Cave and Karst Systems of the World

Series editor

James W. LaMoreaux, Tuscaloosa, USA

For further volumes:
http://www.springer.com/series/11987

William B. White
Editor

The Caves of Burnsville Cove, Virginia

Fifty Years of Exploration and Science

A Contribution of the Butler Cave Conservation Society

Springer

Editor
William B. White
Department of Geosciences
The Pennsylvania State University
University Park, PA
USA

Additional material to this book can be downloaded from http://extras.springer.com.

Cave and Karst Systems of the World
ISBN 978-3-319-14390-3 ISBN 978-3-319-14391-0 (eBook)
DOI 10.1007/978-3-319-14391-0
Springer Cham Heidelberg New York Dordrecht London

Library of Congress Control Number: 2015933618

Printed on acid-free paper

Springer International Publishing AG Switzerland is part of Springer Science+Business Media
(www.springer.com)

This book is dedicated to the memory of friends and members of the Butler Cave Conservation Society who are no longer with us.

Carl Butler
Nevin C. Davis
Thelma Davis
Richard J. Ganjon
Lester V. Good
I. Kennedy Nicholson
Ronald W. Simmons
John A. Stellmack
Fred L. Wefer

Preface

Way back in the 1940s, William E. Davies, pioneering cave geologist, best known for writing *Caves of West Virginia*, casts about for the best locations in the Appalachians where exceptionally large caves might be found. He announced that the most potential was to be found in a small synclinal valley in west central Virginia called Burnsville Cove. The Tonoloway and Helderberg limestones reach their maximum thickness in the Cove, or the order of 800 feet, before they facie into sandstone farther to the south. The geologic structure wraps the limestones around the anticlinal and synclinal axes in such a way as to permit extensive cave development. Most obvious, and requiring less intricate geological reasoning, was the observation that Sinking Creek, the main drainage of the Cove, is underground for most of its route.

A number of mostly small caves were known in Burnsville Cove in the 1940s, but an exception was Burnsville Saltpetre Cave, later known as Breathing Cave, certainly a very large cave by the standards of the day. Cavers from the Nittany Grotto at the Pennsylvania State University became interested in Breathing Cave in the mid-1950s and took on the project of producing a high-quality map. One of the Nittany cavers, George Deike, decided to write his masters thesis on a geological interpretation of Breathing Cave. While Deike's thesis research was underway, I. Kennedy (Ike) Nicholson, his sons, and their friends were searching the Cove for Davies' hypothetical giant cave system. In May of 1958, as the Breathing Cave work was drawing to a close, the searchers found air blowing from beneath a sandstone ledge on the side of a large sinkhole. The Butler Cave-Sinking Creek System had been discovered. Nittany cavers teamed up with the Nicholson family and their group to produce a map and further explore this spectacular new discovery.

The sinkhole containing the only natural entrance to Butler Cave lies on the flank of Jack Mountain, half a mile west and on the opposite side of a ridge from the public road in Burnsville Cove. Access was always somewhat problematic and depended on the good will of land owners whose property must be crossed to reach the cave. There was also the question of controlling access and thus protecting the cave as its size and location became more widely known among the caving community. To solve both of these problems, the Butler Cave Conservation Society was formed in 1968, formally incorporated as a legal entity in the State of Virginia. The BCCS could then lease the cave from the owner and legally install a gate to control access.

As time went on, BCCS purchased the Butler farm and became involved with the exploration of other caves in the Cove. There were further major discoveries on the Chestnut Ridge side of the valley, and in due course, additional properties were purchased. Thus, the year 2008 marked the 50th anniversary of the discovery of

Butler Cave and the 40th anniversary of the founding of the Butler Cave Conservation Society. To celebrate the occasion, the BCCS set out to compile this account of the accomplishments of the past half century. The intent had been to complete the book for distribution at the celebration, but new discoveries outran the editorial process. The discovery of Helictite Cave, then the Subway Section of Water Sinks Cave, and then the large cave called the Wishing Well all required new chapters for the rapidly growing document. In the spring of 2012, BCCS purchased a property containing the sinkhole called Robin's Rift, the collapsed entrance has been dug open and stabilized, and now, exploration and survey is underway. But it is necessary to draw a line and publish the story to date. This book is a milestone in the exploration and study of the caves of Burnsville Cove; it is not the end of the story.

The book has three parallel themes. One is to tell the story of the explorations in Burnsville Cove and the discoveries that have resulted. The second is to document the BCCS organization itself, how it operates, and how it has succeeded in being a good steward for the properties that it owns. The third theme is scientific. It recounts the knowledge that has been gained by the study of the caves in Burnsville Cove.

Curiously, after 50 years of exploration, remarkably little has been published about the caves of Burnsville Cove. In 1982, BCCS members put together a collection of scientific papers that were published as a special theme issue of the *National Speleological Society Bulletin*. At that time, very little was known about the caves under Chestnut Ridge so the scientific discussion focused almost entirely on Butler Cave. Scientific papers are an important means of sharing knowledge, but the editorial custom in scientific writing makes the papers dry as dust for the general reader. Exploration stories are much more interesting. Although the BCCS has published an annual newsletter containing write-ups on many of the exploration trips, the circulation is limited. There is value in telling the exploration stories in more detail and in a more systematic fashion. Further, these chapters emphasize the human interest side of exploration. Butler Cave is big and therefore challenging, but it is not intrinsically difficult. There are few places in Butler Cave that are beyond the abilities of the moderately competent caver. The caves of Chestnut Ridge are an entirely different story. The exploration of the Chestnut Ridge System required strength of body and strength of will that very few cavers were capable of attaining. These exploration stories, therefore, take on a special significance. One may also note the increasing effort required to get access to the caves in the first place. Breathing Cave, the first to be mapped and studied, has a walk-in (or crawl-in) entrance. Opening Butler Cave required moving only a few rocks. Contrast these with the digging efforts required to open the Wishing Well to exploration.

The earlier chapters tell the exploration stories partly in historical and partly geographical order. Butler Cave was the first to be discovered and is documented in Chap. 2 which also includes some of the history of Breathing Cave. Butler Cave focuses attention on the caves associated with the springs which are described in Chap. 4. Then, the tale moves to Chestnut Ridge with chapters on first Bobcat, then Blarney Stone, then Burns, and finally what are referred to as the "pancake caves." Together, these make up the Chestnut Ridge Cave System. Barberry Cave (Chap. 10) forms a link between the Chestnut Ridge System and Butler Cave, a link that is tantalizingly close to completion at the time of this writing. The most recent impressive discoveries are in the downstream end of the Sinking Creek Valley and so the final exploration chapters describe caves found in the Water Sinks Depression and the most recent breakthrough into a large cave through the dig known as the

Wishing Well. Some of the chapters in this book are original. Others are based on previously published material (as indicated in the chapters). However, most of the previously published material has been edited, revised, and updated to various degrees. The maps of the large cave systems are included in electronic form so that all of the fine detail can be displayed without the necessity of large sheets of paper.

Acknowledgments

This book is a group effort of many BCCS cavers, many more than those whose names appear as chapter authors. It is an attempt to display, between one set of covers, what has been discovered over the past 50 years and also the very human effort that was expended to bring those discoveries into being. We must acknowledge the contributions of all members and friends of the BCCS, past and present.

Also acknowledged are:

The late Carl Butler who permitted access to his farm for many years and was willing to lease the cave entrance to the BCCS.

Landowners who have permitted BCCS members to prospect for caves on their land. Pancake weekends would not have been possible without these friendly landowners.

The National Speleological Society for permission to reprint sections of this book that have previously appeared in the *NSS News* and the *NSS Bulletin*.

The photographers, especially Philip C. Lucas, Nevin W. Davis, and the late Ron Simmons, for allowing their images to be used to illustrate the volume.

October 2014 William B. White

Contents

Burnsville Cove

1

William B. White

Abstract

Burnsville Cove is a small limestone valley in west-central Virginia, Highland and Bath Counties. The presence of Silurian/Devonian limestones and the complex geological structure has permitted the development of at least 97 caves in the Cove. A geographical description of the Cove is provided as well as a short history of the settlement of the Cove over the past several centuries.

1.1 Introduction

Beneath a small valley in the rugged Appalachian Mountains of west-central Virginia lies one of the most remarkable cave systems in the United States. The discovery and exploration of these caves has occupied a large number of highly qualified cavers for more than 50 years. Exploration, survey and photography provide a physical description of the caves. Scientific investigations have revealed much about the processes that formed the caves and deposited the speleothems that now decorate them. Our objective in this book is to tell the entire story—the exploration, the discoveries, and the new scientific knowledge gleaned from the discoveries.

The chapter that follows introduces Burnsville Cove. It describes the landscape and a bit of the geologic framework in which the exceptional caves of Burnsville Cove developed. It provides the backdrop against which the cave exploration takes place and explains why this has been such a time-consuming and challenging experience.

1.2 Location and Geologic Setting

Burnsville Cove is located in the Valley and Ridge Province of the Appalachian Highlands about 35 miles west of Staunton, Virginia in Highland and Bath Counties (Fig. 1.1). The villages of Burnsville and Williamsville are situated near the southern and northern limits of the Cove respectively. The northern hydrologic boundary of the area discussed here is the Bullpasture River, a tributary of the James River, which is part of the Atlantic slope drainage.

Burnsville Cove is a very rural part of west-central Virginia. To reach it by road or on maps, locate US Highway 250 which picks its way eastward from Elkins, West Virginia to Staunton, Virginia, more or less perpendicular to the long, parallel ridges of the Appalachian Mountains. Route 250 enters Virginia and crosses US Highway 220 in Monterey. Continuing eastward across Jack Mountain, the critical intersection is the village of McDowell in the valley of the

With a contribution by Judy A Marks.

W.B. White (✉)
Department of Geosciences, The Pennsylvania State University, University Park, PA, USA
e-mail: wbw2@psu.edu

W.B. White (ed.), *The Caves of Burnsville Cove, Virginia*,
Cave and Karst Systems of the World, DOI 10.1007/978-3-319-14391-0_1,

Fig. 1.1 Appalachian topography of the Valley and Ridge Province in west-central Virginia showing Burnsville Cove. Image from Google-Earth referenced to Williamsville, VA

Bullpasture River. East of McDowell, Route 250 crosses Bullpasture Mountain, the Shenandoah Mountains and reaches the Shenandoah Valley at Staunton. From McDowell, Virginia Route 678 follows the Bullpasture River southwestward to Burnsville Cove (Fig. 1.2). A key intersection is that of two narrow blacktop roads where route 609 branches from route 678. The continuation of route 678 goes through the Bullpasture Gorge to Williamsville. The gorge forms the northern boundary and is the location of the springs that drain the caves of Burnsville Cove. Route 609 goes south through Burnsville Cove to the village of Burnsville (Fig. 1.3). The southern limit of the Karst area is about a mile south of Burnsville.

Burnsville Cove is a synclinal valley underlain by the Tonoloway–Keyser–Helderberg group of Silurian–Devonian limestones. The Burnsville Cove karst is an interesting, perhaps unique, example of major cave systems developed in these limestones. There are three main cave-forming groups of carbonate rocks in the Appalachian Highlands: The largest caves are in the low-dip Mississippian limestones of the Allegheny and Cumberland Plateaus. There is extensive karst and cave development in the folded and faulted Cambro-Ordovician limestones in the Valley and Ridge and Great Valley Provinces. The third group is more restricted. The Silurian–Devonian limestones are generally less than 300 feet thick. The stratigraphic relation of the limestones to the overlying Oriskany Sandstone determines that the limestone crops out as narrow, sinuous bands along the flanks of secondary ridges. Thus, although many caves are known in the Keyser and Helderberg Limestones in Pennsylvania, West Virginia and Virginia, many of them are small, there is little surface expression of karst, and developments of large, integrated underground drainage systems are rare. It is the structural setting of Burnsville Cove with synclinal and anticlinal folds as well as the increased thickness of the limestone that permits exceptionally large cave systems to develop.

The synclinal and anticlinal folds that form Burnsville Cove plunge to the northeast. The cavernous limestones rise to the southwest and have been destroyed by erosion just southwest of Burnsville, thus

Fig. 1.2 Topography and drainage in the Burnsville Cove and immediate vicinity. Extracted from U.S. Geological Survey Staunton sheet, 1:100,000 map series. *Note* this map has a metric scale with a 50 m contour interval

Fig. 1.3 The Village of Burnsville. View to the east. WBW photo

terminating the possibilities for cave development. To the northeast, the plunge of the structure takes the limestones below impermeable shales and sandstones. The deep gorge of the Bullpasture River cuts into the limestones and provides outlets for the water as a series of springs. The result is that the downstream reaches of the cave systems are below the regional water table and can be explored only by divers.

1.3 Topograhy and Drainage

The dominant landforms of the Valley and Ridge Province of the folded Appalachians are long, roughly parallel mountain ridges with intermediate strike-oriented valleys. Figure 1.2 shows the arrangement of topography and stream patterns in the immediate vicinity of Burnsville Cove. The highest ridge top elevations are in the range of 3200–3800 feet; valley floors range from 1800 feet at the Bullpasture River to 2500 feet at the Burnsville divide.

The principal surface stream in the region is the southwest-flowing Cowpasture River. One of its tributaries is the Bullpasture River, which flows southwest from its headwaters near Doe Hill along the axis of a shale-floored valley until it abruptly turns southeast, cuts its deep, narrow gorge through Tower Hill and Bullpasture Mountains (Fig. 1.4), and joins the Cowpasture River near Williamsville (Fig. 1.5). The gradient of the Bullpasture is to the southwest while the gradient of Burnsville Cove is to the northeast. The Bullpasture maintains a well-developed flood plain throughout most of its length. The flood plain is at an elevation of approximately 1800 feet at the point where the river leaves the valley to enter the gorge. The Bullpasture channel deepens very rapidly and is a steep-gradient, rough-run stream on a boulder/cobble bed through the gorge until it emerges at grade with the Cowpasture River at an elevation of about 1600 feet. The Cowpasture River also has a well-developed flood plain at this elevation. Figure 1.6 shows the flood plain elevation of the two principal rivers, the summit lines of the mountains and intermediate ridges and the approximate gradients of the Burnsville Cove drainage.

Burnsville Cove is bounded on the west by Jack Mountain, which forms a continuous wall with no breaches in its Clinch Sandstone cap. Streams rising on the eastern flanks of Jack Mountain flow down into the Cove and many sink at the contact with the Helderberg Limestone.

On the east, the boundary is Tower Hill Mountain, also capped with the Clinch Sandstone. The northern hydrologic boundary is the Bullpasture River where it

Fig. 1.4 The Bullpasture River in Bullpasture Gorge during low flow conditions. WBW photo

Fig. 1.5 The Cowpasture River near Fort Lewis, several miles southwest of Williamsville. WBW photo

cuts through the Bullpasture Gorge. To the southwest there is a drainage divide separating drainage flowing northeastward into the Cove from drainage to the south along Dry Run. Beneath is also a ground water divide marked by the boundary between the Silurian Tonoloway Limestone and the underlying Wills Creek Shale. Several miles southwest of Burnsville, the complex folding caused by the arching of the Sinking Creek Syncline axis, forms Warm Springs Mountain which has a north-facing nose directed into the Cove. The southern or Burnsville divide is considerably modified by karst processes.

North of Burnsville, Sinking Creek drains to the northeast and the valley thalweg joins the valley of the Bullpasture just upstream from the Bullpasture Gorge. Much of Sinking Creek is now underground although

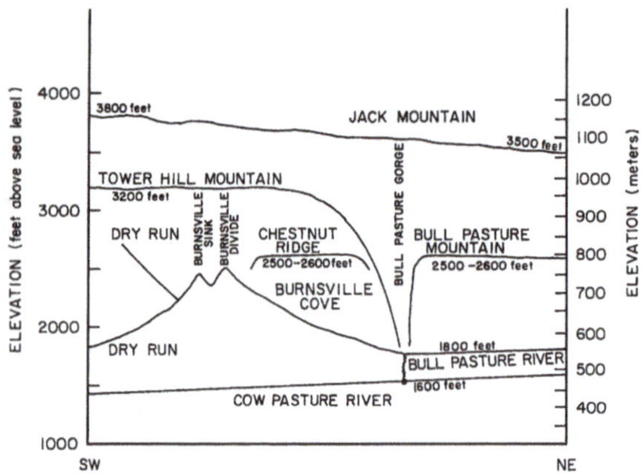

Fig. 1.6 River elevations, mountain summit elevations and valley profiles of Burnsville Cove and Dry Run Valley. From White and Hess (1982)

there is an irregular surface channel that is active during flood flow (Fig. 1.7). The surface valley profile is broken at Water Sinks where Sinking Creek and a major tributary from the west end in a blind valley. South of Burnsville, Dry Run curves around the nose of Warm Springs Mountain, flowing first north and then near Burnsville, flowing south to join the Cowpasture River around the southern nose of Tower Hill Mountain.

Burnsville Cove is divided by Chestnut Ridge, formed by the Oriskany Sandstone where it is brought to the surface by an intermediate anticlinal fold. Sinking Creek flows north along the west side of Chestnut Ridge; the valley on the eastern side between Chestnut Ridge and Tower Hill Mountain is drained by White Oak Draft which also heads near Burnsville. Like Sinking Creek, White Oak Draft is mainly a dry channel.

Figure 1.8 shows the main surface features of Burnsville Cove. There is a surface divide in the form of a pronounced saddle that crosses the cove about a half mile north of Burnsville. However, the large closed depression of Burnsville Sink collects all surface runoff from an area extending to the line of hills across the valley south of Burnsville. The catchment of Burnsville Sink forms the headwaters of the underground streams in the Sinking Creek System. South of Burnsville Sink, Daggy Hollow and the next unnamed stream sink and emerge at Cathedral Spring. Further south, tributaries of Dry Run flow onto Silurian clastic rock, and there is no underground drainage.

Northeast of the Burnsville divide various tributary streams on the flanks of Jack Mountain flow into Sinking Creek. The large closed depression that contains the BCCS Homestead and the entrances to Butler Cave is located along the Jack Mountain flank (Fig. 1.9). Without exception these streams sink during dry weather along the limestone contact, and these and many smaller tributaries without surface expression form the various streams seen in the caves.

Fig. 1.7 The valley of Sinking Creek with dry stream channel. WBW photo

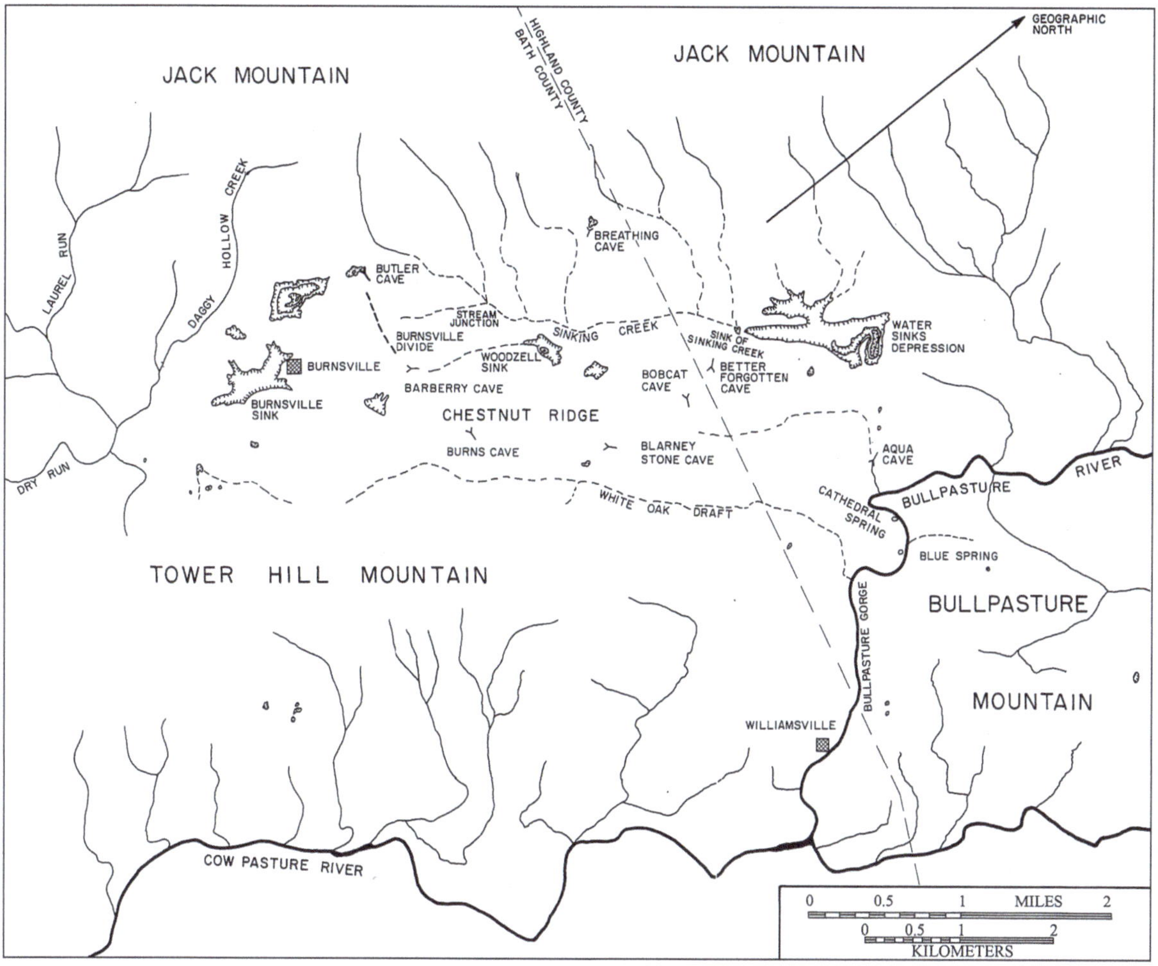

Fig. 1.8 Burnsville Cove, based on U.S. Geological Survey Williamsville and Burnsville 7.5 min quadrangles. Adapted from White and Hess (1982)

Fig. 1.9 **a** The closed depression containing the BCCS Homestead. View to the northeast from the Homestead. **b** View to the southwest from the hill at the northeast end of the depression. WBW photos

The surface channel of Sinking Creek, however, is maintained for a distance of three miles along the axis of Burnsville Cove (more or less the axis of the syncline) to the ultimate sink point at the southern edge of the Water Sinks depression. Other tributary streams flowing into Water Sinks from the west also go underground at this point. There is no surface channel downstream from the Water Sinks depression. The surface channel of Sinking Creek carries water only during periods of high runoff—spring snow melt or exceptional storms. Most seasons of the year the main stream bed is dry throughout its length.

On the east side of Chestnut Ridge, White Oak Draft also flows northward as a tributary of the Bullpasture River. This is a well-defined valley with a partial stream channel, but the entire upper reach of White Oak Draft is a dry or underdrained valley. The surface channel is degraded and the course of the valley is marked by a line of sinkholes.

Extension of the valley profiles of Sinking Creek and Dry Run suggests that most of the area now occupied by closed depressions near Burnsville formerly drained to the south through Dry Run. Development of underground drainage to the north has pirated this section of the Dry Run Basin and made it into the upstream catchment area for a subsurface tributary of Sinking Creek. The piracy was doubtless enhanced by the dip of the shaley Lower Tonoloway Limestone. The Tonoloway crops out along the southern margin of Burnsville Sink and acts as a lithologic funnel causing all of the internal drainage of the sink to follow the syncline to the northeast.

1.4 Karst Geomorphology of Burnsville Cove

The surface expression of karst in Burnsville Cove is mostly in the form of closed depressions on various size scales. There are a few exposed limestone ledges which have been carved by dissolution but most of the area is soil-covered. There are no quarries or extensive road cuts to expose the epikarst so the soil/bedrock interface is generally obscured.

The most prominent surface features are the large closed depressions such as Burnsville Sink, Woodzell Sink, and the Water Sinks Depression along with a number of somewhat smaller sinks. The large closed depressions, shown in Fig. 1.8, have irregular ground plans and seen close-up are indistinguishable from surface valleys (Fig. 1.10). They range in depth from 50 to several hundred feet and in diameter from 300 to 3000 feet. There are many smaller sinkholes, typically 10–30 feet in diameter and often fairly deep in

Fig. 1.10 View into Woodzell Sink show irregular topography of the sink floor. WBW photo

Fig. 1.11 Small sinkhole in the valley of White Oak Draft. WBW photo

proportion to their widths (Fig. 1.11). The smaller sinks have often been chosen as digging sites and excavation of some have revealed underlying caves.

1.5 The Caves of Burnsville Cove

Considering the number of caves that nature has endowed on Burnsville Cove, she has been very stingy with entrances. Very few caves have classic walk-in openings on the hillside or on the river bank. In fact, many of the caves in Burnsville Cove have no natural entrances at all. Without continuous and intensive digging exercises, many of the caves would never have been discovered. One of the few exceptions, with a traditional entrance, is Breathing Cave. As a result, it is one of the caves that has been known for the longest time and was the only significant cave in the Cove at the beginning of the time period described in this volume. Some of the significant caves were known, but because of tight crawls and near-blocked passages, only a minor length of passage near the entrance had been explored.

There was no systematic compilation of cave description data in Virginia prior to the 1958 discovery of Butler Cave and the ensuing exploration. The first such compilation was that of Douglas (1964). The Douglas compilation was augmented by Holsinger (1975). Since then, cave data have been collected by the Virginia Cave Survey. Table 1.1 lists 97 caves in Burnsville Cove that were known as of October, 2013. This list, like all cave lists, is transient. New discoveries and extensions of known caves are continually adding to the list.

Cave names have been changed from time to time over the years, a practice of great concern to Holsinger who, as a cave biologist and zoogeographer, wanted to be sure that biological data, particularly species distributions, were firmly locked to reliable cave names and cave descriptions. The names given in Table 1.1 are those used by the Virginia Cave Survey and should be considered the definitive names. The previously used names are listed in the table for reference to earlier literature but the obsolete names should not be used.

In the period up to about the mid-1980s, most of the caves of interest were on the Sinking Creek side of Burnsville Cove. Breathing Cave is there and the main development of the Butler Cave–Sinking Creek System is along the northwest axis of the valley. Few caves were known on Chestnut Ridge and White Oak Draft was just a line on the map. All of that changed in the mid-1980s with the discovery of what became known as Bobcat Cave. Bobcat was only the first and

Table 1.1 Caves in Burnsville Cove updated to October, 2013

Cave name	Length (feet)	Length (m)	Previous names
Almost Borehole Cave	50	15	
April Fools Pit	65	20	
Aqua Cave	9492	2893	Lockridge Aqua Spring
Armstrong Cave	1160	354	
Audreys Cave	22	7	
Baby Bottle Borehole Cave	39	12	
Backyard cave	400	122	
Barberry Attic Cave	30	9	
Barberry Cave	18,053	5502	
Basswood Cave	710	216	
Battered Bar Cave	7431	2265	
Bear Den Cave	60	18	
Better Forgotten Cave	4100	1250	
Bicentennial Cave	110	34	
Big Deal Cave	36	11	
Birthday Cave	148	45	
Black Cherry Cave	130	40	
Backdoor Cave	40	12	
Blackroot Cave	195	59	Busycon Cave
Blind Faith Cave	3723	1135	
Blue Spring Cave	48	15	
Bob Robin's Cave	100	30	
Bone Cave	87	27	Nevin C. Pit
Boneyard Cave	20	6	
Boundless Cave	1800	550	Burnsville Sink#2
Bowers Shaft#1	90	27	
Bowers Shaft#2	100	30	
BowWow Cave	96	29	
Box Turtle Cave	185	56	
Breathing Cave	32,525	9914	Burnsville Saltpetre Cave
Bridge to Nowhere Cave	64	19	
Brier Fox Cave	100	30	
Buckets of Smoke Cave	543	165	
Buckwheat Cave	2196	669	
Bullpasture Gorge Animal Den	13	4	
Bullpasture Gorge Rock Shelter	18	5	
Burnsville Sink Cave	20	6	
Butler-Sinking Creek System	88,234	26,903	
Butternut Cave	550	168	
Bvideo Pit	651	198	
By-the-Road Cave	2085	635	
Carpenter's Cave	500	152	
Cathedral Spring Cave	952	290	
Cave 609	450	139	
Chestnut Ridge Cave	15	5	
Chestnut Ridge Cave System	111,250	33,909	
Counterfeit Pit	Shaft		
Coyote Crevice Cave	40	12	
Crevice Pit	40	12	
Cycle Sink	110	34	
Disappointment Hole	78	24	
Divined Disappointment Cave	30	9	
Dragon Hammer Cave	76	23	
Fractured Falls Cave	1713	522	
Fuhl Paradise	450	137	
Gags Gig Cave	15	5	
Haroufs Hole	150	45	
Helictite Cave	38,516	11,740	
Hill Top Cave	110	34	
Jackson Cave	1100	335	
Judy's Find Pit	85	26	
Julian Burns Cave	25	8	
Knotts Cave#1	340	104	
Knotts Cave#2	250	96	
Leap Yer Pit	160	49	
Lockridge Water Sinks	15	5	
Mighty Sarlacc Death Pit	94	27	

(continued)

Table 1.1 (continued)

Cave name	Length (feet)	Length (m)	Previous names
Moravian Cave	50	15	
My Favorite Cave	474	144	
Nevin C. Pit	Shaft		
Owl Cave	3145	959	
Pancake Disappointment Cave	55	17	
Pill Box Cave	19	6	
Pond Drain Cave	40	12	
Rat Hole 1180	499	152	
Rat Hole 1181	Shaft		
Rat Hole 1182	80	24	
Robin's Rift Cave	1787	545	
Round Hill Shaft	65	20	
Sandstone Surprise Cave	338	103	
Sinking Creek Cave	66	20	
Snow Melt Hole	15	5	
Stephenson's Cave#1	175	53	
Stephenson's Cave#2	75	23	
Stephenson's Cave#3	180	55	
Stephenson's Cave#4			
T.S. Pit	35	11	
Thirty Foot Pit	30	9	
Twisting Sister Cave	30	9	
Un-Noticed Cave	10	3	
Water Sinks Cave System	11,885	3623	Siphon Cave#2
Waterfall Cave	110	34	
Wildcat Cave	166	51	
Wishing Well Cave	27,067	8250	
Woodzell Ledge Cave	20	6	
Woodzell Pit	60	18	
Woodzell Sink Cave	35	11	

List compiled by Philip C. Lucas

there have been ongoing new discoveries strung out along Chestnut Ridge. Then some of the caves were connected internally so that instead of individually named caves, these became entrances into the Chestnut Ridge Cave System. Although many of the other caves have not—at the time of this writing—been connected, they are clearly related geologically. It makes sense to talk about the greater Chestnut Ridge System although many of the caves are disconnected fragments strung out along the ridge.

Two names that appear in the Douglas and Holsinger reports are Burns Chestnut Ridge Cave and Chestnut Ridge Blowing Cave. Both were small caves consisting mostly of tight passages and stream crawls and appeared to be of no great consequence. Indeed the original name given by the Nicholson's to Chestnut Ridge Blowing Cave was "Rat Hole 1179." Cavers in the 1980s pushed Chestnut Ridge Blowing to map out the half mile descent through the tight, miserable entrance series and discovered major cave below (Chap. 5). To mask the location, the discovery was referred to as "Bobcat Cave". Later Blarney Stone Cave was discovered and in due course connected to Bobcat forming the Chestnut Ridge Cave System (Chap. 7). Burns Chestnut Ridge Cave was explored independently over a more than 20 year period and again an extremely difficult entrance series broke through into significant cave (Chap. 8). The most recent effort has connected Burns Chestnut Ridge Cave to the Chestnut Ridge Cave System. There is now a single cave system with three entrances: Bobcat (or Chestnut Ridge Blowing), Blarney Stone, and Burns (or Burns Chestnut Ridge). The length listed in Table 1.1 is the aggregate length of the three interconnected caves.

The principal caves, as they were known at the time of writing are shown in Fig. 1.12. How these came to be discovered and how they are linked together will be described in the next set of chapters.

1.6 The People of Burnsville Cove by Judy A. Marks

In the early 1700s the main area of population in the country was along the coast from New England south. As early as 1727 the lands west of the Blue Ridge

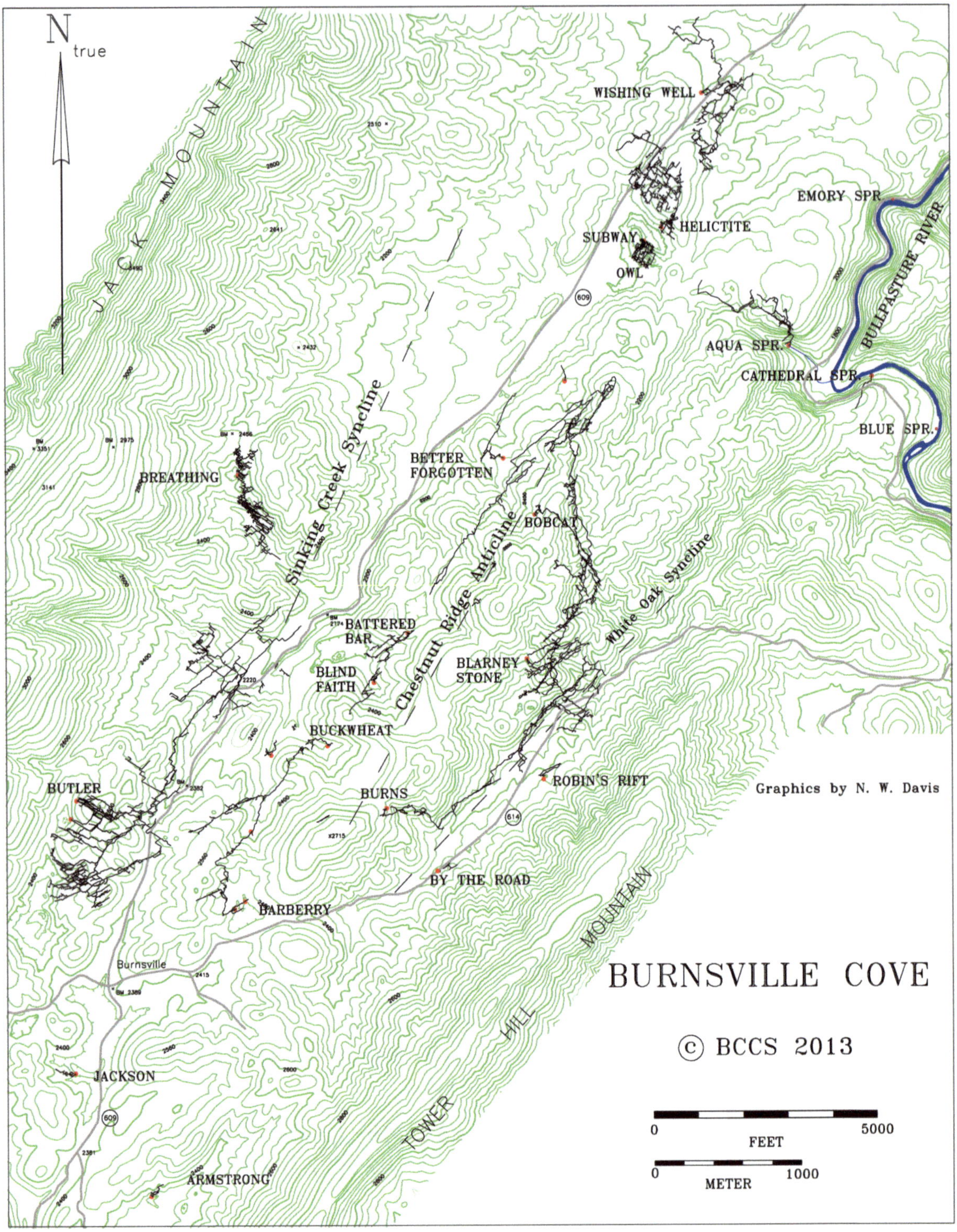

Fig. 1.12 The principal caves of Burnsville Cove. Map compiled by Nevin W. Davis

were virtually uninhabited. Historically the area was used by the Indian nations as hunting grounds and as a military highway. Shawnee from the west, Cherokee from the southwest, Seneca and Delaware followed the well-worn paths that followed the winding rivers. The influx of German and Scotch Irish immigrants in the 1730s helped to open the settling of the Bath County area. Most of the immigrants who came to the area entered it through Panther Gap about 30 miles southwest of Staunton in the Shenandoah Valley.

In 1738, Augusta County was formed from Orange County and named for Princess Augusta, wife of Frederick, Prince of Wales. They were the parents of King George III. The act of formation provided that the new county should remain a part of the county of Orange until there were sufficient inhabitants for appointing officers and organizing a county government. Following this ruling, the county of Augusta was not fully organized until 1748 and extended 240 miles along the Blue Ridge and westward to the Mississippi. Bath County was formed from Augusta, Botetourt and Greenbrier counties in 1791. Highland County was formed from Pendleton County and Bath County in 1847.

The life of the early settlers in the lands west of the Blue Ridge was not an easy one. The settlement at that time was part of Augusta County and it was there that Staunton, the county seat, was located. Any business, legal or otherwise had to be conducted at the county seat. Considering that the main mode of transportation in the early years of settlement was by foot or horseback, this made for very long treks across the mountains. Livestock had to be driven over the mountains to market, including the turkeys and cattle that are today trucked over the mountains. For the most part, people in this area have had to be self sufficient, raising their own food, making their own clothing and logging to build their homes.

A Google search on Burnsville, Virginia, reveals that it is a community or populated place with latitude 38.178, longitude 79.648, and elevation of 2379 feet.

It lies adjacent to the Bullpasture Valley, and not far within the Bath County line. It was originally known as the Red Holes, or Burnsville, settlement. The earlier name is derived either from the reddish loam exposed to view in the sinkholes or from the artificial licks, made by driving stakes into the ground, withdrawing them, and then filling the holes with salt. David Frame patented a tract here that nominally covered 1150 acres. But when it was sold to Elisa Williams, John Burns, and James and Daniel Monroe in 1792, the lines proved to include 1363 acres.

John Burns moved into the Red Holes Valley (now Burnsville) about 1782 with his wife, Mary Shipe, and raised a family of five children (another source says that the year was 1791). Their son Joseph put up a log home across from the present location of the Burnsville Cemetery. For another 25 years or so rooms were added as needed and then in 1907 the present front of the house was built by Charles Wesley Burns for his large family. John Burns I died in 1805 at Red Holes. Sylvia Burns Sanger bought the property in 1945 and her son Julian used it for a summer home.

With the elections of 1860 concluded, the subject of secession became the main subject of meetings throughout Bath County where the residents were divided in their opinions.

> At Red Holes (now Burnsville) citizens met Jan. 23, 1861, for the purpose of 'giving expression to their views with regard to the dangers now threatening this glorious republic and the means best calculated to meet or avert these dangers. After much discussion the following resolutions were adopted:
>
> We, a portion of the citizens of the aforesaid county seeing no remedy in secession for the aggressions of the North against the South, believing that the constitution gives ample security to all sections of our country and needs to be obeyed rather that amended, be it therefore…
>
> RESOLVED, that in our belief simply declaring our withdrawal from the General Government will not stay the hands of the Black Republican party in case they wish to oppress the South and trample it under foot.
>
> RESOLVED, that we will take the Constitution as our guide and maintain our rights in the Union against the Aggressions of the North and any other section from whence they may come.
>
> RESOLVED, that we direct our representative in the coming convention to do all in his power to prevent the secession of Virginia.
>
> RESOLVED, that we deem it expedient that the action of the said Convention be referred back to the people for ratification or rejection.
>
> RESOLVED, that we believe that every state having passed the ordinance of secession should speedily reunite themselves with the General Government and maintain their rights under the Constitution at all hazards.

The Virginia General Assembly, nevertheless, adopted an ordinance of secession April 17, 1861, which was ratified in a state-wide vote May 23, 1861.

Although no actual battles or skirmishes took place in Burnsville during the Civil War, there is a report that in November, 1862 Brigadier General Robert

H. Milroy was scouting in western Bath County and came across the cavalry company of Captain William H. Harness, 17th Battalion Virginia Cavalry in the Burnsville area and several of his troopers were taken prisoner.

Burnsville at one time was a thriving community with an elementary and high school, church, post office, and over the years, three general stores. Hevener's Grocery, which was also the local post office, was located at the intersection across from the high school (now the Burnsville Civic Center). Noah's Ark, which was operated by Wanda and Roscoe Roberts, was located on the south side of the intersection and housed the post office after Mr. Hevener retired. Robert's Grocery, operated after 1948 by Twila and Robert Burns, closed in the late 1980s and had been in either the Burns or Roberts family since it was begun in the early 1900s.

As was common in most small communities, the general store was the community gathering center. It was here that information was exchanged, directions given to travelers, checks cashed, and credit extended on faith. It served as the polling place, and sometimes, housed the post office. The Burnsville post office began operations in 1851 and closed in 1985. Postal patrons now are now served by the post office located in Williamsville. One of the first cars in Burnsville was bought by W. Henry Swadley in 1914.

At one time both Williamsville and Burnsville had elementary and high schools. These schools were later transferred to Millboro. Today, children from Burnsville travel to Millboro and Valley elementary school and to Bath County High School in Warm Springs for high school.

The first Burnsville Elementary school was a 2-room school. At some point, one of the rooms was moved on skids by horses to the Swadley Farm on route 614, which was later owned by Mrs. Willie Mackey. The structure for the one room school remains in Burnsville beside Robert's Grocery. The second school that was built was the high school first named Intermont Elementary School and its first principal was Mr. Seth Bennington.

It is believed that church services began as early as 1799 in the Burnsville area and were led by both Presbyterian and Methodist ministers. Services were first held in homes and eventually in a log school house along the Williamsville road near the present cemetery. About 1850–54 the schoolhouse became too small for the congregation and a new building called New Hebron was built on land owned by C.W. Burns and John Burns. Forty years later, in 1890 the decision was made to erect a new building rather than repair the old one. In 1893 C.W. Burns deeded the land to the church. In 1894, the bell tower and bell were added. In 1983 two Sunday school rooms, bathrooms and a basement which served as a fellowship hall were added. In 1986 a kitchen was added to the fellowship hall. In 2003 a larger fellowship hall was added and was dedicated in 2010. A note to the caving community: "In 1929 a huge hole appeared in the ground directly in front of the church in the center of the church yard. It was an attraction for the children of the church who had fun climbing into the hole before it was finally filled up." This was remembered by Lottie Eagle Jackson in Hugh Gwin's book *Historically Speaking, True Tales of Bath County, Virginia.*

As of 2009 the population of Bath County was estimated to be 4482. Today, the population of the Burnsville area is approximately 150.

It is important to note that the majority of Burnsville area residents are direct descendents of the first settlers. Others who have moved to the area are retired, seeking peace and tranquility in the mountains and rivers. Some things have not really changed over the years. Residents still have to plan ahead for times when it's not convenient to run to the local grocery store. Most do regular shopping in Covington, Roanoke, Staunton or Harrisonburg, still traveling over the mountains to get to town. Neighbors are neighbors in the true sense of the word, helping out in any way that they are able; being there when they are needed.

Down the road a piece, about 3 miles north of the Bath County line along the Burnsville Road. (Route 609), are remnants of another small community. Originally called Poverty, the name was later changed to Flood in honor of Virginia Democratic Representative Henry D. Flood, who served from 1885 until his death in 1921. According to conversations with local residents, I was able to determine that at one time, although not at the same time, there were three churches, a school and a post office.

Franklin School, also known as Flood Colored School was not always a colored school. Originally it was attended by white children, and later by the colored community. Land for this school, according to, *The New History of Highland County, Virginia*, 1983, pg. 19. "There is a deed from Hamilton and Wright for

Fig. 1.13 Franklin School/ Flood Colored School on route 610. Photo by Judy Marks, April, 2011

Fig. 1.14 Black Methodist Church on Route 610, now a hunting camp. Photo by Judy Marks, April, 2011

land in 1897. This was a colored school in the time of segregation. In 1948–49 it was decided to pay tuition and $20 a month board for any child who would have gone to school there. The enrollment was too small to keep the school open. Most of the children went to Staunton and the families followed (Fig. 1.13)."

As you travel on Burnsville Road (Route 609), the white church on the hill is the Lockridge Memorial Methodist Church. This church was served by the Burnsville charge and was built by Lee and Pinckney Lockridge as a memorial to their parents. It was dedicated about 1922–23. An earlier church in Flood was the Prospect Methodist Episcopal Church and it was located about a quarter mile from the Lockridge Memorial Church on land currently owned by R.D. and Sandra Robinson. Mount Zion Methodist

Fig. 1.15 Flood Post Office—current home of Joseph and Betty Jean Lockridge. Photo taken roughly 1900. Photo courtesy of Betty Jean and Joseph Lockridge

Episcopal Church was a black Methodist Church and is located on Job's Hill Road (Route 610); it currently is a hunting camp (Fig. 1.14).

The Poverty Post Office (first suggested name was Mulberry) was located in a log building about 10 miles from McDowell which would locate it near the present property of Jerry and Betsy Fairclaw. The Post Office at Poverty was established in 1889 and was to serve about 125 people. When the name was changed from Poverty to Flood in 1915, it was to serve about 40 people and it was relocated to the home of Lee Lockridge (the current home of Joseph and Betty Jean Lockridge) who was postmaster until he was called to service in WWI. It was then moved to the home of Bud and Annie Hamilton. On April 6, 1935, service was discontinued and mail was then sent to Burnsville (Fig. 1.15).

The historical review is based on Bath County Historical Society (1991), Gwin (2001), Highland County Historical Society (2000), Martin (1969–1985), and Martin (1990, 1996). Some information in this chapter was obtained from personal conversations with R.D. Robinson, Joseph and Betty Lockridge, Barbara Hall, and Dempsey and Joyce Hevener.

References

Bath County Historical Society. 1991. *The bicentennial history of Bath County, Virginia 1791–1991*, 46. Missouri: Heritage House Publishing, Marceline.

Douglas, H.W. 1964. *Caves of Virginia*. Falls Church, VA: Virginia Cave Survey. 761 pp.

Gwin, Hugh. 2001. *Historically speaking, true tales of Bath County, Virginia*. Warm Springs, VA: The Bath County Historical Society Inc.

Highland County Historical Society. 2000. *The new history of Highland County, Virginia 1983*. Marceline, Missouri, Fourth Printing: Walsworth Publishing Company.

Holsinger, J.R. (1975) Descriptions of Virginia Caves. *Virginia Division of Mineral Resources Bulletin, 85*: 450.

Martin, Oren. 1969, 1972, 1979, 1969, 1985. *A history of Highland County Virginia*. Baltimore, Maryland: Regional Publishing Company.

Martin, Oren. 1990, 1996. *Annals of Bath County Virginia*, Staunton VA, 1917. Reprinted for Clearfield Company, Inc. by Genealogical Publishing Co., Inc. Baltimore, Maryland.

White, W.B., and J.W. Hess. 1982. Geomorphology of Burnsville Cove and the geology of the Butler Cave—sinking creek system. *National Speleological Society Bulletin* 44: 67–77.

2 Early Exploration: Breathing and Butler Caves

William B. White

Abstract
Prior to 1958, Breathing Cave was the only large cave known in Burnsville Cove. That changed in May, 1958 when air blowing from beneath a sandstone ledge guided explorers to the discovery of the Butler Cave-Sinking Creek cave system. The exploration of the Butler Cave-Sinking Creek system along with nearby Breathing Cave is described in detail. Breathing consists of a large network maze on the flank of the Sinking Creek Syncline. Butler Cave consists of a master trunk passage and underground stream that follows the axis of a syncline northeast to a series of terminal sumps. Connecting with the trunk passage is a series of network maze caves that extend up the northwest flank of the syncline. Overall, Butler Cave contains 16.71 miles of surveyed passages. The excavation of a new entrance in 1998 allowed easy access to the cave and in recent years exploration is being continued.

2.1 Introduction

This and chapters that follow present a detailed history of the exploration of the various caves in Burnsville Cove. The exploration of recently discovered caves such as Helictite and Wishing Well Caves can be described in what might be called "real time"—exploration logs that were written as the exploration proceeded. The present chapter deals with the earliest discoveries and is more of a challenge. It is being written half a century after the events. The historical record is mixed. There is the author's memory—that of a retired professor recalling his activities as a graduate student, there are field trip reports in old caving club newsletters, mostly the *Nittany Grotto Newsletter*, and there is an extensive published history that carried the story from the beginning to the late 1970s (Wefer and Nicholson 1982). In 1976 the Butler Cave Conservation Society (BCCS) began publishing an annual newsletter so from that time forward the historical record is much more complete.

In writing history, there is always the question of where to begin. Burnsville Cove was certainly known as cave country to the early settlers and the few caves with large and obvious entrances were known although it is doubtful if much exploration took place. In the absence of earlier written records, this exploration history begins with the first scientific and systematic surveys that begin in Breathing Cave just after the 2nd World War. This is followed by the discovery and subsequent exploration of the Butler Cave-Sinking

With an addendum by Philip C. Lucas.

W.B. White (✉)
Department of Geosciences, The Pennsylvania State University, University Park, PA 16802, USA
e-mail: wbw2@psu.edu

W.B. White (ed.), *The Caves of Burnsville Cove, Virginia*,
Cave and Karst Systems of the World, DOI 10.1007/978-3-319-14391-0_2,

Creek System by I. Kennedy (Ike) Nicholson, his sons, and his colleagues in the late 1950s.

Oscar P. Estes, Jr. (July 18, 1913–December 14, 1981) was one of the Ike Nicholson's companions in the initial exploration of Butler Cave in the late 1950s. He accumulated a photo file of more than 100 images which passed to the Butler Cave Conservation Society after his death. These provide the best available photographic record of the early exploration. The photographs in this chapter are drawn from that collection. Many of these photos are the work of Huntley Ingalls but the exact photo credits have been lost.

2.2 Breathing Cave

2.2.1 Early History

Organized cave exploration was nucleated in the District of Columbia area in the late 1930s, then remained almost dormant during World War II, and finally sprang to life after the war. Cavers from DC and elsewhere spread out over the limestone valleys of eastern United States searching for caves. Very quickly information was collected for published descriptions of caves such as the books by W.E. Davies on West Virginia in 1949 and Maryland in 1950 and by R.W. Stone on Pennsylvania in 1953. There was no equivalent book on Virginia but cavers were active in the Shenandoah Valley and in the mountains to the west. They reached Burnsville Cove and found Breathing Cave, already well-known locally as Saltpetre Cave, at least as early as 1944.

The entrance to Breathing Cave is at the bottom of a sinkhole on the lower flank of Jack Mountain (Fig. 2.1). The entrance passage slopes steeply downward to end in a breakdown choke. Crawlways to either side lead to the otherwise independent North and South Sections of the cave. There was a strong air current moving through the crawlway leading to the South Section and, to the surprise of the early explorers, the air current reversed direction periodically. This "breathing" phenomenon ultimately gave the cave its name. The DC cavers were intrigued by the breathing phenomena and observed it over a period of years (Faust 1947). Cournoyer (1954) devised a fast response barograph and made at least one quantitative measurement of the air current oscillation. Apparently the DC cavers also began a map of the cave.

2.2.2 The Nittany Grotto Survey

Nittany Grotto, the caving club of The Pennsylvania State University, had been founded in 1948 but had gone through a short moribund period when it was re-invigorated in 1951 by William Devitt III. Devitt was a mining engineering major at Penn State and had, perhaps, a greater appreciation for the importance of accurate maps then many of his fellow cavers. Following his lead, the Nittany cavers were busy mapping many caves in central Pennsylvania. However, Pennsylvania caves tend to be short, small, and muddy so it became a Grotto custom to take occasional multiple-day trips to areas in West Virginia and Virginia where there were larger (and dryer) caves. The first Grotto visit to Breathing Cave was on one of these trips in January, 1954, when the group spent six hours roaming large and dry cave passages. It was George Deike's first visit to the cave and he was greatly taken by it. Mightily impressed by the cave, the Nittany cavers made Breathing Cave one of their primary objectives for their next winter-break trip south (Deike 1955). On this trip they spent an entire day exploring the cave and getting an impression of its immense size and complexity. For the first time, the Nittany cavers had come up against a cave that really seemed to be without end. They went back to Breathing Cave in April, 1955 and spent 17 h only to find more cave opening up before them. By this time they had accumulated about 10,000 feet of sketch map.

What quickly become obvious was that sketch maps were completely inadequate for a cave the size of Breathing. So, by the time of their visit in the spring of 1956, a new, careful, survey of the cave was underway. By the end of the trip, about 15,000 feet of cave had been surveyed. The big push was Christmas break, 1956 (Deike et al. 1957). Three teams of surveyors were assembled for a five-day stay in the cave. Following their usual custom, the group camped in the cave. After dragging their gear through the entrance crawlway, they set up camp using one chamber for cooking and other activities and a deeper chamber, Sand Alley, warmer and with a smooth sand floor, to lay out their sleeping bags. After breakfast, the three teams would split up and each go to its assigned area of the cave to continue the survey. The teams would reassemble in the Camp Room in the evening for dinner, to plot out their mapping accomplishments, and decide on objectives for the next day (Fig. 2.2).

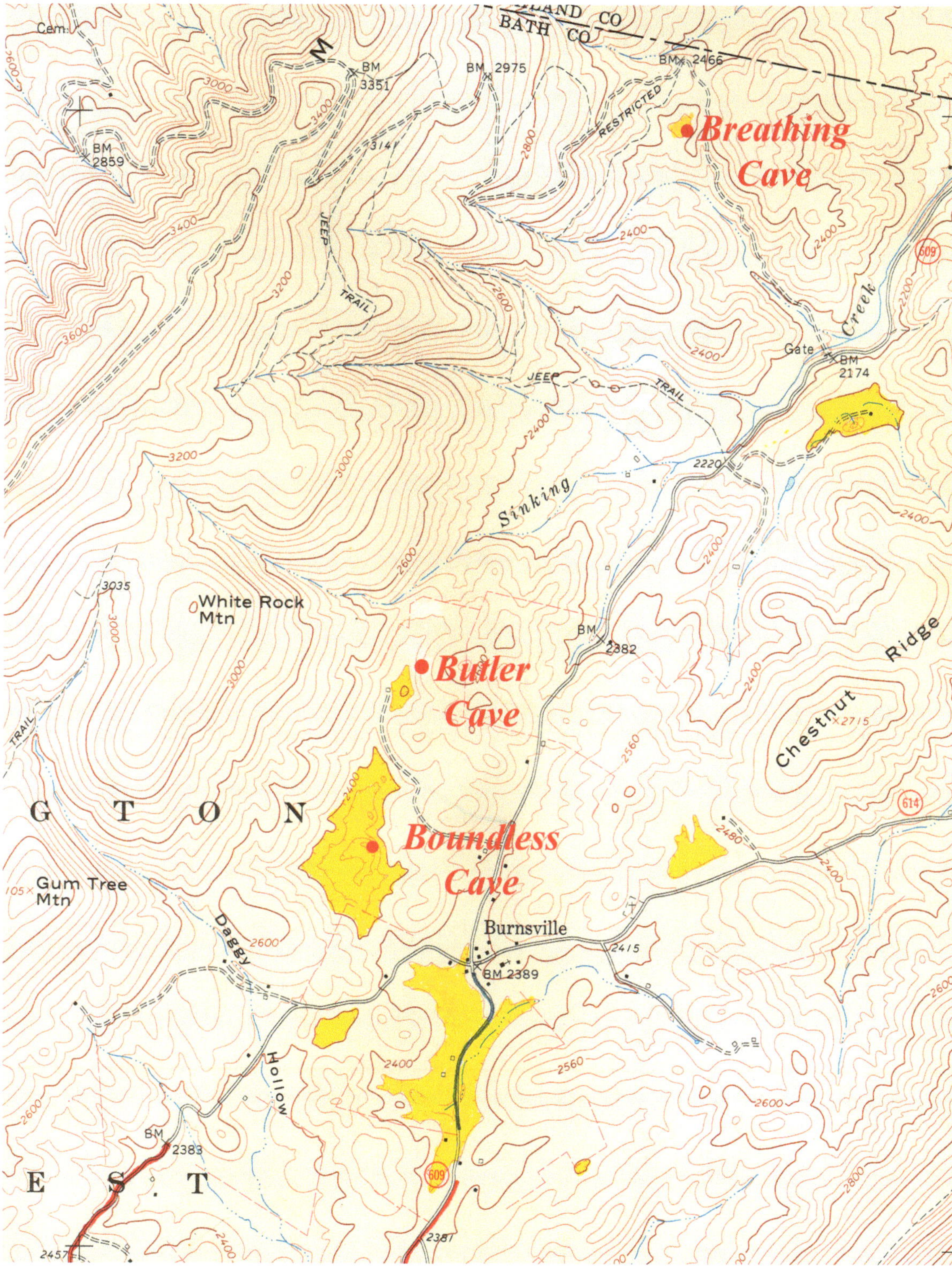

Fig. 2.1 Topographic map of the area near Breathing and Butler Caves. Extracted from U.S. Geological Survey Burnsville, Virginia 7.5 min Quadrangle. *Yellow areas* delineate closed depressions

Fig. 2.2 New Year in Breathing Cave. *Left* to *right* Herb Black, George Deike (concealed), Ken Graves, Jack Stellmack, Larry Matthews, and Harry Johnson. [Nittany Grotto Newsletter, Vol. 5, Number 3, Page 6 (January, 1957)]

George Deike expressed it this way: "Each morning the three parties split up and went into the darkness to their own areas. They might as well have gone a hundred miles away. All day they worked alone, without hearing from each other, as if no one else was in the whole cave. It was odd to reflect as you worked that somewhere in those thousands of feet of passage two other parties were laying their tapes through some seldom seen gallery or pit. At the end of the day it was very warming to come into camp and see a bright lantern and friendly faces."

At the end of five days of mapping, most of the known cave had been mapped. More trips were made over the next year to complete various segments and to correct errors that surfaced when the final map was plotted. The compiled survey data were plotted and passage detail drawn into produce the final map. The total effort had required 14 separate trips made over a period of 6 years, involving more than 50 cavers and more than 3500 man-hours in the cave. The original map was reproduced at 20 feet per inch and resulted in a huge sheet of paper. Reduced copies of the map were prepared and widely distributed (Fig. 2.3). For its day, the Breathing Cave map is an altogether remarkable document. It was certainly one of the earliest large caves to be mapped in detail by project caving—a systematic approach to surveying the same cave on trip after trip.

2.2.3 Geological Investigations

George Deike was a geology major at Penn State and was one of the most persistent Breathing Cave mappers. He received his BS degree in 1957 and moved to the University of Missouri to study for a master's degree. There he convinced the geology department that the geology of Breathing Cave would be an excellent thesis project. This, in itself, was an accomplishment because in the 1950s most geology departments considered cave studies to be completely frivolous. George and his wife Ruth spent much of the summer of 1958 at Breathing Cave completing a topographic overlay of the land surface above the cave, investigating the geology of the region, and examining cave passage cross-sections, profiles, and clastic sediment fills in considerable detail. The results were compiled into an MA thesis that appeared in 1960 (Deike 1960). A summary of these results appears as Chap. 18.

The detailed investigation of Breathing Cave marks something of an historical turning point. The classic papers on the origin of caves that had appeared in the 1930s and 1940s were largely based on intuitive reasoning with very little solid field data. Cave maps were not adequate and field observations were limited. The Deike thesis was one of the first to draw conclusions based on a comprehensive cave map, on detailed

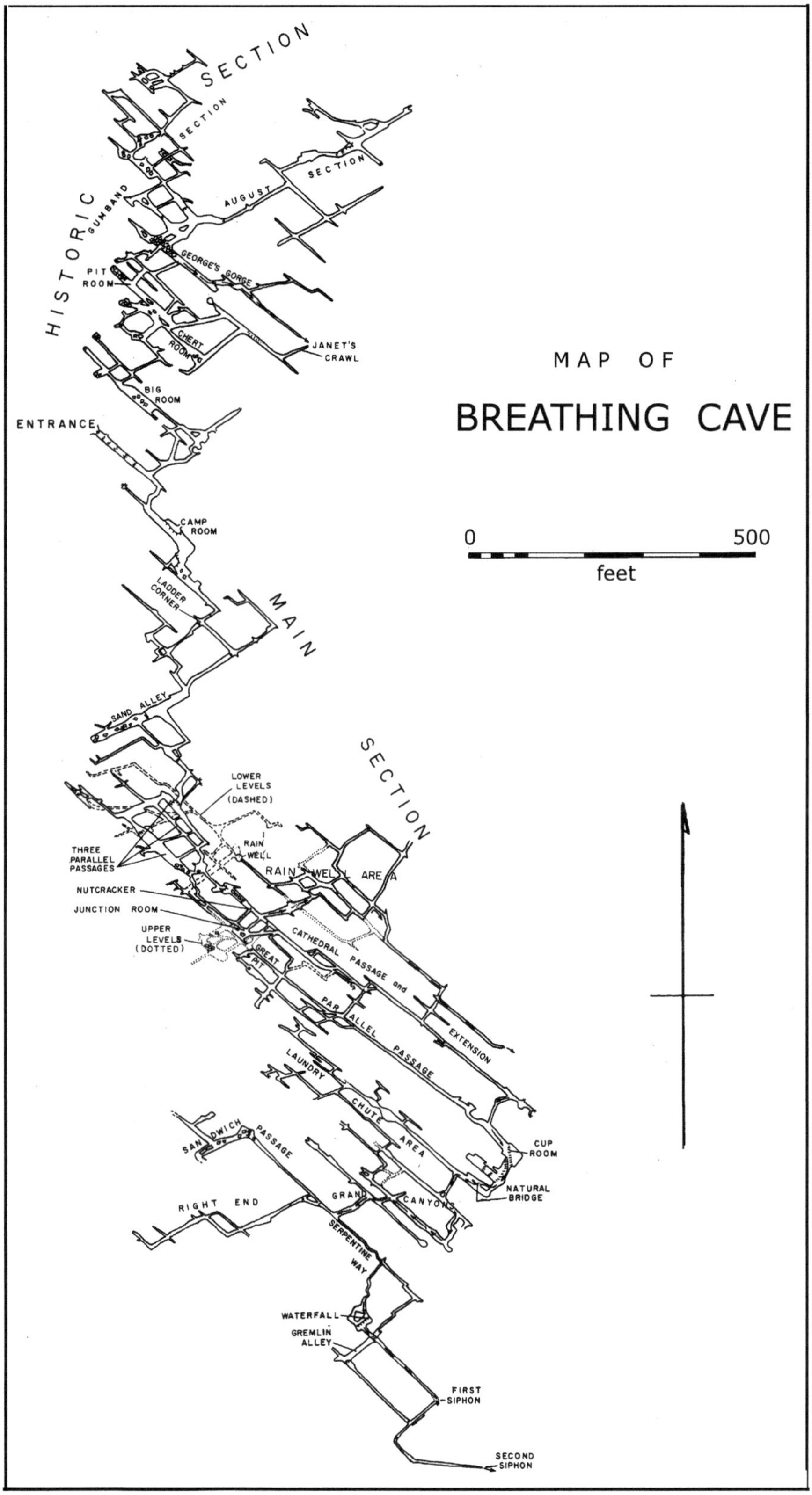

Fig. 2.3 The 1957 Nittany Grotto map of Breathing Cave. This version adapted from Deike (1960)

examination of the cave, and on a solid description of the cave embedded in its local geologic setting. William E. Davies of the U.S. Geological Survey organized a symposium on the origin of caves at the December, 1959 meeting of the American Association for the Advancement of Science. Breathing Cave was one of the star presentations. The Deike thesis also won an NSS Certificate of Merit in 1961.

Breathing Cave is unusual. It is a network maze and most network mazes are nearly horizontal. Instead, Breathing Cave follows the dip of the bedding and so there are elevation differences of several hundred feet between the highest points in the cave and the lowest. A most important discovery of the thesis work was that Breathing Cave is sandwiched between two sandstones. In spite of the overall relief of the cave, it is constrained to about 77 feet of limestone. These aspects of the geology appear again in other caves in the Cove and are important for the interpretation of the overall history of cave development.

2.2.4 Later Explorations and Surveys

As luck would have it, Deike's investigation of what was thought to be the longest cave in Virginia was just being wrapped up when a much longer cave was discovered only a mile and a half up the valley. With the discovery of the Butler Cave-Sinking Creek System, interest in Breathing Cave faded and it was visited mainly for sport caving for many years. There was a short period of intense re-investigation in the late 1960s and early 1970s as explorers searched for a connection between Breathing Cave and the Butler Cave-Sinking Creek System. Then Breathing went back to sport cave status until a major re-mapping effort was undertaken early in the present century. The cave was completely re-mapped by The Gangsta Mappers in the 2000s. The length of the cave is 6.74 miles. The depth is 512 feet.

2.3 The Butler Cave—Sinking Creek System

2.3.1 Discovery

While the investigations of Breathing Cave were underway, other cavers were prowling Burnsville Cove. Surely one big cave implied another. It was also noted that the main drainage in the Cove, labeled Sinking Creek on the map (Fig. 2.1), did indeed sink and the main channel down the Cove was completely dry most of the year. The D.C. area cavers had a cabin along the Bullpasture River which served to draw cavers to the area. One of these was I. Kennedy (Ike) Nicholson and his sons Michael and David. With their friend Oscar Estes from Staunton, Virginia, they spent a great deal of time combing the hills for possible caves. It wasn't entirely without success. One of their companions, Beven Hewitt, did a SCUBA dive in Aqua Spring and discovered Aqua Cave. They discovered a small cave high on Chestnut Ridge that they called Rathole 1179 (later renamed Chestnut Ridge Blowing Cave) that, unknown to them, would later become of great importance as the Bobcat Entrance to the Chestnut Ridge Cave System. They discovered two small caves in the large closed depression near Burnsville, Burnsville Sink #1 and Burnsville Sink #2. Burnsville Sink #2 was renamed Boundless Cave but its small silt-clogged passages seemed to go nowhere.

The breakthrough came in the spring of 1958. Two local boys (see addendum at the end of this chapter) told Oscar Estes of wind blowing from beneath a ledge on the Carl Butler farm on the same flank of Jack Mountain as Breathing Cave but about a mile and a half to the southwest. Oscar passed the word to Ike

Fig. 2.4 The original entrance to Butler Cave as it appeared in the late 1950s. From the Oscar Estes collection

Fig. 2.5 The original entrance to Butler Cave as it appeared in the late 1950s. From the Oscar Estes collection

and on Memorial Day, 1958, Ike was at the ledge (Figs. 2.4 and 2.5). The ledge with air blowing from beneath it was an outcrop of the Upper Breathing Cave Sandstone. Ike pulled out a few rocks and was soon able to squeeze in to find himself looking down a 35-foot pit (Fig. 2.6). He climbed down the pit and found the wind whistling up through a crevice in the floor. Ike climbed out of the pit and went back to the cabin for additional help. Later in the day, the crevice, now known as the Glop Slot, was dug open sufficiently for one of the smallest explorers to slip through (Fig. 2.7). The cave descended rapidly, over several down-climbs and a pit called the God-Is-My-Copilot climb before breaking through another sandstone layer into a large room. The small entrance series had traversed the entire 77 feet between the Upper and Lower Breathing Cave Sandstones. The huge passage opening up was below the lower sandstone.

Fig. 2.6 The entrance pit. From the Oscar Estes Collection

2.3.2 Early Exploration

On 14 June, 1958 a 7-person party led by Ike descended the entrance pit, the Glop Slot, and emerged at the top of the breakdown slope in the first big room. At the bottom of the slope, they discovered that the huge passage seemed to end in a silt wall. An opening on the side of the passage part way up the wall gave access to a set of passages that descended eventually to a stream, Difficulty Creek, where they stopped at a water crawl.

Fig. 2.7 The Glop Slot. From the Oscar Estes collection

On 21 June, 1958, the explorers were back. It was possible to cut steps along the edge of the silt bank that blocked the first large passage (Fig. 2.8). At the top, the passage continued downslope following the bedding. Near the lower end of the large passage it was possible to descend into a canyon that trended roughly perpendicular to the main passage trend. A few hundred feet along the canyon, the floor disappeared although a narrow ledge continued. It was necessary to rope down 40 feet to the stream level below. They then explored upstream, climbed Rotten Rock Waterfall and eventually were able to climb up into a dry passage, Dave's Gallery, that sloped back down along the bedding to an intersection with the Rimstone Pool Passage and a passage leading to a silt bank where they terminated their exploration for the day. The explorers returned to the Rimstone Pool Passage on July 5, 1958 and continued down the main passage, through a small linking passage they named 90-Ugh Crawl, and emerged into the main trunk channel at Sand Canyon.

With the discovery of Sand Canyon the cave opened up dramatically. Sand Canyon is in the main

Fig. 2.8 A blank wall of silt is bypassed along a high ledge in the first big room. The Window Passage opens along the wall. From Oscar Estes collection

trunk passage of the system which extends for several miles along the Sinking Creek Valley following the axis of the Sinking Creek Syncline. The Butler Cave section lies on the flank of the syncline so that the bedding and the passages all slope toward the trunk channel. On the initial discovery trip, the explorers hiked down stream more than a mile, discovered the underground route of Sinking Creek, the dry sumps, Sneaky Creek, and went all of the way to 10-foot high waterfall at the 6th of July Room. The lower reaches of the cave were wet, sometimes with chest-deep water. Although the distance from the entrance to Sand Canyon is only a few thousand feet, the route taken by the explorers was rough and complicated, requiring rapelling down a drop, climbing a waterfall and negotiating a variety of canyons and crawls so that just reaching the unexplored areas required a great deal of time. An underground camp seemed to be the answer.

Sand Canyon is a flat silt terrace above the usually dry overflow route of Sinking Creek and an ideal camp site (Figs. 2.9 and 2.10). August 8, 1958 the group of seven cavers hauled gear through the Butler Cave section and set up camp. For the next week they explored both upstream and downstream along the trunk channel. Upstream they discovered Huntley's Cave, another side cave on the flank of the syncline upstream from Butler Cave, the natural bridge (Fig. 2.11), and a complex maze section upstream from the Natural Bridge. Downstream, they pushed beyond the 6th of July Room to discover the Rat's Doom Sump (called a siphon on the map and in earlier reports) and Dave's Lake (Fig. 2.12). Nearer to Sand Canyon they discovered the Moon Room (Fig. 2.13), the Crystal Craters, and the Crystal Passage, all upper level dry passage above the main stream-way. By the end of the week they had discovered and reconnaissance mapped about 15,000 feet of new cave.

The exploration route from the entrance to Sand Canyon was complex and time-consuming (Fig. 2.14). The Nicholson's and their friends had been secretive about their new find in the summer of 1958 and of course, other cavers became extremely curious. One of these, Cliff Forman, decided to follow them into the cave. He followed their footprints through the entrance series, down the first major passage and into the cross-canyon. On the ledge where the canyon crossed the Bean Room he lost the trail. Sitting down to have a cigarette, Forman notice the smoke swirling up through the breakdown over his head. Climbing into an upper passage, Forman followed it and quickly reached Sand Canyon, completely bypassing Rotten Rock Waterfall and Dave's Gallery. Later, a small crawlway, the Rabbit Hole, was found to connect the first big passage to the second big passage near the top of the fill bank that was the initial barrier to the first exploration. These connections provided a direct route and reduced the travel time to Sand Canyon from

Fig. 2.9 Sand Canyon camp. The passage in the background is the connection with Butler Cave. From the Oscar Estes collection

Fig. 2.10 Camping in Sand Canyon. From the Oscar Estes collection

Fig. 2.11 The Natural Bridge. From the Oscar Estes collection

hours to less than an hour. No further camping expeditions were necessary.

Two more exploration trips in September found the Last Hope sump and the stream passage later called Slippery Creek. There were now three streams in the system, Sinking Creek, Sneaky Creek, and Slippery Creek, all flowing more or less parallel before disappearing into sumps. There were also tributary streams such as the Huntley's Cave stream, Rotten Rock Creek, and Difficulty Creek flowing down the west flank of the syncline. If all of these streams were tributary to some master drainage, the junctions were not visible in the humanly-accessible part of the system. Overall, the 1958 explorations by the Nicholson parties had produced more then six miles of new cave as revealed by very fast reconnaissance surveys. Clearly, a more careful and detailed survey was needed.

2.3.3 Surveys

The first Nittany Grotto Survey was undertaken on Thanksgiving Weekend of 1958. The two alternate routes from the entrance to Sand Canyon were surveyed producing a figure-8 that allowed for an assessment of the closure and therefore the accuracy of the surveys. The survey also established that Sand

Fig. 2.12 Downstream passage. From the Oscar Estes collection

Fig. 2.13 The Moonroom. From the Oscar Estes collection

Canyon was 300 feet below the entrance. On May 30, 1959 the objective was to lay survey along the trunk passage. One team made a rapid trip to the July 6th Room and began surveying upstream. The second team began surveying downstream from Sand Canyon. They met below the dry sumps, each team having set more than 50 stations and with 6844 feet of survey in the book. This survey demonstrated that the July 6th Room was 420 feet below the entrance. There were four additional mapping trips between June, 1959 and July, 1960. On Thanksgiving weekend, 1960 it was possible to field four mapping teams for two days each working in many different parts of the cave. A progress report (White 1960) noted an overall total of 5.3 miles of survey which covered most of the cave as it was then known. Although mapping continued, the next major mapping effort was August 16, 1964 when five teams were fielded. John Haas led a trip to Marlboro Country, the purpose of which was to explore the Candle Room area but no new passages were found. Haas had noticed

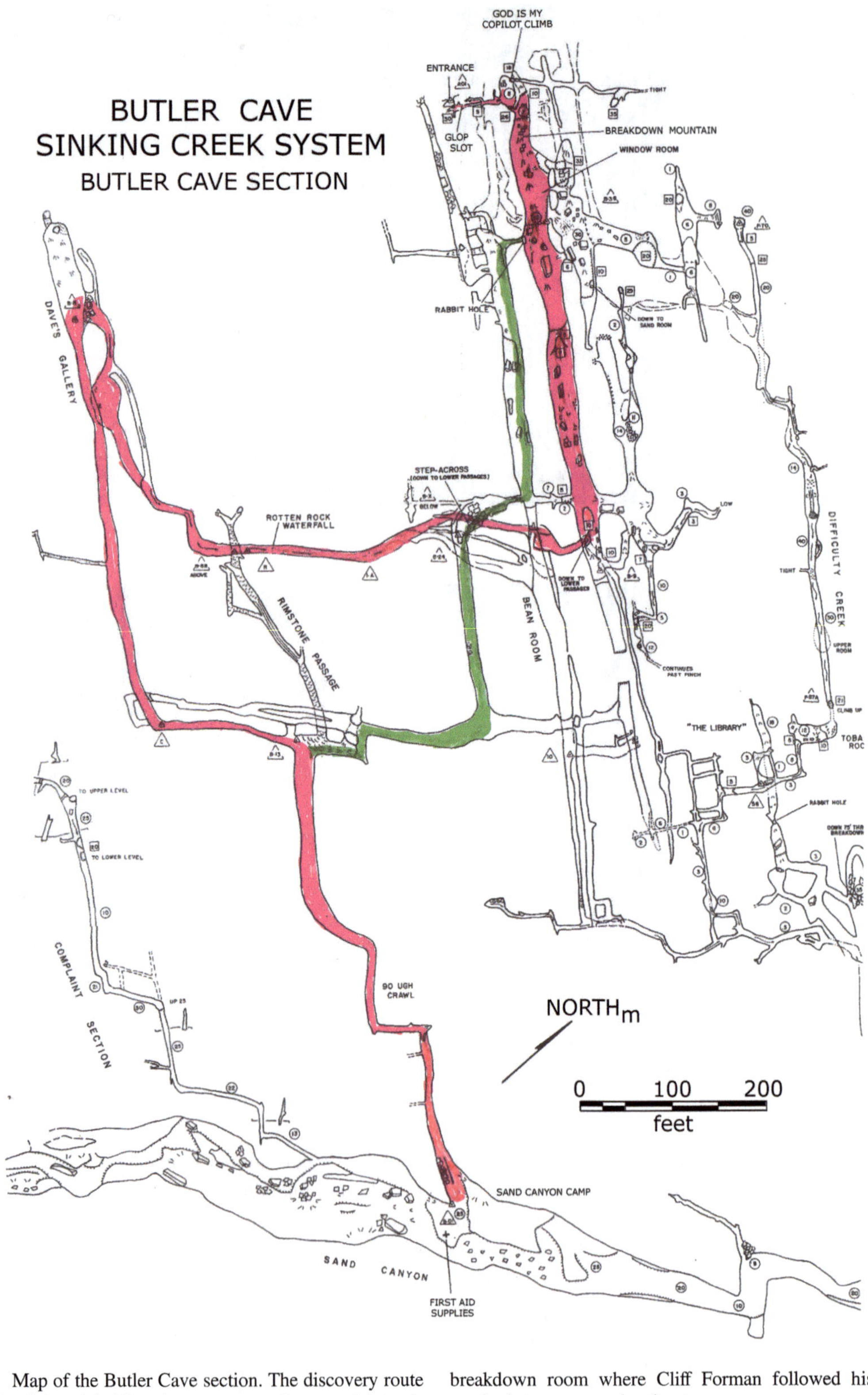

Fig. 2.14 Map of the Butler Cave section. The discovery route is highlighted in *pink*. The shortcut route discovered later is highlighted in *green*. The *pink* and *green* routes cross at the breakdown room where Cliff Forman followed his cigarette smoke into an upper level passage

that the trunk channel went up through the lower sandstone at Dry Sumps. Thus the Downstream Loop is also between the two sandstones. In crawling down Crisco Way, they were crawling on the lower sandstone, which was penetrated at the 40 foot-pit. Marlboro Country is below the lower sandstone at the same stratigraphic horizon as most of the upstream areas. The expedition also deployed mapping teams in Huntley's Cave, the Natural Bridge area, the Crystal Craters Section, and the downstream Trunk Channel The Trunk Channel team discovered what came to be known as the Pat's Room Section. Several thousand feet of passage were added to the map. Wefer (1982) describes the details of the surveying methods.

There were dozens of trips to Butler in the late 1960s and early 1970s, most of them added something to the growing map. By the late 1970s the main features of Butler Cave were known and most of them mapped. The mapping, however, had left many dangling ends and much of the decade from the mid 1970s to the mid 1980s was required to survey these odds and ends. Further, when an attempt was made to systematically evaluate all of the survey data, poor closures and some blunders were discovered, necessitating a certain amount of re-survey. By 1985, Lester Good was able to compile sectional maps of Butler Cave that showed nearly all of the known passages. Good's map folio was scanned and is included with the on-line collection of maps attached to this book. The index sheet which also gives an outline map of the entire cave is shown in Fig. 2.15.

Although the Butler Map was stabilized by the release of Les Good's sectional maps in 1985, clearly these maps were not the final word especially with regard to the length of the cave. There were redundant surveys, some of which got added into the aggregate cave length, there were still some errors and confusions that needed rectification. There was some new survey. The survey data were moved into digital files by Tony Canike so that new survey and needed re-survey can be easily identified. With redundant surveys removed, the length of Butler Cave, as of October, 2013, is 16.71 miles.

2.3.4 Pushing the Boundaries

The Butler Cave-Sinking Creek System had been shown to consist of a master trunk passage roughly two miles long that followed the axis of the Sinking Creek Syncline. It headed in a maze of smaller passages at the upstream end and ended in sumps at the downstream end. Tributary to the trunk passage were a series of side caves, mostly network mazes that extend up-dip along the western flank of the syncline. The cave grew mainly by filling in more and more detail of the side caves. In addition, however, were two other significant discoveries, one upstream and the other downstream.

Upstream, the stream passage above the Natural Bridge emerges from what appears to be a sump. In an early Nittany Grotto trip Karl Francis and Dick Kutz pushed the sump, now called Penn State Lake, found that although very wet, it had sufficient air space to proceed, and found open cave on the other side. The cave beyond Penn State Lake was explored in the early to mid-1960s and explorers found new cave by climbing up into a sequence of narrow fissure passages and some rooms. They had crossed the Lower Breathing Cave Sandstone, which makes up the roof of the trunk passage at the upstream end, and were in the Breathing Cave horizon.

In the summer of 1964, Mike Nicholson brought a young aborigine from New Guinea, the subject of an anthropological study by Dr. Carlton Gajdusek of the National Institutes of Health, to Butler and his explorations beyond Penn State Lake. Mbagintao (pronounced "bog-in-taw") proved an excellent explorer and the passages explored on that trip were collectively named Mbagintao Land. These passages are included on the Butler Cave master map.

Downstream, in June, 1963, Mike Nicholson and Joe Faint followed a small stream into a mud-coated crawlway they called Crisco Way. After 700 feet, the crawlway ended at the top of a 40-foot pit. Later, Mike and Dave Nicholson and Dave Head again crawled down Crisco Way, descended the pit, and discovered Marlboro Country, an extensive part of the cave with large galleries. Marlboro Country lies under much of the lower sections of the Sinking Creek System separated from the upper passages by the Lower Breathing Cave Sandstone. The pit connecting the two parts of the cave penetrates the sandstone. There were three streams in Marlboro Country, stream #1 being a continuation of Sinking Creek beyond its sump in the upper cave. The discovery of Marlboro Country added considerable passage length to the system. These are shown on the sheets 10 and 11 of the Butler Cave map set in the electronic file. It is a long and difficult trip to

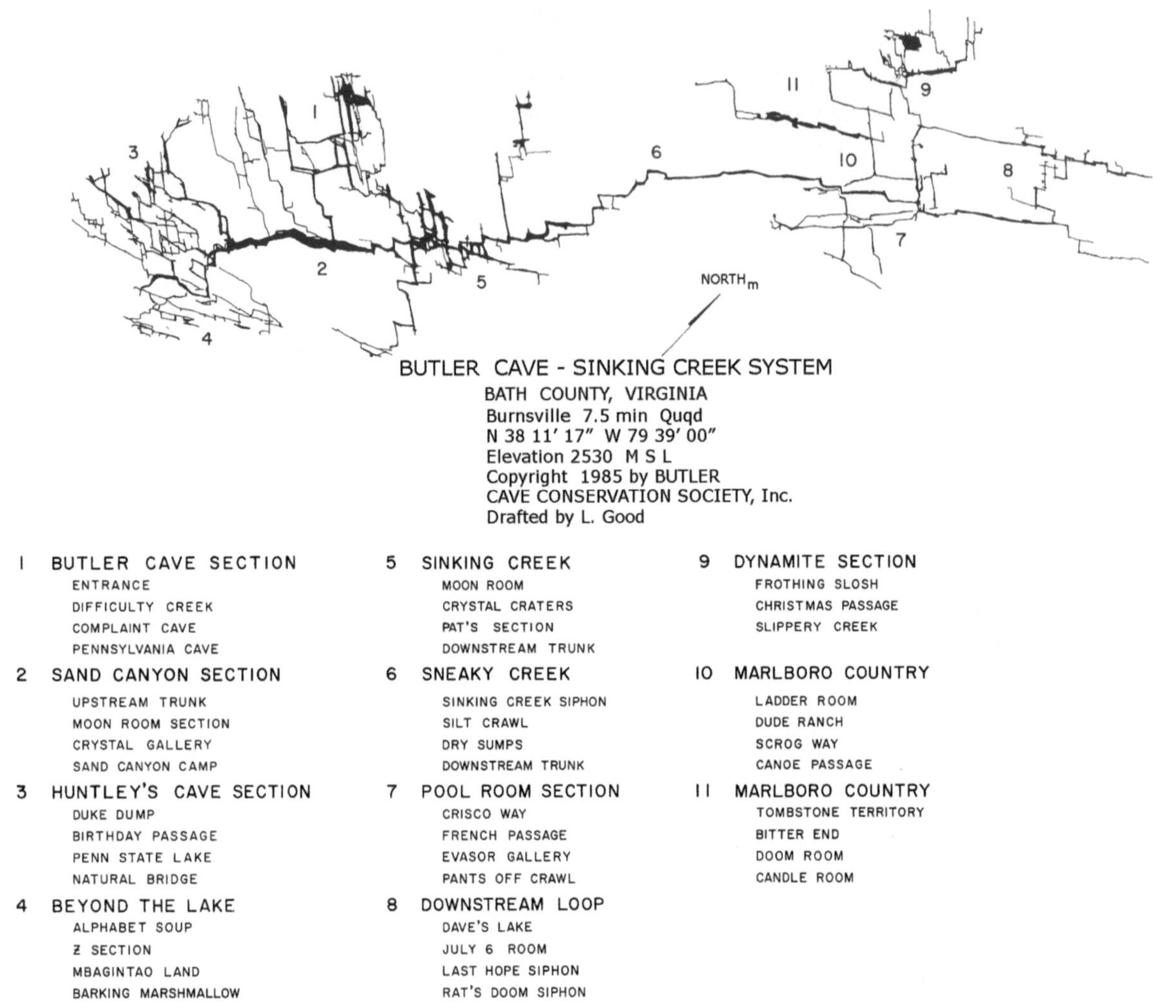

Fig. 2.15 Index map for the Butler Cave-Sinking Creek System with major place names. Full maps sheets are in the on-line map files

Marlboro Country and those contemplating one might well read Fred Wefer's account of the return from one such trip (Wefer 1979).

With the exploration of Marlboro County downstream and Mbagintao Land upstream, the Butler Cave-Sinking Creek System had developed a most peculiar pattern. The Sinking Creek Syncline plunges to the northeast, down the valley, toward the Bullpasture River. The trunk channel sloped downstream also to the northeast but at a lower angle than the plunge of the limestone beds. As a result, the trunk channel actually crosses the lower sandstone at the Dry Sumps so that the upstream portion lies below the lower sandstone while the downstream portion is above the lower sandstone. At the upstream end, Mbagintao Land is an upper tier of cave above the lower sandstone while Marlboro country is a tier of cave below the lower sandsone.

2.4 Related Caves

2.4.1 The Long-Sought Breathing-Butler Connection

The discovery and exploration of Butler Cave created an entirely new perspective on Breathing Cave. Butler Cave is a trunk channel with a collection of side caves, most of which are network mazes on the flanks of the syncline. Breathing Cave is a network maze on the flank of the syncline. Therefore, Breathing Cave should be just another side cave of the Butler Cave-Sinking Creek System. Furthermore, the final downstream sump in Breathing Cave and the downstream sumps in Butler Cave were not very far apart. It was just a matter of finding the connection.

The late 1960s and early 1970s saw almost a frenzy of efforts to locate the connection. The extreme downsteam end of Breathing Cave was searched and searched again, hoping to find the obscure crawlway that would take the explorers past the final sump and into Butler Cave (*Nittany Grotto News*, **15**(3), 49–56 (1967); **16**(5), 89–91 (1968); **16**(7), 138–139 (1968); **17**(1), 9–17 (1968)). No connection was found. Earlier surveys suggested that the final Breathing Cave sump, the "pseudosiphon" was less than 500 feet from Butler. The successful dive of Last Hope Sump and discovery of the Good News passage should have closed the distance substantially. Then more accurate survey linked to points on the surface by cave radio dashed the hopes for an immediate connection. The relationship between the caves is now known accurately (Fig. 2.16) and no humanly passable connection seems likely.

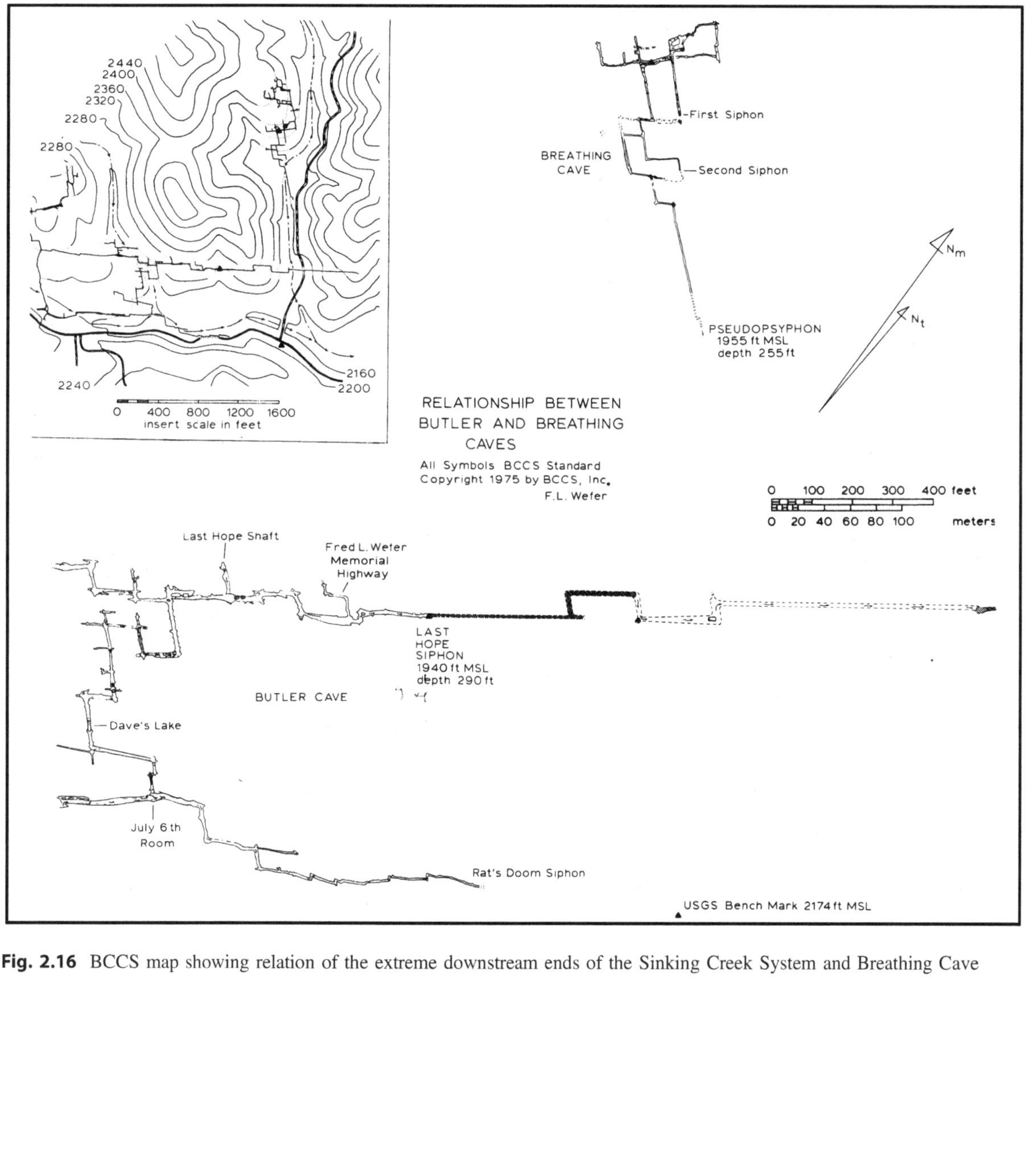

Fig. 2.16 BCCS map showing relation of the extreme downstream ends of the Sinking Creek System and Breathing Cave

2.4.2 Boundless Cave

Access to Butler Cave was tightly controlled after its discovery, first by secrecy, and later by formally leasing the cave from Carl Butler. Other cavers were busily looking for an alternative entrance. The leading candidate was Boundless Cave (Burnsville Sink Cave #2) which was quite close to the upstream maze in Butler Cave. Boundless is a maze cave with narrow passages that are often silt filled. Because the cave takes storm flow from Burnsville Sink, the silt tends to get rearranged and previously opened passages are later found to be plugged. In spite of numerous attempts by different groups of cavers, no one succeeded in finding a way through to Butler. Then, in early 2014, BCCS cavers succeeded in digging through silt-choked passages in the upstream maze and connected Butler Cave to Boundless Cave. Although, technically, Butler now has a third entrance, the connecting passages are very tight and frequently closed by silt.

2.4.3 Better Forgotten Cave

The streams in the Sinking Creek System all end in sumps but the sumps are several miles up the valley, southwest, from the known resurgence at Aqua Cave Spring. There must exist a large amount of unknown cave passage between the sumps and the spring and there is at least the possibility that some of them are air-filled. Thus there was a search for other caves in the Sinking Creek Valley downstream from the sumps. Better Forgotten Cave was one of the discoveries.

The entrance to Better Forgotten Cave is on the western flank of Chestnut Ridge on property now owned by the BCCS. The cave was first explored in 1959 through a tight entrance series to a 100-foot pit. Descending the pit proved very difficult and no leads were found at the bottom so the cave was declared "better forgotten" as indeed it was for nearly 10 years. In the fall of 1969 there were six exploration pushes into Better Forgotten Cave sparked by Jack Hess and Nevin Davis (Hess 1970). On these trips, the groups successively passed the ladder drop, the 100-foot pit, the vertical crawlway, and other obstacles to break out into a major stream passage. Upstream, the stream emerged from a sump but there were several thousand feet of passage downstream before the stream reached its final sump. The first sump encountered had a bypass passage. The second and final sump is the present terminus of the cave. Rather then a backdoor to downstream Butler, Better Forgotten Cave revealed yet another stream apparently independent of the Butler streams but also draining to the Aqua Cave Spring. Explorers returned to Better Forgotten Cave in the 1980s and 1990s to complete an accurate survey and produce the map shown in Chap. 24 and in the electronic map file. The downstream sump is 411 feet below the entrance.

2.5 More Recent Events in Butler Cave

2.5.1 Sump Diving

The first attempt at diving the Last Hope Sump was in June, 1960, when Hank Hoover penetrated 200 feet into the sump and reported it tight but continuing. In the August, 1975, Sheck Exley, one of the country's most experienced cave divers, had a try and penetrated 500 feet into the sump before running out of line. In October, 1975, Sheck was back with more equipment. Nevin Davis tells the story in the BCCS Newsletter (February, 1976).

> At the siphon we struggled with the equipment, assembling every thing and checking its operation. When everything was ready, Sheck walked out into the pool and while still moving, put his flippers on, pulled down his face mask, and disappeared. With the double tanks, he had the capacity to penetrate 2000 feet of submerged passage and have a duration of one hour with an hour reserve if the dive was shallow. After he left we sat around periodically checking a watch. In one hour we started to worry; in an hour and a quarter we were beginning to wonder if he would return. In about one hour and twenty minutes there was a glow which rapidly grew brighter; then splash, he was back. Sheck removed his mask, smiled, and said. "Well, I have some good news and some bad news. Which do you want first?" Someone said, "The good news." Sheck said that the siphon came up into air-filled passage about 100 feet beyond the end of his dive on 31 August. There he removed his tanks and explored about 950 feet of passage mostly 4 to 5 feet high and up to 20 feet wide in one area. The passage got as wide as 20 feet. At the end of his downsteam trek was the bad news, a second siphon. It was too difficult for him to carry his tank down the passage alone to dive the second siphon. There was one side passage which quickly became too narrow to follow wearing a dry suit.

Measurements with compass and length of dive line provided a sketch map (Fig. 2.17). Sheck Exley included a first hand account of his dives in Last Hope Siphon in his memoirs (Exley 1994). Later David Whall and Karen Wark attempted to dive the Bad News Sump and found it clogged with mud.

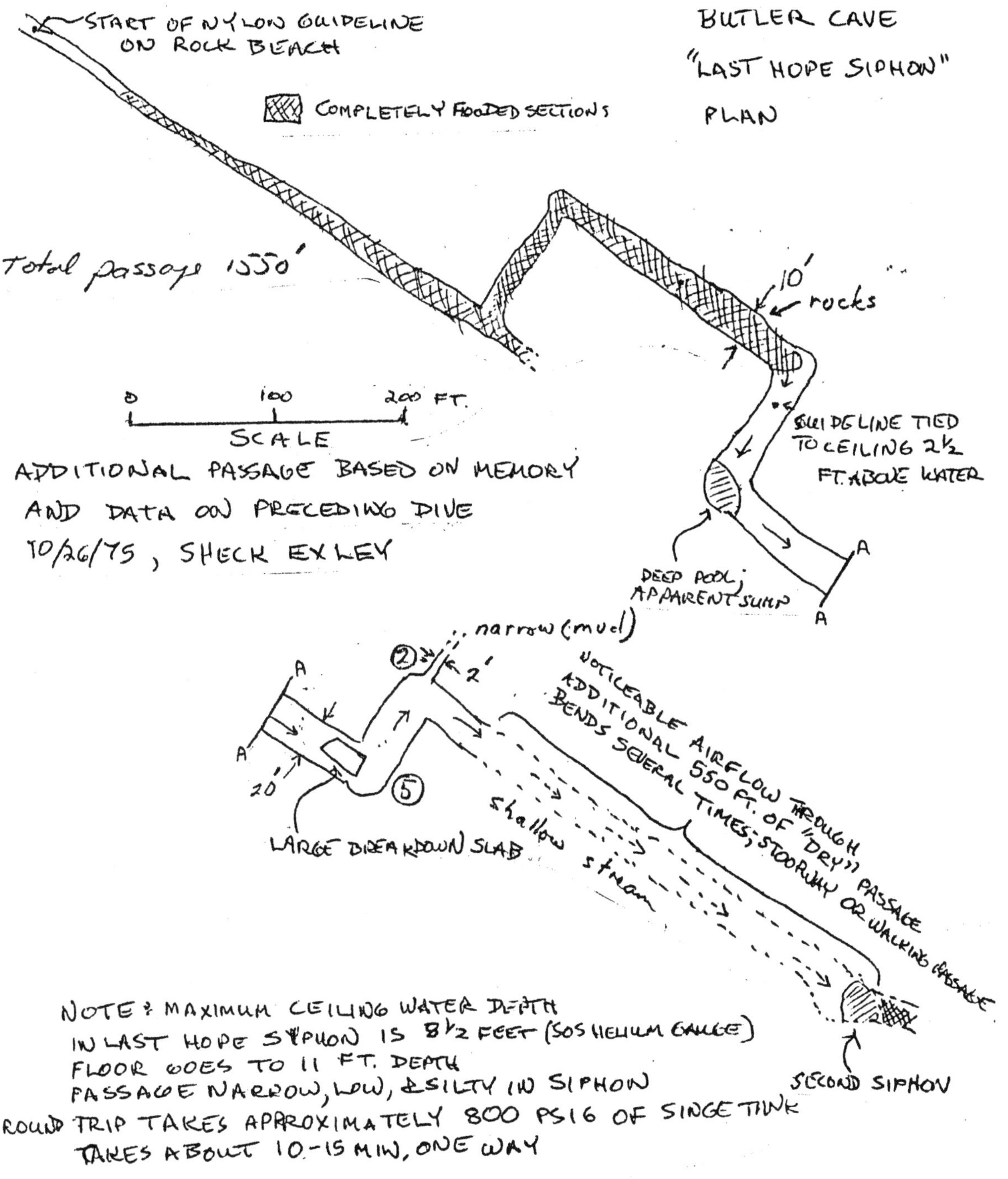

Fig. 2.17 Sketch map of the Good News Passage beyond Last Hope Siphon. From BCCS Newsletter, February, 1976

2.5.2 The SOFA Entrance

By far the most significant event in the recent history of Butler Cave was the construction of the SOFA entrance in late October, 1998 (Marks 1999) It had long been known that Dave's Gallery terminated against a breakdown and gravel fill in the walls of the sinkhole just below the Butler farmhouse. However, various desultory attempts at digging had produced nothing. Several days work with a track hoe and a bulldozer were necessary. Once breakthrough had been achieved, a 5-foot culvert was slid into the opening and the pit backfilled. The culvert was fitted with a door, steps, and a handrail. Later, a protective roof and rock walls were added to protect the door from the continuous sloughing of loose rock from the cliff overhead. Instead of a tedious series of downclimbs from the original entrance on the hillside (now known as the Nicholson Entrance), one opens the door, descends easily though the culvert and steps out into Dave's Gallery. Sand Canyon is then only minutes away.

Did BCCS really need the SOFA Entrance? Certainly, the trip from the Nicholson Entrance was not exceptionally difficult for experienced cavers. The 300-foot difference in elevation between Sand Canyon and the Nicholson Entrance was an aggravation for tired cavers returning from some activity deep within the system. Safety was improved because the time and difficulty necessary to extract an injured caver was greatly lessened. The main virtue of the SOFA entrance is that it provides access to the cave suitable for class field trips and other educational activities in which the participants may not be experienced cavers. These activities have included Outdoor Adventures arranged through the Smithsonian and also geology class field trips from Penn State and from the College of William and Mary.

2.5.3 Current Exploration

Following intensive exploration efforts in the 1970s, new discoveries in other part of Burnsville Cove diverted attention from the Butler Cave-Sinking Creek system. Most of the important leads had been checked and mapped and although there were educational and sport caving trips, few new discoveries were made. A new interest in Butler has been underway for the past five years. Exploration takes the form of digging out silt-filled passages and using bolts and climbing poles to probe potential leads in the cave ceiling. The long-suspected connection to Boundless Cave was one of the accomplishments.

An important discovery, made in late 2010, was an air connection between Backyard Cave and Evasor Gallery in Butler. A fan with a reversing motor was placed at the entrance to Backyard Cave and a recording anemometer was placed in Evasor Galley. The air pressure signal generated by the fan was clearly recorded in Evasor Gallery with about a 15 s delay.

The end of the story cannot be written. There is still more cave beyond the terminal sumps. The connection to Breathing Cave may or may not be a lost cause. There is a real possibility of a connection to Barberry Cave if the right silt-filled passage is dug open. New discoveries are often made where they are least expected and the best guidance for cave explorers is to just keep looking.

2.6 2008 Addendum: Placing a Bronze Plaque at the Nicholson Entrance by Philip C. Lucas

All discoveries are preceded by a sequence of events

I'm going to pull together those events, those things, those people that ultimately led to the discovery of Butler Cave, one of the most significant caves in Virginia and in the US.

To begin, the Burnsville Cove, the valley from Burnsville down to the river was much then as it is today—very rural with mostly farms and forests. This spot here was open pasture land—there were no trees except for that old snag of an ash tree down there. Fifty years ago, it was in its prime. The roads have been paved but mostly everything looked pretty much the same as it does now.

Not much was known about caves around here but that was about to change in a big way. The catalyst for that was the Burnsville Saltpeter Cave used for saltpeter mining during the Civil War. This was a huge cave thought to be the biggest in Virginia. with about 3.5 miles of passages. Its name got changed to Breathing Cave when it was discovered that it breathed with inflowing and the out-flowing air currents. People from all over would come to explore Breathing Cave. It was because of Breathing Cave that an interest in finding more caves in the Burnsville Cave began to develop.

Now I need to talk about people and those folks who called themselves cavers. The NSS was a youthful organization in the fifties. There was an early NSS cabin along the Bullpasture Gorge where cavers would congregate. Two of those cavers who would see the potential for finding a big cave were Ike Nicholson and Oscar Estes. Ike lived in Maryland and Oscar in Staunton. Oscar and Ike were good friends and caving buddies and they were on a mission. Oscar's talent was people and he made many contacts gaining information about caves. Ike was the organizer, the dreamer of what might be, and the scout checking the ridges and valley for leads. They were a great team. When Lockridge's refrigerator spring was dived and Aqua Cave discovered—the hunt was on to find where all that water was coming from.

Some of Oscar's cave explorations had been with two young men who lived near Burnsville; Tommy Burns and Jimmy Puffenbarger. Oscar knew that landowners were well acquainted with their property and whether there were caves or possibilities of caves. So Oscar asked Tommy and Jimmy to keep their eyes and ears open for caves. And so it was that in the spring of 1958 Jimmy and Tommy were sitting on the porch of the store in Burnsville when they saw Oscar driving by in his jeep station wagon. One of those guys, Tommy Burns, is with us today and so Tommy, tell us about what happened next.

Tommy (Emory) Burns then explained how he and Jimmy Puffenbarger took Oscar Estes up to the Butler Farm and up the hillside to where they had seen the blowing hole beneath a limestone ledge. He said that Oscar crawled into the cave a short ways until he came to the top of a pit. He dropped rocks down to estimate the depth of the pit. Not having come prepared that was all he could do that day.

What exactly happened next is lost to history but I think the conclusion is inescapable that Oscar quickly told his caving buddy Ike about this new pit and the first opportunity for Ike to check it out was Memorial Day Weekend. Oscar being an ex-Navy man was probably taking part in some aspect of Memorial Day celebrations back in Staunton. So Ike, following Oscar's directions, finds the cave and after descending the pit finds a nearly choked fissure too tight to get through but blowing air. He quickly determines that he needs help. So he goes back to his cabin along the Bullpasture River and secures two young volunteers and they go back up to the cave that afternoon. One of those young men was Donnie Miller and he is with us today. Donnie tell us about your efforts that day.

Fig. 2.18 The plaque at the Nicholson entrance to Butler Cave. WBW photo

Donnie Miller explained how Ike had come down to the Roller cabin excited about the cave and asked for help in digging to try and get through the fissure. He explained that he agreed to go and later that day the fissure had been opened to the extent that he could wiggle through. Donnie then said that he went some distance into the cave and that it got bigger so he finally decided he had gone far enough and crawled back through the fissure.

So they retreat from the cave and it is nearly two weeks before Ike returns with his two sons, Mike and Dave for an exploration trip. Mike is here and I'm going to ask him to describe his memories of that first cave trip.

Mike then explains that they went through the fissure (now called the Glop Slot) and down the steep slope into the first big room in the cave. They then went through the window in the Window Ledge and explored a lot cave passage in the Sand Pit area of the cave. He then told a little about subsequent trips including one where he fell from a climb on top of his brother Dave when a rock had broken cutting his brother's forehead. Tommy Burns suddenly exclaimed that he had been on that trip and remembered the accident. Mike went on to explain that his brother was not badly injured.

And so with this it was realized that an immense cavern had been discovered. Later that summer, a week-long expedition revealed further the huge size of the cave. Even though we are fifty years down the road, discoveries are still being made in Butler. Just last year a new species of crustacean was found. And so with this fiftieth anniversary the BCCS has placed a bronze plaque to commemorate this occasion. I will read what is written on this plaque (Fig. 2.18).

References

Cournoyer, D.N. 1954. The speleo-barometer. *National Speleological Society Bulletin* 16: 91–93.

Deike, G.H. 1955. Thirteen Yankees invade rebel caves. *Nittany Grotto Newsletter* 3(3): 1, 11–17.

Deike, G.H., H. Black, and J. Stellmack. 1957. Breathing Cave, VA. *Nittany Grotto Newsletter* 5(5): 1–7, 10–13.

Deike, G.H., III (1960) Cave development in Burnsville Cove, west-central Virginia, with special reference to Breathing Cave. MA Thesis, University of Missouri, Columbia, MO, 155 pp.

Exley, S. 1994. *Caverns measureless to man*, 156–161. St. Louis, MO: Cave Books.

Faust, B. 1947. An unusual phenomenon. *National Speleological Society Bulletin* 9: 52–54.

Hess, J.W. 1970. The exploration of better forgotten cave, Virginia. *Nittany Grotto News* 18(2): 48–55.

Marks, F. 1999. Butler Cave walk-in entrance. *BCCS Newsletter* 24: 4–9.

Wefer, F.L. 1979. Return from Marlboro Country. *BCCS Newsletter* 5: 16–19.

Wefer, F.L. 1982. Surveying the Butler Cave-sinking creek system. *National Speleological Society Bulletin* 44: 64–66.

Wefer, F.L., and I.K. Nicholson. 1982. Exploration and mapping of the Sinking Creek System. *National Speleological Society Bulletin* 44: 48–63.

White, W.B. 1960. Progress report on the Butler Cave project. *Nittany Grotto Newsletter* 9(3): 46–49.

BCCS: The Organization

3

Fred L. Wefer and Keith D. Wheeland

Abstract

The Butler Cave Conservation Society (BCCS) was created as a non-profit scientific, education, and conservation organization in November, 1968. In 1970 the BCCS was incorporated in the Commonwealth of Virginia. The purpose of the organization to own and manage caves as a conservation strategy, initially in Burnsville Cove, but not limited to the Cove. The organization is guided by a seven-member Board of Directors which includes the officers. Membership is by invitation. Lists of members, board members, and officers are provided. An addendum provides a description of the 50/40 celebration of the discovery of Butler Cave and the founding of the BCCS.

[**Editors Note**: Fred Wefer's contribution to this chapter was based on his article "A Profile of the Butler Cave Conservation Society, Inc." *Proceedings of the National Cave Management Symposium*, p. 179–189 (1991).]

3.1 Introduction

The Butler Cave Conservation Society, Inc. (BCCS) is a non-profit scientific, education, and conservation organization incorporated in the Commonwealth of Virginia. It was formed in November of 1968 and is the oldest such organization in the United States to employ ownership of wilderness resources as a major element of its conservation strategy. In 1970 it was incorporated in the Commonwealth of Virginia.

The BCCS was originally formed to manage and conserve the Butler Cave-Sinking Creek System commonly known as Butler Cave. Butler Cave was discovered in May of 1958 so that by 1968, the cave had been known for ten years. Although access during that period had been generally limited to exploration and survey trips, unauthorized "tourist" and "orientation" trips were beginning to have an effect on the cave in the form of spent carbide dumps, litter, graffiti, and breakage.

Several related events contributed to the formation of the BCCS. Approximately three years after he opened Butler Cave, I. Kennedy (Ike) Nicholson expressed in a 1961 letter to his friend and caving buddy, Oscar Estes, that he would like to form an organization that would be dedicated to protecting the cave. At the 1956 National Speleological Society (NSS) Convention, the Nittany Grotto, an NSS chapter, had successfully petitioned the NSS to sponsor a project to survey and map Breathing Cave which is

With an addendum by David Kohuth.

Fred L. Wefer—Deceased.

K.D. Wheeland (✉)
2192 Mountain View Avenue, State College,
PA 16801-7214, USA
e-mail: kwheeland@psualum.com

W.B. White (ed.), *The Caves of Burnsville Cove, Virginia*,
Cave and Karst Systems of the World, DOI 10.1007/978-3-319-14391-0_3,

located near Butler Cave. The Nittany Grotto members, primarily from Pennsylvania, were finishing up the Breathing project when Butler Cave was discovered. They were invited by Ike Nicholson to apply these same skills in Butler Cave and by 1968 were leading these activities. The Nittany Grotto cavers not only had considerable skills in the surveying and mapping of large caves, they also possessed some unique experience in managing wild caves.

Hosterman's Pit Cave (located in Centre County, Pa.) had been discovered in 1961 (Davis 1963). Concerned about liability, the owners requested that the Nittany Grotto gate the cave and limit access to qualified cavers, a process begun in 1963. By 1968 the Nittany Grotto members had Five years of experience at managing a gated wild cave. The experience clearly demonstrated that gating and strict access control were viable (if unpopular) conservation techniques. It was obvious to the founders of the BCCS that they could apply the same conservation techniques to Butler Cave that had worked well in Pennsylvania, that is, conservation via access control. The early involvement by Nittany Grotto explains why many of the BCCS members are from Pennsylvania.

While most of the BCCS activities have been concentrated in the area of west-central Virginia called the Burnsville Cove, it should be noted that neither the BCCS constitution nor its bylaws limit its activities geographically. The BCCS has sponsored expeditions to Hawaii, Iceland, the Dominican Republic, and Mexico. See Appendix 1.

The BCCS maintains a policy for access to the caves that it owns and manages. See Appendix 3. Open expedition weekends are scheduled each year. Over the years the number of weekends has varied, but now stands at five. Cavers may also ask any BCCS member for permission to visit a cave at any time. One day of the Memorial Day weekend expedition has traditionally been set aside to work on the grounds and buildings. The special weekend of the regular annual membership meeting includes the Pig-Out, a combination cash meal and pot luck. A special expedition to prospect for new caves—Pancake Weekend—is scheduled annually. See Chap. 9.

The BCCS has published a newsletter since 1976. Copies of the newsletters plus other cave related materials are maintained in the BCCS library which is located in the home of a member who lives near Butler Cave. The core of the library came from the widow of Fred L. Wefer who donated Wefer's private collection. The library has been cataloged and a title and author index is available on the BCCS website. The NSS maintains a link from it's website to the BCCS website.

The BCCS has conducted various community outreach programs in Burnsville and Williamsville beginning in 1976. All members of the immediate surrounding area are invited to a gathering usually held in a local church or community center. Slides of local caves, and programs on conservation, scientific techniques, and water quality have been presented. In 2006, Phil Lucas had a showing of cave photographs at the library in Monterey. These are usually accompanied by a social where food and beverages are served.

3.2 Organization

As stated above, the BCCS is a non-profit, scientific, education, and conservation organization and is a recognized Conservancy of the NSS. It has been accepted by the IRS as a tax exempt organization under section 501(c)(3) of the Internal Revenue Code. Details of the structure of the BCCS are contained in its articles of incorporation [see Stellmack et al. (1970) and in its bylaws Sproul (1972)]. The existence of the BCCS was brought to the attention of the general caving community by Stellmack (1971). Hess (1976) discussed the BCCS at the 1976 National Cave Management Symposium. His paper also included a copy of the articles of incorporation.

The three recognized elements of the organization are the BCCS Board of Directors, the BCCS Members, and the BCCS Friends.

3.2.1 BCCS Board of Directors

A seven member Board of Directors (BOD) runs the BCCS between regular annual membership meetings. The BOD consists of the three Society officers (president, vice president, and secretary/treasurer) plus four directors at large. All seven members of the BOD are elected by the members each year at the regular annual membership meeting. The only formal requirement is that members of the BOD must already be members of the BCCS. See Appendix 2.

The BOD has the major responsibility of managing membership policies. New members of the BCCS are elected by the BOD, not by the existing membership. A strict limit is provided in the bylaws on the total number of members, a limit that can changed only by a unanimous vote of the seven members of the BOD. This bylaw was specifically designed to empower the minority, as described in some detail by Wefer (1980).

3.2.2 BCCS Members

There are no formal membership procedures. By design, anyone is eligible for membership in the BCCS. In contrast to the NSS, there is only one class of membership in the BCCS. The payment of dues can be done annually or for life. For additional information on membership in the BCCS, refer to Wefer (1978a, b) and Williams (1978).

Membership in the BCCS is viewed as a long-term commitment. New members typically have demonstrated this commitment by past support, agreement with Society goals, compatibility with the existing membership, acceptance of possible financial obligations, possession of superior caving skills, and considerable patience. Membership in the BCCS is not touted; there are few benefits, participation is possible without being a member, and membership only means responsibilities. See Appendix 2.

BCCS members are responsible for helping to define the goals of the Society, helping to develop policies that support the goals, providing guidance to the BOD in making major decisions, and providing the necessary resources for achieving the Society goals, namely ideas, labor, and money.

3.2.3 BCCS Friends

Friends of the BCCS are non-members who participate in the activities of the BCCS. This participation may be attendance at expeditions, financial contributions, or other activities such as helping to maintain the grounds and buildings. Friends may also attend BCCS meetings and participate in discussions, but have no vote. BCCS Friends may elect to receive the annual *BCCS Newsletter* that chronicles the year's activities. BCCS Friends number between one and two hundred, the number varying from year to year.

3.3 Appendix 1: Out of the Cove Activities and Expeditions

12/87-01/88 Cueva De Diamonte, Mexico.

10/15/88 1989 Expedition to the Dominican Republic approved as an official expedition.

10/14/89 1990 Expedition to the Dominican Republic approved as an official expedition.

04/21/91 1991 Expedition to the Dominican Republic approved as an official expedition.

10/12/91 1992 Expedition to the Dominican Republic approved as an official expedition.

10/2/93 1994 Expedition to the Dominican Republic approved as an official expedition.

2/21/97 1997 Expedition to the Dominican Republic approved as an official expedition.

5/30/99 The Hawaii expeditions were approved as official expeditions

2001 Expedition to Mexico approved as an official expedition.

10/10/02 Expedition to Iceland approved as official expedition.

3.4 Appendix 2: BCCS Members, Past and Present Through 2012

Name	Life	Date elected to membership—resignation
Arnswaldt, Frederich W. Van		1970–1975
Artz, Michael M. (Mike)	L	1988
Brady, Joseph M. (Joe)	L	1969
Canike, Tony	L	2007
Carter, Peter G. (Pete)		1980
Christenson, Keith		1996
Clark, Lawrence		1970
Clemmer, Gregg	L	1981
Cooper, Brad	L	2010
Cunningham, Paul A.		1970–1973
Davis, Nevin C.	L	1968, Charter, Deceased
Davis, Nevin W.	L	1968, Charter
Decker, William J. (Bill)	L	1969
Deike, George H. III		2010

(continued)

(continued)

Droms, Yvonne		2009
Dyas, Michael D. (Mike)		1991–1994
Faint, Harry		1968–1974
Farrar, Nathan	L	2010
Ficco, Michael J. (Mike)		1995
Ganjon, Richard (Rich)		1977–2005, Deceased
Gibson, Nancy	L	2000
Gilman, Lee B.	L	1976
Good, Lester V. (Les)	L	1978–2002, Deceased
Grimm, Alfred D. (Al)		1978
Haas, John	L	1970
Haverly, Kathryn (Kathy)		2000
Hess, John W. (Jack)	L	1968
Igoe, John W. (Jack)	L	1969
Jones, D. Scott		1982
Jones, William		2006
Kehs, Edward C. (Ed)		1996
Kistler, Michael R. (Mike)		2002
Kohuth, David A. (Dave)		2001
Lillestolen, Jon		2012
Longenderfer, Jay R.		1988
Lucas, Philip C. (Phil)	L	1971–1980, 1990
Lucier, Molly		2012
Madison, Holly		2002
Marks, Franklin J. Jr. (Frank)	L	1969
Martin, Duane		2002
Maxwell, G. Maret	L	1975
McAdam, Hope E.		1991
Miller, Ronald	L	1968–1985
Minton, Mark		2009
Morrow, David B. (Dave)		1989–1997
Nicholson, I. Kennedy (Ike)	L	1968–1999, Honorary Life, Charter, Deceased
Nicholson, Michael K. (Mike)	L	1968, Charter
O'Holleran, Tom		1978–1983
Olson, Scott T.	L	2001
Rigg, Richard H. (Rick)	L	1970
Rosenfeld, John H.	L	1986
Schrumph, Ronald L.		1970–1980

(continued)

(continued)

Schwartz, Benjamin (Ben)		1995
Shifflett, Tommy		1988
Shuster, Evan T.		1975–1976
Simmons, Ronald W. (Ron)		1988–1992, Deceased
Sproul, Mason		1970–?, Deceased
Stellmack, John A.(Jack)	L	1968–2009, Charter, Deceased
Sweet, John R.		1998
Swezey, Christopher	L	2006
Thompson, Glenn	L	1968
Trees (Williams), Toni L.	L	1978
Uhl, Jeffrey W. (Jeff)	L	1989
Vargas, Jean S. (nee Hartman)		1997
Walter, Nathaniel (Nate)		2001
Ward, Rockwell P. (Rocky)		1971
Wefer, Fred L.	L	1968–1999, Deceased
Wheeland, Keith D.	L	1969
White, Elizabeth L. (Bette)		1991
White, William B. (Will)	L	1968
Wilson, John M.	L	1975
Winter, Paul A.		2012

Officers

President	
Clemmer, Gregg	2001–present
Davis, Nevin W.	1971–2000
Stellmack, John A.	1969–1970

Vice President	
Clemmer, Gregg	1995–2000
Hess, John W.	1974
Ficco, Michael J.	2001–present
Marks, Franklin J. Jr.	1975–87
Nicholson, I. Kennedy	1969–71
Wefer, Fred L.	1971–72, 1988–94

Secretary-Treasurer	
Davis, Nevin C.	1969–71, 1975–2001
Gibson, Nancy	2002–present
Thompson, Glenn	1972–1974

Directors	
Artz, Michael M.	1989, 1996–97
Brady, Joseph M.	1975–76
Clemmer, Gregg	1985, 1987–94
Davis, Nevin W.	2001–present
Ficco, Michael J.	1998–2000
Ganjon, Richard	1980–86
Good, Lester	1980–83, 1986–91
Grimm, Al	1992–2000
Hartman, Jean S.	1999–2000, 2002
Haverly, Kathy	2005–2009
Hess, John W.	1971–73
Igoe, Jack	1977–79, 1981
Kohuth, David A.	2004–2005, 2009–2010
Lucas, Philip C.	1999–2003
Marks, Franklin J. Jr.	1974
Maxwell, Maret	1984–87
Nicholson, I. Kennedy	1972–80
Nicholson, Michael K.	1982–83
Olson, Scott	2010–present
Rosenfeld, John H.	1995
Schwartz, Benjamin	1998, 2001–2003
Shifflett, Tommy	1990–91
Stellmack, John A.	1971–80
Wefer, Fred L.	1971, 1977–78, 1981, 1984
Wheeland, Keith D.	1982–98, 2001, 2003–2006
White, William B.	1971, 1975–76, 1979, 1988, 1992–97

Newsletter Editors	
Canike, Tony	2009–present
Christenson, Keith	1997–2001
Gibson, Nancy	2002
Haverly, Kathy	2003–2008
Igoe, Jack	1976
Wheeland, Keith D.	1977–1996

Expedition Leaders	
Clemmer, Gregg (Bobcat)	
Good, Lester	1979
Grimm, Al	1991
Hess, John W.	1971–74
Igoe, Jack	1975–76
Longenderfer, Jay	1988
Maxwell, Maret	1983–87
O'Holleran, Tom	
Stellmack, John A.	1973
Uhl, Jeff and Wilson, John	1990
Wheeland, Keith D.	1989, Temporary 1987

BCCS Member's Awards and Recognition
Davis, Nevin C. and Thelma 1983 BCCS Certificate of Appreciation
Davis, Nevin W. 1992 NSS Lew Bicking Award
2003 BCCS Limestone Award
Deike, George H. III 1961 NSS Certificate of Merit
Dyas, Michael D. (former member) 1991 NSS Lew Bicking Award
Lucas, Philip C. 1990 NSS Certificate of Merit
2001 NSS Lew Bicking Award
2007 BCCS Limestone Award
Shifflitt, Tommy 1996 NSS Lew Bicking Award
Simmons, Ron (former member) 1998 NSS Lew Bicking Award
Stellmack, John A. 1971 NSS Honorary Member
Wheeland, Keith D. 2005 BCCS Limestone Award
2008 NSS Outstanding Service Award
2012 NSS Certificate of Appreciation
White, William B. 1958 NSS Certificate of Merit
1975 NSS Outstanding Service Award
1994 NSS Science Award
2001 Karst Waters Institute Award
2005 NSS Certificate of Merit
Wilson, John 1997 NSS Outstanding Service Award

3.5 Appendix 3: Cave Access Policy

The BCCS conducts expedition weekends at the Butler Cave Homestead. The number of weekends per year is decided at the prior annual regular membership meeting.

Competent cavers are welcome to participate in the continuing exploration, survey, and other cave related activities in Butler Cave and other nearby caves.

1. Only experienced and fully qualified cavers will be accepted for caving trips on the expedition weekends.
2. There will be no tourist trips during the expedition weekends. All trips will be work trips, i.e. survey, dig, exploration, or research.
3. There will be trips of varying duration and difficulty which cavers can state their preference.
4. Trips will be made into Butler Cave as well as other nearby caves. There is no guarantee that a participant will be assigned a trip into Butler Cave.
5. Tourist trips (as well as work trips) may be arranged by contacting any of the members of the BCCS. Trips may be scheduled at a time mutually agreed upon by all parties.

3.6 2008 Addendum: Milestones in Limestone by David Kohuth

[**Editor's Note**: This addendum is an edited version of an article that appeared in the *BCCS Newsletter* Volume **33** for 2008. Photos that appear below were taken by the editor.]

Just in case somebody missed the hoopla, the Butler Cave Conservation Society achieved an important milestone on May 24, 2008. Oh, the "milestones" weren't just in limestone, but also in the hearts of all those brothers, sisters, friends and neighbors who have been involved with this long, strange trip in the Burnsville Cove. The 50/40 celebration honored the dedication of Society members and friends who have contributed to "50 years of discovery; 40 years of stewardship". From Ike Nicholson's 1958 discovery of Butler Cave, through the Butler Cave Conservation Society founding in 1968, to the current day activities of the BCCS, the commemoration was well deserved. The accomplishments of the Society have made it a role model for the entire caving community as well as the general population. By Official Proclamation of Virginia's Governor, Timothy Kaine, May 24, 2008, was declared Butler Cave Discovery Day across the Old Dominion.

I remember it was a few years ago that my Engineering brain ran the numbers and targeted a day "out in the future". I mentioned the dual anniversaries to Gregg Clemmer and wound up with a job. The initial proposal to the Society happened almost immediately so that we could get ready. That was back in 2004. Whoever thought that the big day would arrive.

And arrive it did! The event was described by Society President Gregg Clemmer as follows:

"First, the weather was perfect. Clear, sunny, zero humidity. Registrants officially numbered 142. Saturday evening's BBQ loaded up nearly 200 plates. Campers found the grounds clean, dry, accessible, and mowed. Action verbs popped up from the Homestead like the sandstone float we once dug there: Caving. Photographing. Joking. Eating. Imbibing. Reminiscing. Meeting. Laughing. Hiking. Exploring. Jawing. Playing. And always, Grinning. Burnsville Cove is one of those special places we harken to when we "need to get away" (Fig. 3.1). Some of us have been doing this for decades. A few of us finally moved there. Know that all of you are good stewards of the land. Above and below ground."

And what items will I always remember?

Tremendous weather—blue and white sky for a totally fabulous day.

A throng of smiling people huffing up the hillside to the Butler Cave Nicholson Entrance while...... Duane and Jean's failure to open the locked steel door on Butler Cave just minutes before the Entrance Dedication ceremony and Nevin Davis arriving in the nick of time to open that door, peer inside, and greet an old friend (Fig. 3.2).

Phil Lucas' rendition of the discovery of Butler Cave, highlighted by the personal remembrances of Tommy Burns and Donnie Miller (Figs. 3.3 and 3.4)

A stupendous feast prepared by Master Barbequer Kenny Davis and his crew (by any other name) Hog Wild (OK, Dylan, Austin and Duane). If "barbequer" ain't a word, it is now (Fig. 3.5).

Rockwell Ward's description of "**a thin place**" beloved by us all. Tommy Shifflett reads the Commendation from the State of Virginia Legislature (Fig. 3.6).

Fig. 3.1 The homestead on a bright spring day, May 24, 2008. WBW photo

Fig. 3.2 The crowd waiting for the plaque dedication ceremony at the Nicholson Entrance. WBW photo

The placement of the limestone BCCS monument (carved by Ed Kehs, Sr.) on the Homestead's rock wall (Fig. 3.7).

"Burnsville Bock" crafted by Mike Kistler. Brewed in the Cove with water from the Lockridge Aqua Cave resurgence, a pinch of grass from the Butler meadow, hops from Pennsylvania, bottled on Easter Sunday 2008 and conditioned inside Dave's Gallery of Butler Cave.

The look on Jeff Uhl's face when someone finally told him, without a doubt, that one of the half kegs

Fig. 3.3 Ackie Lloyd, Charlie Moss, and Tommie Burns. WBW photo

Fig. 3.4 Donnie Miller at the Plaque. WBW photo

was indeed Dos Equis. The next look on Jeff's face when that very same keg kicked about 2 min later.

A roaring campfire in the "amphitheater" fire ring and Mike Ficco's raucous fireworks display.

The campfire stories shared with friends over the back-up keg of Sierra Nevada Summerfest.

Late Saturday night, after many had retired from the campfire, Mike Ficco broke from the group and

Fig. 3.5 The barbeque setup. WBW photo

Fig. 3.6 Rocky Ward delivering the dedication of the monument. Tommy Shifflett waiting to read the Virginia legislature commendation. WBW photo

Fig. 3.7 The women of BCCS. Clockwise: Bette White, Jean Vargas, Hope McAdam, Holly Madison, Nancy Gibson, and Kathy Haverly. WBW photo

headed off into the meadow alone. On the way he stopped to answer a simple question with a one word answer spoken very softly: "loud".

Great people and a great time

References

Davis, N.W. 1963. The discovery of hosterman's pit cave. *Nittany Grotto News* 12: 33–40.

Hess, J.W. 1976. The butler cave-sinking creek system and the butler cave conservation society. *National Cave Management Symposium Proceedings*, Mountain View, AR, 26–29.

Sproul, M.M. 1972. Bylaws of the Butler Cave conservation society, incorporated. *BCCS files*, 27 June 1972.

Stellmack, J.A., Davis, N.C., and Nicholson, I.K. 1970. Articles of incorporation of the Butler Cave conservation society, incorporated. *BCCS files*, 25 March 1970.

Stellmack, J.A. 1971. Butler Cave and the Butler Cave conservation society Inc. *NSS News* 29: 30–31.

Wefer, F.L. 1978a. BCCS membership policy. *BCCS Newsletter* 4: 19–22.

Wefer, F.L. 1978b. Membership in the BCCS. *BCCS Newsletter* 4: 22–25.

Wefer, F.L. 1980. The Ike clause. *BCCS Newsletter* 6: 37–40.

Williams, T.L. 1978. BCCS membership, a new member's perspective. *BCCS Newsletter* 4: 25–26.

Exploration of the Outlet Caves

4

William B. White

Abstract

Most of the drainage from Burnsville Cove is underground. Surface streams are few and most flow only during periods of high rainfall. The drainage from the Cove re-emerges from four springs in the gorge of the Bullpasture River. Emory Spring emerges at river level from a rubble pile beneath a highway and nothing is known about the feeder system. Aqua Spring emerges from a large cave at the head of a tributary valley 100 feet higher than the river. The cave streams rise from sumps that extend to unknown depths. Cathedral Spring rises on the river bank from a water-filled cave that has been explored by divers to a length of 950 feet and a depth of 150 feet with further exploration limited by diving technology. Blue Spring is fed by a sand boil at the bottom of a 50-foot deep water-filled pit with no further exploration possible.

[**Editors Note**: Ron Simmons died in a cave diving accident on February 14, 2007. He had agreed to write a chapter describing his underwater explorations of the springs where the water of Burnsville Cove discharges into the Bullpasture River and had provided an outline. What follows has been drawn from various sources including edited excerpts from accounts that Ron published in the BCCS Newsletter. The chapter is, therefore, somewhat of a patchwork but it's the best we can do in memory of an important friend of Burnsville Cove. An interview with Ron appeared in the January, 2002, issue of the *NSS News* (vol. 60, pp. 29–30) and an obituary appeared in the May, 2007 *NSS News* (vol. 65, pp. 30–31).]

4.1 Introduction

The Bullpasture River has its headwaters near Doe Hill some 30 miles northeast of Burnsville Cove. The river flows on shale southwest along the Bullpasture Valley until, at the mouth of Burnsville Cove, it makes an abrupt turn to the southeast through a deep gorge to its confluence with the Cowpasture River at Williamsville. In carving the Bullpasture Gorge, the river has sliced across the complex structure of the Cove, exposing the limestone, and providing outlets for the ground water system. There are four springs draining

In Memory of Ronald W. Simmons. With Contributions by Ron Simmons, John Rosenfeld, Fred L. Wefer, I. Kennedy Nicholson, and Nevin W. Davis.

W.B. White (✉)
Department of Geosciences, The Pennsylvania State University, University Park, PA, USA
e-mail: wbw2@psu.edu

W.B. White (ed.), *The Caves of Burnsville Cove, Virginia*,
Cave and Karst Systems of the World, DOI 10.1007/978-3-319-14391-0_4,

the Cove. From upstream to downstream, these are Emory Spring, Aqua Spring, Cathedral Spring, and Blue Spring (Fig. 4.1).

Except for Emory Spring, the spring orifices are of sufficient size for human entry, but all require SCUBA diving for exploration of the caves that lie behind the springs. Aqua Spring is also the outlet for an extensive air-filled cave, sumped at the upstream end. Cathedral Spring and Blue Spring are entirely water filled and pose significant problems in exploration.

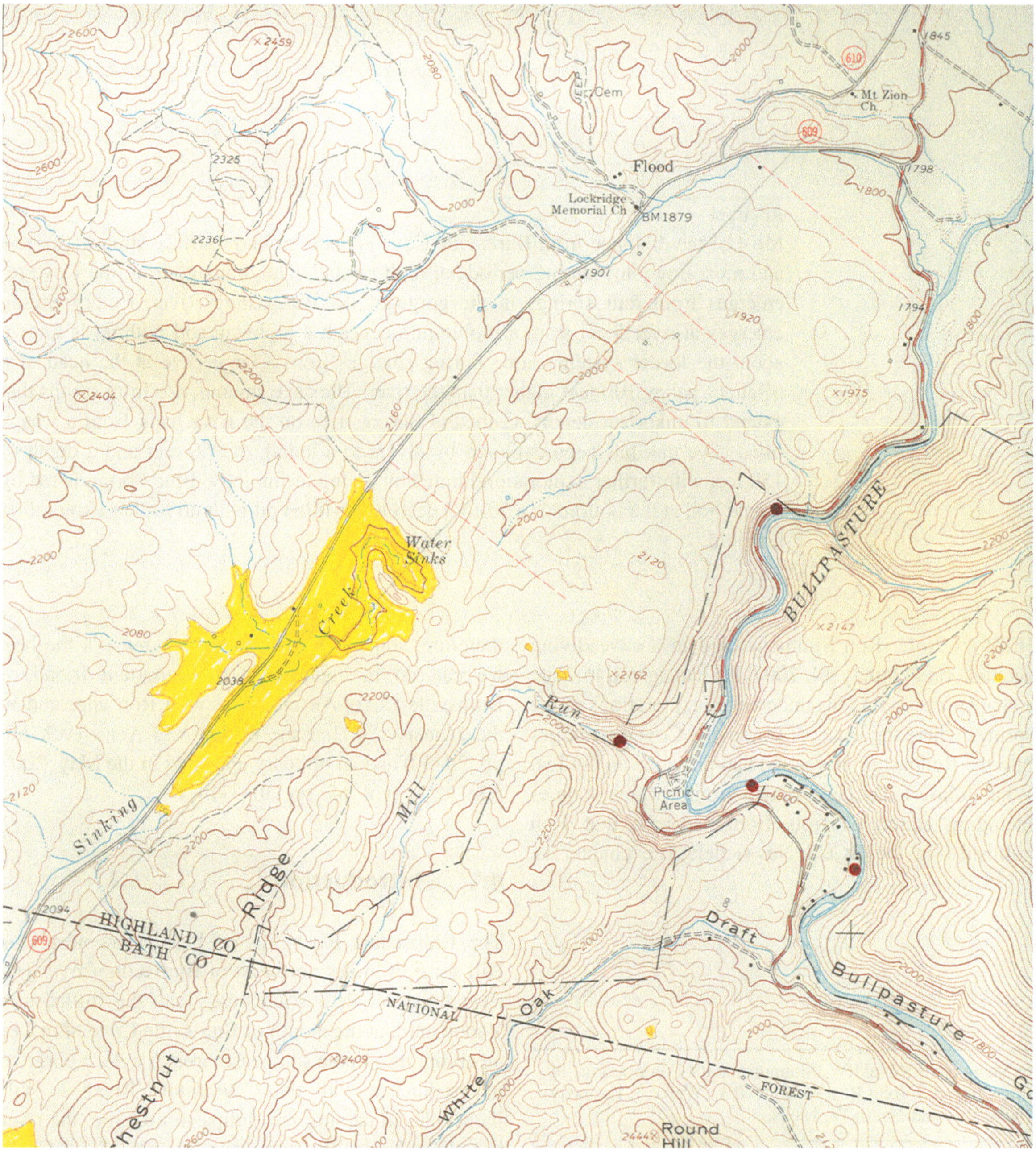

Fig. 4.1 Topographic map of the Bullpasture Gorge taken from U.S. geological survey Williamsville 7.5 min quadrangle. Springs are shown as *red circles*. From north to south, these are Emory Spring, Aqua Spring, Cathedral Spring and Blue Spring. The *large highlighted area* is the water sinks depression

4.2 Emory Spring

Emory Spring emerges from a rock fill beneath Route 678 in the upper Bullpasture Gorge (Fig. 4.2). A map showing the boundaries of the Emory Spring drainage basin is given in Chap. 17. It is a large basin and, based on discharge characteristics, must be draining through a conduit system. Depending on storm distribution, the Bullpasture River may run muddy while Emory Spring remains clear or, when the rainfall is in the Emory catchment, the spring may discharge cloudy or muddy water into the clear water of the Bullpasture (see Chap. 17). There is a steep bluff in the Licking Creek Limestone across the road from the spring but no cave entrances have been found. Likewise, there seems to be no record of the appearance of the spring in its natural state before the construction of the highway.

4.3 Aqua Spring and Aqua Cave (Mill Run Spring; Refrigerator Spring)

Aqua Spring differs from the others in that it does not emerge at river level. The spring—really a cave entrance—is 1200 feet up a tributary hollow at an elevation of 1770 feet and about 100 feet higher than the river (Fig. 4.3a, b). Aqua Spring forms the headwaters of Mill Run and discharges a tremendous quantity of water during flood (Fig. 4.4). Mill Run drains an estimated area of 7.7 m^2.

4.3.1 The Early Exploration by Fred L. Wefer and I. Kennedy Nicholson

[**Editor's Note**: The following is extracted from Wefer and Nicholson's article "Exploration and Mapping of the Sinking Creek System"].

The upper course of Mill Run is usually dry and overgrown but carries an intermittent stream. About a quarter of a mile upstream from the mouth of Mill Run is an abrupt rise of 30 feet in the stream bed. Here, Mill Run Spring issues from an underwater opening 2 feet high and 9 feet wide, feeding a permanent tributary to Bullpasture River. At the suggestion of I. Kennedy Nicholson, Bevin Hewitt donned wet suit and SCUBA gear one weekend in July 1956 and slid into the 51 °F water of Mill Run Spring. About 35 feet in and 6 feet below the elevation at which he had entered, Bevin looked up and saw his air bubbles breaking the water surface. He was soon able to clamber out of the stream onto some breakdown and remove his fins. Further on, the ceiling rose to a height of 60 feet; the passage continued 30 feet wide for as

Fig. 4.2 Emory Spring under normal flow conditions. WBW photo

Fig. 4.3 **a** The head of Mill Run and the entrance to Aqua Cave. WBW photo. **b** Close-up of the spring and cave entrance. WBW photo

far as he could see. He quickly returned to the outside to tell the support party what he had found.

Two weeks later, a party of three spent 4 h exploring the new "Lockridge's Aqua Cave" as Bevin had named it. Big streams were found to run through most of the cave. One lake they discovered was over 40 feet wide and 15 feet deep. A new entrance was created later that summer which allowed entry without

Fig. 4.4 Aqua Cave Spring in flood. Photo by Philip C. Lucas

SCUBA gear. This was done by removing rocks from the spring outlet to lower the water level and by a careful application of dynamite to a hole to the left of the outlet. In the early fall, Mike Nicholson made a 6 h solo trip; the notes he made resulted in the first map of Aqua Cave. On Thanksgiving Day, 1956, a six-member party carried one set of SCUBA gear to the Siphon Room at the end of the right hand, or B, Passage. Mike dived into the gently downward-sloping underwater passage. The slight current was sufficient to carry off the mud as soon as it was raised, thus visibility was quite good. At the end of his 80-foot safety line, the passage was a broad avenue, still sloping downward. Because the Nicholson's had become involved in other activities in Burnsville Cove, the next attempt at diving in Aqua Cave would not be made until May, 1960. This attempt to dive French Lake at the end of the left hand, or A, Passage was foiled by unusually high and turbid water. The first mention of life in Aqua Cave appears in the report of a trip made that September.

In 1962 there were 3 more dives in Aqua by Hank Hoover and Dick Kutz. First to be attacked was the Siphon Room. The diving was ideal; 30 feet of visibility and enough current to carry away any mud. The underwater passage was 5 feet high, 10 feet wide, and curved toward French Lake. The total length was 150 feet at which point it ended in breakdown too small to crawl through. Next to be attempted were the sumps beyond French Lake. The First and Second Siphons were easily negotiated. The Third Siphon was 50 feet deep where the end of the 180 foot safety line was reached. The passage continued in very clear water with a cross section of 10 by 10 square feet. The third dive was again directed at Third Siphon, beyond French Lake, this time with a much longer safety line. The diver turned around 400 feet into the passage and 80 feet down; the 5 foot high by 12 foot wide passage was still going. The return was not without incident; a knife, the safety line reel, and both fins being sacrificed in the process. During these two diving trips, a Brunton compass and tape survey of the cave was also completed (Fig. 4.5).

4.3.2 Later Exploration by John Rosenfeld

[**Editor's Note**: Below is the historical part of "Lockridge's Aqua Cave" that appeared in the *BCCS Newsletter*, **12**, 11–15 (1986). The length and depth were brought up to date by Nevin W. Davis, June, 2011.]

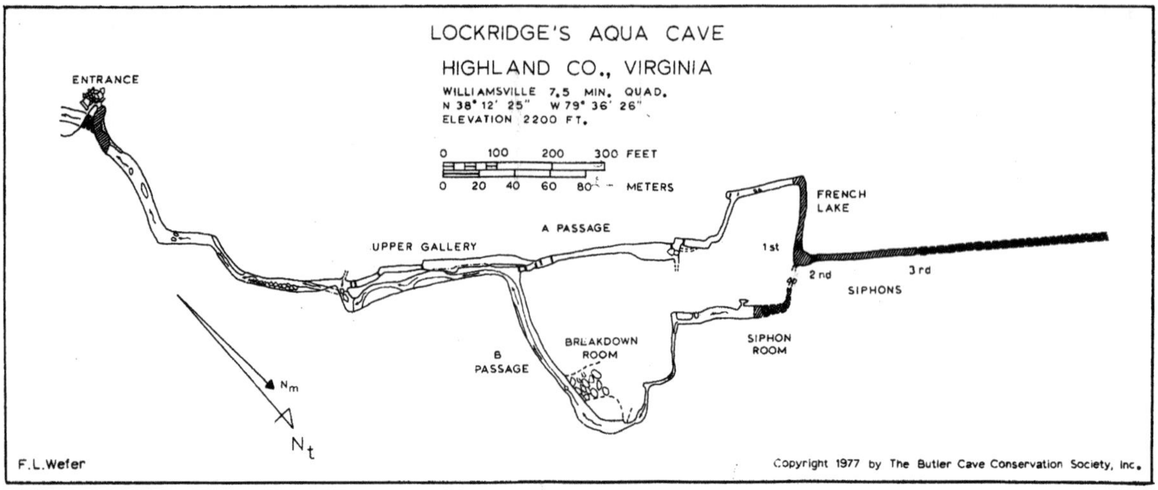

Fig. 4.5 The Hoover-Kutz map of Aqua Cave

Aqua Cave was visited mostly for recreational purposes from about 1962 until 1980, with a few notable exceptions. In 1972, Allen and Doris Haarr attempted the first resurvey of Aqua Cave. They mapped from the entrance to the beginning of the Upper Gallery. This resurvey effort was never completed. In 1975, Sheck Exley led a group from the NSS Cave Diving Section in a dive of French Lake. The divers were unable to locate the long underwater passage beyond French Lake that Hank Hoover had explored nearly 20 years earlier, but they did note a "mud crawl" which they apparently did not push. Additional dives were done in 1980 by Karen Wark, Dave Whall, Dave Morrow, Toni Williams, and Maret Maxwell. The Siphon Room was extended nearly over to French Lake and this sump was surveyed. At this time it was recognized that the end of the Siphon Room sump was located in French Lake (according to the existing surveys). In addition there was no vertical control in the existing surveys. It was thus proposed by the BCCS to undertake a carefully-done resurvey of Aqua Cave in order to correct these problems.

The resurvey effort was begun on April 23, 1983. Partly because it was desirable to solve the superposition problem with the sumps at an early date, the resurveying began at the rear of the cave. The cave was mapped out towards the entrance and not in from the entrance. During these first couple of survey trips, a number of leads were noticed that were high on passage walls or in ceiling domes that certainly had never been checked. These leads were especially intriguing because there is usually a strong wind in Aqua Cave's entrance and its source was unknown. Indeed, it would seem to be a contradiction to observe strong airflow in a cave that ended in sumps after a short distance (Fig. 4.6).

It was suggested that scaling poles could effectively be used to check these high leads. The first scaling pole trip occurred on July 2, 1983. Most of this trip involved erection of the pole and exploration of passages off of the B-Passage and the entrance stream passage. About 200 feet of new passage were discovered on this trip; most of it was found in the entrance stream passage above a short pole climb into a narrow stream crawl which was named Hillary's Climb. One additional intriguing lead was found that could not be reached with 20 feet of 5-foot pole sections.

The second pole trip to Aqua Cave took place on Saturday evening, May 26, 1984. A 25 foot pole was erected at what is now called the Nervous Breakdown Climb. This immediately led to the discovery of the Nervous Breakdown Room. The survey of this new upper level cave began immediately, but it was shortly realized that there was more cave then could easily be mapped in one evening. Thus the pole climb was not derigged and surveying of the upper level as well as tie-in surveying to stream-level stations was continued the next day.

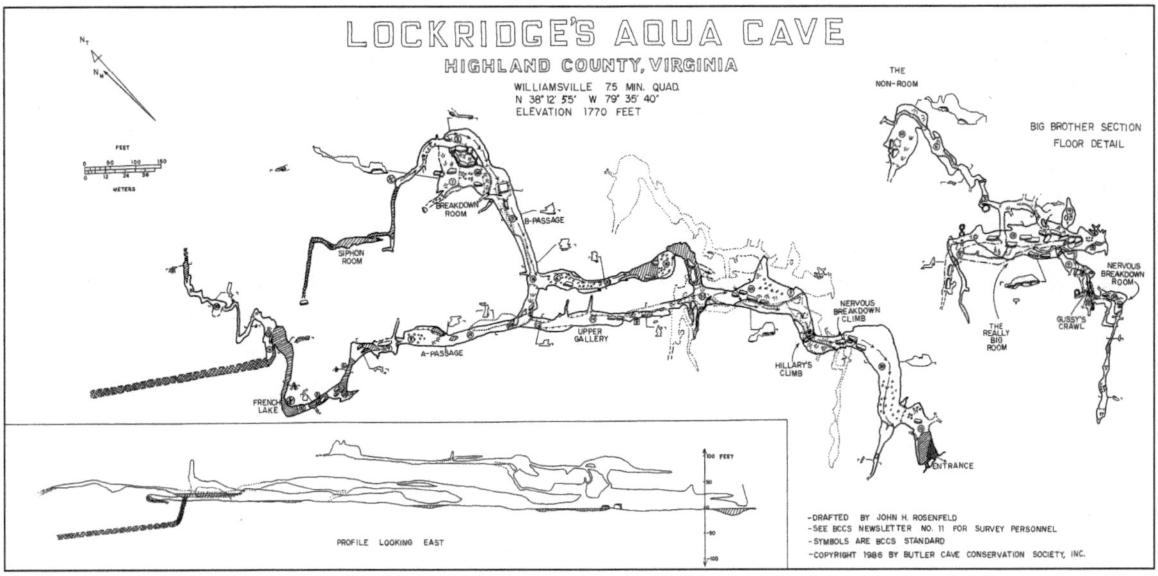

Fig. 4.6 Map of Lockridge Aqua Cave. Map provided by John Rosenfeld

During the trip on Monday, May 28 to de-rig the pole climb, a crawlway was located which immediately led to the discovery of the Really Big Room and some related passages. More passage and another large room named the Non-Room were discovered during a survey trip on July 1, several weeks later. This new upper level was named the Big Brother Section because it was found in 1984 and because it had been watching cavers walk under it for many years. Both main upstream terminations were found to end in blowing dig leads.

A separate find also made in 1984 was the discovery of air-filled passages beyond French Lake. This find was made by Nevin W. Davis and Maret Maxwell. It resulted in about 600 feet of additional survey. This passage was probably the same one noted by Sheck Exley's dive crew in 1975.

Including the estimated survey of the French Lake Sump, the resurveyed length of Lockridge's Aqua Cave is 10,536 feet. This is more than three times the cave length reported by Holsinger. Three thousand feet of this difference is due to the discovery of the Big Brother Section. Six hundred feet of this difference is due to the discovery of the Davis-Maxwell extension beyond French Lake. The remainder of the difference is due to the incompleteness of the first survey of the cave. The known vertical extent of Aqua Cave is 334 feet. The deepest known point is the French Lake Sump and the highest known point is the ceiling of the Non-Room. The sump elevations with respect to zero datum at the entrance are summarized below.

French Lake	28.8 feet
Siphon Lake	25.0 feet
Perched sump beyond French Lake	48.1 feet

4.3.3 Aqua Cave Diving Attempts by Ron Simmons

[**Editor's Note**: Aqua Cave ends in sumps along all of the main upstream passages. The short account that follows was extracted from Ron's trip accounts in *BCCS Newsletter* **14**, 43 (1988) and *BCCS Newsletter* **14**, 53 (1988).]

May 29, 1988. The objective was to dive the sump on the far side of French Lake. With the assistance of D. Scott Jones, Alan Staiman, Jeanne Wetterling, Nevin Davis, and Tommy Shifflett, dive gear was hauled to the sump and I attempted the dive. French Lake was sumped for about 45 feet. I dove through to the breakdown pile upstream of French Lake. I carried my gear to what I believed to be the main upstream sump. The surface of the pool was still with no sign of flow. I penetrated about 25 feet till the visibility went down to zero. This pool can not be the main sump. I think it is a backwater pool from high flow periods. The silt was still settling out. A foot below the surface, the silt formed a level surface that had not been disturbed by any flow of water. French Lake is also still water.

With this new information coupled with that of the last dive, I now think I know the location of the main sump. On the first dive, I found a hole of the right wall of French Lake that loomed into a parallel passage. I think this is the area in which the main stream sump is located.

August 21, 1988. My intention on this trip was to dive what I thought to be the side passage off of French Lake. Assisting were Keith Goggin, Kathy Rosenfeld and Mark Reihart. Due to the drought, the water was way down. My side passage turned out to be part of the French Lake Passage. After much checking, I could find no underwater passages out of French Lake. There is no flow there. This is a perched pool. Keith even found some calcite rafts forming on the pool. The pool just beyond French Lake and the breakdown is also a perched pool. I went out the passage on the other side of French Lake to the sump at its end. There does not look to be any divable passages off of the pool. It looks like a gravel sump. By the looks of the gravel bank in front of the pool, I would say that this is a lift tube from a lower passage. It blows the gravel out in flood stage and fills back up when there is no flow. I believe that French Lake is kept full by this sump overflowing plus some drip infeeders from the surface. The whole French Lake section is an overflow during floods when the main stream cannot handle the volume. After these observations, I did not dive in French Lake.

We hauled the gear back to the main stream and went upstream. The last 300–400 feet of passage before the sump is mostly hands and knees crawl with some low air spaces (down to 6–8 inches in some spots). The upstream sump can only be gotten to during low water periods.

The sump had fabulous visibility. It was at least 20 feet and could have been 30 feet. The water looked blue with my 20 W sump light. Since the 1985 flood, visibility has been low because the water was milky. It was now crystal clear.

When I started into the sump, I found an old dive line. I didn't go far until it was buried under gravel. I laid 122 feet of line to a spot where the ceiling had collapsed. One big block filled the passage (at least 4 feet by 20 feet). I was able to go over it and continued on another 40 feet of so. The passage had much breakdown and was tight, but I could see much larger tunnel ahead. It may open up in another 30 feet or so, but it was too tight to continue. I had to back out the last 10 feet of exploration. Visibility was very good even when in contact with the floor or breakdown. The flow is high; the passage roars from the flow.

[**Editor's Note**. As is apparent from the descriptions above, the extent and interconnections of underwater cave beyond the Aqua sumps is ambiguous. Nevin Davis culled through Ron Simmons' dive notes and constructed a map of the sump area. His notes follow.]

The map (Fig. 4.7) is based on sketches that Ron made. The map was drawn with the same north orientation as on the Rosenfeld map (Fig. 4.6) and if printed at the same scale should adjoin it perfectly. The numbers 1799 feet above mean sea level are the surveyed elevations of the points taking 1770 feet above mean sea level as the entrance elevation. There are probably errors in the elevations. The Siphon Room pool level is 1791 feet while the elevation of French Lake is 1799 feet and there is another sump at 1815 feet. One of these should be at the same elevation as the flowing stream at the Siphon Room. Numbers with a bar over them are depths of water below the local water surface.

The Habatub is a decompression chamber fabricated by Ron from a large, tough garbage container. It is secured in place by cables and contains a way for the diver to secure himself in it mostly out of the water for long decompression times.

The dive upsteam by Dick Graham and Ron with Nevin watching occurred on 20 August, 1988. The first sump survey was on 14 October, 1995. This survey was not accurate and was not used. The second sump survey occurred on 19 September, 1998.

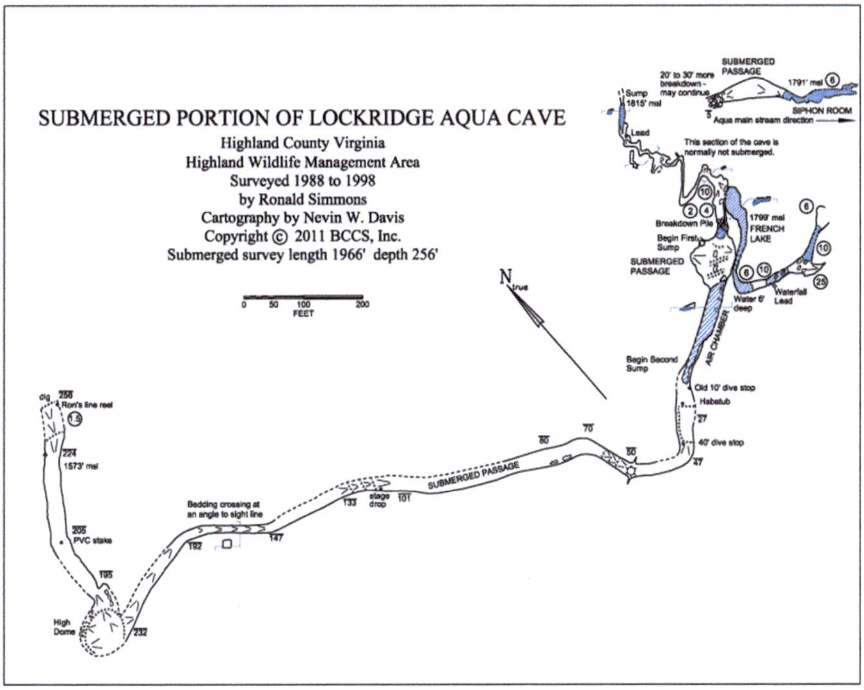

Fig. 4.7 The submerged portion of Aqua Cave. Map compiled by Nevin W. Davis

4.3.4 Description of Lockridge's Aqua Cave by John Rosenfeld

[**Editor's Note**: "Lockridge's Aqua Cave" is a lightly edited version of the cave description that appeared in the *BCCS Newsletter* **12**, 15–21 (1986).]

The entrance coordinates are N38° 12′ 25″, W79° 35′ 40″ and elevation 1770 feet msl. The streams in Butler, Breathing, Boundless and Better Forgotten Caves as well as those in the north end of Bobcat Cave all drain through Lockridge's Aqua Cave to Mill Run Spring. The cave is not significant for its length but for its relationship to these other caves.

From the entrance, the cave continues upstream as a large stream passage for 600 feet to an intersection with a dry parallel upper level called the Upper Gallery. There are several relatively short side passages leading off from the Entrance Stream Passage. This passage is one of the largest in the cave. It is up to 60 feet high and wide. The Upper Gallery parallels the stream passage upstream from the junction for about 300 feet and rejoins it at the upstream end. The stream passage is low and wet in this area and is not visited very often for this reason. There is an enlargement in the passage at the point where the Entrance Stream Passage intersects the Upper Gallery. A short upstream side crawl leads off of this area. There is some airflow in this passage but digging in the mud/cobble choke at its end has not yet produced an extension. Five cubic centimeters of ethanethiol scent tracer was released into this passage during cold weather when the airflow direction was inward. No positive scent detection on the surface was established during subsequent ridge walking of the Mill Run Valley.

At the upstream end of the Upper Gallery, two separate passages continue upstream to sumps. The A-Passage trends for 500 feet to French Lake. Several short upper level loops and side passages are developed along its length. A typical passage dimension is 15 feet wide by 20 feet high. French Lake is a sizable passage that is mostly filled with water and contains a deep divable sump at one end. It is possible to float across this lake in times of low water although for much of the year the entire passage is sumped. At the far side of French Lake, the cave divides into two continuations. The lower continuation is a sump which has been penetrated for 400 feet to a depth of 80 feet; at this point the passage was still continuing at comfortable dimensions. The other passage leading off from the far side of French Lake is the Davis-Maxwell Extension. This passage meanders for 600 feet to an undived sump perched 16 feet above French Lake.

The other passage leading off from the far end of the Upper Gallery is called the B-Passage. It consists of a large 30 foot diameter stream passage trending north for 250 feet to a large room called the Breakdown Room. It is possible to climb up into this room for 200 feet until a breakdown collapse prevents further progress. The stream passage continues upstream for another 300 feet to the Siphon Room. This room contains a sump which has been penetrated to a location very close to French Lake.

The upper level of Aqua Cave joins to the lower level cave at two climbs. Both of these connections occur in the Entrance Stream Passage. The closer of the two to the entrance is called the Nervous Breakdown Climb. It is at the top of a dome located about 250 feet from the entrance. The other connection occurs at a long narrow ceiling rift about 325 feet upstream from the entrance. This rift, Hilary's Climb, can only be accessed by a scaling pole at least 20 feet long.

At the top of the Nervous Breakdown Climb there is a large airy room called the Nervous Breakdown Room. Several passages lead off from this room. One passage loops around to the north of the room and rejoins it as a balcony high on its west wall. This passage also continues southwest from the balcony for 300 feet and terminates as a flowstone choke.

A low crawl in the passage to the north of the Nervous Breakdown Room trends for 160 feet north to a large room called the Really Big Room. This room is up to 25 feet high, 60 feet wide and 300 feet long. It is possible to climb down through the breakdown in the floor of this room and follow a sinuous, tight, popcorn-encrusted stream crawl for about 300 feet to the top of Hillary's Climb. The west end of the Really Big Room continues as a stoop way for a short distance to a down climb 20 feet high. At the base of this climb a cobble-floored fossil stream passage is intersected. This passage can be followed "upstream" for a short distance to a consolidated cobble dig. The airflow in this passage is notable. The Really Big Room contains several "tombstones" embedded in deep sand as well as one notable anthodite cluster.

A crawlway leads off the north side of the Really Big Room for 500 feet to another large room called the Non-Room. This room is 120 feet long, 60 feet wide and 35 feet high. Air blows out of this room during the warm months. The source of the airflow is undetermined. The passage that leads to the Non-Room also contains anthodite clusters.

Lockridge's Aqua Cave lies at the northeast end of Chestnut Ridge. The Chestnut Ridge Anticline in this area is a broad, folded anticline that plunges gently to the northeast. These folds tend to complicate the geology. The cave is developed along a well-defined lineament that trends perpendicular to Chestnut Ridge. This lineament can be identified from features shown on the topographic map. Examination of local land features shows that Water Sinks, the lower part of Mill Run Valley, Aqua Cave, and the Bullpasture Gorge are roughly in line with each other.

At some time in the geologic past, the drainage from Sinking Creek Valley was diverted from its northeasterly course. Water was channeled underground to the southeast along the lineament and into the developing Bullpasture Gorge. Evidence for the approximate time of this event may possibly be obtained from study of Aqua Cave. There are cobble banks in the upper level of the cave that were left behind as passages were cut farther down into the Jersey Shore Limestone. A few large anthodite clusters also exist at this level in sizeable passages. This part of the cave is now 100 feet above active stream passages.

[**Editor's Note**: The original write-up had Aqua Cave correctly placed in the Keyser Limestone but because of the extensive revision of both stratigraphic nomenclature and also the relation between caves and strata (see Chap. 16) the following paragraph has been re-written to be consistent with current nomenclature.]

Aqua Cave is developed mainly in the Keyser Limestone. The stream level passages are developed in the upper Keyser Formation with Clifton Forge Sandstone exposed along the stream passage. The existence of impervious layers of sandstone has had an effect on the development of the cave. The big Brother Section is developed partly in the upper part of the Keyser Limestone and possibly extends into the overlying New Creek Limestone. A portion of the passage development seems to be at or above the contact between the Keyser and New Creek Limestones. A few places in the upper level also pass through the New Creek and Corriganville Limestones into the cherty Licking Creek Limestone. This generally occurs in domes or in tall narrow fissure passages.

Fig. 4.8 The largest of the sandstone boulders in Aqua Cave. These have apparently been rolled up the lift tube through the Aqua sumps. Photo courtesy of Arthur N. Palmer

The upper level of Aqua Cave has passages that are floored with non-carbonate cobbles up to one foot in diameter with several that are extremely large (Fig. 4.8). Since these cobbles are not generally found in passages in Burnsville Cove that lie under Chestnut Ridge, it may be assumed that they had their origin on Jack Mountain to the northwest of the cave. This would then require that they had traveled for a considerable distance underground in order to exist at their present locations. The Water Sinks, located 3000 feet to the northwest, is the closest currently existing point from which these cobble could have been washed underground.

4.4 In Search of the Cathedral System by Ron Simmons

[**Editor's Note**: This section is a lightly edited version of the article from the *BCCS Newsletter*, **16**, 1–5 (1990). What we know about the downstream end of the large Cathedral Spring drainage basin is Ron Simmons' account. It is written in the dive log format. He never had the opportunity to update what he wrote in 1990.]

Cathedral Spring is located along the bank of the Bullpasture River, about 2000 feet downstream from Mill Run and Aqua Cave. The drainage basin of Cathedral Spring runs parallel to the Aqua Basin and extends about five miles to the south. By all indications there should be a large cave in the Cathedral Basin, as there is in the Aqua Basin. Part of the Cathedral System has been found in the Chestnut Ridge caves. The present downstream end of Bobcat Cave is about 8300 feet from Cathedral Spring.

The spring emerges from among boulders across a wide area of talus slope, in front of a large cliff of Licking Creek Limestone (Figs. 4.9 and 4.10). Under the talus slope are what appear to be large blocks that have fallen from the cliff face. There is a small entrance between two of these blocks, just an alcove with water on the floor. Just below water level is a wide low horizontal passage going into the hill. This is what I saw when I first visited the spring. The following article is a log of my efforts to explore the Cathedral System.

Fig. 4.9 Cathedral Spring. Photo by Philip C. Lucas

Fig. 4.10 Water from Cathedral Spring draining into the muddy Bullpasture. Photo by Philip C. Lucas

4.4.1 May 20, 1989

When I came back for a serious attempt, the water level had dropped about a foot. This exposed the horizontal passage under the ledge. The passage was only about a body length long and opened into a small room, big enough for a couple of people. The room was floored in water several feet deep. On the side of the room, going into the cliff, was a narrow slot in the floor. A large percentage of the flow of the spring appeared to be coming out of this slot. The slot was a little over a foot wide but had many thin projecting ledges, which made it impassable.

On my first dive I used a single 60 cubic foot tank. The first part of the dig was essentially open water in that I could get back to air in a single breath. The work involved removing projections from the walls of the slot. In reality this slot was not in solid bedrock, but was made from the sides of huge blocks that had come off the cliff. The projections were thin chert and limestone ledges. It only looked like a passage in solid bedrock. Using a large rock, I was able to remove enough so that I could move forward on my side. The flow in the slot was quite strong and any silt I stirred up was quickly swept away. The clarity of the water was also excellent. It was the clearest water I had seen on a cave dive outside of Florida.

At this point I brought a second tank. The slot I had opened up took a sharp right hand turn at a depth of about 5 feet. I tried backing around the turn feet first, dragging my tanks behind me. This didn't work very well. So I went in head first with my tanks in front. This was very awkward, but I managed to get around the corner and into a big enough space to turn around. There was a passage there running parallel to the cliff. The ceiling was the bottom of a huge block and the floor was all breakdown. The main flow was coming out of a hole, with breakdown piled in front of it. I managed to get my head down the breakdown slope far enough to see a good solid passage beyond. This was far enough for this dive. It took a long time to work both tanks and me back out to the little air-filled room although it was probably not more than 20 feet. I was only wearing a wet suit with no harness, fins, helmet etc. And the passage was still very tight. Obviously this dive site would need some more work before I could get into and explore the passage I had glimpsed.

4.4.2 June 3, 1989

The next dive called for a new strategy. I strapped an 18 cubic foot pony bottle to a steel 60 so I could move my redundant air supply as a single unit. The pony was to be used as a bail-out bottle if anything happened to my main tank or regulator. The big tank was my main air supply for digging. I had a three pound hammer on this dive and used it to remove some more of the ledges in the first part of the passage. At the breakdown slope I spent some time moving rocks. I moved enough so that I could back down the slope into the passage at the bottom. This was a real passage with solid bedrock ceiling, floor, and walls. It was 15 feet wide and 3–4 feet high. There was no silt on the floor! The high flow keep it swept clean. I ventured down the passage a short distance. It got larger and kept on going, but I wasn't set up for swimming. So I returned to the breakdown slope at the end of the entrance restriction and continued removing rocks. The entrance into the tube was still a very tight squeeze. On this dive I spent over an hour in the water and I was starting to get cold in my ¼-inch wetsuit.

4.4.3 June 11, 1989

On the next dive I decided to try and get full dive gear into the passage so I could explore. The entrance section is far too tight to wear my gear in, so I decided to drag it in and put it on when I got into the bigger tube. I wore my side mount harness with buoyancy compensator and battery pack attached. I also wore my helmet. I used two steel 72 tanks on this dive. One had a seven foot hose on the regulator which turned out to be very handy. The passage was too small to move both tanks at the same time, so I had to sort of shuttle the tanks using the seven foot hose. At the breakdown slope the extra bulk of the harness, buoyancy compensator, and battery pack made or an extremely tight fit. My helmet wedged several times. I finally took it off, making one more thing to keep up with in the low visibility. The flow was considerably down on this day and I was stirring up a lot of silt under the breakdown. My light battery pack also got pulled off.

Finally, I got all my gear and me into the bigger passage. It took over half an hour to cover 40 feet of cave. I also had used quite a bit of air. At this point I decided not to explore, but save all my air for the exit. Considering all the trouble I had getting in, I didn't want to risk running out of air trying to get out. Who knew how many times I would get stuck on the way out? Obviously I was going to need a new approach to this dive, if I wanted to get anywhere (Fig. 4.11).

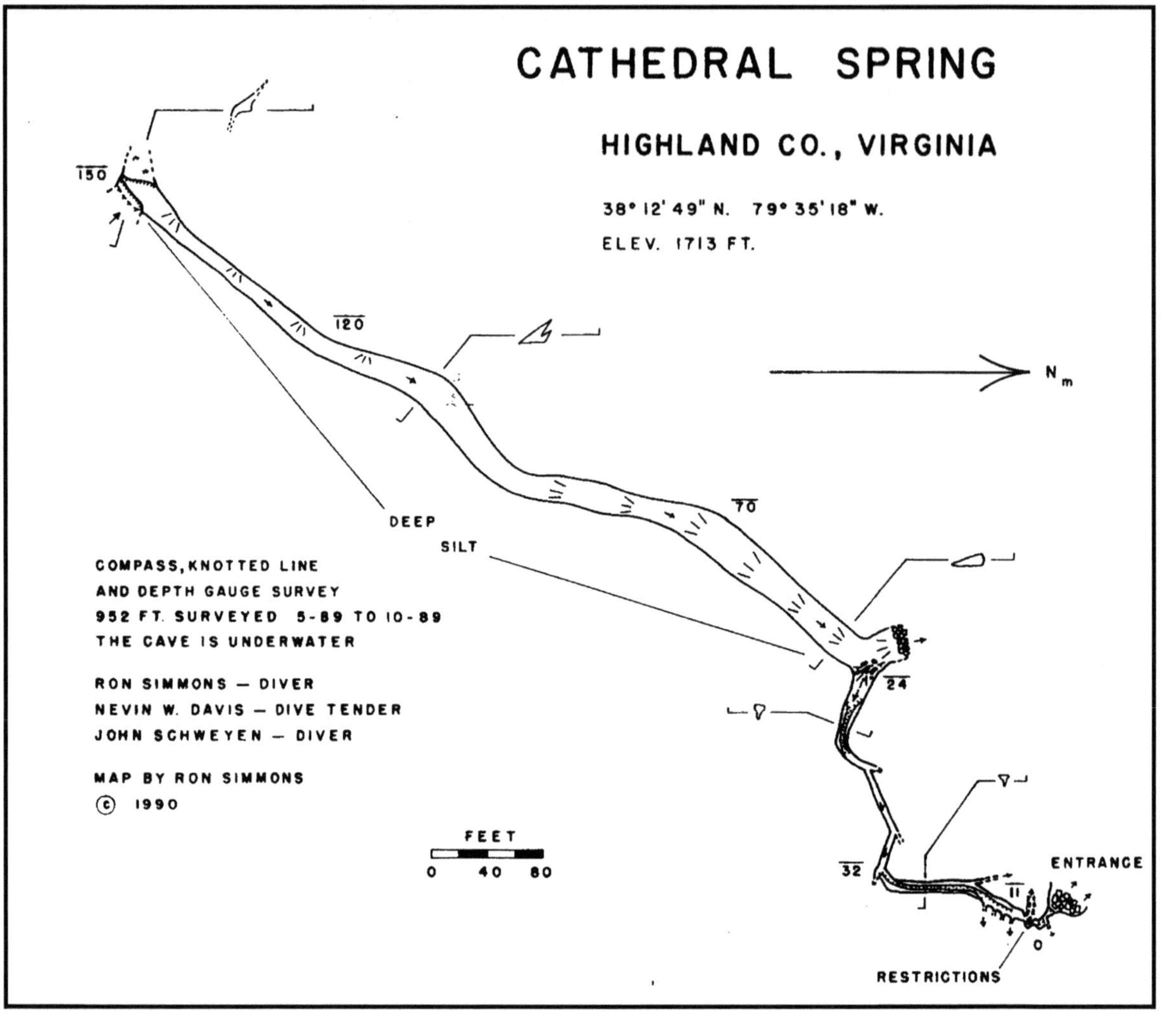

Fig. 4.11 Map of Cathedral Spring prepared by Ron Simmons in 1990

4.4.4 July 16, 1989

What I came up with was a regulator with a 50 foot hose on the second stage. Using this I could place a tank at the entrance and use the air in it to get my other tanks and gear into the passage where I could put them on. In this way I saved all my air in my side mount tanks for the dive. One unexpected drawback of using the hose was rubber-hose breath. The hose flavored the air a bit.

On the next dive I used this new technique. I put a steel 95 tank at the entrance; I wanted plenty of air in case I got really stuck. My main tanks on this dive were two steel 60s. I wore a wet suit with side mount harness and buoyancy compensator attached. Everything else, lights, helmet, fins, reel, etc. was strapped to the two tanks. In this way I could make two trips into the main tube using the 50 foot regulator. This worked much better. With only one load at a time to deal with I could slide through the restrictions much easier. My main diving tanks served as backup if anything happened to my 50 foot hose.

Once into the main tube I used air from my outside tank while I kitted up. The 50 foot hose was just long enough to reach a spot in the main tube big enough to put my gear on. The passage at this point is a bedrock tube with no silt on the floor. Nevin Davis, who had been helping me haul gear to and from the spring, now helped by serving as my dive tender. He put on his wet suit and came in as far as the little air-filled room.

He belayed the 50 foot hose in and out as I hauled my tanks and gear in. Otherwise I would have gotten wrapped up in my work, so to speak.

Now I would finally get to see what this passage did. The spot where I put my gear on was at a depth of 10 feet. From here the tube sloped gently downward to about 30 feet deep, where it became a joint-controlled canyon. It did not look phreatic at all. More like any Virginia maze cave. The passage was mostly bare rock due to the high flow, but there was still silt in protected pockets, easy enough to stir up. On this dive the visibility was down from the usual 15–20 feet to about 5–8 feet, because of rain a couple of days before. The passage width varied from 3 to 5 feet and the height was 4–10 feet. This went on for about 300 feet.

Then I came to a breakdown slope and the passage got wider. The slope led up to a low spot. I had to move a few rocks to get through. On the other side was a larger passage; the depth was only 24 feet. To my right the passage ended quickly in silt and breakdown fill. By surveying, Nevin later determined where this spot is on the surface. It is a place under about 100–120 feet of talus along the cliff, upstream from the main spring. When Cathedral Spring is really pumping out a lot of war, water also comes out along the base of the talus at this spot. Apparently at one time there was a large passage that came out to the Bullpasture River here. Now the main flow of the spring goes through the newer and smaller canyon to emerge at the main spring.

To the left of the restriction a large passage headed off. This passage looked phreatic. It had an oval cross-section and was about 8 feet high in the middle and around 20 feet wide. I started down this passage. It was descending at a gentle grade. At a depth of 70 feet and a little over 600 feet from the entrance I tied the line off and started to survey out. The survey out was a cold one. I had a hard time holding the compass still enough to read. My ¼-inch wetsuit was starting to seem thin during such a long exposure. The water temperature is 51 °F. A complicating factor was low visibility, down to 1–2 feet. Any silt that I stirred up in the larger passage took much longer to clear than at the entrance.

Back at the end of my 50 foot regulator I took my gear off and packaged it back up to haul out. Visibility was essentially zero going out. I had to feel my way out of the entrance restrictions. I was real glad that I have done this section a number of times before. This time I was going on memory. There was no line in the first 30 feet of passage. It was just too tight and the danger of entanglement was too severe. If I got tangled in the line I might not even be able to get a hand near the tangle. But it's very hard to get lost in such a small place. My total time on this dive was an hour and 25 min.

4.4.5 July 23, 1989

I had been able to do the last dive with the entrance section the way it was, but it had been very tight. So I decided to put in some time improving the entrance.

Nevin again helped me both in carrying gear down to the spring and also in serving as my hose tender. This time I wore a dry suit to keep warm. I set up my 50 foot regulator on a 95 tank and took a pony bottle in with me as a bail-out. If anything happened to the 50 foot regulator, the pony would give me enough air to get out. On the first dive of the day I went into the breakdown slope to remove rocks. The first restriction was about as good as it was going to get. I had removed all the projections down to bedrock. At the breakdown slope there was nowhere to put the rocks without taking them out of the cave. So I pushed the down the slope into the main passage. There I started building a rock wall along one side of the passage. Moving large rocks underwater proved to be fairly easy. I wish it were this easy in air-filled cave. I must have moved between ½ and 1 ton of rock. I was even able to remove part of the floor at the restriction into the tube.

On my second dive I removed more rocks. I also extended the survey out to the entrance from the beginning of the permanent line. The entrance section was now an awful lot better. Instead of a 30-foot long squeeze, I had a 10 foot restriction followed by a 5 foot restriction with a small room in between.

4.4.6 August 20, 1989

This time Kathryn Rosenfeld would serve as my hose tender since Nevin was on a Bobcat camp. On this dive I was going to explore the passage further. I used two steel 95 tanks. At the entrance restrictions I did a little more clean-up work before dragging my gear in and kitting up. After kitting up, I did some sketching of the passage and also cleaned up a few spots where the line was badly belayed. I tied off at the end of exploration and started laying more line. The passage character was changing. I was now in a more strike-oriented passage. It had a slanting profile at the angle of the bedding, about 30°–50°. The silt was getting really deep. Line tie-off's resulted in zero visibility around the tie-off. The passage was still going down at a steady rate, angling across the strike of the bedding. I seemed to be going down one limb of an anticline.

I reached the end of my line but couldn't find a good spot for a tie-off. The silt was too deep for a drop weight. So I reeled back up the passage until I found a tie off. Since I had accumulated a lot of bottom time and had hit a depth of 135 feet, I did not survey out. I had a long enough decompression waiting for me already. As it turned out, I was underwater for 2 h and 24 min. Only 44 min was actually spent on the dive. The rest of the time was spent getting in, moving rocks, decompressing and in getting out. At least decompression benefited in one way. It gave time for the silt I stirred up during the dive to clear out.

4.4.7 August 27, 1989

This dive was to be yet another dig. I spent an hour and 13 min removing more rock. I was able to peel up some large slabs on the floor of the second restriction and improve it considerably. The entrance was now almost reasonable. Once again I used my 50 foot hose and a pony bottle for digging—a real nice arrangement for underwater digs within 50 feet of an entrance.

4.4.8 September 10, 1989

Since the visibility of Cathedral Spring is so good (especially for a sump dive), I decided to try to get some photos. But to take photos I needed a model. John Schweyen agreed to help me. Nevin once again helped out in hauling gear and as hose tender.

Logistics became more complicated for two divers. I went into the cave first with my tanks, using the 50 foot hose. Then I came out. We waited for the silt I had stirred up to clear. The flow was down and this took a while. When it had cleared I went in first with a pony bottle and my camera gear. John then came in bringing his dive gear using the 50 foot hose. I was able to get some good shots of John going through the restrictions. I did have a back up air supply; one of John's tanks was always within easy reach, since he pushed them ahead.

Once into the bigger passage both John and I kitted up. We then went down the canyon. The canyon was not very wide and there were quite a few shelves with silt on them. So silting was a problem especially with low flow.

After I finished taking photos, I dropped my camera gear at the beginning of the big passage. John continued on down to lay some new line. He laid about 150 feet of line and reached a depth of 150 feet. Since his bottom time was already fairly long, he just laid the line and did not survey it. I followed him for a way but soon turned to head out. It would be best for both of us if we did not reach the entrance at the same time.

By the time I had finished my decompression at 10 feet, I could see John's light at his 20 foot stop. The water was still quite silty. Visibility was down to 1–2 feet during my exit. By the time John had finished his decompression visibility was somewhat better. I had been in the water almost 2 h and John was in the water over two and a half.

4.4.9 October 15, 1989

To improve safety I set up a valve system for the end of my 50-foot second stage hose. I hooked up outputs from two first stage regulators to valves and then to the fifty-foot second stage. In this way there would always be air in the long hose even if a tank needed to be changed. This also allowed for the addition of pure oxygen for the 10 foot decompression stop, adding a bit more safety to the dive. This was not a dive to have any problems with decompression sickness. The nearest recompression facility is at Duke University in North Carolina. Nevin again served as my tender.

On this dive I wanted to survey the line that had been laid on the last two exploration dives. I used my steel 95 tanks again. On the 50-foot hose I had air and oxygen. Since at dive times of over 2 h even my dry suit gets cold, I tried something new. I used a separate pony bottle to inflate my dry suit. In the pony I put a gas mixture of carbon dioxide and argon. The mix is one commonly used in welding and is readily obtainable. The use of this mix makes the suit warmer; as it does not conduct heat away from the body as fast as air.

I had become more efficient going into the passage. It only took me 16 min to get the gear in and kit up. It took another 16 min to get to my last survey station, seventy feet deep and about 600 feet from the entrance. This is where I began to survey. Things were going fine until there was an implosion and everything went black. The lens on my light had shattered which in turn broke the bulb. I was using a modified Wheat Lamp hooked to a pack of high capacity Ni-Cd batteries. This had been such a good trouble-free light that I did not think about its maximum depth. The glass lens on the front can only stand the pressure down to 120 feet. I switched on my other primary light. Since I had plenty of backups, I continued surveying. I reached the end of the line where John had tied it off to a lead weight, at a depth of 150 feet and 952 feet from the entrance. Just beyond the end of the survey, the bedding became more vertical and looked like a pit. The top of the pit was a restriction, with deep silt on the floor around it. This cave didn't want to let up. The survey had taken 10 min, and getting back to my first decompression step took me another 10 min. In order to do the 10 min of surveying I now had to plan an hour of decompression before it would be safe to exit the cave. By the time I had all my gear out of the cave, I had been in the water for about 2 h.

During decompression I found a flat spot on the ceiling and pumped my dry suit up with the carbon dioxide-argon mix. It really was a lot warmer. It was comfortable to just hang there and let the time go by. Decompression is boring. At the 10-foot stop where I could use the 50 foot hose, I used pure oxygen. Nevin had switched the hose over after I stopped using that regulator on the way in.

This last dive was the tenth that I had done into Cathedral Spring. Of those ten dives, six had been used to dig out the first 40 feet of passage and make a reasonable entrance into the larger passage. One of the dives had been a photo trip. On the other three I laid line or surveyed. John Schweyen also laid line on the photo trip. Cathedral is not going to give up its secrets easily. No one expected the passage to go so deep, but it did. If the passage keeps going down it will be very difficult to go much further. A depth of 150 feet is enough to cause concern about nitrogen narcosis, the impairment of metal activities that is caused by breathing nitrogen at greater than atmospheric pressure. To go much deeper on air is not a real good idea considering that the cold water and the extremely silty conditions can add to the narcosis effect. There is also the problem of decompression. By going so deep, the times for decompression go up rapidly. Long

Fig. 4.12 Blue Spring showing spring run draining out to the Bullpasture. Photo by Philip C. Lucas

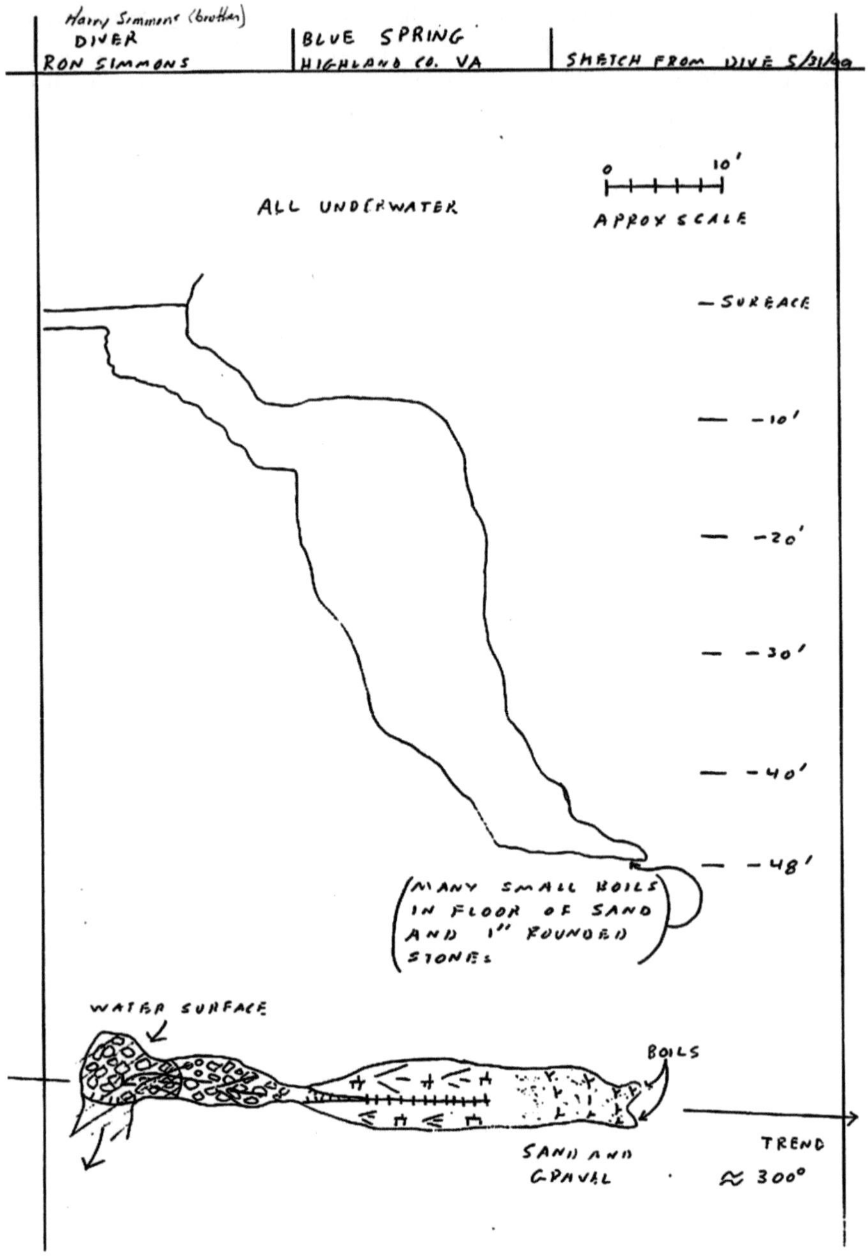

Fig. 4.13 Blue Spring. Previously unpublished sketch map by Ron Simmons

decompressions in 50° water are not going to be comfortable, dry suit and argon or no. Switching from air to a special gas mixture of helium and oxygen would solve the problem of nitrogen narcosis but is very expensive and actually makes the decompression times longer.

At present the cave stands at a surveyed length of 952 feet and a depth of 150 feet. Cathedral Spring will not likely be pushed much further, at least at this time. Some time in the future either myself or someone else will need to use more advanced technology to explore this cave.

I want to add a special note of thanks to Nevin W. Davis. He helped on all of my dives at Cathedral except one. Nevin has not only helped haul gear to and from the spring but has proven to be an invaluable tender for my 50 foot hose. I wouldn't trust just anyone with taking care of my air supply. I also want to thank the many other people who have helped haul my gear to and from the spring.

4.5 Blue Spring

Blue Spring is the farthest down-stream of the four springs on the Bullpasture River. The spring is only a few feet above river level and rises from the edge of narrow flood plain. The gorge is wider here where the valley of White Oak Draft enters from the south. It is connected to the river by a short spring run (Fig. 4.12). The spring is in an area of open fields and cabins and has been used as a water supply.

The water issues from a solutionally-widened fissure in the Tonoloway Limestone and spills over the lip into the spring run. According to Douglas in his 1964 report on the Caves of Virginia, Beven Hewitt dove the spring in July, 1956 and found the fissure to open into a dome-shaped room that bottomed out at 50 feet. The next dive did not occur until May 31, 1999 when Ron Simmons explored the spring. His exploration essentially confirmed Beven Hewitt's report of 43 years before. There were no leads from the water-filled room. The room slopes downward and at the end water rises from sand and pebbles as a boil on the passage floor (Fig. 4.13).

The Discovery and Initial Exploration of the Chestnut Ridge Cave System

5

Gregg Clemmer

Abstract

Chestnut Ridge Blowing Cave, later known as Bobcat Cave, proved to be the first entrance into the large Chestnut Ridge Cave System. The discovery, however, required five years of patient and extremely difficult exploration to conquer the half-mile entrance series into the big cave that awaited below. A detailed account is presented of the obstacle-by-obstacle advance into what would become one of Virginia's longest cave systems.

> A strong current of air blows steadily out of a small hole in the slope. The cave was opened with a crowbar. The entrance opens into a large flat room 50 ft. wide and 100 ft long with a ceiling 2–12 ft. high. At one point there is a shaft which can be descended for about 30 ft. into another large room. From this room there are several fissure side passages. Most of the air comes from a fissure heading NE which drops down to a tortuous stream passage 800–1000 ft long which gets too small for a man. The current of air is still strong at this point.

Thus reads Ike Nicholson's description of Chestnut Ridge Blowing Cave as published in Henry Douglas' *Caves of Virginia*. Yet Chestnut Ridge Blowing was just one of more than 40 known caves peripheral to the Butler Cave-Sinking Creek System, then the Old Dominion's longest known cave. By the 1980s, the exploration of Butler was thought largely complete. To be sure, leads remained and there was always mop-up survey to be done (or redone), but the main features in the system were considered in place. Still, cavers suspected Burnsville Cove harbored much more cave than had been found. Hydrology studies hinted at large and lengthy underground conduits—with quick transit times—for the subsurface streams that coursed beneath the Cove. Nevertheless, despite these strong indications of potential discovery and two plus decades of active caving, new discoveries in Burnsville Cove remained rare (Fig. 5.1).

Then in the Summer and Fall of 1983, a large cavern complex located in the heart of Chestnut Ridge was entered for the first time. The story that follows recounts the exploration of the Chestnut Ridge Cave System, an intense, often tortuous project that at times challenged the physical and mental limits of all who sought its secrets, an exploration that ultimately climaxed in the bottoming of one of the most physically demanding caves in the United States.

5.1 Just Another Mudhole…?

When they first entered the cave in March, 1979, Shenandoah Valley Grotto members Kent Seavers, Paula Casale, Doug Molyneaux, and Gregg Clemmer found Chestnut Ridge Blowing Cave to be little more than a thousand feet long. At its known end, a small muddy stream gurgled into a low, sloppy crawl,

G. Clemmer (✉)
25 Silverton Court, Hunt Valley, MD 21030, USA
e-mail: yorkst@aol.com

W.B. White (ed.), *The Caves of Burnsville Cove, Virginia*,
Cave and Karst Systems of the World, DOI 10.1007/978-3-319-14391-0_5,

Fig. 5.1 The Bobcat Cave entrance. Nathan Farrar, Jon Lillestolen, Molly Lucier, Scott Wahlquist, Paul Winter, and Brad Cooper. Compare the team about to enter with a team after exit (Fig. 5.12). Philip C. Lucas photo

disappearing after 200 miserable feet into an impenetrable slot six inches high and eighteen inches wide (Fig. 5.2). The only indication to the foursome of more cave beyond was the persistent breeze in the passage. A dig trip in wetsuits to this point a month later proved an exercise in futility.

Interest was rekindled when Clemmer talked Molyneaux and Tom Shifflett into a return visit in April, 1982. According to long-time Burnsville Cove caver, Nevin W. Davis, an obscure, tight crack above the low stream crawl led to an upper level bypass that eventually intersected a domepit several hundred feet beyond the 6-in. slot. At the time of his survey trip in 1973, Davis had noted a stream crossing the floor of this dome and disappearing into breakdown. Rick Rigg had hastily checked this and pronounced it “no go”.

Fig. 5.2 Tommy Shifflett starts into the historic water crawl watched by Fred Wefer, Nevin Davis and Gregg Clemmer. Photo by Ron Simmons

Knowing how things can change over time, Shifflett, Molyneaux, and Clemmer elected to have another look. Could this stream be the continuation of the one disappearing into the slot? Could they make it go?

The trio reached the domepit after worming through very tight, slimy crawls and descending into a series of nasty canyon cracks. Clemmer stayed at the top as belayer while Shifflett and Molyneaux dropped the 19-foot pit. Just as Rigg remembered, the stream disappeared into muddy breakdown. There was no obvious continuation, yet Shifflett and Molyneaux believed that removing one large rock might reveal more cave. Resolved to return, the trio slimed their way back to the entrance, exiting well after dark, taking more than two hours to travel less than 400 feet. This was going to be a tough "little" cave to follow (Figs. 5.3, 5.4 and 5.5).

With Molyneaux unable to participate, Shifflett and Clemmer entered the cave a month later and blasted the rock choke, bringing down a considerable portion of the ceiling. With debris still crashing erratically to the floor 30 minutes later, they exited with no real idea of the results.

In late June, Molyneaux led a survey of the cave accompanied by Paula Casale, Joe McKenney and Dave Hall. Shifflett, Clemmer, and Kent Seavers went ahead to check the blast. The now shattered fissure obliterated any leads seen in April. Yet the cave's breeze continued through the collapse; moving rocks revealed a small lead. This opened into wider space. A narrow, jagged slot, blowing air, sloped down and away. After removing more rocks and mucked gravel, Shifflett tried to force his way through, but got stuck. Clemmer attempted to clear the projecting rocks by smashing them with his feet but the rocks only wedged tighter. Finally, with Shifflett's lamp sputtering and Clemmer yelling encouragement, Seavers forced his body—minus helmet—down the slot, wedged his shoulders into a narrow corner and then popped out of sight. Shifflett's lamp sputtered out as Seavers shouted back from beyond, "It goes!"

In several minutes he was back and began enlarging the slot—now known as the Snake Hole—from the other side (Figs. 5.6 and 5.7). Shifflett and Clemmer soon pushed through, the team mapping 100 feet of sharply dipping, irregular stream canyon. Although the passage continued with a good breeze, a greasy, barely passable, plunging crack stopped their efforts for the day. Bobcat Cave, as the team was now calling their new project because of the cat seen near the entrance, had grudgingly yielded its first virgin passage in a decade.

Fig. 5.3 Dave Morrow pushing his camp duffel through one of the flowstone restrictions in the Historic Crawl. Photo by Ron Simmons

Fig. 5.4 Gregg Clemmer in one of the tightest spots of the Historic Crawl with his duffel. Photo by Ron Simmons

Fig. 5.5 Tommy Shifflett starting down the 19-foot Historic Drop. Photo by Ron Simmons

Fig. 5.6 Tommy Shifflett descending through the Snake Hole with his camp duffel. Photo by Ron Simmons

Fig. 5.7 A view of the Snake Hole from the lower side

5.2 Poisoned Limestone

The initial enthusiasm of the group soared on reports of virgin cave. Yet subsequent trips achieved little as many first-time participants lost their ardor in the slippery mud, the tight, steep canyons, and the slimy, chilly crawls. With the cave already 107 feet deep at the Snake Hole and now plunging steeply into the ridge on its eastern side, it seemed that with each succeeding trip that summer, any zeal for this nasty, unrelenting, but still going cave easily melted away.

In July, Shifflett led a trip that pushed downward to a sizeable chamber. The room quickly shut down to a single, extremely tight lead. Joe McKenney, in an attempt to climb down a narrow chimney 25 feet above the stream, dropped his lamp into a series of deep plunge pools. Duly christened the Lost Lamp Room, the cave now had cut its sinuous path to 270 feet below the entrance. Yet the barely negotiable crack that swallowed the stream still blew air. Seavers, Shifflett and Clemmer returned on a follow-up trip to push this only lead out of the Lost Lamp Room. Seavers stayed behind because of a bruised knee injured on the way in. He gave the pair 30 minutes to explore ahead. Shifflett and Clemmer immediately encountered a tortuous stream canyon. Up, over, down, across—the passage seemed to end half a dozen times only to present a small, barely passable slot in the most obscure place just ahead. The contorted crack carved its path downward, providing little resting space. Often it required 20 feet of up and over caving to advance a mere five. After 200 feet of such gnarled cave passage, the stream entered a comfortable dome—High Lead Dome. Although their time was up, Clemmer and Shifflett grunted ahead another 50 feet, stopping in a deep pool at the bottom of still another twisting, narrow climb-down. The cave continued, but with no sign of easier going ahead, they turned back to Seavers. The trio exited the cave well after midnight, taking four hours to exit.

Until this trip, most of the project members had figured the cave would soon break wide open or end in a miserable, low, gravel choke. No one had really considered that the cave might continue as a plunging, tight, muddy crack for several thousand more feet. After the trip beyond the Lost Lamp Room, this bleak possibility began to rear its head. Several in the group quietly, yet understandably, opted to ignore Bobcat for more "enjoyable" caving. To most, Bobcat, while still a going cave, had never really been pleasant. Quiet but sincere concerns were raised about safety and the wisdom of continuing the survey in such a hypothermic nightmare. And yet one undeniable lure remained: *where did that persistent breeze beckon from*?

Shifflett and Clemmer, unable to recruit anyone and not really relishing a return survey beyond High Lead Dome, nevertheless entered the cave in late August. They mapped over 300 feet of continuously tight, convoluted, muddy stream canyon, plunging steadily down the inside of Chestnut Ridge. Though they finally ended their lonely survey exhausted, the canyon continued mockingly and monotonously ahead, now over 355 feet deep. It was on this trip, in light of the recent Tylenol/cyanide murders in Chicago, that Shifflett and Clemmer dubbed this "poisonous place," Cyanide Canyon.

Yet the fact that Cyanide Canyon "still went" recharged some of the group. Frustrated at the slow progress of the survey and perhaps even subconsciously angered at the cave's persistent nature to "defeat us," Shifflett and Clemmer launched a fresh effort to wrap up the survey in early 1983. On March 19, Shifflett, Molyneaux, and Clemmer entered the cave in wetsuits and after a very wet, scratchy, confining trip to the bottom, surveyed only 52 feet before being stopped by a tight, impassable slot. Although it could be widened with a chisel, the newest obstacle seemed to darken their hopes. Old questions resurfaced. *Why continue such a miserable and unrewarding project?* With knees of hamburger and muscles of jelly, the dejected trio started the long, chilling ordeal out of the cave, their cotton coveralls pounded in glop, yet all three fully aware that the cave's breeze from the unknown promised better things to come. Like a siren from the deep, it would lure them back (Figs. 5.8 and 5.9).

It took four months. Heralded as a "do or die effort," Shifflett, Seavers, Molyneaux, and newcomer Dave Morrow returned to the tight slot at the end of July and chiseled through. Now 366 feet deep, the cave guided the four into another impassable spot that stubbornly yielded to the chisel. Lying in water, Morrow managed to inch a few feet ahead in the very narrow crack. A third fissure, this time only the size of a fist, seemed to spell the end. *Was this the final "capsule" of Cyanide Canyon?*

Shifflett eased up to and past Morrow for a look. Beating and knocking, small bits of rock fell with his efforts and he eventually squeezed past this third constriction and found a fourth. Kent pressed ahead at this point with his thinner frame, but could not advance. He wedged his body into a higher slot, but still no go. Shifflett, underneath Seavers, began to work with the chisel. The cave stream, by now a cold,

Fig. 5.8 Gregg Clemmer coming through the Chisel Slot in Cyanide Canyon

Fig. 5.9 Dave Morrow with camp duff in Cyanide Canyon. Photo by Ron Simmons

muddy slop resembling intestinal gruel, poured around and across Shifflett's flattened body, exiting from both sleeves of his coveralls. Yet he worked steadily with the chisel, fortified by an occasional curse, until he was able to slither past even this nasty spot. Beyond, the cave could be seen to open up, but another very tight spot remained. More chisel work, angled in wet contortions and ample hard words, finally brought success.

Twisting, watery crawlways led for several hundred miserable feet to a ten foot drop with a pool at the bottom. Fired by a strong desire to "make the cave go," the foursome considered belaying the drop with packs tied together end-to-end, but yielded to better judgment. Finally, Morrow managed to free-climb the thing and Molyneaux followed. Not finding a promising lead, Dave told Doug to give it a go. Molyneaux followed the narrow, slithering stream crack to a point where he had to shove rocks and muck aside to continue. Part of one wall collapsed on his arm in the process.

Worming through the sleeze after digging himself out of this mess, Molyneaux noticed he could crawl upward to a wider area. He wallowed up a slope and continued, now on all fours. *Was the cave changing character?* The stream six feet below, wound around the corner out of sight. Several feet farther, the passage widened to twenty feet—and then he glanced up into a darkness that swallowed his lamp's beam!

5.3 Breakout

Jubilant whoops brought the others to where Molyneaux had climbed. A quick peek revealed several large, going leads into both stream passage and dry, fossil cave. With time a definite consideration, Molyneaux, Morrow, Seavers, and Shifflett turned back, excited, but tired and unsure of exactly what they had found. They exited the cave at 3 a.m., 15 and 1/2 h after having entered. News of this breakthrough spread quickly among the "Bobcat Crew."

Nevin Davis, who had decided to chance a trip during the pushes in Cyanide Canyon and who had then retired from the group's efforts with the sage suggestion to "call me if you find something better," received a smug notice that it might be time for another look. With Seavers, Molyneaux, and Clemmer committed to the push trip, Davis would have plenty of company—but the day of the trip, amid tales of grim going, stories of two and three day recuperations after a trip, all amid the smirks and gallows humor of the group, he opted to sit out the trip in favor of a headache and upset stomach. "Maybe next time, fellows." He would be held to his word.

It was decided that Molyneaux and Clemmer would continue the survey at the Chisel Slot. Seavers and Morrow would scout ahead, looking for a possible bypass over Cyanide Canyon. After many contorted shots (that have been glazed into this writer's memory as images of Molyneaux's white teeth framed in lots of mud), the survey broke out of Cyanide Canyon and up into what is now known as Terminal Canyon, finding Seavers and Morrow returning from their scout. They had discovered a superb bypass (now known as Bypass Dome) at the top of a dark dome that Molyneaux and Clemmer had just surveyed. Future trips would no longer have to grovel through Cyanide Canyon's final mud slurry. Reunited, the foursome continued their survey to the south, finding the passage averaged 20 feet wide and 80 feet high.

Climbing up into a dry passage over the stream, the survey bent to the right. Then several stations later, it ended abruptly in a blank wall. Floor pits hinted at more passable cave below. Returning to Terminal Canyon, the group climbed into a lead 40 feet above the floor and surveyed north. Dry as a bone, this passage—Tombstone Alley—carried good air (Fig. 5.10). But after 700 feet, the bottom fell out of the passage, and yet another muddy, gloppy drop seemed the only way to continue. With almost 1300 feet added to the survey and the cave now over 380 feet deep, at 10 p.m. the cavers began the long, wet, cold agony of exiting. The urge to sleep several hours after fighting the mud and slop of Cyanide Canyon was almost overwhelming. Drained of reserves, the team exited the hillside at 6 a.m. after 19 hours of intense caving. Two days later, one of the crew mentioned that he felt as if he had been in an automobile wreck.

5.4 The Trip We'll Never Forget

The crew, strong, eager, and prepared, entered the cave at 11 a.m. on October 8, 1983. It was as if they had forgotten the nightmares of previous trips. In minutes, however, Molyneaux trashed the zipper in his

Fig. 5.10 The beginning of Tombstone Alley off of Terminal Canyon. Note figure for scale nearly hidden in the *center of the picture*. Photo by Ron Simmons

coveralls. He left the cave and returned clad only in his wetsuit. Nothing would stop him this day. Four and one half hours later, (entry times were beginning to improve), the five cavers entered Terminal Canyon. Davis seemed a bit disappointed by his first look, but joined Clemmer and Molyneaux as they headed south to survey the stream below the dry upper passage mapped in August. Shifflett and first-timer Maret Maxwell were to check some side leads off Tombstone Alley. Possibly one might intersect Cyanide Canyon.

Clemmer and Molyneaux quickly got wet amid Davis' chuckles as the stream survey ended in an impossible slot with only 30 feet surveyed. Shifflett and Maxwell were finding "tracks"—Morrow's and Seavers'—in what was supposed to be virgin cave. But it didn't go anyway. After 90 minutes of this unproductive exercise, the two teams collectively ended up at station A20, the survey station of Tombstone Alley ending at the edge of the gloppy pit. Enthusiasm was ebbing in favor of an early exit. Memories of warm clothes, hot food, and sunlight—so far away—began to erode morale. Conversation drifted—to any subject except the one at hand. To his credit, Shifflett quietly began rigging the pit at A20. "We might as well use the rope since we brought it this far." The others listened to the rope hum as he descended, first one drop, about 25 feet, then a second, this one a very sloppy 20 feet.

"Off rope." Only Shifflett's boots cracking in the mud broke the silence. Then, "BOREHOLE! BOREHOLE! IT'S BOREHOLE DOWN HERE!" Those at the top initially thought this a poor attempt at hype, but then Molyneaux got curious and went down. "YOU GUYS BETTER GET DOWN HERE! IT'S BIG CAVE!"

The remaining trio scrambled to see who would be next on rope. And, indeed, it was borehole! Blackness roared off to the north and south as Tombstone Alley emptied into this large trunk passage. Maxwell, agog with such a find, headed north with Shifflett. They surveyed 400 feet of large trunk before being stopped by a wide, deep trough. Beyond, on the other side, more darkness cloaked the cave's secrets (Fig. 5.11).

Davis, Molyneaux, and Clemmer headed south. Dry, tall canyon wound through the depths of Chestnut Ridge, now more than 500 feet above. Soon 1000 feet had been mapped and the group's only concern was that Clemmer's mechanical pencil might run out of lead. He was instructed to write *very lightly.*

Their survey seemed not to end, the cave unfolding before them. Then, as Davis climbed a rocky slope and pulled tape for the next station in a small alcove, he exclaimed, "I can't see the walls. It's getting black out there." The 20 × 30 foot cave passage had just given way to 100 by 60 foot passage…with no end in sight! They shot a few stations more and then erected a rock cairn. Far below, a stream gurgled. Leads were everywhere! And mingled with the excitement, the elation, and the satisfaction that only such a trip can bring, came the strange, almost melancholy feeling of the remoteness that permeated this chamber.

The trip out was another muscle-stretching nightmare, a repeat of the "Death March" in August. It took all night to make the entrance, and when they surfaced, all five cavers, dead tired from exertion and lack of sleep, collapsed into a stupor on the leaf-carpeted hillside, wet, muddy clothes and all (Fig. 5.12). No one moved for minutes until two individuals became

Fig. 5.11 Tommy Shifflett at the bottom of Scoopers Drop. Photo by Ron Simmons

Fig. 5.12 Emergence (or is it Deliverance?) The typical appearance of an exploration party emerging from the Bobcat Entrance: Tommy Shifflett, Nevin Davis, Gregg Clemmer and Dave Morrow

physically ill and one vomited. The sun's early rays that Sunday morning were never more welcome, but it was becoming obvious that this sorry, pitiful group of cavers was going to have to find a better and safer way to explore Bobcat Cave (Fig. 5.13).

5.5 A Better Way

If there is anything that characterizes the true caver, it is a short memory and a vivid imagination. While physically hammered after the October discovery, everyone burned in speculation over "what the cave might do." Yet given the five hour transit time to traverse the 1800 foot entrance series and the resultant inefficiency in getting people to a mapping station, it was obvious that the team would have to adopt some other method to continue the project. But what? Wintertime speculation around fireplace and armchair succumbed to the lure of virgin cave, blurring the trip-trashing muck crawls and rotten rock walls of the previous October.

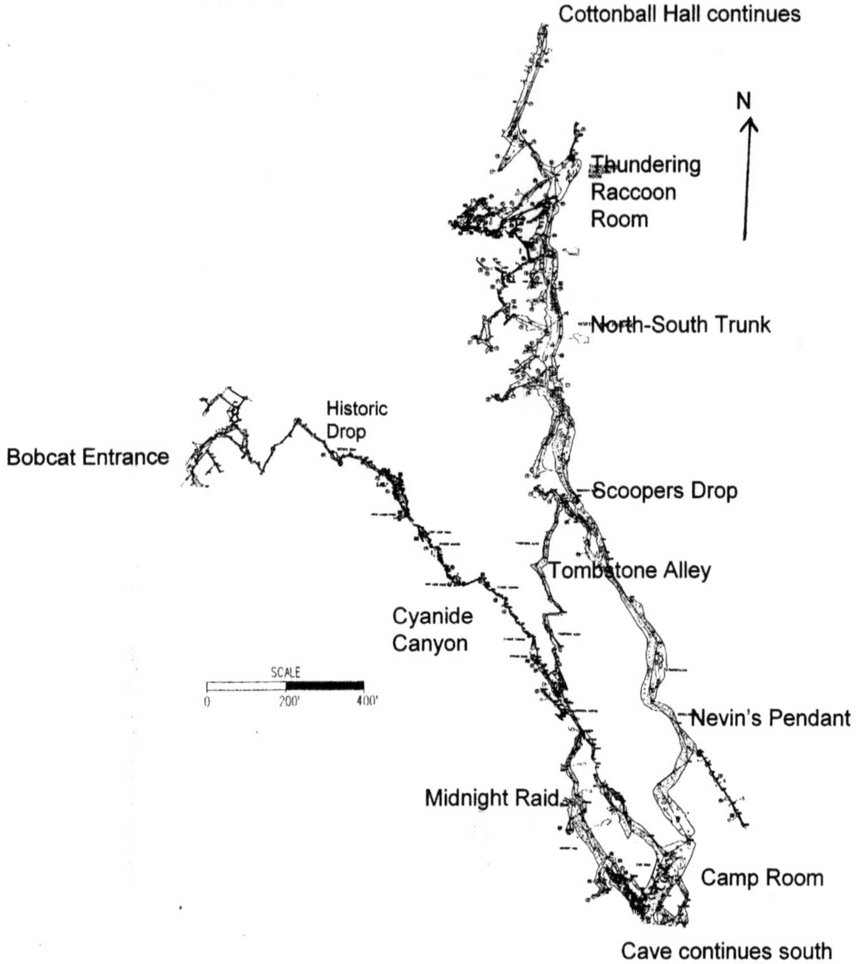

Fig. 5.13 A segment of the Chestnut Ridge System map showing the entrance series, Tombstone Alley, the descent into the North-South Trunk at Scooper's Drop, and the Camp Room

On March 17, 1984, Molyneaux, Clemmer, and Morrow entered Bobcat, followed by Seavers, Shifflett, and first-timer Ron Simmons. The first three would continue the survey due south from the rock cairn, while Shifflett, Seavers, and Simmons would survey east from the same point. Molyneaux's team mapped 800 feet of large rambling passage, ending their survey underneath the cairn room. All leads seemed to end in more pits.

Meanwhile, Shifflett, Simmons, and Seavers were in for a treat. Their passage headed east, immediately plunging down a very steep slope. After 700 feet it broke into a large dome room, named Shamrock Dome for the March 17th discovery. They pressed the mapping ahead for several stations, but by this time, Molyneaux's team had joined the effort and there was talk of leaving. Several members of both teams were getting cold—and yes, entrance fever loomed. Even in going cave, the dreaded Cyanide Canyon continued to cast a pall over the trip.

The exit through Cyanide Canyon was not as bad as it had been in October, but the trip still impressed all with the need to adopt a better "method" of cave exploration. This trip had taken six months to organize. Clearly, the last two efforts had taxed everyone's abilities, and yet the surveying had been in very large, easy passage. What would happen once the easy stuff was mapped?

Shifflett and Simmons suggested camping. Both were experienced expedition cavers with multiple underground camps to their credit. Some of the crew pointed out that their experience had been in Mexico where caves were 60 plus °F. Simmons and Shifflett countered by saying that with proper gear, even Bobcat's 49 °F temperature could be made comfortable. With no other visible options, everyone listened.

Cotton coveralls and other garments disappeared in favor of nylon, DaMarts, polypro and other synthetics. Freeze-dried food replaced Beenie-Weenies and gorp. Simmons made everyone a camp duff to haul their gear into camp, including light-weight, yet very warm, biker's sleeping bags. Therm-a-Rest pads underneath added immense comfort. By the end of May, nine cavers were ready to give this camping idea a try.

With this frantic activity to prepare for the "great cave camp," Shifflett had been busy crunching numbers from the last several surveys. What he discovered astounded the group. At the final station down next to Shamrock Dome, the cave had plunged to 648 feet deep! A quick check of deep caves in Virginia had found only Butler to have a depth of more than 600 feet to −623 feet. Bobcat was now the Old Dominion's deepest!

5.6 Seven Hundred Feet… and Deeper

Camp One entered the cave on May 18, 1984 and did not exit until 100 hours later. Nearly 4000 feet of virgin cave were mapped, following the trunk north near the axis of Chestnut Ridge. To the south, however, beyond Shamrock Dome, Shifflett led Maxwell, Simmons, and Davis into a terrible slop hole, one that quickly and tightly became vertical. It plunged to a sump in one of the more desolate sections of the cave. Reports put the mud there as having the consistency of magnetic axle grease. This station, however, charted the cave to a depth of 705 feet, something no one had expected. The following year, in an adjacent passage almost as slimy but just as treacherous, Shifflett, Clemmer, and Morrow mapped to a lower point, this time documenting the cave to 722 feet below its entrance on Chestnut Ridge. Bobcat was now deeper than any cave in Virginia or West Virginia. But more importantly, this first Bobcat camp had worked far better than expected. The team had a viable plan to continue the cave's exploration and mapping: subsequent exploration would be via underground camps.

The Bobcat Camps

6

Gregg Clemmer

Abstract

Once the large cave system accessed through the Bobcat Cave entrance series had been discovered, there was the problem of how to explore it. The entrance series required 5 h transit time each way, leaving little time and stamina for exploration. The solution was to camp in the cave and to conduct exploration from the camp over several days. Camping techniques are described followed by detailed logs of the first 32 camps. Exploration eventually revealed a large, complex, and deep cave extending over much of the northeastern end of Chestnut Ridge.

In the aftermath of the May 1984 cave camp, the Bobcat crew contemplated their next such venture deep into the system. With more than 4600 feet of virgin cave "in the book" and everyone safely in and out of the cave after the four day camp, confidence soared. But no cave in the United States had ever been continuously pushed and mapped in such a manner. Dare they try it again?

To be fair, few caves offered the isolation and daunting physical challenges that justified camping underground. Primary concerns centered on how to thwart hypothermia, obtain adequate nourishment, maintain endurance, and still safely and efficiently get the cave competently explored and mapped. Custom nylon coveralls were the first big change initiated by the Bobcat campers. Until the early 1980s, experienced American cavers—with rare exceptions—went underground clothed primarily in cotton and wool. "Farmer John" or Montgomery Ward coveralls ruled the day. Wet suits when worn were tight and uncomfortable but battled hypothermia and provided protection far better than blue jeans and corduroy jackets or the wool sweaters and flannel shirts of earlier times. Comfort counted and the eternal cold of soggy cotton and smelly wool when still miles from the entrance begged for garments promising warmth, agility, and the ability to dry.

6.1 Techniques and Logistics

Nylon and other synthetics did just that. Participants in the first Bobcat camp purchased yards of the fabric, adapted coverall patterns provided by Ron Simmons–aided with groundbreaking repair work by cave gear pioneer Mike Artz who would join the project on Camp 7—and sewed their own. The bulky, battered metal towing cans of the 1954 C-3 Expedition and 1973 Project Simmer never even came under consideration. Instead, long, cylindrical, flexible nylon duffel bags—the first batch sewn by Ron—performed admirably. With a tether on one end for upright attachment to seat harnesses when ascending or descending drops, a handle in the middle for grasping in crawls and crevices, and back straps for carrying over long distances, the "camp duff" proved invaluable for getting gear and

G. Clemmer (✉)
25 Silverton Court, Hunt Valley, MD 21030, USA
e-mail: yorkst@aol.com

W.B. White (ed.), *The Caves of Burnsville Cove, Virginia*,
Cave and Karst Systems of the World, DOI 10.1007/978-3-319-14391-0_6,

food into camp. Double or triple thicknesses of trash bags protected food, clothes, and sleeping bags from devastating leaks. Sucking out the air from such a packed bag before tying it off provided additional space. In today's caving, nylon packs and coveralls are the norm and even exude their own fashion statements, having spawned a cottage industry in custom cave gear, vertical rigs, and personalized repair using a variety of incredibly durable fabrics. Changing into dry, warm camp clothes upon reaching camp boosted morale, especially if one's body heat aided the process; thus, wickers and polypropylene replaced cotton and wool undergarments (Fig. 6.1).

Sleeping underground, though, had always been a prolonged struggle against chill and dampness. Cotton or down sleeping bags proved dismal failures in the netherworld, but lightweight fiberfill or synthetic bags worked nicely when laid on a foam or Therm-a-Rest pad atop a reflective ground cloth. A stocking cap kept head and ears warm all night. Some campers even wore gloves. A dry change of socks, bound up in small plastic bags, assured dry feet even when moving about camp in wet, muddy cave boots. An extra polypropylene top and bottom, properly bagged, provided the luxury of a pillow. Hammocks, although favored by some on Mexican expeditions, were quickly abandoned after a fitful night tossing in the damp, 48° chill of Bobcat Cave. The camp site itself needed to be relatively level, spacious enough for sleeping quarters and a community kitchen and eating area, and fairly close to reliable water. A drop of iodine per gallon of water accomplished water purification. The latrine was located in respectable proximity to camp, dug into a clay bank (Figs. 6.2, 6.3 and 6.4).

Fig. 6.1 Packing gear. Photo by Ron Simmons

Fig. 6.2 The dining area. Photo by Ron Simmons

Fig. 6.3 Sleeping accommodations. Photo by Ron Simmons

Fig. 6.4 The camp room. Photo by David Morrow

Eating revolved around breakfast and dinner, supplemented during the day by personal preferences (energy bars, gorp, cheese, candy, pre-made sandwiches, and in one case, a baked potato). Freeze-dried food covered most menus, being far tastier than the wretched examples of the past and significantly lighter than canned foods. Tea, coffee, sugar, salt, oatmeal, dried fruit, pepper and other spices, and even luxury condiments were easily stuffed in zip-lock bags and buried in the depths of packed sleeping bags.

Aside from ropes, climbing and survey gear, group gear on the initial trip included a small white gas stove with repair kit; several full, secured fuel bottles; a cooking pot for hot water; three or four collapsible plastic gallon jugs; first-aid kit; and trowels and toilet paper, all divided among the participants. Outside of replenishment items, these were secured in the cave from camp to camp. Luxury items ranged from washcloths and personal journals to cards and a harmonica. In contrast to today's advances in LED lighting, carbide provided 90 % of the illumination with candles around camp adding an intimate touch and saving acetylene.

Every Bobcat expedition entered the cave as a small, self-contained team. Never did more than nine cavers (three teams of three) participate; six proved the average. No surface crews lounged in administration tents fielding phones attached to miles of wire strung through near-virgin cave. Instead, one or two individuals (usually Judy Davis, Nevin's wife) agreed to be the surface contact person awaiting the safe return of the team at camp's end. Despite the exploration and survey success of 28 safe, rescue-free expeditions into the far reaches of Bobcat Cave and more after the 1994 connection with Blarney Stone Cave establishing the Chestnut Ridge Cave System, cave camping remains a seldom-used tool of American cavers.

6.2 Bobcat Camp Chronicles

The chronology that follows is a short synopsis of the first 32 cave camps in Bobcat Cave according to date, participants, and significance. The photo documentation is provided from the collection of Ron Simmons and from more recent photography by Nathan Farrar.

6.2.1 Camp 1. 18–22 May, 1984

Paula Casale, Gregg Clemmer, Nevin Davis, Maret Maxwell, Doug Molyneaux, Dave Morrow, Tom Shifflett and Ron Simmons

This initial five-day camp incorporated three days of survey between the entry and exit days. On day two, Davis, Maxwell, Shifflett, and Simmons continued the going North Lead survey into the unknown, charting just over 1000 feet of roomy, 15–25 foot wide × 20–30 foot tall trunk passage, an extending gallery sprinkled with aragonite trees and stunning walls of helictites. Their real surprise however, was the discovery of hundreds of what appeared to be trampled raccoon prints, but which later proved to be *fisher* tracks, an animal not documented in Virginia in the last two centuries. Beyond, this Thundering Raccoon Room narrowed a bit but showed good air flow. Davis commented at the time that this discovery was better than any commercial cave he had ever seen! On their first day of survey, Molyneaux, Casale, Clemmer, and Morrow headed south from the Camp Room, mapping into a maze of heligmites off an area they called the South Lead.

Day two saw Casale, Clemmer, Molyneaux, and Morrow pick up the mapping beyond the Thundering Raccoon Room, discovering the pristine beauty of Cotton Ball Hall and then the immensity of 100-foot-wide, 100-foot-high SVG Hall (Fig. 6.5). Molyneaux and Morrow managed to climb 70 feet up the face of SVG Hall's western wall before the team headed back to camp.

Davis, Maxwell, Shifflett and Simmons took their second day beyond Shamrock Dome, mapping the cave to more than 650 feet deep. That evening back in camp, Clemmer, Shifflett, and Maxwell surveyed 660 feet in the Midnight Raid just off the Camp Room.

Fig. 6.5 Molly Lucier in SVG Hall. Photo by Nathan Farrar

Fig. 6.6 Aragonite speleothems. Photo by Nathan Farrar

Fig. 6.7 Nevin Davis at Rainbows End. Photo by Ron Simmons

Maxwell and Davis exited the cave May 21 while Shifflett, Molyneaux, Simmons, Clemmer, Morrow, and Casale headed back to climb to the dark lead at the top of SVG Hall. Shifflett and Morrow succeeded, finding a going lead headed north with good air 99 feet above the others. This first camp garnered 4649 feet mapped and deepened the cave to an estimated depth of 650 feet (Figs. 6.6, 6.7, 6.8 and 6.9).

6.2.2 Camp 2. 23–26 August, 1984

Gregg Clemmer, Nevin Davis, Maret Maxwell, Dave Morrow, Kent Seavers, Tom Shifflett, Ron Simmons and Fred Wefer

Seeking to mop up loose ends around the Camp Room, this camp netted 3000 feet over two days, mainly cut-arounds, including a tie-into the Midnight Raid, pushing leads off Shamrock Dome and beyond

Fig. 6.8 Natural bridge in North-South Trunk south of Scoopers Drop. Photo by Ron Simmons

Fig. 6.9 North-South Trunk going toward camp. Photo by Ron Simmons

the Texas Whore Drop, and finding what they thought was a raccoon skeleton not far from camp.

6.2.3 Camp 3. 10–13 January, 1985

Gregg Clemmer, Nevin Davis, Mike Dyas, Maret Maxwell, Dave Morrow and Tom Shifflett

Shifflett, Maxwell, and Morrow mapped 435 feet in and around Satisfaction Junction. Dyas, Davis, and Clemmer mapped some 800 feet, first into the Q Survey, then chimneying into a large piece of borehole above the Heligmite Room, ultimately looking down into the South Lead. The next day, Clemmer, Shifflett, and Morrow surveyed going passage beyond Shamrock Dome, climbing down to Satisfaction Junction to find the Cynic's Attic, discovering even more anthodites. Davis, Dyas, and Maxwell returned to SVG Hall, climbing to the top and mapping into the air-blowing lead they named the Porpoise Passage.

6.2.4 Camp 4. 14–17 March, 1985

Pete Carter, Gregg Clemmer, Nevin Davis, Maret Maxwell, Dave Morrow and Tom Shifflett

On the first of two days of mapping, Morrow, Shifflett, and Clemmer pushed the limits of the South Lead from the Camp Room, extending the cave beyond an anatomically accurate speleothem in what is now known as Paula's Penis Passage. Despite continued good air, their survey ended in a collapse at the far south end of the South Lead. (This would be where Clemmer, Ficco, Mike Futrell, and Ben Schwartz would return in 1994 to make the connection to Blarney Stone with Davis, Andrea Futrell, and Shifflett.) Returning to camp, the threesome trekked up the North Lead to Scooper's Drop to finish mapping the Bypass. Still wanting to survey, they added another 270 feet to the every growing Midnight Raid passage off the Camp Room.

Carter, Davis, and Maxwell jumped into the Porpoise Passage above SVG Hall, getting up the Mud Piton Climb and mapping to the edge of a deep pit. They surveyed up and around it, netting 500 feet for their day.

Day two saw Morrow, Shifflett, and Clemmer return to the far reaches of the South Lead, finding even more helictites, but pushing obscure leads and breakdown in what can only best be termed a mop-up survey. Carter, Davis, and Maxwell, now armed with rope, went back to their virgin 86 foot pit which they named Damart Drop. This shaft dropped into a falling canyon that soon edged to a second pit (Polypro Pit), topping out on a large borehole going in two directions. Davis postulated at the time that this almost certainly drained to Aqua Cave.

6.2.5 Camp 5. 30 May–2 June, 1985

Nevin Davis, Doug Molyneaux, Dave Morrow and Tom Shifflett

This foursome returned to the top of Polypro Pit, rappelled to the bottom and elected to survey the virgin leads to the north, discovering a stunning array of cave cotton, gypsum formations and selenite needles. Some 350 feet into their survey, the passage split and they continued left, mapping a very dry, but going cave all the way to what they would term Big Bend, a twist of passages that turned them back in the direction of their original approach. They ultimately mapped 3157 feet, charting Nylon Hall above Polypro Pit on their way back to the Porpoise Passage. They arrived in the Camp Room at 3 a.m., downing a meal at 4 a.m. Yet with Morrow suffering a stubbed toe and the other three complaining of nausea and appetite problems, the team stayed in the Camp Room the next day (Figs. 6.10, 6.11, 6.12 and 6.13).

6.2.6 Camp 6. 5–7 July, 1985

Gregg Clemmer, Nevin Davis, Maret Maxwell and Tom Shifflett

In the first of what would be dubbed the "weekend warrior" camps—in Friday, survey Saturday, and out Sunday—Clemmer, Davis, Maxwell, and Shifflett descended Polypro Pit on Saturday, July 6, and continued the survey south, ignoring the large, decorated lead to the north. They mapped into a huge, sediment floored borehole extending more than 500 feet. This 6th of July Room averaged 50 feet wide by 40 feet high. Continuing their survey into adjacent leads, the crew of Camp 6 eventually mapped 2695 feet, putting Bobcat over the five mile mark (Figs. 6.14, 6.15, 6.16 and 6.17).

Fig. 6.10 Brad Cooper at Nevin's Pendant in the North-South Passage. Photo by Nathan Farrar

Fig. 6.11 Nevin Davis in North South Passage south of Nevin's Pendant. Photo by Ron Simmons

Fig. 6.12 North-South Passage before Thundering Raccoon Trail looking toward Thundering Raccoon Trail. Photo by Ron Simmons

Fig. 6.13 Bottom of Shamrock Dome. Photo by Ron Simmons

Fig. 6.14 Gregg Clemmer doing the Texas Whore Drop. Photo by Ron Simmons

6.2.7 Camp 7. 24–28 July, 1985

Mike Artz, Gregg Clemmer, Doug Molyneaux, Dave Morrow and Fred Wefer

Breaking into two teams, everyone headed out the North Lead to SVG Hall, then up the 99 foot wall and out the Porpoise Passage to rappel Damart Drop and Polypro Pit. Artz, Molyneaux, and Wefer took the first big lead north of Polypro Pit, discovering more helictites, anthodites, and cave cotton in pleasant walking passage. Morrow and Clemmer headed into a lead to the northeast that Davis had probed on Camp 6, mapping more than 600 feet of "pipeline" that eventually corkscrewed into part of the Rocky Mountain section, tying into station GM16, a neat closure of several thousand feet, linking the Sahara Pipeline, Devil's Tower, and Big Bend sections of Bobcat Cave (Fig. 6.18).

Fig. 6.15 A section of the Porpoise Passage. Photo by Ron Simmons

6.2.8 Camp 8. 13–15 September, 1985

Gregg Clemmer, Nevin Davis, Dave Morrow, Tom Shifflett and Ron Simmons

With so much virgin cave discovered in just 15 months, Simmons and Clemmer photo-documented the North Lead all the way to SVG Hall, shooting various angles in the Porpoise Passage, Damart Drop and Polypro Pit. Close ups of the cave cotton, anthodites and sprouting helictites in newly named Jewel Cave just north of the bottom of Polypro Pit grabbed the photographer's attention but the highlight of the effort was Simmon's ten-flash, ghost image of the 6th of July Room, a photograph that would be featured on the 1995 NSS Convention Guidebook—and a double treat for Ron, since it was his birthday (Figs. 6.19, 6.20, 6.21 and 6.22).

Fig. 6.16 The Mud Piton Climb at the end of the Porpoise Passage. Gregg Clemmer on rope. Photo by Ron Simmons

Fig. 6.17 Damart Drop viewed from the bottom. Photo by Ron Simmons

Shifflett, Morrow, and Davis mapped out Maret's Lead off the 6th of July Room, surveying a total of 2227 feet to the 622 Sump, so designated as the sump is −622 feet below the entrance. Here, they ended their survey for the day at an intersection with another stream coming in on the left, blowing air! Only later would they discovery that Bobcat had broken 6 miles.

More importantly, the *weekend warrior* two-night camp was proving effective. Most of the group had now gained considerable confidence in negotiating the constrictions of Cyanide Canyon, several saying they felt they "had it wired." Others expressed the idea that such Friday to Sunday expeditions in the future also offered a double advantage: no impact on vacation time from work, while also not impacting the cave as hard. Indeed, now that all going leads were far beyond Scooper's Drop, day trips seemed unfeasible, if not ludicrous. Until another entrance was found, expedition camps—and those, increasingly two-nighters—seemed the only way to continue.

6.2.9 Camp 9. 8–11 November, 1985

Gregg Clemmer, Nevin Davis, Maret Maxwell, Doug Molyneaux, Dave Morrow, Tom Shifflett and Ron Simmons

This team entered the cave just days after rains from Hurricane Juan deluged Bath County with the "flood of the century." Roads throughout Bath and Highland Counties, most notably in Bullpasture Gorge, were washed out and impassable. Huge deposits of sand coated the stream banks a hundred yards beyond the resurgence of Aqua Cave. Locals said they had not seen such devastation since a memorable flood back in 1913.

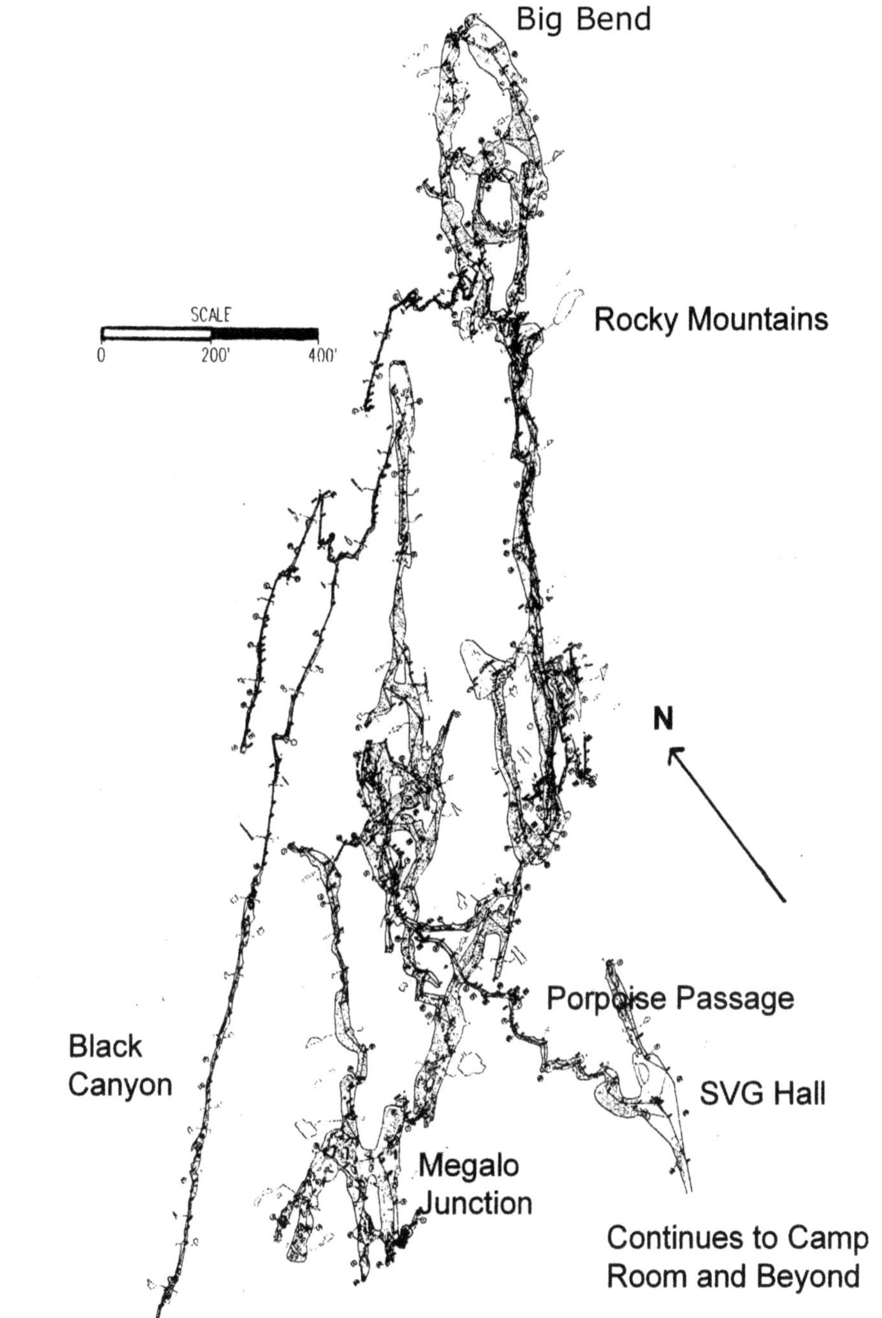

Fig. 6.18 The northeastern end of the Bobcat Section. The southeastern corner is the northern end of Cottonball Hall. The southwestern corner continues as the Black Canyon to its connection to the Burnsville Turnpike

Remarkably, the crew noted only minimal evidence of high water going into Cyanide Canyon, probably because of its elevation above the flood zone. Lines on walls, deeper pools, and washed gravel seemed the most obvious changes.

Molyneaux, Maxwell, and Clemmer pushed through a tight crack at the south end of the 6th of July Room into newly-named Megalo Junction, but netted only 160 feet. Shifflett, Davis and Morrow pressed on to the 622 Sump, then started their survey up the

Fig. 6.19 The main passage at the bottom of Polypro Pit. *Note* rope on extreme right of the photo. Photo by Ron Simmons

Fig. 6.21 Elkhorn anthodite with foot-long stalactite. Photo by Ron Simmons

Fig. 6.20 Aragonite dendrites on flowstone in Jewel Cave passage. Photo by Ron Simmons

Fig. 6.22 Crystals in the Sahara Pipeline Passage. Photo by Ron Simmons

air-blowing stream discovered on Camp 8. Tight but going, they mapped 2505 feet up Black Canyon, turning around in still-going, blowing, wet cave. Bobcat Cave approached 7 miles in length. Everyone exited the cave on Sunday, November 10, with some spending Monday in Burnsville Cove to examine the flood damage.

6.2.10 Camp 10. 27 February–2 March, 1986

Mike Artz, Gregg Clemmer, Nevin Davis, Ben Johnson, Maret Maxwell, John Rosenfeld, Tom Shifflett and Ron Simmons

Artz, Johnson, Simmons, and Davis entered the cave on Thursday evening and made good time to the Camp Room. Yet they got a late start the next day, departing camp about 1 p.m. Their destination: the known end of Black Canyon where Davis, Morrow, and Shifflett had mapped a half mile of virgin cave in November, turning around in a sometimes tight, air-blowing stream passage that still beckoned into the unknown.

After nearly a 3 h trek from the Camp Room, they broke out their survey gear and started mapping Black Canyon's going stream passage, wading pools up to 2 feet deep. With 200 feet in the book, Davis took the tape up a small waterfall, turned and stared into a tall stream trunk heading to the southwest! They named it the Burnsville Turnpike, now regarded as one of the largest, most remote, rarely visited cave passages in Virginia. All noted the going side leads, especially the enticing high leads in the left wall. At one point, the trunk curved, and up the bank on the left, a large, 50 × 50-foot breakdown chocked passage seemed to continue to the north. They deemed this an excellent future camp—flat and sandy with water nearby. After 1/2 mile of surveying borehole and now farther from the Camp Room than anyone had ever been, the foursome turned around in going blackness, leaving the next crew with an incredible 130 foot wide by 75 foot high continuation (Figs. 6.23, 6.24 and 6.25).

Meanwhile, Maxwell and Rosenfeld had started into the cave on this Friday afternoon, working they way through Cyanide Canyon and stopping in the Lost Lamp Room to replace a very worn rope. After ten camps, it was time. Clemmer and Shifflett brought up the rear several hours later, getting to the Camp Room in less than 3 h. They found the rope at Bypass Dome exhibiting considerable wear and tied it off.

When Team One returned sometime after 3 a.m., they couldn't contain their excitement, waking the newly arrived foursome with news of their find. Vivid imaginations battled sleep the rest of that night, Team Two, eager, but in wonder at what awaited them as they would be soon headed into what was now possibly the most remote section of cave in all of Virginia.

A brief description illustrates the journey out there. From the Camp Room, Team Two made their way out the North Trunk, essentially broken trunk passage, past Nevin's Pendant, Rainbow's End, and toward Scooper's Drop. Continuing out the North Lead, they scrambled past aragonite flowers, over boulders, and two dicey pits and climb downs, then gingerly stepped through the Minefield and edged around the Thundering Raccoon Passage. A sloppy, unnamed crawl and the two nuisance drops preceded Cotton Ball Hall. Yet as they stalked into large SVG Hall, they were

Fig. 6.23 Molly Lucier climbing out of Black Canyon into the Burnsville Turnpike. Photo by Nathan Farrar

Fig. 6.24 A view of the Burnsville Turnpike. Photo by Nathan Farrar

only one fourth the distance to the Burnsville Turnpike.

Up the roped 99 foot west wall at SVG Hall, they pushed into the Porpoise Passage, feeling the breeze in their faces. After the Mud Piton's 20 foot ascent, another 100 foot scramble got them to the edge of Damart Drop—an 87 foot rappel. Far below, a short scramble of 150 feet eastward slid all to the lip of Polypro Pit's 35 foot intersection into canyon passage.

Fig. 6.25 Another view of the Burnsville Turnpike. Photo by Nathan Farrar

Heading south down this canyon brought them into the 6th of July Room. Averaging 60 feet wide by 40 feet high, it stretched 500 feet west to what would be named Megalo Junction. Halfway into 6th of July, they entered Maret's Lead on the right, the way on to the 622 Sump. For a half mile, all climbed up and slid down mud mountains—some with challenging exposure—descending through dark, dank, sloppy cave as they approached the 622 Sump.

Once there, they turned left—a sharp left—and climbed up into Black Canyon. Advised to plunge in and not worry about getting their feet wet, they squeezed through the Tollgate—a nasty, tight, muddy—exposed piece of passage—until finally popping through into some enjoyable stream canyon. This bore essentially straight ahead for half a mile, beautiful flowstone walling behind islands of helictites.

At the waterfall they found what Team One had mapped virgin the day before: stunning borehole. Fifty foot shots became the norm with the Turnpike occasionally vaulting 145 feet wide and more than 100 feet high.

Team Two mapped 1600 feet of the Turnpike's continuance, until stopped by a breakdown choke busting in from the ceiling. Curiously, all noted small flies near a trickle soaking across the breakdown. Mapping an uptrending lead secured another 200 feet in the book before also ending in breakdown. The rock-choked Turnpike stream, however, appeared followable, but only in low water. Back at the large, breakdown filled side lead noted on the way in as a future camp site, Team Two cooked some food, then continued their survey here, netting 700 more feet in two north trending leads headed straight toward distant Megalo Junction.

The discovery of the Burnsville Turnpike on the 10th Bobcat Camp—putting a record 5009 feet into the book—documented the cave past Unthanks Cave as Virginia's third longest at 7.4 miles or 39,146 feet. Yet given the extreme logistics experienced, some now began asking if establishing a second, deeper camp might not be more feasible.

6.2.11 Camp 11. 18–20 April, 1986

Mike Artz, Gregg Clemmer, Nevin Davis, Doug Molyneaux, Dave Morrow, John Rosenfeld, Tom Shifflett and Ron Simmons

While Simmons and Artz spent their day attempting to capture the Burnsville Turnpike on film, Shifflett, Morrow and Molyneaux continued the survey of the Turnpike's breakdown-chocked side passage headed north back toward a possible connection with Megalo Junction. After mapping everything to massive breakdown, they joined Simmons and Artz at the far south end of the Turnpike, three of the five visually confirming the dye trace Judy Davis had earlier placed in Woodzell Sink.

Davis, Rosenfeld, and Clemmer mapped a dead bottom pit near the Mud Piton Climb then dropped Damart Drop and mapped richly decorated leads in Nylon Hall, ultimately pushing a narrow crawl to a connection with "Station 34," near the top of Damart Drop, bypassing 90 % of this drop. Despite all these efforts, only 909 feet were surveyed (Figs. 6.26 and 6.27).

Fig. 6.26 Fibrous gypsum in the Sahara Pipeline. Photo by Ron Simmons

Fig. 6.27 Gypsum fibers along bedding planes in Sahara Pipeline. Photo by Ron Simmons

6.2.12 Camp 12. 30 May–1 June, 1986

Nevin Davis, Dave Morrow and Tom Shifflett

This was the first "three man" cave camp. Tommy and Dave dug out an air blowing lead in the South Lead that did not go but the team still managed to map 400 feet of new cave at the bottom of the slope leading to the Heligmite Room. Despite the discovery of Blarney Stone Cave still five years in the future, Nevin would write, "I feel that the strong air entering the south end goes out through the floor breakdown on the south end room," unknowingly identifying the connection point of Blarney Stone with Bobcat that would not be discovered for another eight years.

6.2.13 Camp 13. 11–13 July, 1986

Gregg Clemmer, Nevin Davis, Steve McCampbell, Maret Maxwell, Dave Morrow and Tom Shifflett

Shifflett, Morrow, and McCampbell pushed leads off the South Lead, mapped the Water Hole area, and ended their day with 272 feet. Davis, Maxwell, and Clemmer headed for the far south end of the Burnsville Turnpike, mapping 175 feet into terminal breakdown beyond H257. After pushing the low, wet, flowstone/breakdown-chocked continuation southward, Clemmer left it, unable to go farther but noticing a small breeze. Their total survey—including an upper, parallel passage—netted 1258 feet, bringing the cave's total footage to 42,128.

Eager to get back to camp, not only because it was his birthday, but to answer nature's call, Clemmer pressed ahead of Maxwell and Davis, soloing down Black Canyon, focusing up and out Maret's Lead, and ascending Polypro Pit. To speed his return, he climbed up into Nylon Hall, crawled through the new bypass and jugged up the last dozen feet of rope at Damart Drop, bypassing 90 % of the pit, and ultimately arriving in the Camp Room 90 min ahead of his compatriots.

6.2.14 Camp 14. 8–10 August, 1986

Gregg Clemmer, Nevin Davis, Rod Morris and Ron Simmons

This foursome mopped up survey in and around Scoopers Drop, then headed out the North Lead. The day took a decided turn when Davis discovered a virgin lead—with air—off to the left of the Mine Field. They later dubbed a second, deep stream lead, also angling left but nearer to the fisher tracks, the Mystery Lead, getting 824 feet surveyed at day's end and putting the cave over 8 miles.

6.2.15 Camp 15. 26–28 September, 1986

Mike Artz, Pete Carter, Gregg Clemmer, Nevin Davis, Les Good, Dave Morrow, John Rosenfeld, Tom Shifflett and Ron Simmons

Davis, Good, and Clemmer, reaching camp by 3 p.m. on Friday, headed out to the Mystery Lead to set bolts and work on the two approach nuisance drops into SVG Hall. The next day, Carter, Good, and Rosenfeld mapped 420 feet off the Texas Whore Drop while Morrow, Shifflett and Artz surveyed over 750 feet up a slimy, narrow stream passage they speculated might head toward recently dug open Butternut Cave. Simmons, Davis, and Clemmer dropped the 15 foot pit into the Mystery Lead, mapping a series of tight, clean-washed, blowing cracks, but found no definitive ends or breakouts. Despite a maximum of nine campers, Camp 15 only put 1765 feet into the survey book, bringing the cave to 44,718 total feet mapped.

6.2.16 Camp 16. 7–10 November, 1986

Pete Carter, Gregg Clemmer, Nevin Davis, Steve McCampbell, Maret Maxwell, Doug Molyneaux, Dave Morrow and Tom Shifflett

Despite eight campers, this was another mop-up survey with efforts concentrated in the Mystery Lead north of Scooper's Drop, a lead at the top of Damart Drop, and in the lower Shamrock Dome area. Despite two intense days pushing the Mystery Lead and 500 feet mapped beyond Shamrock Dome, Camp 16 netted just 909 feet of virgin cave.

6.2.17 Camp 17. 6–8 March, 1987

Gregg Clemmer, Nevin Davis, John Ganter, Ben Johnson, Steve McCampbell and Tom Shifflett.

With Johnson, Clemmer, and Shifflett chasing more mop-up off Shamrock Dome, Davis, McCampbell, and Ganter headed to Nylon Hall to clean up several leads. There, an obscure, small crack surprised them, revealing a large room above Nylon Hall that led to two pits including an 81-footer taking lots of air. Despite the cave's arduous challenges, first-timer Ganter noted a week later, "Well the slimy memories are fading, and the ones of that 81-foot pit (not to mention the drop into the canyon, which we kind of forgot about in the excitement!) are growing stronger." Indeed, such remarks rather summed up the motivational chemistry fueling Bobcat's cave campers. With the passage of a few weeks, the stirring lure of discovery always seemed to dull the sharp recollections of their bruised and bone-weary exits back onto Chestnut Ridge and the outside world.

6.2.18 Camp 18. 1–3 May, 1987

Mike Artz, Gregg Clemmer, Nevin Davis, Maret Maxwell, Dave Morrow and John Rosenfeld

Artz, Morrow, and Clemmer checked leads in the South Lead beyond Satisfaction Junction and pushed water leads beneath the Camp Room but mapped little virgin cave. Davis, Maxwell, and Rosenfeld bottomed the 81-foot drop found on Camp 17 only to discover it connected near to Bobcat Dome, a feature on the way to the 622 Sump and Burnsville Turnpike.

6.2.19 Camp 19. 7–9 August, 1987

Gregg Clemmer, Nevin Davis, Ben Johnson, Doug Molyneaux, Dave Morrow, Tom Shifflett and Peter Shifflett

Motivated to return to the Burnsville Turnpike after a year of mop-up expeditions, two teams ventured to Bobcat's most distant corner. Yet Davis, Shifflett, Johnson, and first-timer Peter Shifflett, armed with "instant cave," failed to get to the far south end of the Turnpike after a rock fall "redirected" their intentions. Morrow, Molyneaux, and Clemmer once again pushed into the Turnpike's stubborn northern breakdown near the proposed second camp area but found no way through, mapping just 125 feet for their efforts.

6.2.20 Camp 20. 6–8 November, 1987

Gregg Clemmer, Nevin Davis, Phil Lucas and Tom Shifflett

This team continued mapping upper leads off the South Lead and Heligmite Room, including an interesting, 60 foot deep pit, putting 444 feet into the book with extended pushes at the South Lead Terminus. VSS President Lucas, on his first camping trip in Bobcat, mentioned that the entrance area needed a Significant Virginia Cave sign.

Learning that the BCCS was negotiating with Bobcat's owner to buy the entrance and a parcel of surrounding acreage, Lucas decided to approach the owner to see if he would be willing to sell additional land. Phil and Charlotte Lucas ultimately purchased 82.6 acres that included the Water Sinks, Siphon # 1 Cave (Owl Cave), and Siphon # 2 (Water Sinks Cave), but their move to Burnsville Cove also opened the door to the stunning discoveries of Helictite Cave (1996), The Water Sinks Subway (2007), and Wishing Well Cave (2010).

6.2.21 Camp 21. 25–27 March, 1988

Gregg Clemmer, Nevin Davis, Mike Futrell, Dave Morrow and Tom Shifflett

Just east of Nevin's Pendant, Clemmer and Davis dug open a small slot on Friday evening, the team surveying 280 feet of new cave the next day. Futrell and Clemmer continued to push this dig, but Morrow, Shifflett, and Davis could not make the pit in the Heligmite Room go. As Bobcat closed on 9 miles of mapped cave, and the BCCS closed on buying the Bobcat entrance and 84 acres of Chestnut Ridge, the hot lead list seemed to be cooling.

6.2.22 Camp 22. 29–31 July, 1988

Mike Artz, Pete Carter, Nevin Davis, Mike Futrell, Lester Good, Ben Johnson and John Rosenfeld

These two teams revisited small, but going leads off Black Canyon and the Sahara Pipeline, detailing the decorated passages out to Big Bend, but finding no surprising discoveries.

6.2.23 Camp 23. 28–30 October, 1988

Mike Artz, Gregg Clemmer, Nevin Davis, Mike Futrell, Dick Graham and Ron Simmons

While Simmons, Graham, and Davis did an extensive photographic study of Jewel Cave and the nearby cave cotton, anthodites, and selenite needles, Artz, Futrell, and Clemmer pushed the low stream Gun Barrel passage off of Black Canyon beyond the Toll Booth, mapping 695 feet of very low crawls and putting Bobcat Cave past 9 miles.

6.2.24 Camp 24. 6–8 May, 1989

Nevin Davis, Mike Futrell, John Rosenfeld and Tom Shifflett

This single survey team mapped additional passages beyond the Sahara Pipeline and into Big Bend.

6.2.25 Camp 25. 11–13 August, 1989

Mike Artz, Pete Carter, Gregg Clemmer, Nevin Davis, Mike Futrell, John Rosenfeld and Tom Shifflett

This camp sustained the only "rescue" when Futrell became sick on Saturday morning and was "escorted" to the surface without incident by Clemmer and Rosenfeld. The rest of the team pushed a series of climbs above the lower South Lead beyond Frustration Junction, but mapped nothing.

The next day, Clemmer and Rosenfeld aided Ron Simmons' cave dive at Cathedral Springs. With Simmons having mapped Cathedral to a depth of 160 feet below the Bullpasture River, with no bottom in sight, any connection of Cathedral with Bobcat or Burns would push the system's depth close to 1000 feet.

6.2.26 Camp 26. 10–12 November, 1989

Gregg Clemmer, Andrea Dakowski, Nevin Davis, Mike Futrell, John Rosenfeld, Tom Shifflett and Jeanne Wetterling

Digging and blasting through the Mystery Lead by Clemmer, Shifflett, and Davis netted the mapping of 150 feet of sleeze. Rosenfeld, Dakowski, Futrell, and Wetterling photographed and took casts of the unknown animal prints in Thundering Raccoon Passage which were later identified by caver Fred Grady of the Smithsonian as fisher prints.

6.2.27 Camp 27. 20–22 July, 1990

Gregg Clemmer, Nevin Davis, Ed Divine and Tom Shifflett (Doug Molyneaux pulled his gear while taking Dave Hubbard on a day visit during Saturday.)

This was yet another trip to the Burnsville Turnpike to blast the flowstone breakdown in the south end, an effort that took two attempts to explode, but ultimately could not be pushed because of hanging fumes. With leads getting harder to find, push, and map, and camper attendance languishing, Camp 27 marked the end of the first Bobcat Camp series. Noted Clemmer at the time, "it is not productive to continue mapping/exploration in this fashion. Camp [site] 2 looms as the best alternative. Details on such a venture will have to be considered carefully."

6.2.28 Camp 28. 23–24 July, 1994

Gregg Clemmer, Ben Schwartz and Tom Shifflett

Almost precisely four years later, Clemmer, Shifflett, and first-timer Ben Schwartz returned to Bobcat's Camp Room to coordinate with the day team of Nevin Davis, Jeff Uhl, and Phil Lucas who had entered nearby Blarney Stone Cave to make a possible connection. Despite a voice and smoke connection, Uhl became ill and had to exit the cave, ending the connection attempt. Clemmer, Shifflett, and Schwartz spent the night in Bobcat's Camp Room, pondering the implications of what now appeared to be an imminent connection.

A month later, on 20 August 1994, Clemmer, Schwartz, Mike Futrell and Mike Ficco entered Bobcat in an attempt to link with Nevin Davis, Tom Shifflett, and Andrea Futrell who had entered Blarney Stone Cave. The two teams' successful connection and the Bobcat teams subsequent through trip that same day established the Chestnut Ridge Cave System as Virginia's second longest known cave system.

6.2.29 Camp 29. 28–31 December, 1999

Gregg Clemmer, Mike Ficco and Tom Shifflett

The millennium's final Bobcat camp—now in the newly named Chestnut Ridge Cave System—sought not only a return to the Burnsville Turnpike but the establishment of a second camp out there. Yet no one had been out to the Turnpike in nine years, and with just three committed to the trip, moving camp proved, as Clemmer would note afterwards, "a losing proposition, a fool's errand."

After a taxing trip in, Shifflett, Ficco, and Clemmer camped near the bottom of Polypro Pit, "a miserable, muddy camp, and not worth the extra effort to haul gear out there as compared to the relative comfort of the Old Camp Room." Continuing on to the Turnpike the next day, they surveyed 344 feet up a narrow in-feeder coming in the east wall of the Turnpike. The lead ultimately climbed 100 feet toward the surface, ending in a 40 foot dome going up into better limestone and what looked like going cave. But this venture did not trigger another round of underground camps. Indeed, it would be another eight years before Camp 30 got underway.

6.2.30 Camp 30. 7–13 October, 2007

Pete Carter, Ed Kehs, Sr., Ed Kehs, Jr., Jon Lillestolen and Tommy Shifflett

This week long camp saw two trips to the Burnsville Turnpike. The first, over October 8–9, pushed a 32-foot-high, 4 bolt lead at the Turnpike's beginning. Once climbed, the team surveyed a breezy, crawl/walk tube some 650 feet to dense breakdown that might ultimately bypass the Turnpike's air-sucking breakdown side passage to the south and possibly continue to Aqua Cave. The next two days—rest days—saw leads mapped off the Thundering Raccoon Passage and Sahara Pipeline where Carter at one point wormed his way down a steep bedding plane to the edge of a pit, hearing water below.

The team returned to the Turnpike on Thursday, October 11. Although one member elected not to go to the Turnpike because of a painful shin injury, Shifflett still sensed "there was a feeling of trepidation" amongst the others "returning to the Turnpike despite the curiosity of what the opposite lead might do."

This second, "opposite lead" simply reconnected with the Turnpike, and the team surveyed only 92 feet for the day, despite more arduous pushing of the breakdown. Netting 1300 mapped feet for the entire camp, Shifflett could only remark that "hopefully, the younger generation that participated on this trip will be motivated to return."

6.2.31 Camp 31. 24–28 November, 2010

Pete Carter, Nathan Farrar, Jon Lillestolen, Molly Lucier and Philip Schuhardt

It took three years before such a "return" was organized. Remembering several good leads from Camp 30, yet recognizing the intense challenge an underground camp presented, Lillestolen recruited four

Fig. 6.28 Burnsville Turnpike Camp 3: Paul Winter maintaining his journal in the kitchen area, Molly Lucier just waking up, and Scott Wahlquist fast asleep. Photo taken at 5:09 p.m. on June 4, 2012 showing the thrown-off schedule of cave camping. Photo by Nathan Farrar

other cavers, include Bobcat veteran Carter for a four night stay in the Camp Room.

On day one, Lillestolen and Farrar set dye traps at the 622 Sump terminus of Black Canyon to coordinate with a surface team's dye dump. Carter took Lucier and Schuhardt back to his bedding plane pit off the Sahara Pipeline and mapped over 500 feet of floor-slanted stream passage, noting good air. This Stompen Borehole, so named because of the mud steps the trio booted into the slanting floor continues to the north-east at this writing—off the map.

The next day, Lillestolen and Lucier did an aid climb above Polypro Pit, surveying some 300 feet while Farrar and Schuhardt checked leads out toward Big Bend. Returning on day three, Lucier, Schuhardt, and Carter continued the aid climb survey—now called *Praise the Lord and Pass the Ammunition*—ending at a deep pit that later was matched to the top of a tall, dripping dome in the 6th of July Room. Farrar and Lillestolen retrieved their dye traps at the 622 Sump and checked leads in Megalo Junction and out Maret's Lead.

Fig. 6.29 Brad Cooper examining flowstone at the extreme south end of the Burnsville Turnpike—the extreme end of the Chestnut Ridge Cave System, two days travel from the entrance. Photo by Nathan Farrar

6.2.32 Camp 32. 2–8 June, 2012 (Write-up by Nathan Farrar)

Brad Cooper, Nathan Farrar, Molly Lucier, Scott Wahlquist and Paul Winter

Farrar had procured a grant from the National Geographic Society to place a camp in the Burnsville Turnpike, two days from the Bobcat entrance. The team of five spent the first and last nights in the Camp Room, but spent the other four nights beyond the Black Canyon, at a newly established camp in the Paleo Turnpike, just off the Burnsville Turnpike trunk passage. It took 16 h for the crew of five to travel from the Camp Room to the Burnsville Turnpike, hauling eight full and heavy camp duffs, along with three coils of climbing rope; granted, there were many photography stops along the way (Fig. 6.28).

Dye was injected into the 622 Sump to investigate a possible hydraulic connection with the Emerald Pool (Water Sinks Cave System) and Aqua Spring. Farther upstream in the Black Canyon, dye traps were set in an attempt to establish a hydraulic connection with the Barberry Cave stream.

On day three, designated a 'rest day', the team set out to photograph the Burnsville Turnpike for the National Geographic Society (Fig. 6.29). On day four, Farrar and Winter surveyed into a lead seen directly above the new camp. After about 250 feet of survey, the passage ended in breakdown collapse. The traverse line through the passage was left rigged, optimistic that Back Door surface dig will connect there. Cooper, Lucier, and Wahlquist had set off to push the collapse piles at either end of the Burnsville Turnpike, to no avail.

Next on the to-do list were multiple bolt climbs. The first, completed by Wahlquist with Lucier belaying, was near the camp and came up empty. On the

Fig. 6.30 Scott Wahlquist climbing an 80-foot wall in the Burnsville Turnpike. Photo by Nathan Farrar

fifth day, Wahlquist began the climb up the impressively tall, east side of the Burnsville Turnpike to investigate a vanishing ceiling channel; Winter belayed. Cooper, Farrar, and Lucier worked on capturing more photographs (Fig. 6.30). Wahlquist ran out of bolts halfway up the wall, perhaps 50 feet in the air. With his improved view of the lead, Wahlquist felt ever more confident that the lead was worth pursuing. Completing this climb, along with further attempts to penetrate the collapse at either end of the Burnsville Turnpike, are the two hottest objectives for the return trip to this section of the cave system.

The team spent the following two days retreating from this remote point to the surface. Rope and a first aid kit were the only items left at the Turnpike camp location. The team, along with the SD cards, exited the cave safe and sound after over 149 h underground.

Interest in the Bobcat Section of the Chestnut Ridge System has picked up in recent years. At the time of this writing (July, 2013) Camps 33, 34, and 35 have been fielded, mostly to check out leads, clean up uncertainties in the survey, and general tidy-up. However, one never knows when that obscure hole behind a breakdown block or that dark shadow at the top of a bolt climb will lead to something big.

The Chestnut Ridge Cave System is currently mapped to 21.26 miles at a depth of 806 feet, the 5th deepest cave east of the Mississippi River and tied as the 29th longest cave in the continental United States.

Finding the Blarney Stone

7

Gregg Clemmer

Abstract
Blarney Stone Cave was discovered by means of an intense digging effort in a small sinkhole where snow had been observed to melt during the winter. A difficult entrance series opened into a large and complex cave system which was explored to both the northeast and the southwest. Blarney Stone Cave is exceptionally well decorated with unusual aragonite speleothems. To the northeast, Blarney Stone Cave approached closely to the southwestern end of Bobcat Cave. After systematic exploration followed by digging to remove clastic fill, the two caves were connected to form the Chestnut Ridge Cave System.

7.1 The Discovery

With discoveries down and enthusiasm for multi-day Bobcat expeditions on the wane, the 1991 Pancake weekend presented an attractive, far less intense option. There was no way John Rosenfeld would miss this date after the Pancakers had dug open a cave on his property the year before. Mike Artz promised to show up, but if the weather turned good…*well, climbers will be climbers.* Fred and Ann Wefer would be there as would Gregg Clemmer and Mike Dyas, secretly up for the NSS Bicking Award at the instigation of most of those present. Coming from Ohio, Andrea Dakoski and Mike Futrell had upcoming nuptials after cave hunting. Tom Shifflett had another "get out of jail free" card and well, Nevin Davis lived there.

G. Clemmer (✉)
25 Silverton Court, Hunt Valley, Cockeysville, MD 21030, USA
e-mail: yorkst@aol.com

Despite moderating weather, most of Chestnut Ridge still had a thin blanket of snow which could only highlight the sought out melt spots. Armed with plenty of tools, the crew made their way to two such bare sites several hundred feet above the entrance of Butternut Cave. The Wefers, Davis, Shifflett, and Clemmer started a small hole next to an exposed slab of massive limestone while Rosenfeld, Dyas and the Futrells plunged into a cobble-filled sink, conspicuously bare in the late winter snow cover.

Lunchtime came and went. The day warmed, melting the last of the white. The first dig quickly found that all of the rock either had to be chiseled, broken by a sledgehammer, or worse. Dyas reported the cobbles in his dig were bottomless. Everyone stayed on task, yet by mid afternoon, the diggers' mood had mellowed. Ann vanished to ride her horse.

"Nevin, what about those couple of features down the east side of the ridge…maybe we could take a look before dark?" No one had any idea what Clemmer was talking about except Shifflett and Davis. "Remember the sinks you showed Tommy and me last winter? We need to get back to those and we just never seem to find the time."

W.B. White (ed.), *The Caves of Burnsville Cove, Virginia*,
Cave and Karst Systems of the World, DOI 10.1007/978-3-319-14391-0_7,

"Well, there is that melt spot I showed you and Tommy on the east side of the ridge," Davis mumbled as Clemmer picked up a shovel and crowbar. Hyping the site to Mike Futrell, he headed down the ridge… encouraged by the leaf-crunching footsteps behind him. A ten-minute walk brought everyone to a small sink with a drain hole. Futrell was not encouraged and after a token probe, the crew walked a hundred feet down the hill to the second site, a small depression no more than four feet across and two feet deep tugging down a gnarled old tree. Limestone bordered the northeast corner of the hole. Some of Nevin's faded orange flagging still wrapped the tree.

Futrell leaned down and scraped leaves out of the depression with an entrenching tool. Clemmer fueled the effort with words while the others plopped down on the hillside to take in the late afternoon sun. Soon, he too grabbed a shovel and began spading dirt from the bottom and sides of the depression.

It was the organic black stuff that grandmothers love for their flower gardens. Soft and easy to dig, Mike and Gregg soon doubled the size and depth of the depression. A crowbar plunged downward revealed more of the same. Fred and Nevin lounged on the ground, talking computers. Tom offered advice but couldn't seem to find a tool that fit his hand. Dyas kept an eye on the bottom of the dig while Andrea kept an eye on her digging man. John just kept asking if anyone was getting tired.

"Not yet," said Clemmer, continuing to spade the black soil out of the hole. Four feet down, solid bedrock defined the north wall. He hacked at some roots, thinking this dig was a lot more promising than the others. *Why hadn't we come here first? Oh, well. My back aches; let's spread the fun.*

"Here John. I need a break." Futrell gardened the edge of the hole as Rosenfeld shoved shovel into soft dirt. On his second spadeful, the shovel plunged into a void! All hillside loungers quickly circled the hole like vultures on roadkill. Fresh, warm air welled up from a dark, dirt-lined hole. Clemmer asked the time: 4:47 on Saturday, the 16th of March. CAVE!

Everyone bounded up across the hillside in excitement. The fatigue of hammering on slab limestone and digging cobbles by the cubic yard had melted with the March snow. They had a cave and tomorrow they'd be back, "armed for bear."

Mike, Avery, and Amanda Artz showed up at the Davis farm at supper time (the climbing had been fantastic, he said) just in time for the big scoop. Everyone devoured chicken dinners Judy Davis brought from the Fireman's BBQ and then savored several hundred cave slides for the rest of the evening.

Sunday morning after pancakes, syrup, and sausage in Williamsville, the dig crew returned to Nevin's and Judy's, grabbed gear, threw in a climbing ladder and an ancient piece of Goldline and roared off for the hillside. Tom won the race to get ready first and started down the hole on belay. When he called back that he had down climbed some 30 feet and was out of the way, Andrea scooted for the rope. Gregg followed when she cleared the bottom.

The newly dug entrance had breached the ceiling of a narrow fissure one to two feet wide, thirty feet deep. Abundant handholds made the climb easy and once down, the threesome squeezed to the far end of the fissure where a small opening on the right continued, emitting a warm, intermittent flow of air.

Shifflett squeezed through first and whispered back to Clemmer, "You better see this. This is hot!" Gregg inched ahead while Andrea waited for Mike to join her. To the left, a black void opened onto the edge of a window looking into a large room. This thing *was* looking good.

Dyas and Futrell joined with the cable ladder and belay rope. At the top of the drop, they rigged the ladder and sent Shifflett down. Eighteen feet later, he stepped off on a landing, finding himself in a sizeable, irregular room some 70 feet long, 25 feet wide and 25 feet high. Beneath the ladder, the pit continued. With Mike Dyas running belay, Tom and Mike Futrell started down this drop, stepping off some 25 feet below into a tight stream canyon. They down-climbed farther and then followed this tiny stream around a series of muddy meanders to the top of a 30 foot dome. Out of rope and heeding Dyas' admonishment to survey what was found, everyone began the exit. Was borehole around the corner?

Outside, the crew pondered their discovery. Tom wanted to name the cave Brendan's Fever because of his son's recent sickness. Mike Artz favored Blechuguilla and so the nonsense ran. Gregg noted that the day was St. Patrick's Day and since they had "to do" the entrance with a rather intimate association against the bedrock before gaining access, how about calling it Blarney Stone? That stuck and as for a return date, no one wanted to be left behind. After much wrangling, the crew picked April 20. What a wait—and the anticipation was worse than Christmas for a six-year old. Yet five weeks later, everyone gathered for the big push, everyone except Dyas who had broken his arm in a bike accident.

The first survey trip into Blarney Stone consisted of two teams. Davis, Mike and Avery Artz, Rosenfeld, and Clemmer started in first to rig the drops. Within an hour, they were at the 30-foot dome. Mike rappelled first and pushed the crack that carried the small stream. It didn't look very promising, so he climbed up toward the ceiling, following very tight slots. Davis and Clemmer meanwhile, wormed into the stream for a second look, angled around a very tight bend and clawed their way through a slot to a series of slide-down crawlways. The cave did not end, but it wasn't exactly borehole either. Rosenfeld followed as best he could, contorting his long frame around the bends. Back at the entrance, Shifflett, Wefer, and the newly-wedded Futrells began the survey.

As long as the cave continued, Davis and Clemmer lamely rationalized they should push the passage. This actually made sense because there could be many trips surveying this virgin slop without knowledge of the tight spots, digs, and other obstructions that lay ahead in the dark. John caught the pair as they pondered their options yet Mike soon shouted through the tube that he and Avery were headed out of the cave.

Indeed, the passage reminded all of Bobcat's Cyanide Canyon, and they soon dubbed it Strychnine Canyon. Narrow down-climbs snaked around cobble-filled corners but the cave did not end. Scraping out an angled slot filled with cobbles, Clemmer squirmed his way into a room decorated with elegant drapery—reminding him of his grandfather's old tobacco barn—and waited for Nevin and John. This Tobacco Barn was far enough for one day and the threesome set a station and started a survey back toward Wefer's group. After a half dozen stations, they hooked up and tied in their stations, but Andrea, Mike and Tom, anxious for their share of virgin cave, surveyed another 100 feet beyond the Tobacco Barn, including a tricky down-climb to a serpentine stream passage. They turned around in a low, slimy, but going tube. With air!

The survey showed that the cave was 302 feet deep at the last station and still going. The next trip was planned for June 1. Rosenfeld, Shifflett and Clemmer would pick up the survey from last time while Andrea and Mike Futrell would join Mike Artz as Team Two, eventually leap-frog and continue ahead.

The stream passage beyond the Tobacco Barn quickly became the tightest yet negotiated and after several stations, it seemed to be coming to a sump. Shifflett crawled ahead and groveled in the cobbles and fill. *No way on!* The air was also stale. *Where was the breeze?* Within moments, Rosenfeld called out that he had something and after a few strokes with the crowbar, Team One crawled up into a going crawlway, headed northeast. The cave seemed to be changing, more sand instead of mud…and it was dryer. Then they hit Tommy's Pinch.

At first it looked impossible; too many irregular surfaces compressing a distorted cross section. Yet it was only a body length long, rising to a stoopway. An echo hinted at larger things. Shifflett squeezed through head first. Clemmer followed, the tightness tangling a few concerns for self-preservation. With careful breathing and attention to the immediate task of the moment, he got through. Rosenfeld followed, his passage greased by whoops of big cave and goaded by his partners' gallows humor.

Big cave did meet them, fifteen feet high and just as wide. But it was based on breakdown and steep, muddy, pitted floors. They surveyed a hundred feet to a lead rising sharply up to the left which they set aside for Team Two. After a break for snacks, they continued ahead into the unknown B Survey.

It was horrible–muddy and tight cave, broken by nasty climbs with bad exposure. After a half dozen stations, all were slimed, but the cave still went. Through a series of angled squeeze-ups and climbs, Shifflett found a way on, but then experienced light troubles. Rosenfeld and Clemmer fumbled in the cold to get to him, but just as they were setting a station over a narrow dome, Shifflett wrenched his lamp from his helmet and lost it down the pit. To retrieve it, they needed Team Two's rope. Calls got their voices and 30 minutes later, Mike Futrell joined with a rope.

It took 2 hours, but finally everyone and everything was extricated from this wretched place. The trip out was one long slog toward the entrance, an exhausting climb, crawl, and squirm out to the night-time hillside. Despite 700 feet mapped, only one lead of note remained; the one up to the left that Artz and the Futrells had surveyed into, finding a five by four foot continuation on the other side of a 30-foot deep pit.

They said it looked good, but everyone now knew the road to it was a "butt kicker."

The 1991 NSS Convention in New York provided a pleasant diversion and the discoverers forgot about Blarney Stone for a few weeks. But Mike Artz "had a feeling" about this cave and said so at Convention. Across the pit, *well, it looked really good,* he said. With some still burned out from the June death march, he still twisted an acceptance out of Rosenfeld and Davis, who had missed the June trip and was curious about what had been found. On July 20, the trio entered the cave equipped to cross the pit. Armed with his Hilti drill, Mike bolted across the top of Artz Attic (Fig. 7.1). On the other side, he disappeared around the corner while Rosenfeld and Davis waited. He returned with a gleam in his eye! "It's big and black and goes in all directions!" he whooped. The whole hill was hollow!

Fig. 7.1 Nate Walter crossing Artz's Attic. Photo by Nevin Davis

The other two clipped on and muscled their way across the Attic. What they found was nothing short of stupendous. Dry walkways angled from the other side of the pit, narrowing to a small crawl. Beyond lay an enormous room 600 feet long, 100 feet wide and perhaps 70 feet high. In the gloom, a white stalagmite rose from the floor, a 10-foot ghost lurking in this ancient chamber. And so they called it Ghost Hall (Fig. 7.2). The trio surveyed north, toward Bobcat, ending at a blowing wall of formations. This was no ambiguous air flow; this was real cave air on its way to *someplace else.*

News of this discovery spread fast among the crew and a return date was set for August 10. Yet new blood was needed for the survey and Hope and Jeff Uhl were prime targets. A baited hook with the usual hype, secrecy, and Machiavellian manipulation did the rest. They bit big-time, hearing only the part about "good air flow," and "big, going cave." This time Davis, Wefer, and Clemmer would go to the blowing formation wall and make a way through a small hole into what had to be going cave. Rosenfeld and the Artzs would map Ghost Hall south into the unknown, as would Shifflett and the Uhls (Fig. 7.3).

It proved a productive trip for all three teams. Wefer, Davis, and Clemmer surveyed into a well-decorated passage of small columns and totems dubbed the Leprechaun Forest that eventually exploded into the Over Forty Passage, a breakdown trunk that continued tantalizingly off to the northeast. They found numerous leads but contented themselves with charting the borehole. To the south, the other teams found a large upper extension of Ghost Hall and a meandering stream with several leads they dubbed the Black Diamond Crawl. Yet the trip out again busted butts. Discoveries had masked any concern for the hour, and fatigue coupled with the overpowering desire to sleep slowed everyone down. But all slogged it back to the entrance safely and this time emerged into the early light of a new day. The cave now stood at almost a mile in length, more than 400 feet deep, and best of all, had edged to within 175 feet of connecting to Bobcat!

The Pancakers spur-of-the-moment dig had become one of the major caves of Burnsville Cove. Eighteen months after the cave's discovery, mappers in Blarney Stone would chart 3.37 miles of passage plunging to a depth of 599 feet.

Fig. 7.2 The ghost in Ghost Hall. Photo by Nevin Davis

Fig. 7.3 Mike Kistler entering the Boondocks extension beyond the Upper Ghost Hall connection. Photo by Nevin Davis

7.2 Exploration

Exploration of Blarney Stone Cave proceeded rapidly over the several year period from its initial discovery in 1991 to the connection with Bobcat Cave to the north in 1994. As it turned out, Blarney Stone Cave actually overlapped Burns Cave to the south but this was not confirmed until the connection trip in December, 2005 (see Chap. 8). An overview of the entire Chestnut Ridge Cave System shows the Blarney Stone section to be a complex maze of strike passages and cross-structure passages on several levels ranging from dry upper levels to low level streamways (Fig. 7.4). A remarkable feature of Blarney Stone Cave is the profusion of aragonite and other speleothems (Fig. 7.5).

7.3 Connection: Bobcat + Blarney Stone = Chestnut Ridge Cave System

[**Editor's Note**: Edited from article of same title which appeared in the BCCS Newsletter 20, 2–11 (1994/95)]

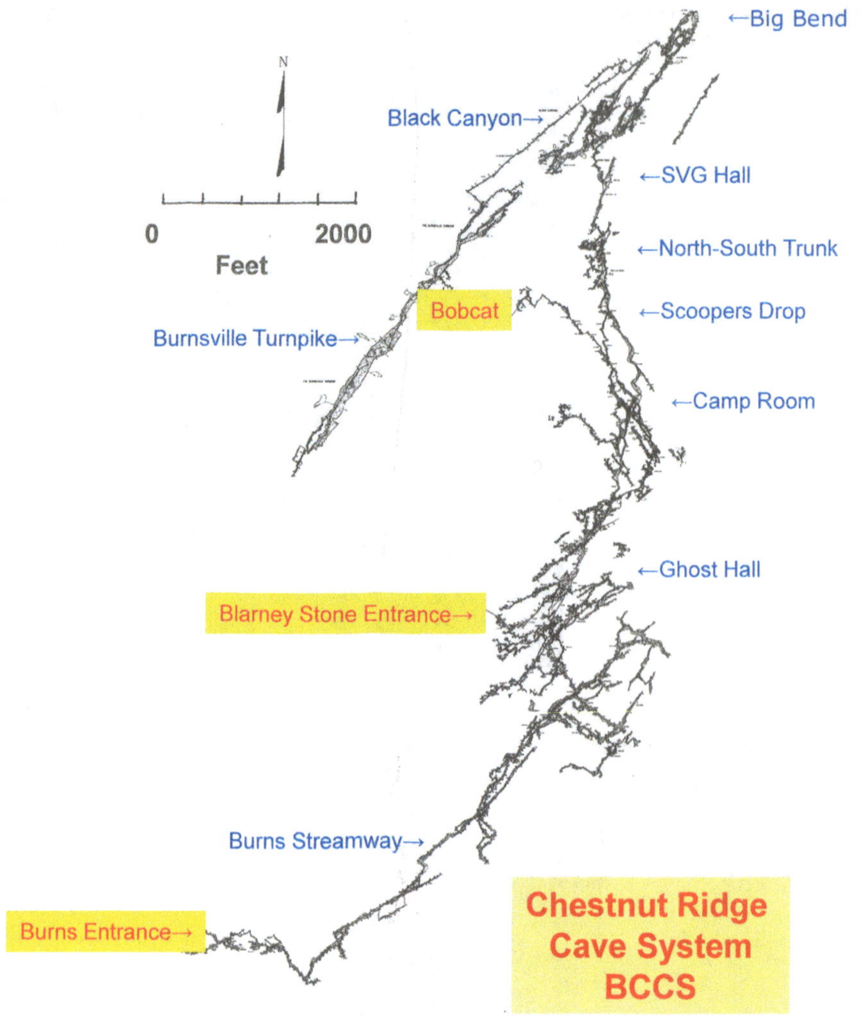

Fig. 7.4 Overall map of the Chestnut Ridge system showing the middle position of the Blarney Stone section

Over the period 1984–1990, Bobcat Cave expanded to more than 9 miles of passage, all mapped from the relative comfort of 27 underground camps. With its discovery on 16 March, 1991, Blarney Stone Cave quickly exploded into several miles of decorated if somewhat arduous cave. Survey trips through the Leprechaun Forest and into the Over Forty Passage discovered a confusing maze of passageways on several levels, all with a distinct trend to the northeast. Gradually, as we pushed and plotted this end of the cave, it became apparent that the South Lead of Bobcat Cave might not be far away through the rock.

In light of the multiple, difficult, still unresolved efforts to connect Breathing Cave to the Butler-Sinking Creek System, perhaps a connection was too much to expect of these newer discoveries. But armed with typical cavers' optimism, plenty of determination, a bit of luck, and a secret weapon, we had to try.

7.3.1 Possibly

By the Fall of 1991, Blarney Stone Cave approached a mile and one half in length. The discovery of Ghost Hall in July by Mike Artz, John Rosenfeld and Nevin Davis, proved that the entrance infeeder did indeed intersect a large trunk segment. When Nevin, Fred Wefer and Gregg Clemmer mapped northeast down this passage in August, we found a voluminous, breakdown-filled gallery with plenty of leads, and one,

Fig. 7.5 **a–d** A selection of images of Blarney Stone Cave. Photos by Nevin W. Davis

that after plotting, trended directly for the South Lead in Bobcat Cave. The bad news was that it ended prematurely in fill and breakdown.

With everyone bringing in thousand foot plus surveys from just about anywhere in the cave each time they entered, there was no overwhelming impetus to push toward Bobcat. Yet Nevin, Tommy Shifflett and Gregg were eager to dive into the breakdown at the end of the Over Forty Passage. We entered the cave on 8 September and yet once out there Nevin and Gregg could find no way on; but while Gregg tinkered with his lamp and Nevin re-examined the geology, Shifflett disappeared.

Digging open an insignificant crack, he pushed a tight, muddy space through the breakdown. There was no air flow, but at best it trended in the right direction. Davis followed in bewilderment and after various crowbar adjustments to the hole's perimeter, Clemmer did too. It was truly a disgusting way for the cave to end and it went nowhere. Yet this little find cautioned us to examine every crack.

We retreated to the so-called Rebel Lead which Clemmer had located in August. It too led northeast but far under the passageway we had just pushed. As Shifflett checked a tight crack, Clemmer dug on a dirt crawl, noticing the good air flow. Two minutes later Gregg was through and they broke out the tape. Three shots brought us to a large junction with a pit in the floor. There seemed to be leads everywhere but after a few stations most pinched. Tommy found one, Spic and Span Pit which needed a rope. A second, a wall climb into the ceiling would need bolts. Thus with two leads, we would have to come back.

Our anticipation simmered for three weeks. This time, Mike Dyas joined us, but try as we might, none of the several floor pits or wall cracks near Spic and Span did anything and while the other three mopped up a lead across the top of the still virgin Spic and

Span, Gregg sat down, somewhat disgruntled that now we only had the ceiling lead. As he contemplated how unfair all this was, he noticed a narrow, steeply dipping slot in the east wall. Hadn't seen this before.

Carefully Clemmer entered and worked his way down and out over what appeared to be a climbable pit. His perch was somewhat confining and the others had no idea where he was. But this looked interesting—yet very muddy. Nevin was suckered into a look-see.

Davis now back from Spic and Span, eyed the slot and ventured forth. Down climbing the pit, he squeezed forward along a squishy slop hole that he described as "very treacherous"—a mud-slickened slot that "if you fall into it you'll never get out". But beyond this horror, he thought it might open up. Doesn't it always?

This was hardly promising—definitely poisonous stuff and in time we would come to call this lead Poison Goes Down. But for now this crack was just another miserable footnote and we were trashed. As we left the cave, we could only wonder just how close we really were to "ole Bobcat".

Despite our optimistic hints that a close encounter with another large cave was in the offing, none of the others mapping booty beyond The Pearly Gates was interested. We quietly set our sights on a return trip in November. This time Davis, Shifflett and Clemmer ignored the Spic and Span area and began mapping a side lead on the west of the Over Forty trunk. They soon found themselves in a veritable rabbit warren of challenging passages and loops, the most significant discovery being a sizeable stream that plunged to the northeast. Calling it Wobblestone Creek for its less than secure footing, they mapped upstream, leaving the tight, somewhat nasty downstream for another time. When they finished, only a bolt climb high above the stream where we had started seemed our best lead.

While we kept up our connection hunt, Jeff and Hope Uhl, Mike Artz, John Rosenfeld and others had continued their methodical survey beyond The Pearly Gates, making the cave grow. Fine. We didn't need a cast of thousands anyway. In December, we would tackle the bolt climb above Wobblestone Creek, a gurgling stream that we now had plotted as an almost perfect match to a bedding plane gusher south of Satisfaction Junction in Bobcat.

On 7 December, 1991, the 50th anniversary of Pearl Harbor, we tried again. Artz and the Uhls opted for leads beyond The Pearly Gates. How appropriate. Well, maybe Davis, Shifflett and Clemmer could get in a pre-emptive strike of their own.

After some effort, Shifflett set two bolts above his perch over Wobblestone Creek. Once up we mapped into a steep, down-trending, large passageway that required a handline. Crawling past a small but exposed pit that dropped 30 feet into the downstream continuation of Wobblestone Creek, we stood up in a large room. Climbing over a nuisance overhang, we found that the passage broke into several segments, all eventually ending in breakdown and muck. Only an obscure lead in the floor, which led to an impressive sediment-lined passage, hinted at more cave.

We turned around at a very tight ceiling crack although Davis managed the thing, working his way through a short series of dripping breakdown squeezes to an obvious end. Dubbed the Earthworks for its similarities to the trenches of WWI, we exited the cave, pondering, but not really knowing just how close we had gotten to Bobcat.

But the cave was going in all directions and with virgin passage in no short supply beyond The Pearly Gates, it was July before we scraped up two teams for a push toward Bobcat. Shifflett convinced Mike and Andrea Futrell to join him at Spic and Span Junction. Not only did the pit need plumbing, other leads and cracks beckoned.

Meanwhile, Davis, Hope and Jeff Uhl and Clemmer returned to Wobblestone Creek. Instead of going up the bolt climb, we dropped into the flowing water. Downstream was virgin and by our "calculations" it almost had to be the upstream continuation of the tight bedding plane squirter Tommy, Ben Johnson, and Gregg had found in Bobcat during Camp 17 in March of 1987.

Nevin promptly worked his way down to the stream through a narrow slot he and Jeff pounded open. Mud glopped from the walls, finding its slimy way down necks and into ears. Sleeze city…too tight…no way on here.

Jeff and Gregg pushed the tight canyon while perched some twenty feet above the stream. Very close but possible. Seventy-five feet brought us to an even narrower cross-section passage dreadfully filled with slices of sharp limestone just waiting for the first customer. Lacking a foolhardy audacity, we retreated, exploring a small muddy alcove. Hope had meanwhile chimneyed up to a hole in the ceiling, finding the top of the pit that Tommy, Nevin and I had crawled past on the last rip. There were still a few leads up there so we continued our survey topside, finding nothing more exciting than a meandering passageway that led back to the muddy alcove.

Shifflett and the Futrells climbed all over Spic and Span Junction, dropping the pit, but finding only tight impassable drains without significant air. Without bolts, the ceiling lead would have to wait. But there was the "treacherous" slot Nevin had examined in late September. Mike entered, kicking down mud coated walls to widen the fissure. Calling for a back up, he peered into the tightening gloom. A hint of air flow? At least it was heading right.

This lead now seemed the best place to go for the connection. We made our plans for 26–27 September. While Nevin, Jeff and Gregg surveyed tag ends near Spic and Span, providing back-up if needed, Tommy, Mike Artz and John Rosenfeld would go after the last remaining, obvious lead in the Junction—the dark ceiling lead. Lugging his trusty Hilti drill into the cave, Mike quickly mastered the climb with two bolts. Tommy and John ascended right behind him. The passage spiraled upwards through a small slot and into a chamber of gelatinous mud. Thoroughly trashed and soaked, and finding no obvious continuations, the coated threesome bargained with us to survey their lead while they pushed Poison Goes Down.

We agreed, managing to keep reasonably clean while they entered the nasty, tight fissure for the unknown. Tommy took point and seeing an obscure slot in the floor, slithered past the tight spot that had stopped Mike Futrell. With Mike Artz and John Rosenfeld as backup, Tommy squirmed ahead. Mike joined him at the edge of a steep narrow slope. Free climbing this, they found a pit and called for rope.

Down Mike went, greased now by more horrible mud. With perhaps a hint of a breeze, he noted that the cave seemed to continue, but it would take digging to make it go. Others were not so certain and about the only thing we could agree on was that Poison Goes Down was aptly named.

Despite going virgin cave in other parts of Blarney Stone, the momentum now to connect to Bobcat took center stage. Two reasons—one obvious, the other novel—powered our resolve. To this point, we had made all of our attempts for a connection strictly from the Blarney Stone side. But if we put teams in both caves, we would have pushers on two fronts. More importantly, teams in both caves now made a sight/sound/smell connection possible. Simply put, this was the next logical step.

The second, indeed state-of-the art, aspect that fueled our push—indeed our certainty—to connect, lay in Fred Wefer's Alexandria condo. For the last decade, Fred had quietly taken our data from Bobcat and incorporated it into an increasingly sophisticated computer map. Able to rotate a line plot of the cave from all points on the compass, Fred's display gave us a startling perception of the cave passages and their relation to one another.

And when we added the Blarney Stone data to the map, then rotated the entire mix in real time, the cave magically tricked our eyes into a three dimensional perception of what really lay beneath Chestnut Ridge—and how the two caves related to each other. They were indeed close. Wobblestone Creek almost certainly was the same stream that fed past Satisfaction Junction. And the South Lead was only some 50 feet through the rock from the Earthworks. Even with the expected error in our ten-year old, ten-thousand foot survey loop—in Bobcat, Blarney Stone, and between the two entrances, the caves were certainly one.

That, perhaps, was the optimistic part. But with some determination—and that proverbial bit of caver's luck—we figured we stood a good chance.

On November 14, 1992, John Rosenfeld, Tommy Shifflett and Gregg Clemmer entered Bobcat prepared for a long day trip. Even after a decade of camping in the cave, day trips "to the bottom" still commanded a healthy bit of respect. Nevin Davis, Jeff Uhl, Mike Artz and Eric Anderson comprised the Blarney Stone team. They would make for Poison Goes Down and push all leads. Tommy, John and Gregg would dig and work from a series of tight leads off the stream beyond Satisfaction Junction. Both teams agreed that the "window of Opportunity" would be between four and six o'clock in the afternoon, i.e. both parties could hammer and listen for each other during these times, hopefully finding a way on toward the connection.

Our trip to the bottom was uneventful and although 45 minutes late, we still had plenty of time for a sound connection. With a three-pound hammer and two-foot crowbar, we beat and knocked—then waited breathlessly for a reply. Nothing. Again and again, the same arm-aching effort. And again—silence.

Then, at 5 o'clock, we heard it! Pinc! Pinc! Pinc! Pinc! Pinc! John and Gregg both shouted at once, "Did you hear that?" "Listen, listen!" But after holding our breath for what seemed an eternity, there was nothing more. Tommy had missed it and thought our imaginations were in overdrive.

But again, in the midst of conversation now, the definite cadence of a half dozen "pinc's" tapped through the limestone to our ears. Realistically, we had no idea how far they were from us, and our whistles, massive hammer blows, even shouting, brought no response. Again and again we hammered back, knowing that such blows could travel great distances through solid rock. But there was nothing more and at seven o'clock we quit.

With mixed emotions we started our long journey "from the bottom" emerging safe an hour past midnight into the November dark of Chestnut Ridge. The Blarney Stone crew had gotten trashed in the muck and horror of Poison Goes Down, arriving early and leaving at five o'clock. It was their last taps that we had heard; our hammered responses had never had a chance.

Despite the definite sound connection, the enthusiasm to connect faded. Trips to this cold, wet, tight section of Blarney Stone paled when compared to the virgin cave that waited along Moon River and up the Stairway to Heaven. And frankly, few had stepped forward, willing to do a day trip into Bobcat. It seemed the severity of both caves—perhaps akin to the chemistry and challenges of Breathing and Butler a generation earlier—was rearing its head once again. Connections just didn't happen in the Cove.

And for more than a year nothing happened. Sure, Blarney State expanded, nearing four miles in length. And on Memorial Day weekend of 1993, we dug into Barberry Cave, finding a nice stream passage beyond a miserable piece of claustrophobia called the Briar Patch—a gnarled squeeze of cave so bad we consumed half of 1994 excavating a second entrance.

Yet even with these finds, everyone knew we had some unfinished business. Tommy Shifflett finally set us straight.

Returned from the NSS Convention in Texas, Tommy started pushing to renew the effort to connect. With the Convention in Blacksburg next summer wouldn't it be nice to have a challenging through trip for some of our friends coming to Virginia? They've treated us to their projects when we visited. Shouldn't we do the same?

He had a point and we all knew it. But more importantly, this focused our attention to solving a nagging puzzle. We knew the caves connected. They had to. And there was no better time to prove it.

Again, we reviewed Fred's marvelous cyberspace projections. Magnified rotations of the approximations of Blarney Stone to Bobcat more resembled twisted versions of Watson and Crick's double helix than any mysterious underground passageways beneath Bath County. Again and again, the blue and green and black lines (Fred could make them any color) bent and turned, their movement giving depth to a planar screen. Butler and Breathing had never endured this kind of assault in the sixties.

Ignoring the vertical line for Poison Goes Down and the green lateral meanders southwest of Satisfaction Junction, we focused on the South Lead. When it was discovered in 1984, we had noted strong air flow at several constrictions out there. But we had never found where the air went. Now on Fred's Silly G monitor, the terminus of the South Lead—a large gallery filled with shattered breakdown—apparently lay near and somewhat above portions of the Earthworks in Blarney Stone. But our extensive survey loop almost certainly contained errors. How close they really came could only be determined by going into both caves at once.

7.3.2 Probably

Sunday morning, 23 July, 1994—and not even ten o'clock yet. Officially, it was Bobcat Camp #28; four years had passed since the last one. As Clemmer prepared to descend Historic Drop he thought of all those on the original camps who had drifted from the project. Today, there were just three, Tommy Shifflett, Gregg, and 20-year old Ben Schwartz from Doe Hill.

We ditched our heavy duffs in the Camp Room and spent some time examining our in-cave gear. Nevin Davis on an earlier day trip with Mike Futrell to haul camp gear out of the cave, had found significant corrosion on our pot and stove. The stove would probably still work but after eight years of service, it was time to haul it out of the cave. Our pot was not so lucky, having numerous corroded holes along the sides and bottom. Thank goodness, Tommy had packed in a smaller replacement. This original relic, he said, was going home as a treasured memento of earlier days.

With still plenty of time before our agreed upon window with the Blarney Stone team, we ambled south beyond camp, continuing our tour with Ben. Thankfully, with the Down/Up Pit was still rigged on both sides, we soon were on our way to South Lead Terminus. I noted marginal air at the constrictions as we worked through the last of the breakdown.

About noon, Nevin, Phil Lucas and Jeff Uhl entered Blarney Stone. They were to reach the Earthworks area by the time we arrived at the end of the South Lead. Three o'clock we agreed—and everyone would stay until six.

Shifflett and Clemmer quickly led Schwartz into the small, nearly hidden lateral gallery that constituted the real terminus of the South Lead. A muddy trickle stream dribbled from the left sloping wall and disappeared into a floor crack along the right. Beyond, thirty feet perhaps, the gallery abruptly ended, the floor filled with massive breakdown. Wanting to try out some of their toys, Gregg lit a smoke bomb and tossed it down one of the dark slots.

Nothing. Gone.

"It didn't go off", Tommy said, "Didn't work."

"OK, I'll do it again." Gregg lit another and dropped it to the floor. Quickly, a white, sulfurous cloud filled the end of the gallery, followed by unkind suggestions regarding Gregg's pyrotechnic skills, intelligence, and general abilities. In this white gloom, we started moving rock, hoping for some surprise opening to reveal itself and give us the magic connection. Twenty, maybe thirty minutes later, we all suddenly realized that the smoke had cleared.

At this same instant, Phil Lucas walked toward the rear of the passage from which the Earthworks continued on the left through an obscure hole. "Sulfur," he thought. "That smells like sulfur."

The cloud had indeed worked its way through impenetrable breakdown from Bobcat to Blarney Stone. Excitedly, he related the news to Jeff and Nevin. Now we had a "smell" connection.

On the Bobcat side, we continued our digging. Gregg poked into the muddy dribble that dived under the right wall. Were those dark voids down there? With growing anticipation, he heaved out more rocks until he suddenly considered just how stable or unstable the whole mess was. With discretion the better part of valor, he decided to let his dig "mature." Moments later, as he joined Ben and Tommy, the whole thing collapsed.

They were still working the shallow hole near the end of the galley when they all heard the distant tapping. Ben hammered back—and got a solid reply. Shave and a haircut—two bits! You bet! Now the Bobcat crew was in business.

The rocks flew out of the hole, but despite our enthusiasm, it was slow going. During his turn in the bottom, Gregg carelessly splattered some grit into his eye. He groped his way to the packs to take out a contact lens and doctor the eye. He had taken perhaps a dozen steps when he heard Nevin's voice come from beneath the floor! With such yelling and whooping, it's a wonder Oztotl didn't appear.

Distinctly, we called back to Nevin again and again, trying to get a fix on precisely where he was and what his surroundings looked like. As Gregg fought corneal abrasion, Ben and Tommy started another dig against the right wall above Nevin's voice. So close, yet so far.

From snatches of conversation, Nevin indicated that Jeff was getting ill. There was talk of having to leave the cave. Then abruptly Nevin did leave, saying he would be back shortly. But more than an hour passed and when he did return it was to tell us that they had to exit. Methodically, Nevin had carefully done a one-man sketch and survey of his surroundings. Now, before he left, Gregg asked him to get one more good visual image of the place and take a compass reading toward his voice.

Ben, Tommy and Gregg continued digging for another hour, Ben at one point just avoiding the dangerous shift of a freezer-sized rock that had a mind of its own. But clearly the wind was out of their sails. With no definite point to concentrate their efforts, the Bobcat crew was laboring in vain. Dejected, they headed for the Camp Room…and hoped the stove worked.

7.3.3 Certainty

The acrid smell of the smoke bomb and the nearness of Nevin's voice galvanized us to the task at hand. Thoughts of virgin cave in Barberry or Blarney Stone did not seem quite so important now. Major connection trips rank among the rarest of all caving experiences. It is the ultimate, yes, logical culmination of every activity a caver works for—ridge walking, digging, pushing, surveying, aid climbing, water tracing, and in our case, cave camping. As Clemmer continued to remind everyone, "This was a plum that was only going to be picked once."

Saturday, 20 August, 1994. It just felt right! From the Blarney Stone side we would have Nevin Davis, Tommy Shifflett, and Andrea Futrell. Not only would Tommy bring the knowledge of where the Bobcat team was digging, he would be able to guide us to the best approach from underneath.

Joining Gregg were Mike Futrell, newcomer Mike Ficco, and Ben Schwartz. Futrell was a previous cave camper, well familiar with Bobcat. Ficco was strong, agile, and a veteran of recent pushes in Blarney Stone and Barberry. And as for Schwartz, well after the last trip, he had flatly declared that Bobcat was his favorite cave.

This day we were armed for bear. Yet, as Gregg toyed with his gear in the parking lot of Spring House Farm, it seemed foolish to plan on a camp, silly to be more concerned with comfort than the matter at hand —finding the connection that had to be. Hints and glances told him that there was little enthusiasm for spending the night in the Camp Room. And yes, Gregg did feel a little guilty about splitting up Mike and Andrea.

"Let's bag the camping idea," he suggested. "We're going to do this thing for sure and there's no need to drag sleeping bags, stove, and extra clothes down there. We gotta show a little confidence!"

No one disagreed. Now we could travel light, devoting our time to digging through into Blarney Stone. Even if we failed, Mike Futrell said, it was only a long day trip.

We made what seemed like record time, arriving in the Camp Room at 1:10 p.m. Our tools were a small crowbar, chisel, foldable E-tool, and Futrell's cave pack for hauling mud. Based on good measurements and detailed sketches from both sides and no little bit of cave intuition, by the way, we were fairly certain where to dig.

As we entered the lower extension galley off the South Lead, just before the trickle stream that flows left to right through the unstable rock and ooze, a low space angled downward to the right. By projecting Nevin's description of his surroundings on the previous trip together with the compass reading and the map we had made of the galley, this low space seemed the best, safest place to dig. There was no hint of a hole, no undermining stream, no obvious crack to follow. But several feet down and beyond, we believed, lay Blarney Stone Cave.

Gregg seized the E-tool and began spading the mud, his effort more aimed at creating a comfortable work place than advancing the dig. Enthusiasm was high and soon Ben, Mike, and Mike had relieved him. Age did have its benefits.

7.3.4 Reality

Gregg walked around the corner, crossing the trickle stream to the point where Ben had dug down toward Nevin the month before. The whole area had undergone an ugly collapse. Almost expectedly, Nevin's voice echoed from the slot. Andrea remained behind, he said, admiring the Earthworks. Tommy sounded off, "All right, let's get started."

Tommy and Nevin talked in muffled tones for a moment, deciding precisely where to start their dig from underneath. As Nevin recalled later, "Once Tommy started where I pointed, there was no stopping him."

Hearing that Tommy was ready to dig, Gregg asked Nevin to tell Tommy to make a lot of noise and see if Mike, Mike, and Ben could hear him. Tommy bellowed and the Bobcat diggers yelled, "He's right underneath! He's right underneath!"

Now the entire chemistry changed with furious digging erupting from both sides. A dozen times and more, Mike's full, muddy pack was hauled out and dumped. Fifty pound rocks, some heavier, were rolled and pushed from the work site. Deeper, wider, the hole opened downward.

Amid a banter of encouragement, hype, and silly jokes, Tommy guided us toward the safest point of intersection. Working with a crowbar, he pulled down a number of large rocks. Ben asked if he could see a large rock wiggling? No, he replied but from the dirt trickling through the cracks, it might be best if Ben moved to the right.

Another rock, possibly a 200-pounder—see it? Yes, Tommy yelled, it moves. Leave it alone! Ben reached down and removed some webbing, "Do you see my hand?"

"Yes!" and an instant later the first handshake linked one caver to another, one cave to another.

Fig. 7.6 Tommy Shifflett in Blarney Stone Cave seen through the connection dig from Bobcat Cave. Photo by Mike Futrell

Now the work reached fevered pitch. To the right of the handshake rock, the dirt was softer. Suddenly Ben's bar broke through to a small void which quickly widened.

"Watch the dirt, watch the dirt", Tommy yelled back, spitting and backing out of the way. Carefully Ben stopped for a moment and Tommy edged forward toward a faint light from above. There! Framed by an eight inch opening three hundred feet beneath Chestnut Ridge and more than a mile from either entrance, Tommy Shifflett beamed back a million dollar smile (Fig. 7.6). Wow!

This was too good to miss and pulling a bit of rank, Gregg edged Ben out of the hole. Widening the slot with the bar, he wanted to make sure that this hole was plenty wide. After all, we were going to run a survey up through here in a little bit.

Tommy retreated from Gregg's efforts and a few seconds later, Gregg slid through, joining him in a handshake in Blarney Stone Cave. But it wasn't so any more was it? Something had changed. There was no more Blarney Stone Cave, no more Bobcat Cave. Yes, there would always be the Bobcat Entrance and the Blarney Stone Entrance, but now they were in the Chestnut Ridge Cave System, thirteen miles and counting of charted passageways with over 700 feet of vertical relief.

It was time for a little ceremony and, as everyone gathered in the South Lead, Gregg broke out some napkins and a small nalgene bottle filled with some Tennessee cave water, properly distilled, of course. He carefully poured some of the "cave water" into his trusty St. Andrew's Cross shot glass and offered a toast, inducting everyone into the Secret and Mystical Order of the Black Jack. Duly downed and acknowledged by all, they proceeded to officially join the two caves by compass and tape.

For the four who had entered the system via the Bobcat Entrance, the trip out via Blarney Stone constituted the first through trip in the cave. Yet as Gregg made his way across Artz Attic and into the twisting, muddy passageways of Strychnine Canyon, he thought of all of the Saturdays they had pushed these rat holes on Chestnut Ridge, of all the dead bottom pits, high hopes, and blowing cracks. So many people had helped to make this connection possible and so few were here now to see it happen.

There would be more days to be sure, where they would map other discoveries and there would be plenty of cave for all. But on this night, as Gregg emerged from the ground, it was time to pause, take stock, and savor the satisfaction of solving part of the puzzle.

Burns, Baby, Burns

8

Tommy Shifflett

Abstract

Burns Chestnut Ridge Cave (usually known as "Burns") was discovered in the 1950s. Exploration over the next 20 years produced only a small and difficult cave. Extremely difficult exploration from 1979 to 2003 finally succeeded in penetrating an interminable sequence of obstacles to discover several stream passages that are part of the Cathedral Spring drainage. The final depth was 782 feet below the entrance making the cave one of the deepest in Virginia. In 2005, exploration from the Blarney Stone Cave side produced a connection and made Burns a part of the Chestnut Ridge Cave System which at the time of connection then had a length of 20.03 miles.

I've just squeezed into a small, cramped room. As I contort myself, I find it barely possible to turn around and see the crack I just exited. My breathing is heavy. Sweat drips down my forehead and stings my eyes despite my being soaked from hours of crawling in 49-degree water. I focus on the passage ahead, the way out of here to the surface.

But there is no way on. I'm confused. Nothing here appears big enough to squeeze into. I'm much too tired to back down this wrong turn. What wrong turn? I remember there are no side passages in this isolated stretch of the cave.

Again I look ahead, spotting the small fissure near the floor that appears too small for me to fit. Peering in, a strong breeze blows in my face, cooling the sweat on my brow. Raw fear hits me. With no other cracks staring me in the face, I know this must be the way on.

After several trial and error attempts, I discover just the right angle to fit my body. I push ahead, squeezing against the friction of the cave's walls, straining to push my pack ahead of me. Progress is measured in half inches. The rushing water cascades down the smooth flowstone floor of this foot-wide passage, into my gloves, up my sleeves, and out my coveralls. Fighting an agonizing fatigue, I begin to ponder the next constriction… then the next…and the next…a preliminary mental torture to the all-night, arduous contortions I must pass to get out of this cave. Hours of bone-weary labor lie before me, though I am but two thousand feet from the sunrise of a new day.

This is something of what the handful of cavers who have pushed the limits of Burns Chestnut Ridge Cave (Burns for short) feel when negotiating the countless twisting squeezes and narrow canyon passages of the cave's entrance series. Regarded as perhaps the most difficult cave in Virginia's Burnsville Cove, Burns is located atop Chestnut Ridge near the Bath County/Highland County line. A trip into nearby Chestnut Ridge Cave System via the notorious Bobcat entrance, said Virginia caver Mike Futrell, "feels like a tourist trip", when compared with a venture to the bottom of Burns.

Those who have been into that Virginia discovery know that Bobcat is no tourist's lark. After

With a contribution by Mike Ficco. [Reprintedwith permission from the *National Speleological Society News* **61**, 181–192 (2003)].

T. Shifflett (✉)
224 Barker Lane, Bluemont, VA 20135, USA
e-mail: tommyshifflett@gmail.com

W.B. White (ed.), *The Caves of Burnsville Cove, Virginia*,
Cave and Karst Systems of the World, DOI 10.1007/978-3-319-14391-0_8,

confronting an exhausting half-mile entrance series of windy, narrow, twisting, toothpaste mud-filled cracks and fissures, we opted to map the system's 14 miles of passage—and 722 foot depth—via 28 muddy underground camps in a constant 49-degree temperature over a period of 12 years.

Burns simply takes the Bobcat example to ridiculous extremes. All who travel to the bottom of Burns are beaten by the continuous onslaught of pinches, squeezes, narrow and twisted canyons, exposed climbs, and soaking, cold water crawls. Just the trip in can be exhausting.

But the story to be told here is more than just that of another difficult push trip into previously uncharted cave. The Burns story is about the handful of people on dozens of trips over many years who wanted to make this blowing cave go, who went back again and again against each constriction and terminus, burning (no pun intended) through at least three generations of explorers. This is about the persistence of many Burnsville Cove cavers, who in a saga of digging and dedicated exploration spanning decades from the 1950s into the 1990s, put a "for the record only" cave into the hushed annals of cavers' campfire lore.

8.1 Early Exploration

Burns was discovered in the 1950s by Ike Nicholson. The first description of Burns is in Henry Douglas' *Caves of Virginia*. Douglas noted several 30-foot drops and a series of narrow rooms 30 feet tall. His description brings the cave to an end at a blocked stream passage blowing air about 200 linear feet from the entrance.

This was the terminus when Duke University students began working the cave in the 1960s. Although little direct information had been found on the Duke cavers' efforts, an article in the *Nittany Grotto News* noted that they "blasted several constrictions and pushed the cave's passages to an estimated 2000 feet, following air current." However, from blast debris found in the cave, and Nevin Davis' account of his attempts to push Burns farther, the Duke students actually penetrated only about 600 feet to a depth of 200 feet below the entrance. The same *Nittany Grotto News* article described Davis and Fred Wefer finding a few hundred feet of virgin cave along the way, but being stopped by the same constriction in a narrow stream passage that had stopped the Duke cavers.

In the next *Nittany Grotto News*, Fred Wefer described a fluorescein dye trace of the Burns entrance stream to Cathedral Spring, a distance of 2.7 miles to the northeast and 815' vertical feet below the cave's entrance. A Nittany follow-up noted Duke's return to continue the exploration. Nevin Davis detailed his own June, 1971 visit to check on Duke's progress, especially on the effects of the fuse and dynamite they had "borrowed" from him. With Paul Cunningham to back him up, Nevin traversed the cave to a 20-foot-deep pit 500 feet from the entrance. This same pit, now called Historic Drop, would become the symbolic spot dividing the old known cave from later discoveries.

Historic Drop is reached via squeezes, exposed climbs, and traverses, all through tight, sinuous canyons, making the cave seem a bit longer than it really is. To reach the actual edge of the pit, one must work sideways into a tight, high fissure with plenty of exposure above the pit. The short drop was formerly rigged with a wooden rung, polyethylene ladder backed by a Goldline belay.

Perhaps because of this exposure and the unsure nature of the ladder, Nevin elected to solo the drop while Paul stayed topside. Nevin pushed downstream from the bottom of the pit and into the First Mud Crawl, a two-foot-high by 3-foot wide passage that can best be described as a fight to stay out of quicksand, or in this case, quickmud! Using elbows and knees he crawled above the muck using narrow ledges along the sides to keep from sinking. Beyond the crawl lay the end of exploration and the point of Duke's blasting effort. Shattered flowstone blocked the stream trickle from which cold air poured. Nevin discovered that the constricted passage had been completely collapsed in a zealous effort to make it larger. Poking around, he found a hole in the ceiling 10 feet back from the site. Enlarging it with a hammer, he pushed up and over the blasted-shut passage. After 150 feet of scrambling in virgin cave, Nevin came to yet another too-tight constriction, this one at the rear of a small room less than four feet high. He described this location as "another constriction through which the water flows and the wind whistles."

The next trip came in August of 1972. This time Nevin brought Ron Miller with him. In Nevin's words, "this was another of those long miserable trips to enlarge the stream crawl in the cave." Besides caving gear, the duo brought a hammer, ten pounds of dynamite, and a 120-foot Goldline rope. When they reached

Historic Drop, they discovered that Duke Grotto had also been back, leaving behind their own 120 foot Goldline rope. Both Ron and Nevin used the same rope ladder left behind to reach the bottom. This time the air was sucking, allowing them to inspect the cave after setting their blast. Digging out the debris, they managed to open the constriction only to find "that there was yet another five feet of very tight crawlway before what appeared to be walking passage." Mercifully, they had no idea of how wrong they were.

8.2 A Bitter End

Nevin returned in February 1974 accompanied by Champe Burnley and John Wilson, lugging a polypropylene and aluminum rung ladder to re-rig Historic Drop. Two and one half arduous hours later, fighting the cave in wetsuits, they reached the blast site. The mission was simple: advance the cave with ten sticks of Kinepac, all capped to a timer so everyone could clear the area safely. Yet even without the burden of ladder and explosives, and cushioned by time and distance from the blast, it still took two and one half hours to exit the cave. Caving without a wetsuit Champe got cold during the second half of the trip. Little did anyone know, the physical requirements to eventually bottom this cave would negate any consideration for a wetsuit, despite constantly crawling in cold water (Fig. 8.1).

Nevin and John came back that same month. More and more it seemed, first time visitors to Burns found little appealing in a return trip Nevin noted that the previous blast had enlarged the cave to where he could see "a canyon up to 5 feet high but only 8 inches wide." Setting another charge with timer the pair exited, this time cutting their exit time to 2 h.

But they didn't go back. As Nevin described later, the cave had worn him down and he couldn't find anyone to help him. John Wilson had found better things to do and Duke Grotto had apparently abandoned the project. Thus, the cave returned to silence and darkness, with only the bats and cold breeze passing the narrow 8-inch crack that marked Nevin's deepest advance.

8.3 A New Generation

By 1980, Gregg Clemmer, a new member of the Butler Cave Conservation Society, had developed an interest in the caves of Chestnut Ridge. Gregg had long lost interest in pushing small caves in his native Augusta County, but still yearned to discover something of greater potential in his native Virginia. Burnsville Cove seemed a good bet. Jack Igoe, a member of the BCCS, had sold Gregg on the theoretical Cathedral System, though there was little—excepting a few obscure blowholes in winter—to suggest a big cave lay under the eastern flanks of Chestnut Ridge. Still, eager to get

Fig. 8.1 Ben Schwartz in the entrance series. Photo by Mike Ficco

started, Gregg asked fellow Shanandoah Valley Grotto (SVG) members Doug Molyneaux, Paula Casale, and Buddy Stein to join him on Labor Day 1979 for a visit to Burns. Somehow by using the myopic quadrangle maps in the back of *Caves of Virginia*, they stumbled onto the entrance.

Already hot and tired from searching for the obscure entrance on a humid, late summer day, the foursome ventured only a couple hundred feet inside. A strong breeze sucking into the cave convinced Gregg that this cave, and indeed Chestnut Ridge, could be very rewarding if pursued. No one else seemed interested in the cave—the Duke cavers had vanished and Nevin Davis and other members of the BCCS groaned about the prospects—so Gregg decided to push it. Some in the BCCS tried to steer him away, saying the caves in the Ridge were young in development and would be too tight to follow. Their theory was that a large cave was unlikely in Chestnut Ridge. Paying no heed to their predictions, Gregg sought help from members of his Grotto.

I had joined the Shenandoah Valley Grotto in 1979, and had participated in their exploration and mapping of Walt Allen Cave in Pocahontas County, West Virginia. On two of these trips I had come to know Gregg and as such was exposed to his slick spiel of enticing people to join him in checking out two known leads on Chestnut Ridge: Chestnut Ridge Blowing (aka Bobcat) and Burns. Captivated by Gregg's lure at finding large borehole cave similar to nearby Butler, I joined Gregg and Kent Seavers for a recon trip into Burns. Our goal was to reach the end of Nevin Davis' exploration and find the source of the strong airflow that had been described by previous explorers. To relocate the entrance to Burns, particularly from the west side of Chestnut Ridge, Gregg asked Nevin's help. Nevin had recently moved to Burnsville Cove and he quizzed us why we wanted to go into Burns. He laughed when we told him of our intent to push the cave where he had left off. Nevin truly thought at the time, that after a good Burns ass beating, we would be back 10 or 12 h later, without ever having reached the constriction, declaring that we would never return to such a horror hole. He was wrong! We did reach the constriction, we liked the prospects, and we quickly began planning a return assault to get through the narrow crevice where he had given up.

8.4 The Terminus

The crevice where Nevin had stopped was just a vertical crack two to four feet high. The passage was difficult to evaluate because a three-foot chunk of breakdown blocked the way. To get a decent look over this jagged rock, I had to double twist my body while my legs were still in a tight, five-foot-long constriction, then arch my torso vertically to see up and over the breakdown boulder. I felt like a pretzel and could only stand such a contortion for a brief moment. What I could see was a straight passage, 4–8 inches wide funneling to a blank wall at least twenty feet away. A black hole in the floor at the base of the blank wall offered a dim glimmer of hope. In cross-section the top half of the passage was wider than the bottom, offering a narrow ledge to aid our work in widening the slot. We knew it would be a long protracted effort but we also felt certain we could get beyond the immediate obstacle. The wind and that black hole were our inspiration. And so we began, Gregg, Kent, Doug Molyneaux and myself, gradually chipping away at the crevice, with occasional help from Joe McKenney, Pat Ward and others in the SVG.

Our first objective was to "gravelize" the breakdown boulder. After that obstacle was removed we went after the walls, widening the passage at the agonizingly slow rate of 2–3 feet per trip. With little room to maneuver—the slot remained at chest tight dimensions—we burned a lot of energy squeezing forwards and backwards, grinding ourselves into the sharp wall corners and floor rubble in order to remove busted rock at the constriction. I still remember that after each trip, my bruised chest remained sore for days afterwards. For this and other awful reasons we named the place the Bone Crusher. After several intense trips we had advanced to within four feet of the blank wall.

8.5 A Breakthrough

In October of 1981, two teams entered the cave. One group—Gregg, Joe McKenney and Dave Hall, began a survey at the entrance. The second team—Doug, Kent and myself—headed to the terminus, intent on making a breakthrough. After more rock removal we finally had the Bone Crusher large enough for a skinny person to force it. Or so we hoped. It still appeared suffocatingly tight. But now it was time for Kent to take over. Truly a master of the squeezebox, Kent possessed the necessary confidence and daring to attempt this kind of thing, an obstacle I feared I might get stuck in and die. Kent had long since earned the name "The Snake" for his particular talent. He would later pioneer another breakthrough and tight squeeze in Bobcat Cave, which bears part of his nickname: The Snakehole.

With much effort Kent finally forced his way to, then through, the last of the Bone Crusher, entering a five-foot-wide by six-foot-long dome. We had finally broken through! Upon hearing Kent's yelps of triumph, I had to follow, forgetting my fear of the place. But that last stretch of the Bone Crusher turned out to be the worst, requiring an extremely tight body force on a steep incline, then another, final womb-tight grind and grimace into that small room. Getting through required maximum twisting and contortion of my body. Others would suffer strained back muscles, bruised hips and shoulders, and in one case, cracked ribs. Only later modifications somewhat "tamed" the Bone Crusher.

The only passages exiting the dome were a body-sized slot and a two to three-inch wide crack in the floor taking the stream. Wind blew through this crack, but no one could follow that. There was no air moving through the slot located about six feet off the floor and requiring a bit of wiggling to get over the slick slimy mud. But once over we dropped down a steep, slippery slope to a low room, 15 feet wide and long which we named the Mud Room for obvious reasons. The bottomless nature of the floor generated awe, then fear, that one could get stuck here for a long time. A tight steep inclined ($\sim 30°$!) crawl at the back of the Mud Room continued for 15 feet to what looked like another blank wall. This incline had no detectable air and as a result no one crawled down it. Frantic to make anything go, we "chemically loosened" a considerable amount of rock from the crack in the floor of the dome. Then, cold and soaked by this muddy gruel, we exited the cave without checking the results of our rock removal effort, too drained and beaten to test the Bone Crusher a third/fourth time. I had little confidence that our rock removal effort would get us through. But on hearing of our breakthrough Gregg expressed satisfaction at finally conquering the constriction that had stopped Nevin. We had made progress. Despite our pessimism on the cave going beyond the Bone Crusher, Gregg pushed everyone for a return trip. In his mind there always was a way to follow the air.

Within a few weeks we returned to check the results. Shattered rock lay everywhere, but the crack continued one by two inches as far as our lights could penetrate. It seemed hopeless. Lest we missed something we made one last-ditch attempt to push the cave by crawling down the 30° inclined mud tube at the back of the Mud Room. We had not noted airflow there before, but to absolutely confirm this lead did not go, someone had to crawl down it. With crowbar in hand I wriggled in, or rather glided down the muddy ooze. As I slid along I was more concerned about getting back out than in not finding anything. I felt certain the passage would not go. After all we could see the end and there was no airflow. Sure enough I came to a blank wall. My first thought was simply "nothing"...then "how to get the hell back out." However, before I attempted to scrunch backwards up and out of the crawlway, something compelled me to poke with the crowbar at the left end side of the crawl. I felt resistance for about four inches and then, presto, the crowbar punched through to a void. When I took it out air whistled from a one-inch hole, hitting me in the face. With a few more pokes I soon had a larger hole blowing much more air. Ahead lay blackness and what appeared to be a continuation of the crawl.

Although excited by the hole in front of me, I was angled down on my head at this ridiculous incline. I had to retreat, wriggling backwards up through the slime and ooze. After telling the others what I saw, plus the discovery of the strong airflow, everyone excitedly began to dig. But we quickly found that any digging down there was nigh on impossible. We had no good tools and nothing to haul the mud. What to do? We reasoned that first we would excavate enough mud from the floor to approach the crawlway horizontally. But not this day! I was already exhausted from the sheer effort of backing out of the crawlway. I was getting cold from having laid in the wet slop. And now there was that long slog back to the entrance. We decided to retreat, beaten for the time being, but exhilarated that we still had another chance to crack this cave.

8.6 Another Disappointment

In the fall of 1982 and early 1983, we made occasional trips into Burns to dig out the crawlway beyond the Mud Room. Nearby Bobcat Cave had taken off and now most of the attention was being focused there. But one trip in Burns made great headway. Gregg Clemmer, Bill Howell, and I dug several feet down in the mud to where the crawlway could be entered on the level. This, however, created a water sump, a very sloppy morass we had to lay in that grossly and literally "dampened" our efforts to push the cave. We named the crawlway the Second Mud Crawl, it being the second location where deep, sloppy mud penetrated coveralls and soaked one to the skin. During this period, Joe McKenney, Pat Ward, Bill Howell and others continued to enlarge the Bone Crusher (Fig. 8.2).

In early 1983, Kent Seavers managed to enlarge the hole at the back of the Second Mud Crawl to "adequate dimensions" and slid through. A 90-degree turn to the left with a following 90-degree vertical turn up between the wall and a mud mound nearly thwarted his contortionist's talents. But determined to push it all the way he oozed and grunted his way into a four-foot-wide by seven-foot-long, standing room. At the back, Kent found a four-inch diameter vertical hole in a flowstone choke. When he peeked in, strong air hit him in the face.

8.7 Another Terminus

Joe McKenney, Pat Ward and I returned to the cave in August of 1983. We brought the "usual tools" for negotiating tight places. After pushing and contorting our bodies to the end of the cave we worked on the vertical round hole in the flowstone until it was a near vertical crawl. Peering down, none of us could discern what the passage was doing, because at the bottom it sloped steeply away from view. I dived in head first to take a closer look. Joe and Pat grabbed my feet to keep me from jamming myself at the bottom. From my upside-down vantage point I spotted a bedding plane less then a foot high sloping left to a wall with an opening too tight to enter. Beyond, though, I could see blackness. Surely this passage was going to open up! But before I could get a second look, rubble from the blast shifted and blocked my view. As I struggled to extract myself Joe and Pat yanked on my legs. I yelled in pain as I felt something give in my back. Before exiting the cave Pat set some "chemical persuasion" to enlarge the vertical crawlway. I also used some to remove a portion of the wall at the end of the Bone Crusher, eliminating a very difficult vertical bend. I exited the cave in considerable discomfort, my injured back keeping me from the next big Bobcat trip.

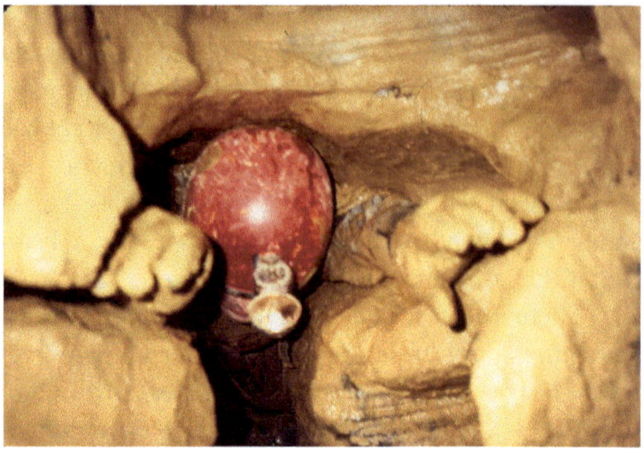

Fig. 8.2 Gregg Clemmer squeezing through the Bone Crusher

The three of us returned in August for more fun and games at this nasty dig. Through our previous efforts, the Bone Crusher was now much easier to negotiate. But Pat's attempt to remove rock in the vertical crawlway at the end of the Second Mud Crawl had failed. Both Joe and Pat inverted themselves as I had done in order to take another look at the lead. Both glimpsed some darkness beyond the rubble, indicating a hint of passage beyond. We had a lot to do and it was not going to be easy. The rubble in the bottom of the vertical crawlway acted like ball bearings, further increasing our chances of getting stuck. It was like going on one's head to scoop out a jar of marbles. Once again this dig was beginning to look desperate.

Joe, Kent Seavers, Dave Morrow, and I waited until March of 1984 before we visited the cave again. We brought our bag of tricks to further enlarge the crawlway, but our efforts seemed futile. The rubble floor and steep approach worked against us. Mud-slimed, sore and discouraged, we retreated with little accomplished.

Bobcat Cave was going great guns. Key personnel needed for Burns' demanding dig were more focused on Bobcat. Joe, Pat and others in the SVG continued to work the end of Burns, but made only marginal progress. The steepness and collapsing nature of the rubble in the dig made rock removal nearly impossible. Worse yet, there was simply no room to work with a hammer and crowbar much less space to dispose of the rock and mud. Just two people could occupy the site and even then, the blast of cold air quickly chilled anyone not actively digging. By the end of the decade, the dig in Burns was shelved for better things (Fig. 8.3).

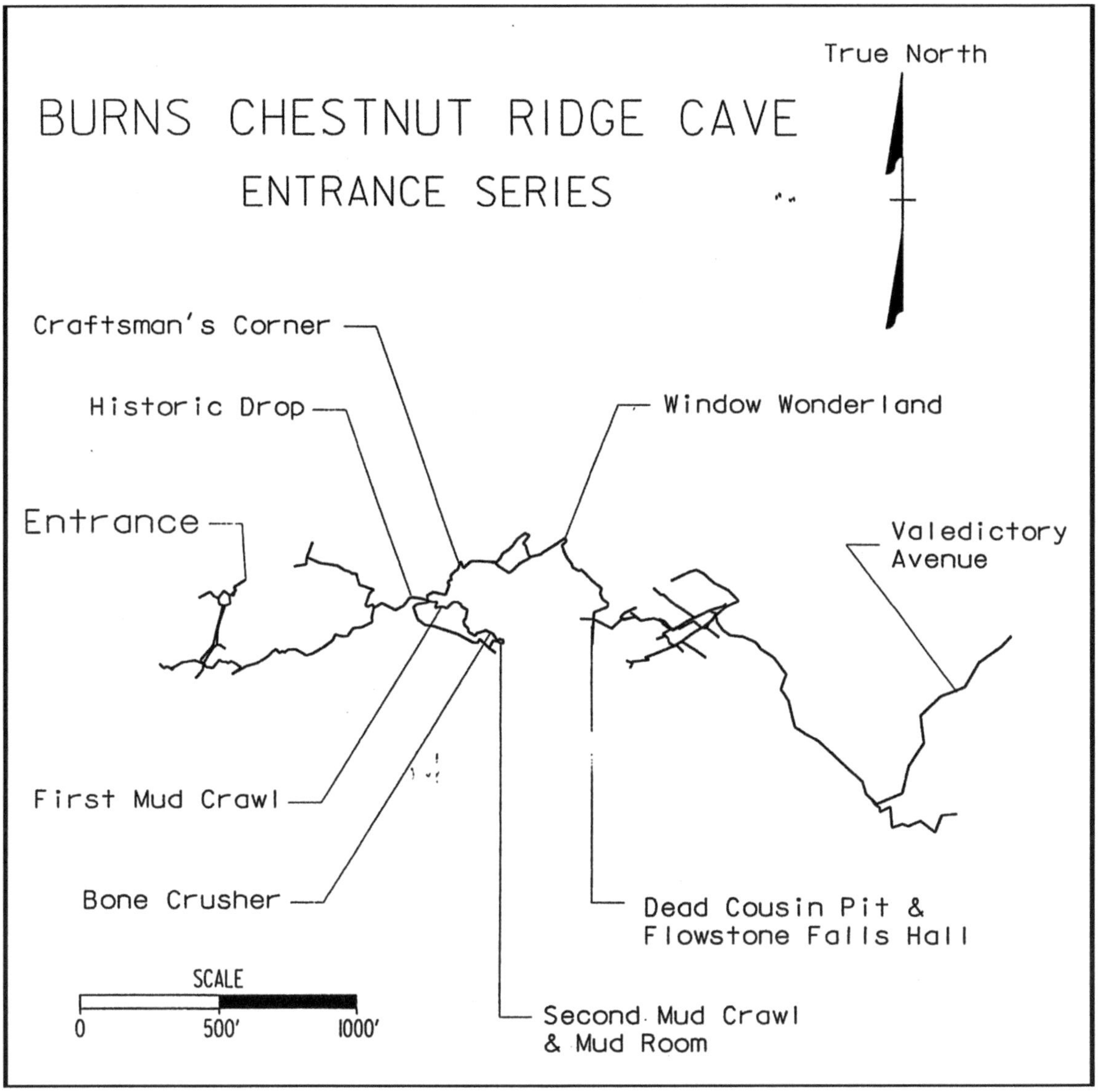

Fig. 8.3 Line map of the entrance series

8.8 Yet Another Attempt

But some of us kept talking about Burns' strong airflow and its potential as both a long and deep cave. I reconsidered our strategy. In October of 1990, we started anew. Accompanying me on this trip was Gregg Clemmer, arguably the most enthusiastic digger I have ever known. Nevin would come for just a short ways, curious to see the second entrance Joe McKenney and other SVG members had attempted to dig open. If they had been successful, this entrance would connect to an upstream canyon lower down in the cave, shortening the travel time by almost an hour. Mike Nicholson also tagged along, revisiting the entrance his father had located in the 50s.

Nevin had been on one of Joe's last efforts to push the end of the cave, and like Joe, had decided the dig was hopeless. Upon reaching the entrance we immediately checked out the potential second entrance located less than a hundred feet away. A raw earth collapse met our gaze, dashing our hopes for an easier way to Historic Drop. Nevin followed Gregg and I for a short distance underground, then turned around to rejoin Mike Nicholson back at the entrance. Gregg and I continued to the known end of the cave.

The Second Mud Crawl seemed larger than we remembered. But the steep rubble slope beyond still looked hopeless. We had no plans to dig this day; instead we needed to evaluate the dig and come up with a viable plan to get through.

> Looking at Gregg, I said, "I believe this will work."
> "Where will we put the rock?" Gregg asked. "There isn't any room in here."
> "In the Mud Room."
> "How?"

The Second Mud Crawl between the small room and the Mud Room was a 15-foot-long, body-tight worm way with deep, sloppy mud and two consecutive, tight 90-degree body bends, one in a horizontal direction and the other vertical, making handing and passing material next to impossible.

"By putting the mud in a bag between two ropes and dragging it through the crawl, one person working on either end," I replied. We figured digging would thus require four people.

We hauled some of the debris out of the crawl and stuffed it in the few crooks and crannies remaining in the Mud Room. Then gathering up old tools and wire left from previous digging efforts, we exited the cave, confident we could get by this next obstacle.

The discovery of Blarney Stone Cave in March of 1991 turned our attention from Burns. It was 3 months before we returned. Both Gregg and I had convinced Nevin of our strategy to get through. Mike Dyas, a caving comrade from the early seventies and a hard-core digger, would join us. We came "loaded for bear:" hammer, crowbar, 50 feet of rope, webbing, bucket and appropriate "chemical persuasion" with all the fixin's. Everyone had a camp pack for use as a haul bag.

Once at the dig we went after the ceiling, taking out rock to create working space. Soon we had a place to sit down in the crawl. No more inching in upside down on one's head. Once we could dig at the blockage, Gregg entered the crawl and started digging debris with the bar, loading the bucket and handing it up to me. I emptied the bucket into a pack held by Mike. Nevin, working from the Mud Room side pulled the loaded pack through the Second Mud Crawl at Mike's signal, Mike having to maneuver the load around the corners or unclog the accumulating muck from repeated haulings. Larger rocks were hammered down to pack size.

After 20+ loads, Nevin was duly exhausted from hauling the mud-globbed pack. But Gregg had dug out enough material for a good look under the ledge. Although still upside down he could see the hole that I had previously spotted before rubble blocked it. But this was no hole in the floor. We were looking through a hole in the wall. We still could not reach it nor could we peer ahead into any going passage. But the howling wind gave us all the inspiration we needed.

8.9 The Final Breakthrough

Blarney Stone and other caves continued to divert our attention. It's truly hard to return to digging and groveling in cold mud and penetrating slop where cave booty is being scooped elsewhere. However, we did return in September 1993. This time only Gregg, Nevin and I made the trip. But now it was Nevin pushing to dig. We had other projects with walking leads and on this same weekend I had a trip scheduled to one of those caves. I felt reluctant to open up yet another cave when I couldn't go underground enough weekends in the month to participate on all the trips my caving comrades were running. But Nevin is a good arm twister when he wants to be, and after all, it was a perfect weekend for the trip—a nice cool, late summer day when the air sucking into the cave would clear the fumes.

We dragged in the usual equipment—a real burden with just three people. As a result, travel to the end of the cave proved slow and arduous. At the terminus, we discovered mud had flowed down into the dig and filled the area to a depth of six inches! It took us a dozen or more buckets to remove this liquid glop, scraping and hauling these dead weight packs out of the Second Mud Crawl. Just three of us filling the packs with the "intestinal gruel" then tugging and pushing the load from its muddy abyss, we could easily have become frustrated. But this day we sensed something electric in the air. No one really said much, but each of us suspected we were getting close to a "view of the virgin."

After digging several large rocks from the walls and floor we finally reached the small tantalizing hole. More rocks needed to come out, a difficult task given the ball-bearing nature of the steep floor that tried to roll and jam me into a bedding plane on one side. Finally by bracing against the far wall with my right arm, I managed to thwart the rolling effect of gravity and the rubble floor. Using only my left arm, I set the "rock removal material" and mud-packed it. Then everyone exited, crawling out the Second Mud Crawl, across the Mud Room and out through the Bone Crusher, a ritual we had done countless times.

To our dismay a warm afternoon outside had changed the dynamics of the airflow in the cave. Instead of clearing the fumes, the cave pulled them by us, then reversed in a curious breathing effect, repeating the entire gagging procedure as we waited. We were ready to abort, except that our packs and other gear were back in the Mud Room. With no other choice, we snaked by through the Bone Crusher for a final look.

I crawled down into the blasted squeeze and noticed that the rock was broken along the perimeter of the small hole in the wall. Pulling out some manageable size hunks helped me finally see around the corner into a crawl of hands and knees dimensions. Going cave! I could not quite see beyond the end of the crawl because of mud dams blocking my view. Pooled water would welcome our tired bodies when we tried to squeeze through. I could care less! Seeing the larger passage beyond the squeeze fired my enthusiasm. I was going for it!

I couldn't fit through on either my belly or my sides because of the steepness of the approach and the height of the opening. The only way was to lie on my back and force it. The pooled water quickly flowed in around my helmet and down my neck, soaking my back as I inched into the squeeze. About halfway, the mud dam snagged me. I pushed and dug at it while on my back, splashing the pool's six-inch depth all over the mud. For two minutes I groveled in this slop before it gave way and I lunged through the squeeze and into the hands and knees crawl. Beyond lay blackness!

We had finally done it. The crawl led to a climbable drop perched 10 feet above a nice-sized room. I explored in the glow of discovery, yelling "borehole," which both Nevin and Gregg distinctly heard in the distance. I could hear them yelling "what's it doing?" but all I could answer back was "borehole!" I was excited, and my shouting just spread the contagion.

The room seemed to end except for an opening in breakdown in the middle. I knew at this point it was going to go. I stopped my exploring and returned to the crawl, digging more out of the mud dams and draining the pool of water. Nevin and Gregg squirmed through. Gregg quickly headed into the breakdown in the floor without first looking around, as if he had been divinely directed where to go to find going virgin cave. After a few minutes, he returned, smiling. Large walking passage was on the other side! We explored this for a few hundred feet to a point where it narrowed to a four-foot-wide canyon with breakdown. A short distance beyond the passage ended except for a hole in the flowstone heading to the top of a pit. Dropped rocks splashed into deep water. We estimated the pit at 35–40 feet deep. A thick layer of bat droppings coated the floor of the flowstone opening leading out to the pit. This was undoubtedly the way on. But we had no rope. This time, as we labored back through the tortuous entrance series, we had reason to be jubilant. Burns was conquered! Or so we thought.

8.10 More Obstacles Ahead

The next month, Nevin, Gregg, and I returned with rope, bolt kit, and mapping gear. To begin our survey, we had to start at the Mud Room with a decade-old survey station. Mapping through the Second Mud Crawl and down the squeeze at the breakthrough was sheer horror. Wet, sloppy mud penetrated everything, and made keeping the book clean a hopeless task. After mapping to the pit we spent an hour rigging, making sure the rope hung free. Upon descending we found the pool at the bottom only a couple feet deep. We continued the survey up and over some massive flowstone mounds, through a squeeze, then into a small room with a too-tight crawlway leading from the floor. Air poured from here. A loose slab of rock above the crawlway offered hope we could enlarge the crawl if we could dislodge it. Gregg had brought along a small crowbar and chisel. We loosened the slab, but to our dismay, a much larger piece fell and completely blocked access. At arms' length we struggled to budge it but the slab was simply too heavy and cumbersome to move aside.

Reluctantly we turned our attention to a two-inch crack above the crawlway. If we got through here we could bypass the blockage. Gregg whipped out the chisel and together with my Petzl bolting hammer we declared war on the limestone. For five hours we pounded, flaking bits of rock and exploiting hairline cracks, cheering with each half inch of progress! Only after taking off six inches and grossly deforming both the hammer and the chisel, was I able to barely scrape through. We named the place Craftsman's Corner in honor of the Sears' chisel we had sacrificed. Once on the other side I managed to maneuver the slab out of the way, enabling Nevin and Gregg to join me. We continued the survey for over a hundred feet, pushing ahead in narrow, clean-washed stream canyon, floored in flowstone. The mud was gone, but the character of the cave stayed the same, squeeze after squeeze, with plenty of cold water rushing off into the unknown.

We picked the coldest day of the year for our next trip. With the thermometer reading well below zero, Nevin fired up his new John Deere tractor to transport us to the entrance. It felt unbearably cold clinging to the tractor as we chugged to the top of Chestnut Ridge. Hoarfrost covered the cave's entrance. Steam billowed up and out of the depths as if a fire was burning inside. For this trip, we changed inside the entrance, despite cramped conditions and having to cling to the steep entrance slope. We had a new chisel but had swapped my bolting hammer for a three-pound sledge. Air screamed through the Bone Crusher, almost blowing out our carbide lamps, but Nevin just grinned, enjoying the lumens of his newly self-built fluorescent light.

Our last survey point was on the wall above a deep pool in a sinuous stream canyon. The water was up and getting there got us soaked to the waist. Gregg took lead tape, and soon began inching ahead on his side. As the passage lowered, water poured into his coveralls and out his sleeves. Progressing further ahead we found the floor begin to steeply dip, but passage width remained claustrophobically tight. The water appeared to be no more than a quarter-inch deep, but its high velocity sheeted over us, soaking everything. But getting wet didn't matter. Just negotiating the passage worked up quite a sweat. Another constriction loomed, then another. And then, the crawlway turned into a tall and narrow canyon only inches wide. Blades of chert and fluted limestone blocked progress. Hammer work got us around a very tight corner, then down into a small room with a pool of water. The only passage out seemed a continuation of the narrow canyon. It seemed to end, but the strong breeze indicated otherwise.

Gregg squeezed through and crawled to the apparent end, then called back that a window, totally unseen from our vantage point, opened to the right. We named it Window Wonderland. But past this point the passage turned nasty again, a tight, low crawl over water, too small to force. Gregg went to work with hammer and chisel and after some time enlarged it enough for me to get through. There was just enough room to stay above the water without total submersion… as if that mattered. To my dismay though, just beyond this obstacle, I came to an opening in solid bedrock too small to pass. The wind blasted through here and soon chilled my soaked body. Nothing more this day!

On the way out we surveyed a side lead that offered a potential bypass to some of the horror we had crawled through. But although dry, it proved longer and tighter than the wet way. We exited after midnight to a bright starry night. The temperature was a brittle 9° below zero! The tractor failed to start, so we hiked/hustled the three-quarters of a mile back to Nevin's house. Fortunately, we had been able to change into our dry clothes in the blowing, 49-degree entrance. Stepping from the entrance "fumarole" into the cold winter night—a dramatic 60-degree temperature drop—capped our collective memory of this hard, brutal trip.

8.11 Yet Another Final Breakthrough

Much happened before the next trip. We made a much-anticipated connection between Bobcat and Blarney Stone caves and a few of us would get trapped by high water in Barberry, an event that made national news. Trying to organize a return trip to Burns, though, proved difficult. The cave had not offered any real breakthrough, despite our advances. The passage continued tortuous and every trip required several days of recovery after the beating. Still, we knew we had to get back "down there." By June of 1995, we were set. Gregg opted out, obligated to attend a funeral for his wife's cousin (He would later confide that he needed a break from this cave). Having been both mentally and physically beaten up by repeated trips into the cave, his absence refueled his enthusiasm by giving him time to reflect.

We replaced Gregg with Ben Schwartz, a new member of the Burnsville team who had proven he could handle the difficulties of Burns by previous trips into Bobcat, Blarney Stone, and Barberry. A trip into Burns would be a true test of not just physical caving ability, but mental stamina: a challenge against one's desire to push onward despite the soaking cold, muddy, bone-weary hammering, deep inside this tortuous, brutal, relentless cave.

I was reluctant to return without Gregg, feeling he was owed the chance to be in on a breakthrough because of all his previous contributions. But Nevin, at age 53, was hot to go back in, as if his caving career would soon come to an end. On Friday night of the weekend trip, I asked Nevin to call Gregg and ensure there would be no hard feelings. Thinking back, it is strange that I would even feel this way, given the previous history of Burns' obstacles. Gregg gave us his blessing to go for it, and we promised to turn around if the passage opened up. So, on Saturday morning, Nevin, Ben and I muscled in "rock removal" gear, hammer and chisel, and a strong plastic bag for hauling mud. Beyond the breakthrough, we had encountered little mud, so we were probably going to have to "carry our own."

The trip in was unremarkable until Nevin discovered he had not brought any cave food. Burns is not a place to forget nourishment, but not willing to turn back, Ben and I offered to share ours. At the constriction I was able to locate a strategic cleft in the right wall for rock removal. It took Nevin and Ben a while to locate some mud for mud packing. Without needed mud for packing, the wished-for results are not nearly as satisfactory, and we didn't want to waste our efforts needlessly. By the time they returned with sufficient mud I had become quite chilled.

I inched back into the lead and finished the packing, then backed away some fifty feet to the only area large enough for all of us to fit. Upon detonation, the shock reverberated through the walls of the cave, telling us the mud had done its job. Strong airflow going into the cave quickly vacated the fumes. Crawling in, I was able to dig away enough chunks of broken rock despite having only one arm stretched in front of me. Popcorn on the walls and ceiling grabbed at my every attempt to inch forward. Each foot gained brought a changing view. Ten feet in I could see that a window opened into total blackness. Looking yet further ahead, the flowstone floor literally poured over into a vast pit.

WOW! This appeared to be the long-sought breakthrough we had been working for. I retreated, letting Ben and Nevin, one at a time in this narrow confine, get a look. We removed more rock, making it more comfortable. Ben, using a hand-held flashlight, determined the expanse below to be a large room. He could see a flowstone floor dropping steeply away, about 40 vertical feet down. Without a rope we were going no farther this day, letting us easily keep our promise to Gregg. We decided to name the pit Dead Cousin Pit after the event that prevented Gregg from being with us (Fig. 8.4).

Fig. 8.4 Tommy Shifflett at the top of Dead Cousin Pit

As for Ben, he had passed the test. His enthusiasm had shown that much. He was already talking about returning before he even reached the entrance. The rest of us always needed a few days recovery before considering such a thing!

8.12 Booty Time

It didn't take long to return. During July 1995, Nevin, Gregg, Ben and I hauled in a bolt kit and rope to bottom the pit, and hopefully map lots of booty. After setting bolts and rigging the rope, Ben descended to the bottom while the rest of us followed with the survey. A standard practice of ours is to survey as we explore, ensuring all in the team share the thrill of discovery, and that most all of the passage seen gets mapped. From the bottom of the pit we could see that the passage we came from was just a small window opening half way up one side of the room, with a waterfall pouring out and over the beautiful canopy of flowstone. Compared to the passage we had come from this room was a monster hall though it measured only 50 feet wide by 90 feet long. We named it Flowstone Falls Hall for the beautiful white flowstone that cascades out of the top of the pit.

This room sloped steeply from one end to the other, leading to another flowstone cascade over breakdown. We mapped down this route over 50 vertical feet to where the breakdown cascade ended in a drop-off of about 20 feet. We rigged this with left-over rope from the first pit. From the bottom of this drop the passage continued as a stream canyon, but more spacious. At two to three feet wide, we WALKED, except for an occasional scramble over breakdown. At a junction room, we explored several leads, noting the stream sumping against one of the walls. To continue, we pushed a tricky climb across a narrow, muddy ledge to a window above the sump. A challenging downclimb got us to the stream where the passage got even wider. More deep pools, breakdown, and occasional low spots blocked our way but we kept on surveying.

Then the stream vanished, but the cave, and the air continued. We intersected upwards into a dry, fossil passage 30 feet wide, three to five feet tall. Sensing we were finally "off to the races," Gregg named it Valedictory Avenue. Soon the ceiling rose to nice walking dimensions.

After several hundred feet, we were again crawling but ahead, the roar of water told us that things were about to get "interesting." Scrambling on through we intersected a tee junction with an impressive stream canyon 15 feet wide and perhaps 40 feet tall. We started cautiously down a "less than straightforward" 15-foot climb and reassembled at the stream. We headed downstream, walking through and climbing over a number of deep pools and rapids. On the left side we passed a walking size lead issuing a stream and a blast of air so strong it nearly blew out our carbide lamps. We were thrilled!

We continued our survey to a 12-foot overhanging waterfall. It appeared unclimbable. Ahead we could see rapids and more waterfalls dropping into the depths. We knew we were very deep, at least for a Virginia cave. We had also spent a lot of energy reaching this point. Recalling the long slog back to the surface of previous trips, we realized that on this trip, we were going to be pushing new endurance limits.

It took us five hours of steady, grueling work to get out, exiting at 4 a.m. We were elated that the cave was going deep. That was good. Having to negotiate the unrelenting obstacles that pounded and bruised our bodies, that was bad. When we reduced the data we found that the top of the last waterfall was precisely 700 feet below the entrance. Our next trip would push downward from 700 Foot Falls.

That came in September with the same personnel. No one wanted to be left out. I had just gotten over the worst part of a cold and felt apprehensive about going underground, but thoughts of booty can sometimes get the better part of judgment. We were motivated enough this time that we actually got into the cave by 9:40 a.m. This was important given the long tiring trip ahead. While heading in and still feeling the lingering effects of my cold, I realized more than before how difficult the cave really is. After a four-hour, quick pace, we arrived at the waterfall lead. Ben performed an acrobatic stretch across the stream canyon intersection with the pit in order to set a bolt away from the waterfall. His legs seemed to be wedged at 180° from his hips! Thankfully he was on belay. With a couple of bolts at the drop's edge, we rigged a tension traverse over to the bolt Ben had placed (Fig. 8.5).

Immediately past 700 Foot Falls we came to a seven-foot drop which we rigged with webbing. A short distance beyond we hit another seven-foot drop. We were out of bolts and there were no natural rig points. By performing a tricky stretch beyond the pit we could access a narrow ledge where we were able to climb down. We mapped 500 feet of passage

Fig. 8.5 Ben Schwartz at the edge of 700-Foot Falls. Photo by Mike Ficco

beyond the last waterfall to a sump. This appeared to end except for a crevice 15 feet above the water. Slight airflow seemed to move toward the crack. Ben and I made our best effort to push it but the crevice was just too tight to follow. We were starting to feel the gloom of shutting down, except we still had the upstream part of the main stream canyon to explore. That at least was walking passage (Fig. 8.6).

After 150 feet this upstream lead ended in a wide sump (Figs. 8.7 and 8.8). We then turned our attention to the side lead that was blowing all that air. Mapping in brought us to a 25-foot-wide by 60-feet-long room floored in massive breakdown. A waterfall poured from the ceiling, perhaps 100 feet up. A large lead appeared to roar off into the unknown 40 feet off the floor. This we felt had to be the source of the air. We mopped up a

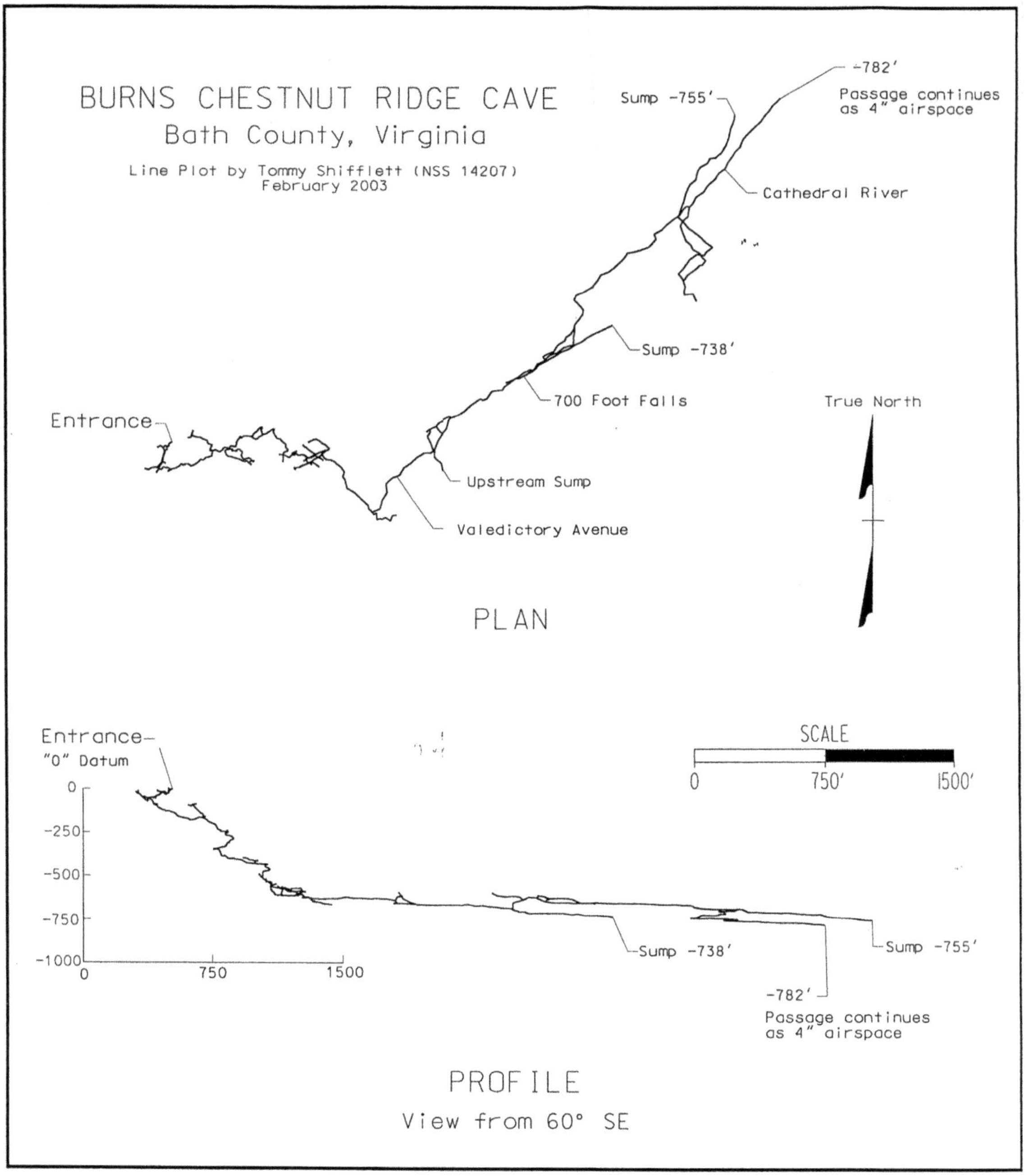

Fig. 8.6 Line map and profile for Burns Chestnut Ridge Cave

Fig. 8.7 Ben Schwartz and Tommy Shifflett in stream trunk upstream from 700-Foot Falls

Fig. 8.8 Mike Ficco by speleothems in stream trunk upstream from 700-Foot Falls. Photo by Mike Futrell

few side leads then headed out. Reduction of the data put the downstream sump at −743 (later adjusted to −739 feet) below the entrance. Burns had become the deepest cave in Virginia.

8.13 Walking on Walls

In June of 1996, we returned with some fresh blood. Gregg Clemmer, Ben Schwartz, Mike Futrell, Mike Ficco and I planned on a two-party survey. Gregg, Ben and Mike Ficco would attempt to push the entrance stream in hopes of bypassing the upstream sump. Mike Futrell and I would attempt to climb to the 40-foot lead and/or map any remaining leads from the same room in an effort to track the incoming air. Nevin, along with other Burnsville Cove cavers, was assisting Ron Simmons with a dive in Aqua Cave.

Again we tried for an early start but only managed a 10:30 a.m. entry. This was a must as Mike Futrell is notorious for not exiting a cave until he has plenty of "booty in the book," as he styles it. We know we were in for a long trip. Our descent went smoothly, the old veterans savoring the new boy's take on this hellish cave and all its countless tedious entrance-series obstacles.

Once we reached the intersection of Valedictory Avenue and the stream trunk, Mike and I headed to the high lead, looking for an easy way up. We found a climbable crevice that intersected the canyon. This led to a balcony overlooking the room, but nothing went. We had lost the air. We turned our attention downstream, finding a passage to the water, but it proved to be a loop-around. By the time the other team caught up they were hungry for booty. Their stream lead had become too low and tight after just a few hundred feet. Things seemed to be shutting down. Where did the air go? With no other plan we started downstream looking for the air and showing the cave to our new visitors.

About 100 feet before 700 Foot Falls, Mike Ficco began to suspect that the air might be going up. In a stream trunk 10–15 feet wide, with nearly vertical walls and soaring to more than 70 feet above us, getting up there was not straightforward. But to our amazement, Mike, followed by Ben, somehow climbed this thing and disappeared out of sight. Mike Futrell attempted to follow but continually peeled off the climb. (After this trip, Mike Futrell made a point to go out and buy the same brand of boots Ben was wearing. His excuse for doing so was that he wanted to be able to *walk on walls* like Ben. Needless to say the boots did not offer that ability.)

It did not take long for Ben to come back and shout down the news that they had found a paleo-passage. We derigged 700 Foot Falls and threw up the rope. Once each of us had ascended we started surveying big, booming, paleo-borehole. Except for a couple of climb-downs this was the easiest passage in the entire cave. After a while we came to a large lead on our right. A dome with a waterfall pouring down intersected our left. The passage ahead still roared off into the unknown, so we continued our survey here, now following a respectable-size stream. After several hundred feet of survey we abruptly came to a sump. The passage had a peculiar V-shape cross-section and was banked in heavy mud (Fig. 8.9). Later calculations showed we had deepened the cave to −755 feet.

Back at the large side lead on the right we mapped up and over a large mud mound, then broke into large walking passage that turned a corner and became a crawl. Ben slid ahead and checked it out, returning with a grin, saying there was a large stream passage behind him. We mapped through to the second largest stream in a Burnsville Cove Cave, the Cathedral River, the main supply to Cathedral Spring on Bullpasture River! We set a survey station both in the upstream and downstream directions. We had 3784 feet in the book, Mike Futrell was happy, and we could leave. We got out at 7 a.m. after more than 20 h underground. Everyone was wiped out, grabbing what sleep we could before the long drive home.

8.14 Visions of Bigger Booty

Despite the aches and bruises, it didn't take us long to organize a return, especially now that we had found the main water to Cathedral Spring. Between the Chestnut Ridge Cave System (Bobcat and Blarney Stone Caves) and Burns, only a small portion of the theorized system has been found. We envisioned the downstream direction taking us toward the Blarney Stone section of the Chestnut Ridge Cave System, hoping for a connection. Upstream and totally "off the map" lies the largest portion of unexplored karst (and

Fig. 8.9 Stream trunk heading towards the −755 foot sump

hopefully many miles of virgin cave) in Burnsville Cove. In August 1996, we went after it with vengeance: two teams, consisting of Nevin Davis, Mike Futrell, Mike Ficco, Ben Schwartz, Gregg Clemmer, and myself. And because we knew that this was going to be a major push, we entered the cave at the unheard of hour of 9 a.m.!

After the previous trip we had come to the conclusion that camping might be the best way to continue exploration in Burns. This would be far more problematic than the many camps we had run in Bobcat in the 1980s. Ten-inch-diameter packs that worked fine in that cave would never do in Burns with all its twisted, gnarly cracks and contortions. We'd never get enough camp gear in on one trip to make it feasible. Even the sheer difficulty of hauling a camp pack of substantial size and weight relegated the attempt to a fool's errand. In an effort to somehow make it work, though, we decided to haul in a few items at a time, establishing a "beachhead" of sorts with a stove, ground cloth, Therm-a-rest, etc. By hauling in gear piecemeal on surface trips we felt a camp could eventually be stocked to launch a major survey and mapping effort, now that the cave had broken wide open. Each participant in his own way had concluded that we were fast reaching the physical limit of surface trips.

But there was no consensus on where to camp, and after hauling gear to the bottom, some felt it perhaps better to use a bivouac system in Burns. Stockpiling an underground camp along Valedictory Avenue was simply too difficult. We confronted the specter of extended surface trips.

At Cathedral River we formed into two teams. Gregg, Ben and I let the others have their pick of leads. They chose upstream, which seemed at the time to be the most promising lead. Going with the water, beginning as a tall, narrow canyon, quickly provided us with a number of pleasant surprises. Wall-to-wall water quickly challenged our surveying skills. Swift water threatened to sweep us from the stations. But after five minutes of mapping we broke into spacious walking passage 20 feet wide. We mapped this dimension for over 1200 feet, finally stopping when the ceiling lowered to four inches. To advance would require total chest immersion with our faces on the ceiling. We could see the water flowing ahead with four to six inches of airspace, so it did not sump immediately. Without wetsuits we were forced to end our survey, putting our last shot on the water surface, and taking a measurement of the stream's depth.

The upstream team quickly surveyed to a sump. They then turned their attention to pushing and mapping side leads, but found nothing that hinted at bypassing the sump. Burns appeared to be shutting down once more. We grimly decided to haul out the camping gear we had brought in, reaching the entrance 17 h after we had started.

Reduction of the survey notes put the cave's depth at −782 feet below the entrance. With a point in the ceiling, near the entrance four feet higher, the total depth came to −786 feet, setting a new Virginia depth record at the time. The vertical profile of Burns' entrance to the level of Cathedral Spring is 815 feet. Considering the 160 foot depth Ron Simmons achieved in Cathedral Spring on past diving expeditions, and his report that the passage is still trending downward, it may be possible to dive over 200 feet below spring level. Hauling tanks to the far reaches of Burns can be considered a near impossibility at the time. But if a connection is ever made Burns could go a thousand feet deep.

8.15 It Still Goes

We ran a few more trips into Burns after this epic survey, all being attempts to find and follow the air. Just before the 1998 NSS Convention in Sewanee, Tennessee, Ben Schwartz, Mike Ficco, and Nevin Davis climbed the dome near the junction in Cathedral River. It proved to be a risky, challenging climb into soft beds of shale and chert. As a result, Ben and Mike were forced to use pitons in the 20-foot ascent. Still some of the pitons literally wedged the rock up, ready to pull out, requiring Mike to quickly place the next piton in a scramble to stay ahead of the anchor failure. At top, he placed a bolt in respectable limestone, but had little time left to explore ahead. Mike believed the passage continued, blowing some air.

To date we have not been back. That's where exploration stands. Perhaps, you ask, the cave has finally pounded us into submission? With other projects going strong, giving us ample opportunity to "put booty in the book" and with razor-sharp memories of the grinding brutality Burns brings to every visitor, it is easy to talk about the next trip without actually ever going back in. But the air still blows down there and this is all the lure true cavers will ever need.

The author would like to thank Mike Ficco and Mike Futrell for offering some of their fine pictures; Phil Lucas for his review of the article and many helpful comments; and a very special thanks to Gregg Clemmer for providing valuable information of events from his cave journal, and his tips on making this article, hopefully, an exciting read.

[**Editor's Note**: Tommy's article brings the saga of the exploration of Burns up to early 2003. As might have been expected, there's more to the story. Burns was indeed entered again but not through the horrendous sequence of obstacles described above. On December 3, 2005, Ed Kehs, Jon Lillestolen and Mike Ficco pushed a lead in the Outer Limits of Blarney Stone Cave and squeezed through the crawl to emerge at the top of Pot Metal Climb in Burns. The final bit of survey connected the two caves.]

8.16 Second Connection: Burns Chestnut Ridge Is Spliced into the System by Mike Ficco

[**Editor's Note**: Edited from a field trip report, *BCCS Newsletter* 30, 10–13 (2005)]

Friday night, December 2nd, 2005, half a dozen of us gathered at the BCCS Homestead in preparation for a Saturday morning trip into the Blarney Stone entrance of the Chestnut Ridge Cave System. The morning dawned crisp and sunny, breakfast was eaten, and we migrated over to the Davis farm for the seemingly mandatory milling around stage of trips into the Chestnut Ridge caves. The plans were for two teams to go into Blarney Stone. Tommy Shifflett was to lead one team to the Pigeon Tooth Dome lead; I was to lead a second team whose object was to push the end of the Outer Limits passage for a possible connection to Burns Chestnut Ridge Cave.

On the trip of January 8, 2005, Ben Schwartz, Ed Kehs and I had previously turned around in Outer Limits at a low sleazy stream crawl that was blowing strongly but too tight to continue. That trip had put a pretty good hurt on us and a return trip had not been high on our list of priorities. However, the lure of a connection to Burns was strong and survey data reinforced our suspicion that we were close to connecting. I felt that the lead resembled one at the top of the Pot Metal Piton climb in Burns where I had turned around during an epic trip in July, 1998. Repressing the less pleasant memories of a trip to that section of Chestnut Ridge, we scheduled a return to push the Outer Limits. Our team, consisting of Ed Kehs, Jon Lillestolen and I, planned to enter the cave prepared to dig/hammer/enhance our way towards the second deepest cave in Virginia.

We entered the Blarney Stone entrance a little before noon loaded for bear. This was Jon's first introduction to Chestnut Ridge caving and regardless of the extent of preparatory warnings/descriptions, one really has to experience these cave's entrance series to appreciate them. He was about to taste his first dose of Chestnut Ridge "poison". Despite our heavy packs, we moved at a pretty good clip down the entrance drops, through Strychnine Canyon, across Artz's Attic and into Ghost Hall. After a brief stop to catch our breath and to allow Jon to take in the sudden increase in passage volume, we proceeded to Upper Ghost Hall and down through the series of dig sites leading to The Boondocks. From here we quickly covered the several thousand feet of large trunk to Duane's Drop, rappelling down to Boulder Dash, where we were pleasantly surprised to find minimal indications of flooding from heavy rains earlier in the week. Had the flooding been more severe, our traverse over the namesake boulders would have been much more treacherous.

Climbing up from Boulder Dash into Opportunity Knocks we found that the wind was blowing strongly into the passage and we prepared ourselves for the hardest part of our journey to the lead. While only half a mile in length, the trip through Opportunity Knocks and The Outer Limits is a non-stop fun-fest of mud, crawling, mud, more mud, and more crawling, with a few vertical drops and climbs thrown in for good measure. In essence it's one long obstacle. We grunted and cursed our way through this obstacle and arrived at our destinations after approximately 6 h of travel.

The lead was as I remembered, a low (one foot high) flowstone shelf with a stream flowing under it into a crawlway choked with cemented stream cobbles, flowstone, and copious quantities of mud, leaving perhaps four inches of open space to the ceiling. Jon had one look at the lead and exclaimed (only half jokingly, I believe) "I didn't sign up for this!" We unpacked our tools of destruction and tried to find solid surfaces to place things, while we sank up to our knees in the muck. I convinced Ed to have the first go at the dig and he dove right in, lying on his back in the stream, loosening material primarily with his hands and a crowbar and placing it in voids along the sides of the dig. He slowly inched forward while a slurry of mud and water flowed into the pant legs of his cave suit.

Meanwhile, Jon, a resident of Tennessee and presumably a wannabe engineer for the TVA, began building a series of dams upstream of the dig. While the intent was to control the flow of water through periodic "scheduled releases", the grand project was short-lived as the series of mud dams slowly failed, sending small flood pulses of spooge in Ed's direction. TVA would have been proud.

After about 20 min of heroic groveling, Ed extricated himself having progressed approximately two body lengths into the dig and reported that the way on looked "gnarly" and was going to require significant hammering. Ed got the full effect of the sleaze factor when he stood up, and all of the accumulated water and spooge drained into those areas that up until that point had managed to remain relatively dry.

Wielding the three pound hammer and crowbar, I then inserted myself into the slot and lying on my back, began pounding on the ceiling, systematically removing small projections which would allow me to force myself inch by inch closer to Burns. After progressing about three feet, my heart sank as I was presented with a lowering of the flowstone ceiling, leaving only an inch or two of open space through which to pass. This was going to be a real problem and I inspected the ceiling for a possible weakness that I could exploit. Seeing none, I gave the hammer a good swing. The result was "thunk", the characteristic sound of solid flowstone, the variety that is just soft enough to absorb the energy of hammer blows without breaking. We had brought additional "tools" of the highly exothermic variety for this sort of situation. However, using it on the ceiling was going to be problematic; once again I paused and scrutinized the now scarred but fully intact barrier.

Then I saw it, a faint hairline crack on the far left side where the ceiling met a small vertical partition. I resumed hammering, now focused on taking advantage of the rock's vulnerability. 'Thunk, Thunk, Thunk—SPLASH!—a 1-foot by 2-foot section of ceiling, 6-inches thick had peeled off and dropped down into the slurry. I quickly maneuvered the slab into a pocket along the right hand side of the crawl and for the first time got a good look at what lay ahead. The good news was that clearing the way for the next body length would be easy, just some rotten doo-doo on the ceiling that had no aesthetic value in a hell-hole like this. The bad news was that beyond that body length, a large flowstone mound extended the entire way across the passage, with what looked like a very tight open slot remaining between it and the ceiling. Half a dozen swipes of the crowbar cleared the way ahead, and I discovered that I had just enough space to roll over from my back onto my belly (it was getting big). I looked through the flowstone slot, and beyond I could see it enlarged to hands-and-knees size. I tried hammering on the mound but with only a few inches with which to swing the hammer, it was futile. The opening appeared to be right at my limit. I test fitted my helmet—it fit with an inch to spare! After a moment of hesitation, considering if this was really such a good idea, I forced my way into the chest compressor. Thankfully, I was fully lubricated by mud, and I squeezed over the smooth flowstone rather easily and scrambled on my hands and knees into the enlarging passage beyond.

A body length past the flowstone mound, the passage T'd into a 6-foot by 8-foot passage that looked vaguely familiar, and suddenly my heart raced as I saw the rope and bolts I had set in the ceiling of Burns seven years prior. WE HAD MADE THE CONNECTION!! I excitedly told the news to Ed and Jon, and with the increased space in which to swing the hammer I completely removed the offending flowstone mound. After scrambling back through the crawl, we got out the already muddy survey gear, found a tie-in station and began the survey. To add insult to injury, Ed was partway through the crawlway/sleazeway when the end of the tape broke off.

The survey continued despite the now non-standard tape length and we were all soon standing in Burns. The only thing remaining to do was to rappel down the Pot Metal Piton Climb and tie into the Burns survey below. The aluminum carabiners that I had been forced to use for rigging at the top of the Pot Metal Piton Climb had deteriorated over the intervening seven years. We hammered them open and replaced them with stainless steel links. The rope was still in good shape so we descended down through the waterfall into the large passage below, found station R204 and connected the two surveys. WHOO…HOOO!

It was 9:00 p.m. and we collectively decided to get the hell out of there, so up the rope we went and through the Sleazeway back to Blarney Stone. Prior to leaving, I checked out an infeeder lead I had discovered on the Pot Metal Piton trip. Consisting of a low, wide crawlway floored in smooth sandstone, I traversed upstream for perhaps 100 feet to an intersection with two walking-size canyons; one carrying the stream, the other dry. These leads are definitely worth returning to and could be the key to unlocking the upstream portions of the Cathedral drainage basin.

On the other side, we packed up most of the gear deciding to leave the hammer, crowbar and the 50 feet of rope Jon had de-rigged from a high dead-end lead. These items are now stashed above the flowstone shelf at the beginning of the Sleazeway. Trying not to dwell on the hard trip we had ahead of us, we began the long slog towards the entrance. Partway through Outer Limits, we assessed a high lead into an apparent large passage/room above that we had seen during the previous trip. A small amount of bolting will be required to gain access to this lead, but I feel that it holds great promise for leading into the upstream Cathedral System.

We exited the cave slowly but without incident, reaching the surface around 5:30 a.m. It took us approximately 8 h to make the trip out from Burns; not a trivial exercise. We were greeted by frigid temperatures and freezing rain, changed out of our gear and drove back the warm Homestead where Scott Olsen has a warm pot of turkey stew waiting for us next to the wood stove. We filled our bellies with warm stew and cold beers, and then hit the sack for a few hours of much needed sleep. What a great way to end a difficult, historic trip into one of the country's great caves.

Data reduction indicates that the connection and additional survey by the two teams pushed the Chestnut Ridge Cave System to 20.03 miles in length, making it the third Virginia cave to break the 20-mile mark.

Pancakes and the Saga of Our Maple Sugar Digs

9

Gregg Clemmer

Abstract

By 1983 most of the open cave entrances in Burnsville Cove had been discovered. The only way to find new caves was to dig entrances. Clues were shallow sinkholes and places where the snow melted in the winter. Beginning in 1986, BCCS initiated a regular annual weekend to dig for new caves. Since the weekend coincided with the Highland Maple Sugar Festival with its fire-company-sponsored breakfasts, this became known as "pancake weekend" and the caves discovered were collectively known as the "Pancake Caves". More than a dozen caves have been opened. Maps and photographs of these caves are given. Most of the caves are relatively small but four have been mapped to lengths of more than a mile.

Whose woods these are I think I know.
His house is in the village though;
He will not see me stopping here.
To dig a hole and make it go.

With apologies to Robert Frost.

Virgin cave in Burnsville Cove has to be earned. The hillsides and valleys have been walked. The obvious entrances have been found. Notable exceptions like Breathing Cave and a few smaller caves like Carpenter's, Jackson, and Owl gave up most of their secrets years ago. So it was that Ike Nicholson's discovery of the Butler-Sinking Creek System while scratching out rocks and loose dirt from a blowhole on the Carl Butler farm in May of 1958 would set the tone for discoveries to come. In short, we were going to have to dig new entrances.

G. Clemmer (✉)
25 Silverton Court, Hunt Valley, MD 21030, USA
e-mail: yorkst@aol.com

But where?

Before Butler, Nicholson and his sons had visited Chestnut Ridge Blowing Cave, digging open this small blowhole to find 800 feet of sloppy, mostly miserable passage. In summer, the cave's breeze vanished down a twisting, narrow muddy crawl. Bevin Hewitt's 1956 diving efforts with the Nicholsons in Refrigerator Spring (Lockridge Aqua Cave), garnered a surprising discovery of more than a mile of walking, decorated stream cave. Entrance modification permitted subsequent access without free diving.

Over the next two decades, surface and in-cave digging by the BCCS, Duke Outing Club, Potomac Speleology Club, and others, burned calories and charted modest finds in such features as Leap Yer Pit,

W.B. White (ed.), *The Caves of Burnsville Cove, Virginia*,
Cave and Karst Systems of the World, DOI 10.1007/978-3-319-14391-0_9,

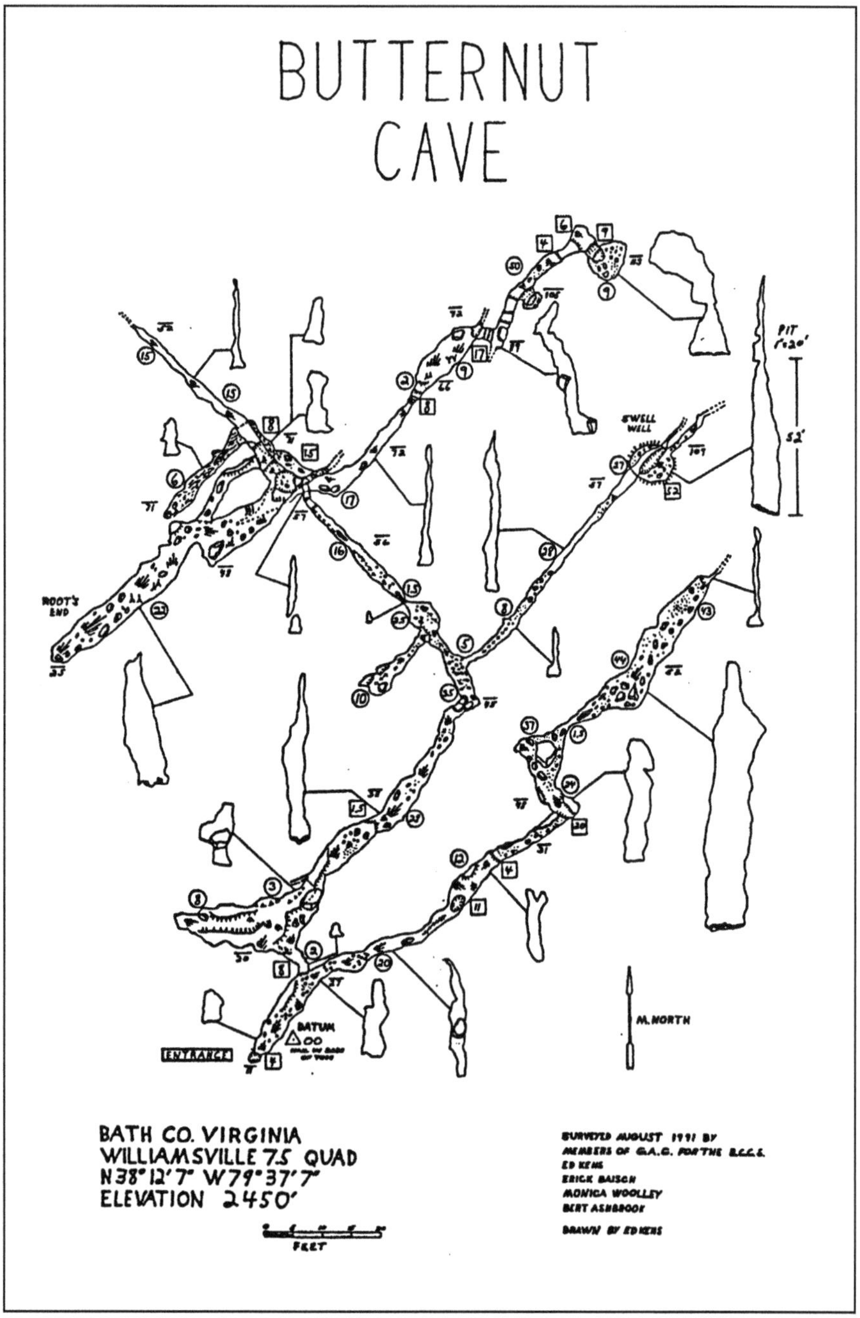

Fig. 9.1 Butternut Cave. Map from the *BCCS Newsletter* **17**, p. 39 (1991)

Counterfeit Pit, Better Forgotten, My Favorite, Boundless, Burns, and Robins Rift. These efforts ranged from individual sojourns to grotto, on again–off again, multi-year projects.

The digging bug bit me in 1978 when I investigated a blowhole on the east bank of the Bullpasture River at the head of Bullpasture Gorge, acting on a vague tip from a customer in my father's hardware store. Subsequent work on this small hole with Shenandoah Valley Grotto members got us into a half mile of virgin cave which we named Fossil-Moss.

After this, digging for caves became one of my passions and as I continued to spend time in Burnsville Cove caving with the BCCS, I realized a great system

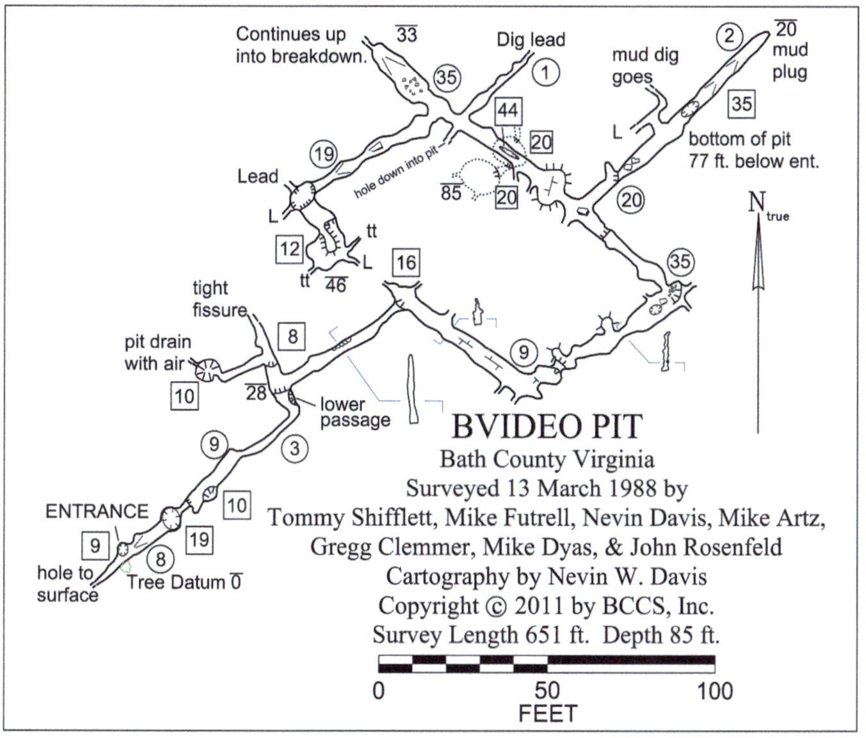

Fig. 9.2 Bvideo Pit. Map compiled by Nevin W. Davis

lay beneath our feet. Not just Butler and hydrologically connected Breathing, but a vast unknown complex, one of the great systems of the world. This was nothing new to the old hands who had already dyed the streams, explored the caves, and ridge-walked the hills for years. But our theorized system boasted few natural entrances, lying deep and difficult beneath Chestnut Ridge and nearby environs.

Dig? Well, yes. But where?

With systematic searches to read the signs and plot the sites, might we employ some muscle with a fair degree of confidence that we'd find cave?

Or was all this a pipe dream?

Occasional digs after Fossil-Moss focused on PSC's re-opening Robins Rift (at least one anxious moment there!), our pushing and clawing past the grim air-blowing terminus in Burns (a visceral multi-decade saga described in Chap. 8), fiddling around (paying our dues) on Bullpasture Mountain with Singing Tree, Seven Shots, and other nameless pud holes, and probing nebulous air cracks above Cathedral Spring. When we tired of these, there was always Nevin Davis and his never-ending Backyard dig. It seemed everyone had a dig going.

By 1983, Nevin's Backyard dig was now Backyard Cave and some of us were fooling around in Chestnut Ridge Blowing Cave. Yet despite good air, there was no easy way on in this sinuous, mud-lathered, interminable crawl. But we kept at it, moving rock, excavating bypasses, chiseling corners, and "gravelizing"

Fig. 9.3 Entrance to Blackroot Cave. Photo by Philip C. Lucas

Fig. 9.4 The intruding tree roots in Blackroot Cave. Photo by Philip C. Lucas

breakdown in Cyanide Canyon on our downward push to follow the breeze (Chap. 5).

And map the unknown

After our breakout in October 1983, we would ultimately map nearly nine miles of virgin cave via 27 underground camps in the "Bobcat" phase of exploration, culminating in the stunning February/March 1986 discovery of the Burnsville Turnpike, one of the largest trunk passages in Virginia (Chap. 6).

The first real Pancake Weekend occurred three weeks later when three of us dug into Butternut Cave (Fig. 9.1). A Sunday morning's excavation in a modest, steep-walled sink high up on Chestnut Ridge revealed more than 700 feet of virgin cave. Eight years after Fossil Moss, it had happened again!

But there was something else that made this trip memorable. Stopped by a constriction in our exploration, Mike Artz and I sent newcomer Rod Morris back to the entrance to fetch a hammer. Outside, Rod encountered "an old man and a dog wondering how we were getting along."

It was Ike Nicholson. Sniffing out a good cave dig, Butler's discoverer had trudged to the near top of Chestnut Ridge. He and his pup were waiting for us when we exited. Back at Springhouse Farm parking lot, Nevin Davis could hardly believe our tale of discovery, amazed at our lucking into new cave in

Fig. 9.5 Passage in Blackroot Cave. Photo by Philip C. Lucas

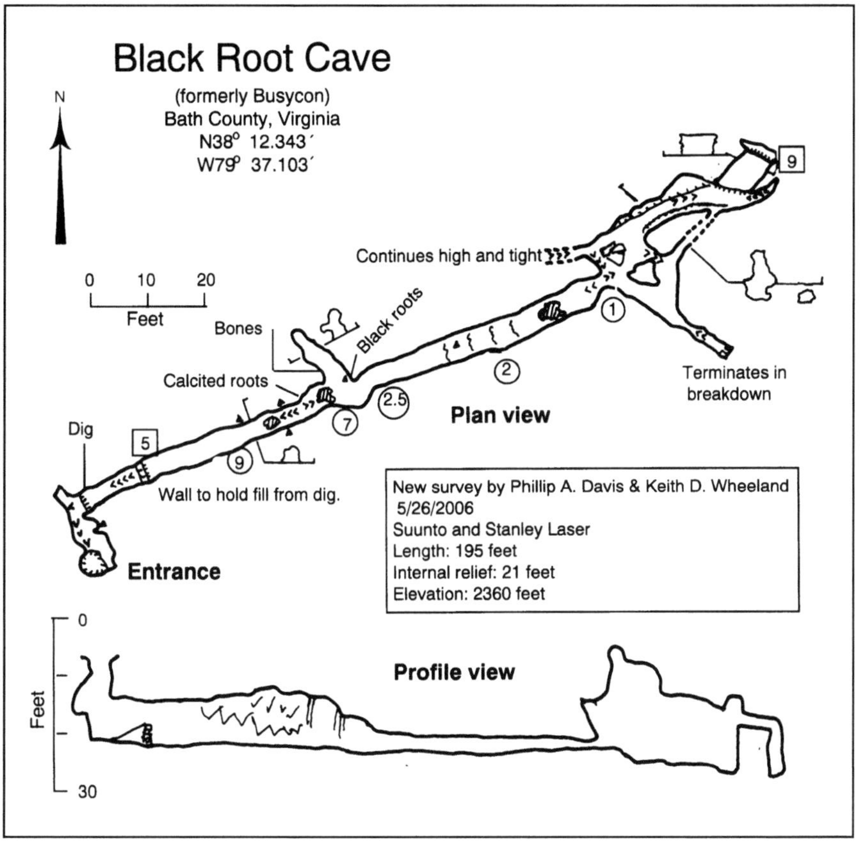

Fig. 9.6 Blackroot Cave. Map by Phillip A. Davis and Keith D. Wheeland

such cavalier fashion. Ike didn't miss a beat: "Nevin, don't you know you have to be anointed for such things!"

And that rather started this Pancake business

Although sore, we'd had a weekend of enormous fun, full of back-breaking work, yet coming home with real cave.

So why not do it again? Indeed, why not hold a regular, systematic "cave hunt," near the end of winter when the weather moderates but before it gets hot and buggy? And coordinate it with the Highland Maple Sugar Festival. It would be an easy date to remember and cavers would be supporting the local fire companies by chowing down pancakes and sausage before heading up the hillsides. Some of us had already been doing this for years. Incorporating an official "dig weekend" into the BCCS calendar just made sense. And although some cavers "cave year 'round," Pancake Weekend seemed a good way to "kick off the year," having an event where cavers young and old could attend and participate.

We had leads aplenty, good landowner relations, and adequate tools wielded by eager diggers with energy and enthusiasm to spare. But would we find any cave? And was Butternut a fluke?

Year two found us digging a thousand feet down the western slope of Chestnut Ridge from Butternut. The sink looked "good on paper" but even with long time caver and C-3 veteran Ackie Lloyd in attendance, we managed only a stale void at the 11 foot level. With no leads. But we'd hit a void at the bottom of a sink. Did that tell us anything?

In February of 1988, Mike Artz, Frank Gibson, and I started a dig on a small blow hole Nevin Davis showed us on the western flank of Chestnut Ridge. As a novelty, I brought my new video camera, filming us digging into an impassable pit. A month later I filmed Tom Shifflett, Mike Futrell, and Mike Dyas dig past the constriction. Aptly named Bvideo Pit yielded

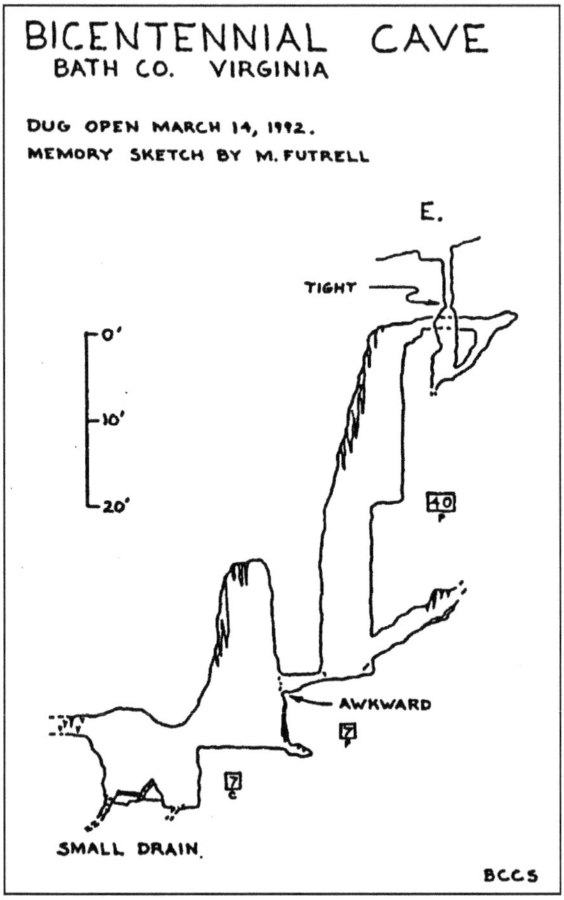

Fig. 9.7 Sketch map of Bicentennial Pit from *BCCS Newsletter* **17**, p. 59 (1991)

more than 700 feet of narrow, northeast trending canyon the next day (Fig. 9.2). Subsequent surveys have added to this total, but the cave for now remains unrelated to any other known cave on the ridge.

The next year, we headed for the top of Chestnut Ridge to dig on Burnsville Cove Lead Map feature #58. Mike Artz, Frank Gibson, Steve McCampbell, Mike Futrell, Frank Marks, Tom Shifflett, Nevin Davis, and I had a decent dig going by noon, lowering the floor and bringing up lots of black dirt. McCampbell got the lucky work tour and popped into a void headed down and away. Artz followed with Gibson muscling behind. They moved more rock but it shut down just as a huge cloudburst hit, driving us off the ridge to Nevin's backporch. Yet once again, we'd hit a void. Was there a pattern here?

It rained the next year, but 1990's bad weather did not deter us. Nevin had found another "melt spot." In his winter hikes, he occasionally identified bare spots in the snow, prime indications of places where warm air came out of the ground, even in the absence of a hole. Mike Futrell and I erected a tarp over the site and started on the thing in the driving rain, soon joined by Andrea, Nevin, and John Rosenfeld who owned the property. We were in virgin cave 3 hours after beginning the dig, finding more than 200 feet of decorated cave for our efforts (Figs. 9.3, 9.4 and 9.5) John called it Busycon Cave. Years later, Keith Wheeland renamed it Blackroot when he purchased the land (Fig. 9.6).

So, after five planned Pancake weekend digs, we'd hit voids every time and found three new caves. I

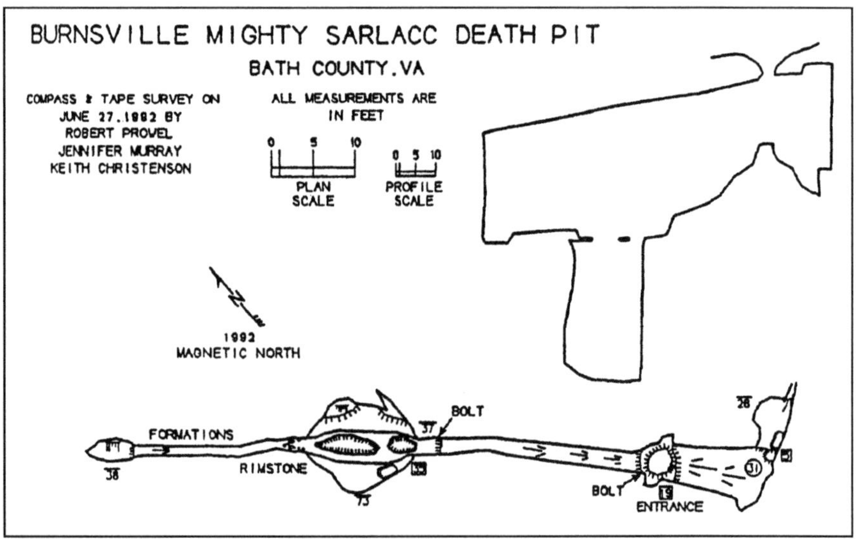

Fig. 9.8 Map of Burnsville Mighty Sarlacc Death Pit. From *BCCS Newsletter*

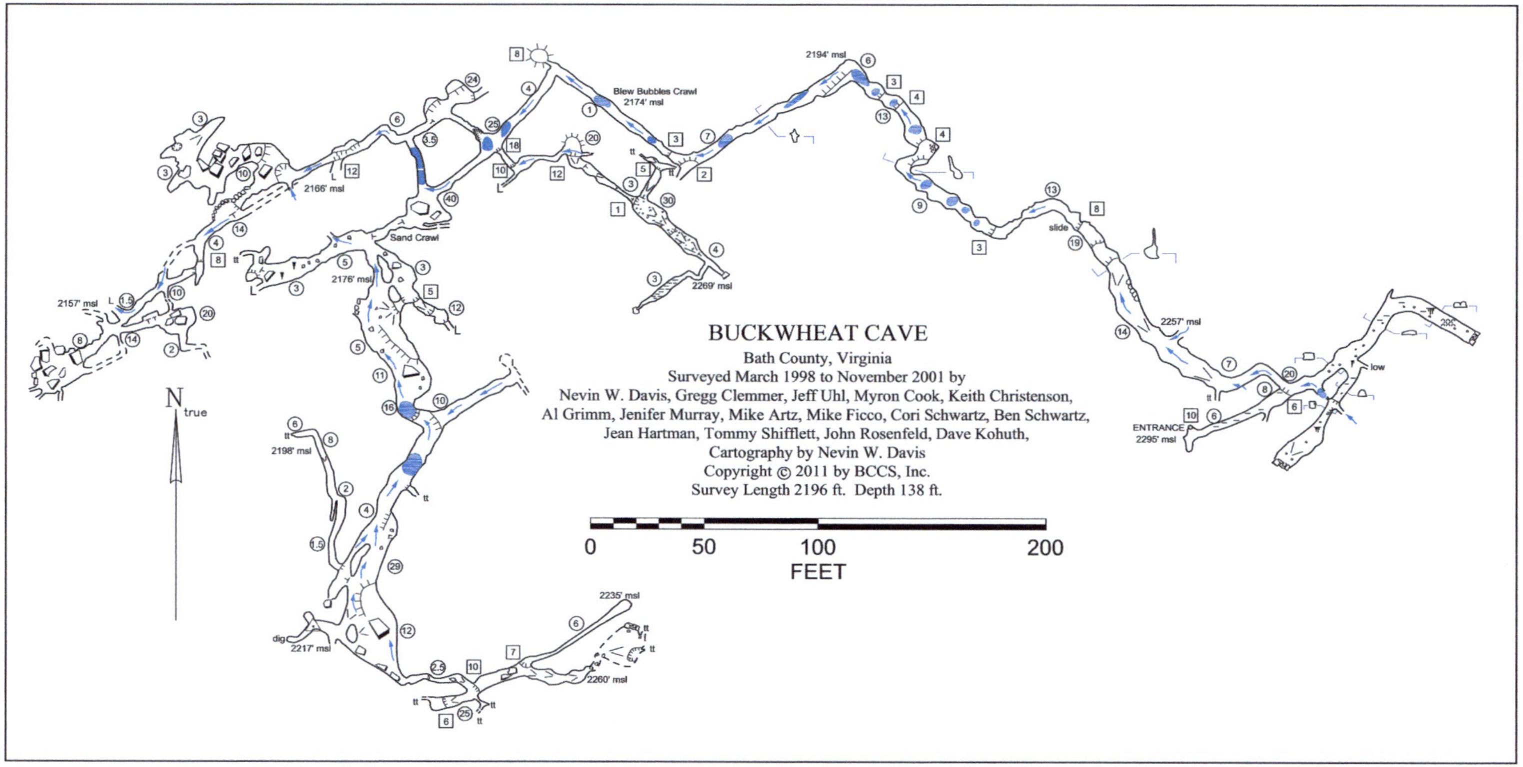

Fig. 9.9 Map of Buckwheat Cave compiled by Nevin W. Davis

hoped most participants rationalized the day-long, back-breaking labor worth the price of discovery. Nothing guaranteed, but...

Besides we could all enjoy the day on the hillside, get caught up on cave gossip, contribute our fair share of cave hype (This dig looks hot! Gonna go big!), and come away with a real sense of...well, having gone after big cave.

As for me, I felt like Tom Sawyer whitewashing that fence: make monotonous drudgery look fun and everyone will scramble to help! I hoped Pancake Weekend would morph from a monotonous, labor intensive, all day slog chasing dreams of discovery into a fun and free-for-all cave hunt...with a decent prospect of finding something!

The next year in less than an hour, nine of us dug into Blarney Stone!

And three years later, we connected Blarney Stone to Bobcat and established the Chestnut Ridge Cave System (Chap. 7)!

But even with our promising find in 1991, cavers came back the next March determined to dig for more entrances. Mike and Andrea Futrell led the charge in 1992 by getting to the bottom of Bicentennial Pit, an essentially dead-bottom pit about a thousand feet south of Blarney Stone (Fig. 9.7).

Thirty inches of snow cancelled Pancake Weekend in 1993. But undeterred, we simply waited for May, opened a crack in a barberry patch a mile east of Burnsville and hit pay dirt again. Surveying through typical Cove entrance horror, (Hemlock Highway and 44 foot Feeder Meeter Pit in this case) got us into some downstream trunk. But it was the legendary underwater digging exploits of Mike Ficco and Ben Schwartz through the all-time sinister Barberric Crawl that revealed the cave's spectacular trunk passage (The Great White WOW among other wonders!) with an enticing, air blowing lead headed for Butler. In search of a better entrance into Barberry, Pancakes-1994 brought us to a sink near the first entrance. Our cave survey plus a radio location indicated digging here might get us a better way in.

It did! And that spring, we installed an inverted gas tank with both ends cut out as our second entrance into Barberry Cave.

The Burnsville Incident of January 1995 (five experienced mappers got trapped by high water for five days) lost us access to Barberry via both entrances, so part of our 1995 Pancake efforts concentrated on finding a third entrance to this decorated "piece of Mexico" lost in Burnsville Cove. We tossed rocks from the bottom of Sugar Bush Cave on the Davis farm knowing this was in the vicinity of the large downstream Barberry trunk. We also dug in a sink over the far northern end of Barberry to no effect. With lots of diggers showing up for pancakes now, we needed more projects, so we plumbed and deepened Hilltop Cave, an obscure, narrow pit cave located

Fig. 9.10 Entrance to Battered Bar Cave. Photo by Philip C. Lucas

Fig. 9.11 Phil Lucas opening the pit entrance

Fig. 9.12 Nevin Davis in the Wormhole, Battered Bar Cave. Photo by Philip C. Lucas

above the far end of the Burnsville Turnpike. Only in December however, did we begin what would be the Big Bucks project on the Davis farm, our third (and most spectacular) entrance into Barberry (Chap. 10).

But were we learning anything?

Well, yes. After ten years of this Maple Sugared nonsense, we knew that sinks, generally, were good places to find caves, especially steep-sided sinks with limestone walls and smaller diameters. But melt spots and blow holes offered the best odds.

And after years of shovel work, "follow the black dirt" had become our mantra. Organic soil, especially dry and soft with old sticks, nuts, and animal bones is the subsidence going into cave. Follow it (and tree roots) and you'll eventually view the virgin. But beware gray or yellow clay which spells death to a dig. The subsurface clues were there, if we just read them. Still, I remembered Mike Futrell's wisdom: "Caves don't wimp, cavers do."

Given our success, digging became the way to find new cave in Burnsville Cove. (And other places we would find out! But that's another story.) We had already seen good efforts in Nevin Davis' Backyard Cave and the Water Sinks' Booger Cave. But Phil and Charlotte Lucas, owners of the Water Sinks, had started another dig in the same sink as Booger. As luck would have it, that big breakthrough came on Pancake Weekend 1996 when Ben Schwartz, Jean Hartman, and Ed and Kim Kehs were the lucky foursome to finally get into Helictite Cave, at seven miles mapped, now the largest cave in Highland County. Everybody, it seemed, was digging, giving us Baby Bottle Borehole, the Burnsville Mighty Sarlacc Death Pit (Fig. 9.8), and Bone Cave.

Fig. 9.13 Passage in Battered Bar Cave. Photo by Philip C. Lucas

Perhaps getting a bit cocky, we took a stab in 1997 at an easier way into Blarney Stone, working on a sink at the eastern base of the hill going up to the cave's entrance. But this dig was more wishful thinking that good geology and the next year we returned to more reliable science, digging in a small, sharp-sided sink near a sinking stream a few hundred yards east of Backyard Cave. In only 2 h, we were into Buckwheat Cave, a narrow but active and winding stream fissure that we would map to within 20 feet—plus or minus—of connecting into Barberry (Fig. 9.9).

The next year, seeing our success in Buckwheat, Ben Schwartz decided to work on a similar sink in the next large ravine coming off Chestnut Ridge to the north. I led a second effort in a similar sink on land owned by Fred and Anne Wefer and Will and Bette White. In late afternoon, my group found a teasing void but no continuation. The news from Ben's crew sounded a bit more promising, as they broke into a low, horizontal passage with air. Subsequent work the next day and in following weeks revealed an extensive cave draining the western flank of Chestnut Ridge, surveyors ultimately putting more than 3300 feet in the book. Ben's good science, hard work, and a little blind faith did the trick to getting us into Blind Faith Cave in 1999.

March of the new millennium had us digging in Eddie Webb's "Wine Gourd," a gentle wooded valley along the western base of Chestnut Ridge sporting numerous sinks overtop Bobcat's Burnsville Turnpike. April Fool's Pit, Hilltop, and Mighty Sarlacc loomed nearby. We visited a number of sites that day, digging in April Fool and a few other pud holes, but Dave Kohuth, John Rosenfeld, Doug and Hazel Medville, and I angled for a small sink atop a narrow saddle 300 yards north of Blind Faith. By lunch, we were looking down a deep pit.

Initially dead bottom, we poked at a narrow slot and felt slight air. More digging, followed by various dynamic processes in subsequent weeks, got us to the top of a 40 foot shaft. Our Battered Bar Cave soon passed a mile in mapped passage (Figs. 9.10, 9.11, 9.12, 9.13 and 9.14), pushing out toward Woodzell Sink and approaching the downstream end of Blind Faith. Slowly, surely, we were filling in some vital blanks under Burnsville Cove.

With these modest successes, Pancake attendance soared. Anne Wefer opened her home to everybody on Saturday night for her Big Enchilada, a scrumptious Mexican dinner *a la fireside*, complimented by good conversation, slide shows, and a simmering hot tub under the stars.

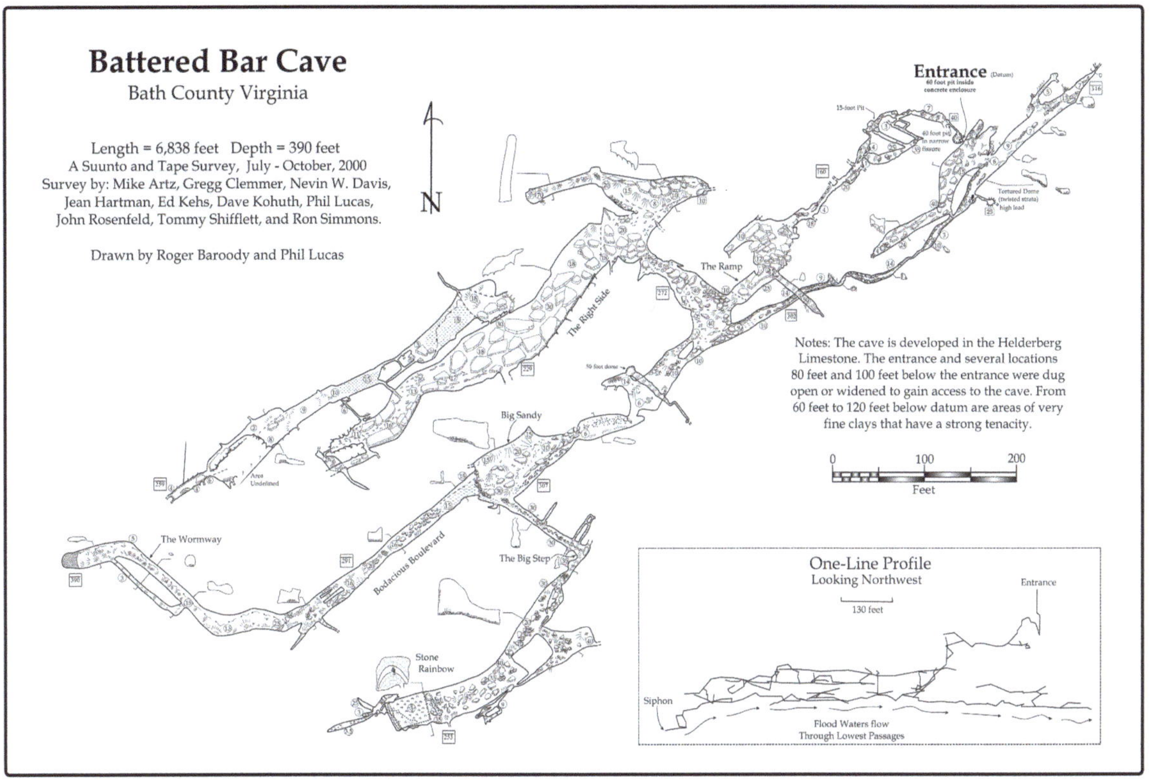

Fig. 9.14 Map of Battered Bar Cave

Fig. 9.15 The excavation of Basswood Cave. Photo by Philip C. Lucas

Fig. 9.16 Installing the culvert in Basswood Cave. Photo by Philip C. Lucas

Fig. 9.17 Nevin Davis in the Little Girls Room, Basswood Cave. Photo by Philip C. Lucas

Fig. 9.18 Tony Canike in the Squeezeway, Basswood Cave. Photo by Philip C. Lucas

Fig. 9.19 The lead. Basswood Cave. Photo by Philip C. Lucas

Fig. 9.20 Lift tube wall, Basswood Cave. Photo by Philip C. Lucas

Fig. 9.21 Calcite crystals, Basswood Cave. Photo by Philip C. Lucas

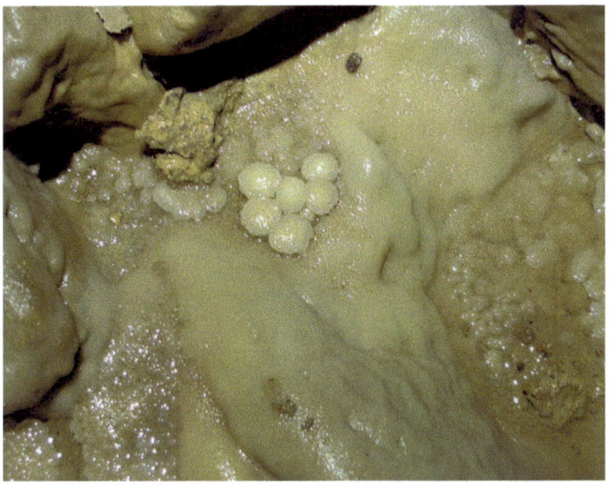

Fig. 9.22 Cave pearls, Basswood Cave. Photo by Philip C. Lucas

The infamous Dimple Sink on Phil and Charlotte Lucas' property corralled our efforts in 2001 while the Kehs' dig team established a one day Pancake depth record of 16 feet on a nearby second hole. Nicely shored, we continued efforts at Dimple into the next year before filling it in, not once seeing any sign of solid limestone. *We cavers had wimped.*

Excluding the Big Bucks Pit entrance to Barberry, our most persistent effort came with the Basswood dig in 2002. With a solid group of diggers (Mike Kistler, Nate Walter, Dick Graham, Mike Ficco, Phil Lucas, Nevin Davis, Dave Kohuth, Duane Martin, Jean Hartman, and others), we started on the bottom of a fairly large sink 200 yards south of Nevin Davis' big barn. Lots of runoff sank in the meadow just west of our work site. Nevin offered tales of a whirlpool spinning out of sight during heavy rains. Because of the sinkhole's diameter and depth, I worried we'd have to dig to China before we hit cave. But we stayed at it and by June, we were down 26 feet, still following the soft, black, dry dirt. Where was this thing going?

After installing a corrugated pipe for shoring, we continued, still following the black dirt. At a protected depth of 36 feet in mid-July, we punched into a void,

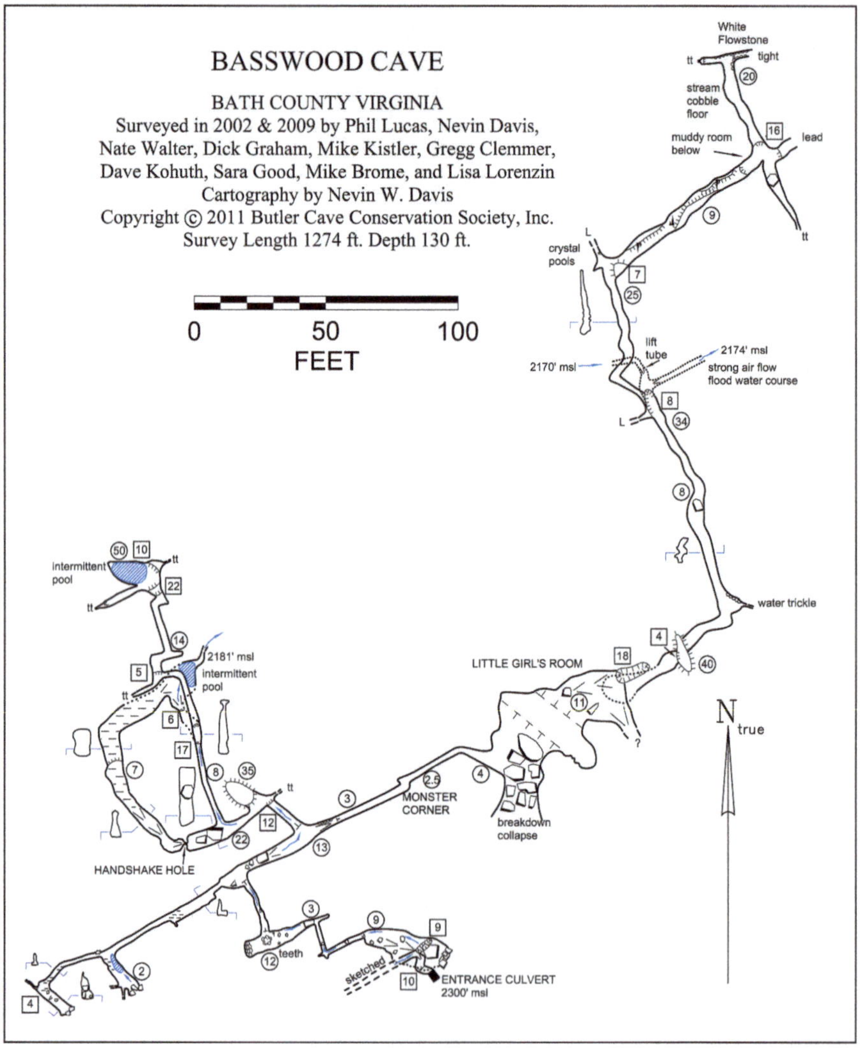

Fig. 9.23 Basswood Cave. Map compiled by Nevin W. Davis

Fig. 9.24 Digging the entrance to Fuhl' s Paradise Cave. Photo by Phillip C. Lucas

Fig. 9.25 The completed entrance to Fuhl's Paradise Cave. Photo by Phillip C. Lucas

mapping 800 feet of multi-level cave and leaving a tight dig lead with a strong current of air (Figs. 9.15, 9.16, 9.17, 9.18, 9.19, 9.20, 9.21 and 9.22). Six years later, we returned, mining this over the next year and a half into a continuation of tall, fissure canyons that ended (for now!) in yet another air-blowing, too-tight crawl (Fig. 9.23).

We tackled a similar dig in 2003 when we returned to the Wine Gourd and started excavating a 4 inch blow hole. Ultimately pushing this prospect to a depth of 44 feet, we stopped work at a 6 inch wide serpentine crack spiraling down. We piped this Backdoor dig off during Pancakes-2006, hopeful that one day the "next generation" will make this a backdoor entrance into the storied Burnsville Turnpike.

Pancake eaters numbering near 30, swarmed Burnsville Cove in 2005. With so many, we went to four sites: a modest sink high on a ridge at Springhouse Farm; a second, smaller sink 150 yards to the north; a melt spot on the Hole Place, first scratched at in 1980; and a melt spot on Keith Wheeland's place.

Keith's crew spent a few hours on their dig then retired out of curiosity to see what the others were doing. Mike Artz and Nate Walter led a strong crew at the Hole Place's melt spot, getting down more than a dozen feet until walls of cobbles hindered their work and fueled safety concerns.

I got a third crew digging out to the sink atop the ridge above Springhouse Farm, while Ben Schwartz, Mike Kistler, and Myron Cook took a fourth team to the lower sink. By early afternoon, following the directional traces of black soil, our crew had poked into a void, finding a hundred feet of cave. A prominent black cherry tree gave the new cave its name.

Fig. 9.26 Fuhl's Paradise Cave. Photo by Philip C. Lucas

Fig. 9.27 Rimstone pools, Fuhl's Paradise Cave. Photo by Philip C. Lucas

Down the hill, Schwartz's crew dug down a dozen feet, then continued up a mud filled phreatic tube, finding cave, but doing it the hard way, digging it out a foot at a time.

Believers in the melt spot/cobble dig returned a few weeks later. After a long day of intense rock tossing and dirt removal, we popped into 450 feet of nicely decorated walking cave. We soon piped Fuhl's Paradise Pit to protect the entrance from collapse, finding that we had opened a bat hibernaculum (Figs. 9.24, 9.25, 9.26, 9.27, 9.28 and 9.29). Fuhl's Paradise Cave is a fragment of trunk passage in the Licking Creek Limestone (Fig. 9.30). There are no obvious continuations.

The Twisting Sister dig grabbed our attention during Pancake-2006. With half our crew pipe shoring the Backdoor dig for another generation, the rest started digging on two small sinks just above the sinking stream in the Owl Cave meadow of Water Sinks. Several weekends of strenuous labor, plus an installed pipe for safety and stability, got us into a low, very muddy room. Despite no air, we lowered the floor and expanded the walls, hauling out huge amounts of mud to *create* cave. Still a dig, a breakthrough here might get us easy access to the far northern reaches of the Chestnut Ridge Cave System.

A recent Pancake gathering saw yet another new cave put in the survey books. Buckets of Smoke, possibly the highest cave in Burnsville Cove, proved to be the typical Pancake cave dig. While not the classic melt spot, this site remained thawed in winter's freeze. An iron bar plunged into the black, soft soil almost to its end. We commenced this dig midst a frigid wind and 3 inches of snow on the ground with more bad weather in the forecast. A campfire, started by carbide lamp, kept spirits up as well as warm.

And down we dug, reaching 13 feet between two fluted limestone walls after two days. A second trip two weeks later got us down 20 feet with hints of a void, even teases of air. More digging on Sunday put our smallest member, Jean Vargas, into real cave. She reported walking passage going in several directions! She was right! A month later with everyone wanting to get in on the fun, our team of seven surveyors mapped

Fig. 9.28 Dripstone, Fuhl's Paradise Cave. Photo by Phillip C. Lucas

Fig. 9.29 Soda straw stalactites, Fuhl's Paradise Cave. Photo by Philip C. Lucas

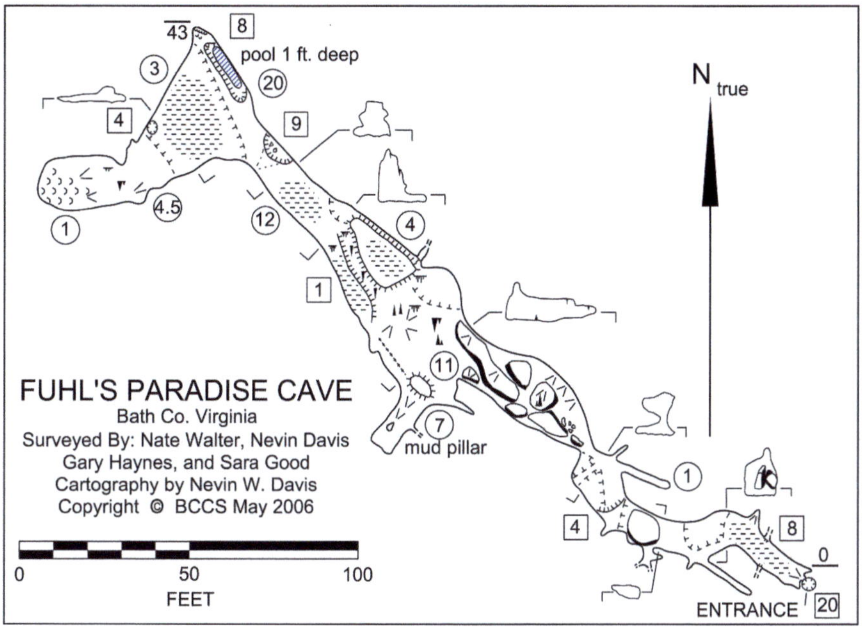

Fig. 9.30 Fuhl's Paradise Cave. Map compiled by Nevin W. Davis

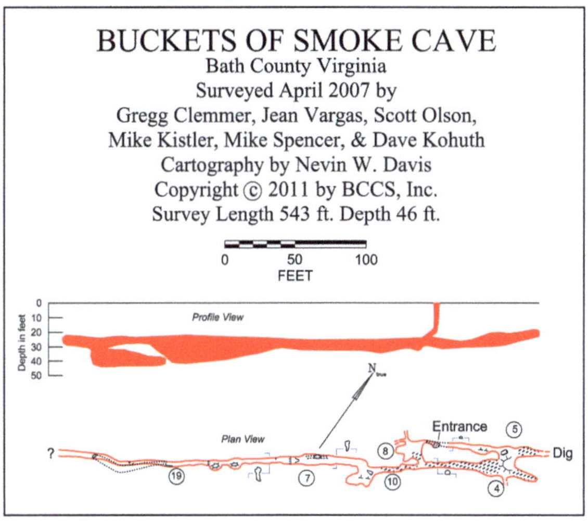

Fig. 9.31 Buckets of Smoke Cave. Map compiled by Nevin W. Davis

over 500 feet of mostly walking passage in this newest Virginia cave (Fig. 9.31). A few tantalizing leads remain.

So after twenty plus years of pancakes, what does all this mean?

For better or worse, we've become aggressive in searching for new cave. Through an annual, somewhat "organized" event, we've been able to open a number of new cave entrances in Burnsville Cove, four of great significance and each of those more than a mile in length. The digs are such that any caver or interested party can participate, regardless of skills. Scheduled during the region's Maple Sugar Festival, the digs offer a great way to usher in spring with a weekend in the mountains, accompanied with the real prospect of virgin cave.

And what does the future hold?

Ponder our prospects in the blowing digs waiting in Buckwheat, Basswood, and my old favorite, Fossil-Moss, or help map in the just-opened Wishing Well, boasting 4.7 miles mapped in its first eight months of exploration. Then ask the question again.

What does the future hold? Discovery!

Exploration of Barberry Cave and the Construction of Big Bucks Pit

10

Benjamin F. Schwartz and Nevin W. Davis

Abstract

Barberry Cave, with a length of 3.4 miles, was discovered by a dig in a sinkhole. The entrance series was tight and time-consuming so a second entrance was excavated. The main part of the cave, with large well-decorated passages, was discovered through a low, easily flooded crawlway. Following an incident of explorers trapped by rising water, a third entrance was achieved by excavating a shaft from the surface into the large passage of the cave. Barberry Cave has some of the largest and best decorated passages in the Burnsville Cove. The Cave is located between the Chestnut Ridge caves and the Butler Cave-Sinking Creek System although no physical connections have been discovered.

[**Editor's Note**: The exploration story is largely extracted from the article by Benjamin Schwartz "Exploring Barberry Cave", *NSS News*, vol. 57, pp. 268–275, Sept., 1999. The inserted description of the construction of the Big Bucks Pit entrance was written by Nevin Davis in the *BCCS Newsletter*, vol. 22, pp. 6–10, 1997.]

10.1 Discovery and Exploration by Benjamin Schwartz

10.1.1 Another Productive Pancake Weekend

Pancake weekend 1993 was delayed due to a blizzard and rescheduled for Memorial Day. The main goal was to open a small swallet in a sinkhole, where on February 19, 1982, Nevin W. Davis had recorded a positive dye trace to Aqua Spring, three miles to the north. With no known cave passage nearby, the crack had considerable potential to reveal a large section of missing cave. Also, due to its location on a low area of Chestnut Ridge, it could intersect large fossil passage, while bypassing the typical difficult and horrible entrance series of caves found higher on the ridge.

The dig began on May 29, 1993, when Nevin W. Davis, Mike Nicholson, and Gregg Clemmer cleared the sink of trash, brush, and animal bones. They uncovered a narrow crack in bedrock that blows air at times and swallows water during heavy rains. The following day, nearly two dozen people gathered for a full day of work. A 70-pound rock drill, an air compressor, and higher powers were used to excavate a 10-foot-deep shaft. This shaft intersects a very narrow vertical fissure that constricts 15 feet farther down. Mike Ficco and I hammered it open just enough for us to squeeze through. Five feet below, it belled out in the ceiling of a small dome. I climbed down, with Mike Ficco and Mike Artz following. After quickly checking a couple hundred feet of small, sloppy passage, we

B.F. Schwartz (✉)
Edwards Aquifer Research and Data Center, Department of Biology, Texas State University—San Marcos, 206 FAB Freeman Aquatic Station 601 University Drive, San Marcos, TX 78666, USA
e-mail: benandcori@hotmail.com

N.W. Davis
12512 Dry Run Road, Williamsville, VA 24487, USA

W.B. White (ed.), *The Caves of Burnsville Cove, Virginia*, Cave and Karst Systems of the World, DOI 10.1007/978-3-319-14391-0_10, turned back at an eight-inch bedding plane with a stream flowing into it. This was not the large passage we hoped for!

The next day, Mike Ficco, Mike Artz, Ed Kehs, and I returned armed with survey gear, hammer, chisel, and prybar. We needed all of it. Squeezing fifty feet into the bedding plane brought us to the first of several constrictions; each was enlarged barely enough to fit through. Most were just beyond shallow pools and forced us to wallow in the sleazy stream. By the end of the day, we hadn't reached an end of the grim horror and our lamp brackets were twisted into pretzels.

A longstanding Burnsville tradition is to name nasty entrance passages after a poison and this one was quickly christened Hemlock Highway. Another tradition is giving caves a "B" name. The name Barberry was inspired by the large patch of these thorny bushes surrounding the entrance.

10.1.2 Digging a Second Entrance and Continuing to Push Downstream

On subsequent trips we discovered a 40-foot pit at the end of the 600-foot-long Hemlock Highway and several thousand feet of passage beyond. The entrance stream is enlarged by several infeeders and flows through a few sections of well-decorated walking-height passage before reaching a near-sump at a three-inch bedding plane. A few nice side passages continued and we had high hopes of finding a bypass around the near-sump. Unfortunately, after more survey trips, none of these side passages were in the desired direction. Since Hemlock Highway is such a nasty trip, taking upwards of one hour each way, and many leads remained, we decided to excavate a second entrance where a passage in the "nicer" part of the cave came close to the surface. A radio-location confirmed the exact locations of the passage under a large, rock-filled sinkhole only 293 feet away from the first entrance. In the flat field, one minute of easy walking traverses a distance requiring one and a half hours underground. On June 28, 1994, we completed the second entrance by installing a 12-by-5 foot tank and a piece of 24-inch culvert.

July 16, 1994, found Mike Ficco and I with nothing better to do, so with full wetsuits and a lack of sense, we attacked the downstream near-sump. Nevin W. Davis, Pete Carter and Mike Ficco had started a dig here in September of 1993. They gave up after 15 feet of difficult progress; the passage ahead looked even worse than what they had just dug open. The near-sump is a passage that averages six to eight feet wide and was originally only three to four inches high with less than half of this being air space. The majority of the airflow in the cave blows through these few inches of air space. Although not a gale of air, it was enough to convince us that something lay beyond. Just before the near-sump is a small room with a water-covered floor. In a fit of inspiration we named this The Pool Room. It became our "camp" while working on the dig. Mike and I devised a new plan for digging that involved lying on our backs and sliding feet first into the passage. That allowed us to press our noses to the ceiling and breathe occasionally. Kicking our heels to loosen the hard-packed gravel floor, we wriggled our bodies and used our legs to push the gravel to the sides of the underwater trench. We dug it only deep enough to allow us to slide in by exhaling and pushing hard. Simply inhaling wedged us as tight as a cork in a bottle. This enabled us to stay in place while we used all our leg strength to kick out ahead and also helped block the waves coming from our feet and legs. Waves were a real concern, with our noses on the ceiling and water up to corners of our eyes. We were unable to wear a helmet, so small dive lights on a string around our necks provided psychological comfort more than anything else. Several times our flashlight batteries died and left us digging in the dark. With only one way out of the dig, this was more of an inconvenience than a problem. Almost all travel and digging was done by feel anyway. Each of us dug for about an hour before trading places. We did this five times that day and made four to five feet of progress per shift. Both of us exited the cave with vivid memories of how horrible a place the near-sump really was. Still no end was in sight and, as on the trip before, the passage ahead looked even worse. In spite of this, we had a good time and made significant progress.

By September 10, 1994, Mike and I had forgotten just enough about the previous trip to look forward to the next. At 12:15 p.m., the two of us crawled through the tank entrance, as Nevin wished us luck and shut the lid behind us. One and a half hours later we reached the Pool Room and began digging. After 45 min of hard digging, I felt a larger space ahead. The floor became softer and easier to move. Soon I slid into a 10-inch high, gravel-floored "room". With my ears full of water, I could hear what sounded like a waterfall ahead. Draining one ear and listening, I realized I was only hearing Mike heat water on our small stove. Beyond this low room, the passage turns a corner to the right and dips into a deep pool where the air space becomes even lower and narrower than before. I slid into the water and very carefully floated for 20 feet, with nothing more than my nose and one eye in the air. For a few feet, the air-space shrank to just one inch high and eight inches wide. Then suddenly the ceiling rose above my head, as I slid onto a small beach and began shouting to Mike that we had just broken through and found borehole passage. I sat up and promptly hit my head on the ceiling. The borehole was actually only two and a half feet high. After seeing a ceiling within one inch of my face for the past hour, 30 inches looked like 30 feet. I carefully sat up again and dug a trench in the gravel floor to lower the water level an additional inch in the second pool. Returning to the Pool Room, Mike and I celebrated this breakthrough with a pot of hot chocolate.

Surveying the horrid thing was our next obstacle. In Burnsville everyone follows the "survey as you go" rule. The nature of the caves is such that, if they aren't surveyed as they are discovered, many passages will never be seen again, much less surveyed. In the near-sump, we accomplished this by taking two shots. The first measured 114.5 feet long and the second 24 feet long. Not only did we do this with a 50-foot tape, but we unknowingly set the survey shot length record for Burnsville Cove. Sliding through the last pool, we managed to re-light our lamps and began surveying real passage. Low stream crawlway continued for several shots until we suddenly crawled out from under a wall and stood up in a real borehole. Passage dimensions averaged 30 feet wide and 40 feet high. Ecstatic, we whooped and hollered for a few minutes before continuing the survey downstream. One thousand feet later, we turned around in big passage and began the trip out. Though the route isn't pleasant, we had broken into a major extension of the cave. Leads were plentiful and the large passage continued. As we climbed out the entrance at midnight, Nevin drove across the field to greet us. I handed him the survey book and we enjoyed the look of disbelief on his face as he flipped the pages.

One week after completing the dig, I woke up with only partial hearing in one ear and none in the other. A visit to an ear doctor produced a teaspoon of fine silt and sand. She had never found anything like this is someone's ear before and called the office staff in to see. When I told her the story of how my ear had filled up with silt, she immediately assumed I was crazy and had come to see the wrong doctor.

Our next trip was planned for only two weeks later, and we convinced Nevin to join us. At the Pool Room, Mike and I spent one and a half hours deepening our trench so Nevin could fit. We all arrived at what is now called The Other Side and hurried downstream to the last survey station. The big cave continued for 100 feet and abruptly ended in massive breakdown. With no hammer for pushing the breakdown, we began investigating the many side leads. One by one, they turned out to be short or simply a side pocket. One well-decorated section of old trunk passage gave us a few hundred feet of survey for the day.

On December 3, 1994, Tommy Shifflett joined Mike and me for a trip to The Other Side. Gregg Clemmer and Nevin Davis accompanied us to the Pool Room where they saw us off before leaving to work on some other leads closer to the entrance. Gregg's last words were "don't feel bad about scooping heavily!" We checked a couple side leads before turning our attention to the downstream breakdown choke. Armed with a hammer, Mike attacked a small crack along the right wall. Fifteen minutes later he squeezed through and soon returned reporting a stream passage large enough to walk down. Excited, we surveyed through the breakdown. After a few more shots it looked like we had cracked the breakdown choke and were off to the races again. The passage enlarged until we were surveying trunk passage with walls 40 feet apart and the ceiling 60–80 feet over our heads (Figs. 10.1 and 10.2). Huge pure-white

Fig. 10.1 Representative passage in Barberry Cave. Photo by Nevin W. Davis

Fig. 10.2 Another example of passages in Barberry Cave. Photo by Nevin W. Davis

dripstone formations, gigantic flowstone mounds, and delicate gypsum formations decorate the route. One room in particular is superbly decorated with an immense white column, tall white totems and dense patches of white and transparent soda straws. We named this room The Great White Wow. Near the end of the day's survey, the main stream disappeared into a small side canyon while large black borehole continued ahead. When the three of us finally turned around with 3550 feet in the book, we decided we had surely "scooped heavily." None of us were feeling guilty about it either. Our last station was set on the slope of a huge breakdown pile. Blackness continued ahead. The only damper on our celebration was the thought of crawling back through the near-sump. When we finally climbed out the entrance, we were greeted by the glow of morning sunlight rising over Chestnut Ridge.

The next day, a line plot clearly showed us that Barberry Cave had become a major part of the puzzle in the Cove and was possibly the key to linking Butler Cave (17 miles long) with the Chestnut Ridge Cave System. The Barberry stream trunk lines up with the Burnsville Turnpike, a huge stream trunk in the Chestnut Ridge System. A side passage in Barberry points in the direction of Butler Cave. Significant airflow at the end of the cave seemed to promise much more cave in that direction.

10.1.3 The Camp Trip and the Incident

A return trip was quickly planned for January 14–17, 1995. This time, we decided to camp in the cave for four days since the near-sump is a major obstacle, and the back of the cave is a 6 h trip, one way. Some of us found the near-sump merely a nuisance while others found it to be downright disagreeable. Whatever the case, it severely limited the number of people who would even see The Other Side by passing through it.

Five of us entered the cave on Friday afternoon: Mike Ficco, Nevin Davis, Tommy Shifflett, Mike Artz and I. Each of us had two huge packs. In addition to our usual camping gear, we carried bolt climbing gear, rope, and equipment for a radio location at the end of the cave. We arrived after a 6.5 h journey and set up camp at a flat spot several hundred feet from the end of the cave. By midnight, all of us had eaten a hot meal and were sleeping soundly. The next morning, we started the day early. Mike Ficco and I chose to survey the side passage leading in the direction of Butler Cave, while the others would continue down the main trunk. Mike and I surveyed 1200 feet of nice walking-size stream passage to a flowstone choke. Since this passage began in The Great White Wow, we called it the Woway. The only way to continue appeared to be a very low and narrow crevice with a stream issuing from it or a dry upper level reached by a climb up through breakdown. Heading back to camp, we met the others. They greeted us with bad news. The large passage ended in breakdown with no easy way through, and only one or two leads had any potential for further progress. They also tried to follow the main stream down a narrow canyon, only to discover a sump after 150 feet. We all trooped back into camp and split up again. Mike Artz and Nevin weren't feeling too peppy and set up the cave radio, while Mike Ficco, Tommy, and I took a hammer and began pushing the best breakdown lead. We spent the next five hours following air, pounding, and digging in horrible, sloppy breakdown. Our efforts produced only 150 feet of unstable passage in the loose rocks with no clear way to continue. Discouraged, we headed for camp, ate dinner, and went to bed.

Sunday morning arrived and we immediately knew something had changed. From the distance, sounds of roaring water and rolling rocks had replaced the quiet hiss from the previous night. We feared that it had rained heavily on the surface and the cave was now flooding. A quick look confirmed our fears as we watched wall-to-wall whitewater foaming into a deep sump pool where a narrow canyon had been the day before. Unable to continue upstream without swimming, Tommy, Mike Ficco and I spent the day surveying and pushing small passages beyond camp. Nevin and Mike Artz were still feeling poorly and turned on the cave radio before resting in camp for the day. On the surface, Phil Lucas and Mike Nicholson received the radio signal at the prearranged time and located a point directly above the cave radio. Although unable to communicate, they knew we were still alive. By that evening, the water level was already beginning to lower a few inches per hour.

The raging whitewater had receded considerably by Monday morning and the deep pools had all but vanished. We packed up our gear and headed upstream to check the near-sump. Arriving at The Other Side, we put on our wetsuits and crawled upstream. It didn't look good. Two hours of digging in the stream produced only one inch of air above very swift water. Deciding it was too dangerous to attempt, the five of us returned to camp. The rest of Monday, Monday night, and Tuesday morning were spent conserving food, checking the lowering water levels and sleeping. One bit of irony is that Nevin spent five days sleeping in a cave when by straight line he was only 600 feet from his house and a warm, dry, bed. So close, yet so far!

By late Tuesday morning, the water level had lowered sufficiently for Mike Ficco and I to dig the near-sump channel open just enough for us to squeeze through. It would still need to be enlarged considerably for the others to fit. Because we were all running low on energy, our group decided Mike and I should exit to eat a good meal, get a good night's rest, and return the next day to help dig out the others. In the event of additional heavy rain, we would bring food and supplies that night. On the surface, though, things were not so simple. A rescue call-out had been organized, and by late Tuesday night more than 100 people arrived to provide support if it was needed. It turned out to be a major event in the history of tiny Burnsville. See Uhl (1994/1995).

On Wednesday morning we returned to dig out the channel, and by that afternoon everyone was safely out of the cave. It turned out that several inches of rain had quickly fallen on Saturday while we were in the cave. Fortunately, we had camped in a dry, comfortable site in no danger of flooding, and our only worry was the possibility of running out of food before the water level lowered enough to exit.

10.2 The Construction of Big Bucks Pit by Nevin W. Davis

[**Editor's Note**: At the time of this writing, access to Barberry Cave is easy and the entrance is accessible and secure. The following is the detailed history of the ordeal of gaining a reliable entrance to Barberry Cave]

In a period of less than 41 months, three entrances to an entranceless cave have been dug. This must be some sort of record. Especially considering that the length of the survey is only 2.72 miles. I need to go back to 28 May, 1993, to explain the chain of events which led to the opening of Big Bucks Pit.

It was on this date that the saga of Barberry Cave began. Mike Nicholson, Gregg Clemmer and I set out to clear the trash and barberry bushes which clogged a little crack at the bottom of a sink on a neighbor's farm 4000 feet south of my house (Clemmer 1993). On this Memorial Day weekend perhaps 70 man hours of effort were expended to reveal the entrance to a cave which over the next 9 months grew to 0.96 miles of survey. It was obvious to those involved that the cave held great potential. After all, one side passage was only 1400 feet from Butler Cave and the air blowing over the Near Sump at the downstream end of the survey would entice even the most confirmed skeptic. There was only one problem. Access to the cave was through the Hemlock Highway. This crawlway required 75–90 min to pass and at the end the explorer was only 420 feet from where he started. That could be up to 3 exhausting hours stolen from a cave trip. Something needed to be done to facilitate the exploration of the cave.

As luck would have it, there was a sink filled with field stone about 100 yards from the entrance. It was over some side passages at the far end of the hemlock highway not far from Feeder Meeter Pit. A cave radio location on Pancake Weekend, 1994 (Rosenfeld 1994) confirmed the location and depth and a second dig was started. For five weekends from 29 March, 1994 to 5 June, 1994, stones were tossed and a 12 foot long tank with its ends cut out was installed to stabilize the excavation. At this point known passage had not been intersected. On 7 June, 1994, after making a careful drawing of the excavation and cave survey, I was able to take my trusty digging bar and plunge it though the wall of the dig at the bottom of the tank into surveyed cave. By 28 June a 2-foot diameter lateral culvert had been installed into known cave and the whole area was backfilled and graded. About 365 man hours were involved in the whole project and it did provide excellent access to the cave. Between 25 June and 4 December, 1994 the cave grew from 0.96 miles to 1.94 miles (Schwartz 1994). The tank entrance was certainly proving its worth.

On the weekend of Friday, January 13, 1995 there occurred the incident described at the end of the Section above (Uhl 1994/95). What followed was the biggest excitement to occur in Burnsville in the last 100 years. The Hot Springs Rescue Squad and all their buddies in surrounding counties along with major news media converged on Barberry Cave. The resulting media event was something cavers can do without. We were ambushed in our own cave project. As a result of the publicity and efforts of well-meaning, would-be rescuers, we were locked out of the cave. My neighbor, the land owner, has not spoken to me or even made eye contact since the "Burnsville Event". Clearly, if we were going to continue exploration of this beautiful and strategically located cave, we needed another plan. Read that—another entrance.

Most of the large passages in the cave underlie my property; Springhouse Farm. At one location the ceiling above a 45-foot high pile of breakdown is only 70 feet from the surface. So choosing this location for our third entrance, on 11 November, 1995, I signed a contract with W.F. Caldwell Well Drilling to sink a 10-inch diameter borehole into the cave. Bill Caldwell's air rotary rig was set up on the "spot" the morning of 8 December, 1995. It took some time to clear the snow from the equipment and get the Detroit Diesel engines running. At about 2 in the afternoon, Phil Lucas and I were counting down the feet until the drill would break through. I said, "He's got about 6 inches to go". The drill inched downward. Then with a small lurch and a change in the sound, he was through. The hole was then reamed to 10 inches. Because of an impending snow storm, Caldwell packed up his equipment and was on the road in 10 min. Phil and I lingered long enough to lower a Marks Products GeoVision Jr video camera down the hole to confirm our location in the cave. It was "bingo" right on the mark. In the process we noticed that you could hear the sound from the 3-foot high waterfall in the cave, just downstream from the well.

On 27 December, 1995, Gregg Clemmer brought his family out to Springhouse Farm for a look at the bore hole. This time, when we lowered GeoVision Jr., it was attached to a working VCR and we videoed the well and the 80-foot high cave at the bottom.

Since we planned to sink a shaft following the well, Caldwell had not cased it through the soil zone. His well log showed a depth to bedrock of 13 feet so Ben Schwartz secured a piece of casing from his father that was 14 feet long and we lowered it into the well. In the middle of January after a thaw and storm I noticed that water was standing only a foot below the surface. Our well had collapsed. A review of the well video revealed that contrary to the driller's log, bedrock was actually more like 26 feet below the surface. Another emotional roller coaster ride was in progress. The well is drilled (the high); the well has collapsed (the low point). This wasn't to be our last disaster.

The information that bedrock was 26 feet down was both good and bad. It meant that there were 13 less feet of rock to sink a shaft through but it also meant that we would have to shore 26 feet of earth before bedrock was reached. While pondering various options, I talked to Paul Cunningham. He suggested that we hire his track hoe to dig a hole and install a 4000 gallon tank (24 feet long × 65 inches in diameter) which had been lying around his farm for 20 years. The ends could be partly cut out and the tank would span 24 of the 26 feet through the soil zone. On 20 March we had the tank, we had the track hoe, we had David Armstrong to operate it, but I had to teach at Dabney Lancaster Community College so Ben Schwartz took over the supervisory role. At 1:00 p.m. I got a frantic call at work that there were rock ledges in the way and I should come home. It seems that the well hole, in traversing the soil zone, had gone parallel to a subsoil cliff, piercing a ledge at 13 feet and continuing on downward to bedrock at 26 feet. We needed to remove some of the ledges and move the bottom of the cliff back a few feet to center the tank on the well. You ask, "How did we know where the well was located, after all it had collapsed?" I had rigged a wire high above the well from which a plumb bob could be suspended. Thus the well could be relocated anytime we wished. With the help of an air compressor and rock drill we made short work of the offending ledges and by the end of the day the tank was set and nearly plumb and probably centered on the well. Two days later, the excavation was completely back filled. Our problem now was 6 feet of water in the tank.

On 23 March, the water had mostly gone away, so Ben Schwartz and Mike Ficco entered the liquid mud world at the tank bottom and eventually uncovered our precious well. The rest of the weekend was spent starting the shaft which would follow the well into Barberry Cave. With three weekends work we were 6 feet below the tank. If you'll remember, it was 26 feet to bedrock and our tank was 24 feet long. We had a 2-foot seam of mud on the tank-bedrock interface. The tank was supported on bedrock at only one point on its circumference. This problem needed to be cured. Most of the weekend of 20 April was spent cutting re-bar and welding it in place between the tank and pins driven into holes drilled into bedrock. Plywood shoring was placed behind this grid. We knew that this was only a temporary solution to the mud problem but we were eager to advance the shaft downward.

By 10 May, 1996, our shaft had advanced to 18 feet below the tank. All the rubble from the excavation was dumped down the well. Pieces too large for the well were broken smaller. Also I should mention that at times during the previous 5 weeks as much as 5 gal/min of water entered the shaft from a soil pipe on the southwest side of the excavation just under the tank. The water made for damp working conditions but in the end just cascaded down the well. On 14 May we had 2 inches of rain. I descended the shaft observing perhaps 60 gal/min of water entering. The sound of the resulting waterfall was audible from at least 200 feet away. On 16 May we had another 1.5 inches of rain and my visit this time was greeted with silence. The tank had filled to within 12 feet of the top with water. Obviously, the well was blocked. I managed to pull up a 50 foot cable ladder which was hanging in the shaft. It was severely stuck but I managed to remove it with large clods of mud sticking to the bottom. The subsoil around the tank had liquified. Our temporary shoring could not take the force and suffered a collapse. There was no option of removing the plug at the well because if we did, draining the water would remove the buoyancy and all of the waterlogged mud behind the shoring would cascade into the excavation.

Any options were soon removed as the tank sank a foot and large cracks began to open at the edges of the track hoe excavation. This was all accompanied by a strange splashing sound coming through the walls of the tank. An emergency call to Phil Lucas at Water Sinks produced all manner of ropes, come-alongs, and slings to secure the tank from falling into the widening collapse. Soon after the tank was secured, the surface on the west side of the tank gave way to a roiling caldron of liquid mud. The 18-foot excavation below the tank was certainly filled with liquid mud. I was depressed.

By 25 May the water had drained from the tank to reveal a large dent in the bottom and the fact that our tank was now an ellipse with a major axis of 70 inches and a minor axis of 60 inches. By this time I had also used my tractor front loader to back fill the collapse on the outside of the tank. We were back to where we were on 22 March except now the well was covered with 18 feet of mud rather than 18 inches. If we were going to continue this madness, we needed to remove the mud. To remove the mud, the water problem had to be solved. Without a drain, anytime it rained the shaft would fill with water. On 25 May, Ben, Gregg and I used compressed air and water to jet a one-inch diameter steel pipe through the mud to the well collar. We even managed to blow air down the well. With this encouragement, Ben and I constructed an 8-inch diameter, 700 pound pipe which we hope could be jetted through the mud to the well. By 23 June with the expenditure of 110 man hours, the pipe was sitting on rocks 6-inches from the bottom of the shaft and not quite over the well. Water would not drain from the 8-inch pipe into the well so this effort was a failure. Back to the drawing board. We had invested so much time and money at this point that project failure was not an option.

Water was a problem we could deal with by pumping and bailing as long as there were no severe storms. The mud had to be permanently walled off. A new plan evolved. We would dig out the mud bucket by bucket 8 feet below the tank. At this point we were 7 feet into bedrock. We would repair the temporary shoring between the tank and the bedrock, wash the limestone walls, lower an 11-foot long, 42 inch diameter culvert through the tank to the top of the mud and fill the space on the outside of the culvert to the tank and the limestone walls with concrete. This all had to be done quickly with all supplies at the site before excavation could begin. By 4 July, we were ready. Between 4 and 9 July, the 8 feet of mud was removed, the shoring was repaired, and the walls were sprayed clean. Of course any water which entered naturally or otherwise had to be pumped or bailed out.

10 July saw Jean Hartman, John Rosenfeld, John Sweet, Ben Schwartz and I installing the 42 inch diameter culvert through the tank. The bottom of the

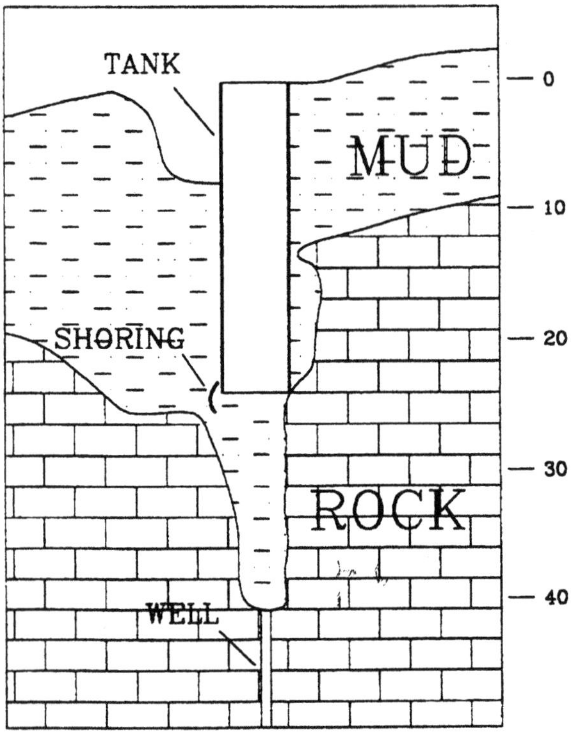

Fig. 10.3 A simplified cross sectional view looking northeast after the collapse in May, 1996. The numbers are the depth in feet below the *top* of the tank

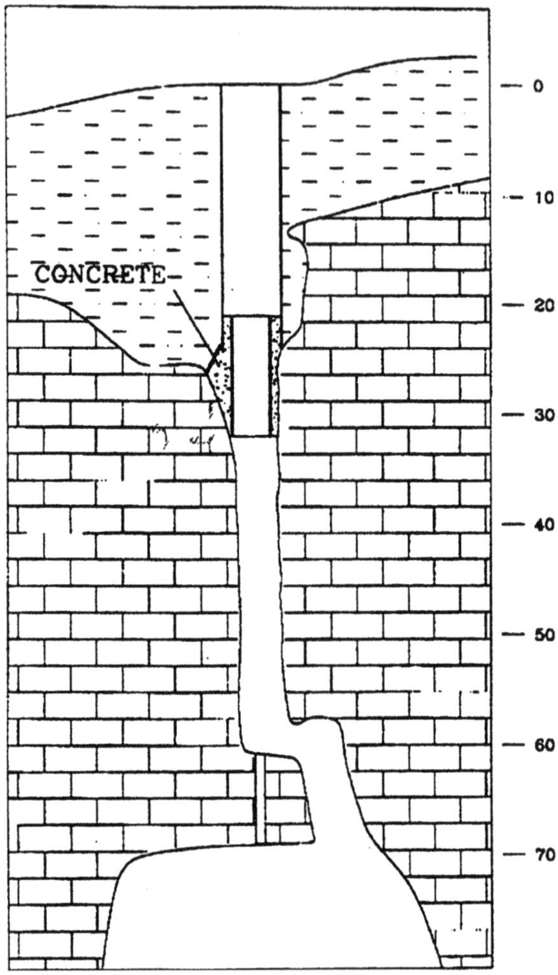

Fig. 10.4 A simplified cross sectional view looking northeast. This represents the finished shaft. The numbers represent the depth in feet below the *top* of the tank

culvert was supported by a wooden platform lying on the mud floor. The tank and culvert overlapped by 3 feet and this space would be filled with concrete locking the two together (Figs. 10.3 and 10.4). The top of the culvert was covered by a tight fitting lid which would prevent the concrete thrown down the middle of the tank from entering the culvert, diverting it instead to the space between the limestone walls and the culvert. The remainder of the day we mixed one cubic yard of concrete with a small mixer and dumped it into our excavation. This tested our forms, our theory, and it served to lock the bottom of the culvert in place for the big pour the next day. Between 8:40 and 10:30 a.m. the next morning, John Sweet, Ben Schwartz and I mostly watched as 6 cubic yards of concrete were poured around the outside of the culvert and into the annular space between the tank and the culvert. When we were finished, the concrete was 11 feet deep. Surely this would wall off the mud.

The next ten days were spent with little things like removing the lid from the culvert and cleaning up. Mike Futrell did his bit by sawing through the platform upon which the culvert rested and removing at least the central part to give us access to the mud below. We had probably walled the mud back but the water was a persistent problem. Every time we did something on the bottom, the first order of business was to bail water. The fact that the well was sealed caused another problem. The air on the bottom became foul after a short time digging. We had to rig a blower to replace the air every few hours. A movable bulkhead was installed above the diggers' heads to provide protection should something fall.

On four different long days, 27–28 July and 4–8 August, 8 people took turns filling, lifting, and dumping 5 gallon mud buckets until at 6:30 p.m. on 8 August, I had the pleasure of removing the last rocks blocking the top of the well. I was greeted with a breath of fresh cave air. For the first time in 84 days our well was open. There was still about ½ cubic yard of mud clinging to the walls, however.

10 August saw me back at the dig. The first order of business was to remove the remaining mud and toss it down the well. Then I removed the bulkhead and, as my final act, washed the entire tank, culvert and 10 feet of limestone shaft with high pressure water until no traces of mud remained. We were now back to where we were on 16 May before mother nature intervened, except that now the subsoil was walled back with steel and 7 cubic yards of concrete. Ingenuity has prevailed on this battle but there remained 28 feet of rock to tunnel through.

Fig. 10.5 Sara Good rappelling down the tank. Photo by Nevin W. Davis

Between 15 August and 18 October a total of 12 people helped with the tunneling operation. We maintained a square shaft about 40 × 40 inches with the well almost centered in the shaft. After 224 man hours on 20 different days we had about 8 feet of well remaining. As luck would have it, I was alone on 18 October as I worked with a pneumatic hammer at the bottom of the shaft. One wall was particularly soft and before long I had mined a 4 × 4 foot hole in the shaft wall into a 4 foot wide ceiling crack in the cave. I could sit on a foot wide ledge and peer into the blackness toward the floor 80 feet below. We had achieved our goal (Figs. 10.5 and 10.6).

I decided to have some fun since I knew Ben planned to help the next day. I called him and asked, "you definitely are going to help me tomorrow?" "Yes," he replied. "Don't bother, let's go caving instead," I said. We did do some work on the shaft on 19 October. Ben used the high pressure water to wash the rock dust and loose rocks from the shaft walls. We then set two 3/8

Fig. 10.6 View from the *bottom*. The drill hole emerged from the center of the ceiling. The climber is approaching the ledge to the *left* of the drill hole. Photo by Nevin W. Davis

Fig. 10.7 The stream passage in Barberry Cave just upstream from Big Bucks Entrance. Photo by Nevin W. Davis

Fig. 10.8 Waterfall over the Williamsport Sandstone that makes up the floor of the stream passage just downstream from Big Bucks Entrance. Photo by Nevin W. Davis

inch bolts and rigged a rope. What a spectacular rappel out of the ceiling past a 40 foot high flowstone cascade. We spent the rest of the day touring what was formerly a very remote area of the Burnsville karst (Figs. 10.7 and 10.8).

The tank now has a locked lid on it and liability releases must be signed to enter. There is a building over the entrance to which electric power has been installed (Figs. 10.9 and 10.10). The cool cave air makes the building a great storage space for apples.

10.3 Exploration from Entrance Number Three by Benjamin Schwartz

Now we had a third entrance, friendly caver-landowners, and easy access to the downstream part of the cave again. At first we concentrated our efforts on the downstream breakdown and the upper levels in the Woway, also known as Juvenile Delinquency.

To date, we have been unable to discover a way through the downstream breakdown. A tremendous amount of air flows out of the breakdown and a very strange phenomenon has been noted. No matter what the outside temperature, barometric pressure, or season, the air always flows in an upstream direction. We have been unsuccessful in tracing it more than 800 feet upstream before it fades away. Possibly the breeze disappears into an as yet unreached ceiling passage in the 80-foot high trunk.

In the Juvenile Delinquency area, survey teams quickly discovered more than 2000 feet of comfortable, dry, upper-level canyon passage. Several patches of very colorful and unusual red, orange, and yellow shale beds were found. Gypsum crusts, balloons, and flowers coat many of the walls and the largest known

Fig. 10.9 The building over the entrance to Big Bucks Pit, December, 2012

gypsum flower in Burnsville Cove is in this area. While attempting to follow airflow and discover a path around the flowstone choke at the upstream end of the Woway, I found a very low crawlway at the bottom of an extremely tight slot. A strong breeze flows through this passage, and I could hear water falling somewhere out ahead, both good signs since this lead later plotted out directly over what would be an upstream continuation of the lower Woway.

To make a long story short, a few of us spent several trips chipping open this unbelievably tight 30-foot long crawlway. Mike Ficco, Cori Schwartz and I finally popped out near the top of an 80-foot dome with obvious large passage leading off the bottom. Fortunately, on the same trip, Tommy, Nevin, and Bruce Dunlavy hammered open the water crawlway in the Woway below us. Tommy pushed ahead for several hundred feet and walked out into the same dome just as Mike Ficco reached the end of the crawlway above. This was fortunate, since getting on a rope in the end of the upper crawlway is nearly impossible. So, in spite of all our hard and painful labor, the water crawl turned out to be the easier route.

Nevin, Cori Schwartz, and I put together the next trip and surveyed 300 feet through the water crawl to the large passage on the other side. We named this junction Heavens Gate (Figs. 10.11 and 10.12). The name seems to fit, since both routes to this wonderful large passage are grim horror. To the left, the passage

Fig. 10.10 Looking down the shaft

Fig. 10.11 The water crawl leading to Heavens Gate and the Cave Pearl Room. Photo by Nevin W. Davis

Fig. 10.12 Cave Pearl Room. The cave pearls are between Nathan Farrar and the wall. Photo by Nevin W. Davis. The place where Mike and Ben emerged from the unbelievably tight crawl is about 80 feet above where the photographer was standing

leads to the dome with the crawlway on top. To the right, we surveyed 1000 feet of pleasant passage. This passage splits up into several smaller parallel leads and all are headed straight toward Butler Cave, 500 feet away. Good airflow is a very encouraging sign in one of these leads. However, several more trips and many hard hours of hammering have produced only a few hundred feet of very tight and wet passage. Typical Burnsville optimism persists, though, and we have yet to give up on the lead.

During Pancake Weekend, 1998, a new piece of the Barberry puzzle was discovered when we dug open yet another new cave. Already 1600 feet long, Buckwheat Cave has been surveyed to within 25 feet of Barberry and will almost certainly be connected in the future. More significantly, it may help us bypass the downstream breakdown choke in Barberry and provide access to a large segment of missing trunk passage.

Despite a few nasty spots, Barberry Cave contains some of the easiest and most spectacular passage in Burnsville Cove. The third entrance has made access to the main trunk a simple rappel, and, while not all is easy walking, it is certainly the most heavily decorated stream passage known in the Cove. With large speleothems, huge flowstone mounds, and rimstone pools, Nevin has compared it to a tropical river cave. Although nothing comes easily in Burnsville Cove, years of exploration and many miles of cave have proven that obstinate persistence eventually pays off. All the difficult obstacles encountered simply make significant breakthrough that much sweeter. Barberry Cave is very sweet indeed.

Work still continues. Several bolt and pole climbs have been completed, but many high leads remain to be checked. We continue to push the downstream

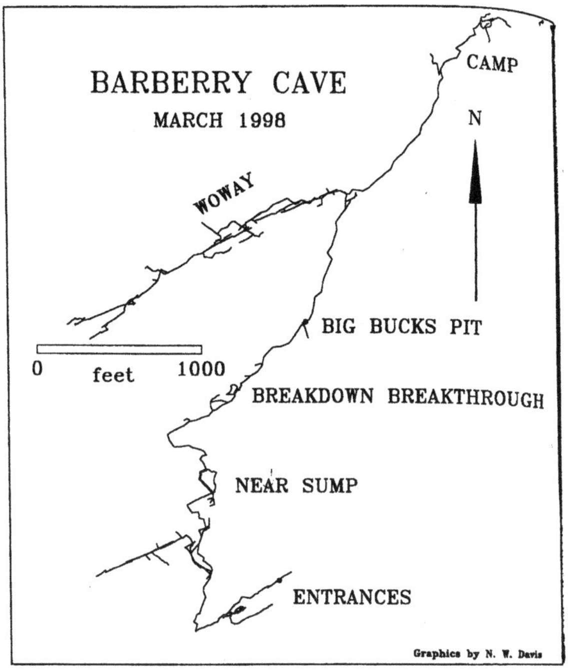

Fig. 10.13 Line map of Barberry Cave, March, 1998

breakdown choke in the hopes of finding a main trunk continuation. As of March, 1999, Barberry Cave has a surveyed length of 3.32 miles and a depth of 330 feet (Fig. 10.13).

10.4 Geological Notes

10.4.1 Passage Layout and Relation to Butler Cave

Barberry Cave may be the link connecting Butler Cave with the Chestnut Ridge System. It contains a curious convergence of large passages. The main trunk passage in Barberry lines up with Buckwheat, Blind Faith, and Battered Bar Caves and through them with the Burnsville Turnpike. However the Woway lines up with passages in Butler. Relatively few passages in Butler extend to the southeast up the flank of the Chestnut Ridge Anticline. The exception is the Crystal Gallery, a set of high, narrow fissure passages that come within 487 feet of the southwest end of the Woway (Fig. 10.14).

10.4.2 Speleothems

Barberry Cave is located at the southwestern limit of Chestnut Ridge where the sandstone caprock has been removed for a long time. This has two effects on speleothem development. First and most obvious, is that there is a direct path for infiltration water from the surface to reach the cave. Secondly, the ridge has a

Fig. 10.14 Stick map showing the relation of Butler and Barberry Caves

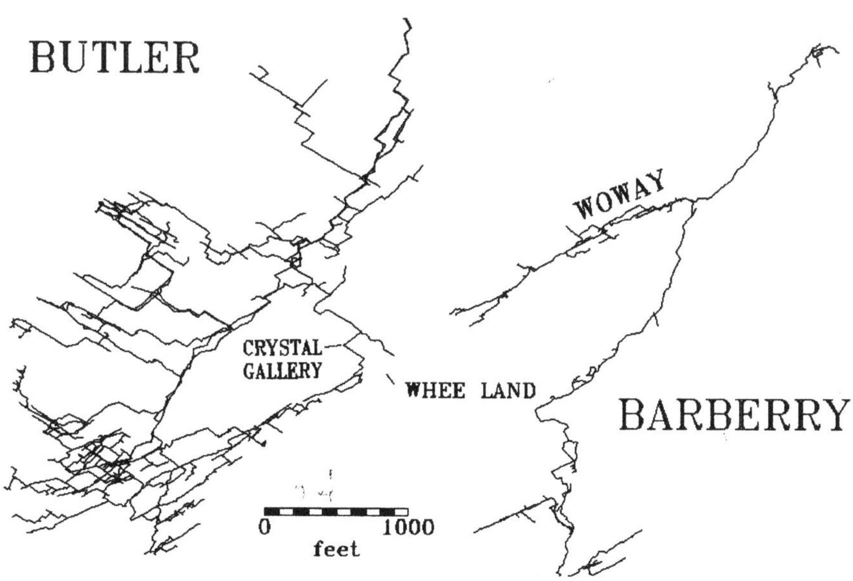

Fig. 10.15 Massive flowstone in Barberry Cave. Photo by Nevin W. Davis

Fig. 10.16 Large stalagmite in Barberry Cave. Photo by Nevin W. Davis

covering of limestone soil which is generally more CO_2-rich than sandstone soils. These two factors are largely responsible for the exceptional development of dripstone and flowstone in the cave (Figs. 10.15, 10.16, 10.17 and 10.18).

10.4.3 Clastic Sediments—The White Clay Deposits

Digging in the Barberric Crawl Bypass uncovered layers of white clay, very different in appearance from the usual brown cave muds (Fig. 10.19). X-ray diffractions patterns of two samples of the material were essentially identical. A major phase is very fine-grained crystalline quartz which produces sharp diffraction peaks. Also present as major phases are kaolinite clay, $Al_2Si_2O_5(OH)_4$, and a fine-grained muscovite mica, $KAl_2(Si_3Al)O_{10}(OH)_2$. Kaolinite was identified mainly by its basal reflection at 12.26° 2θ and muscovite by its basal reflection at 8.86° 2θ. Both of these reflections are weak with respect to the quartz peaks but because of the very small grain size, quantitative comparison cannot be made.

Fig. 10.17 Speleothems in Barberry Cave. Photo by Nevin W. Davis

10.5 The Nitty-Gritty of Big Bucks Pit by Nevin W. Davis

It's difficult to understand how so much effort could be expended on one cave. To put this into perspective consider that over 1670 man hours have gone into the three entrances, not considering the special cave trips, the contract labor, and the planning time, to yield 3.32 miles of survey. I think that works out to 10.5 feet for each man hour of digging. That's a pretty poor average. Not only did the man hours escalate for each successive entrance but the costs also escalated. They went from maybe $30 for the first to about $100 for the second and a whopping $4097 for Big Bucks. I think the reason for the tenacity was that mother nature just kept throwing challenges at us and we weren't going to give up.

I'd like to close this report with a special thanks to the people who made Big Bucks possible. Without the money and labor contributed by those listed below, Big Bucks just would not have been born.

10.5.1 Statistics of the Big Bucks Dig

- 1235 man hours were involved from 11 March until 19 October, 1996.
- Work continued on 72 different days.
- There were a total of 30 people involved excluding contract labor.

Fig. 10.18 The Great White Wow. Photo by Nevin W. Davis

Fig. 10.19 White clay interbedded in cave sediment. Photo by Benjamin Schwartz

- Over 57 % of the man hours were contributed by two people: Nevin Davis with 453 h and Ben Schwartz with 253 h.
- The total cost to date: $4097.
- Total monetary contributions to the project $3465.

10.5.2 List of Persons Involved in the Project (in Order of Their Contributions)

1. Nevin W. Davis
2. Ben Schwartz
3. Fred Wefer
4. Jean Hartman
5. Judy Davis
6. Phil Lucas
7. Gregg Clemmer
8. Mike Ficco
9. Frank Marks
10. John Rosenfeld
11. Les Good
12. Berta Kirchman
13. Jay Longenderfer
14. Cori Schwartz
15. John Sweet
16. N.C. Davis
17. Tommy Shifflett
18. Keith Wheeland
19. Ed Kehs
20. Mike Nicholson
21. Jeff Uhl
22. Al Grimm
23. Ron Simmons
24. Mike Futrell
25. Jack Igoe
26. Myron Cook
27. John Wilson
28. Keith Christenson
29. Scott Jones
30. Rocky Ward

References

Clemmer, G. 1993. Yet another "B" Cave: The ongoing saga of masochists in the Cove. *BCCS Newsletter* 19: 3–7.

Rosenfeld, J. 1994/95. Burnsville Cove pancake weekend. *BCCS Newsletter* 20:17.

Schwartz, B. 1994/95. From the pool room to the other side—new discoveries in Barberry Cave. *BCCS Newsletter* 20:49–53.

Uhl, J. 1994/95. But I don't want a rescue! *BCCS Newsletter* 20:36–46.

Caves of the Water Sinks Depression

11

Philip C. Lucas

Abstract

The Water Sinks Depression is a large closed depression at the northeastern end of Burnsville Cove. It is the final sink point for several surface streams. Two modest-size caves, Owl Cave and Water Sinks Cave, were known earlier. In November, 2007, a collapse revealed a much larger lower level of Water Sinks Cave. The new cave consists of a large master trunk passage, the Water Sinks Subway, and an array of other passages with the characteristics of a floodwater maze. A description of the caves is given along with a detailed description of the exploration of the Subway section.

11.1 The Water Sinks Depression

A thousand feet northeast of the entrances to Butler Cave is a surface water divide which marks the head of a complex blind valley that extends more than three miles to the northeast eventually ending at a closed depression called the Water Sinks. The closed depression is about a mile long and a quarter mile wide. It is bounded on the southeast by Chestnut Ridge which has exposures of limestone ledges and sinks of various sizes along its flank. It is bounded on the northwest by Jack Mountain whose lower flanks are shale and sandstone. Nearly all of the depression on the southeast side of Route 609 is an area called the Pancake Field (Figs. 11.1 and 11.2). It is named this because the mile-long valley here seems to be unnaturally flat. At one point this long flat terrace has been determined to have thick beds of clay (>35 feet) that more than likely represent the bottom of a former lake. Two hills, Cave Hill and Valley View Hill are located at the northeast end of the depression.

P.C. Lucas (✉)
587 Limestone Lane, Williamsville, VA 24487, USA
e-mail: Lucas@VirginiaCaves.org

Two streams, Sinking Creek and Water Sinks Creek, flow into the Water Sinks depression. Sinking Creek enters from the southwest and flows into a swallet in the bottom of a steep walled sink at the head of the Pancake Field (Fig. 11.3). A grass lined channel can faintly be seen meandering across the Pancake Field as an overflow from the swallet. Only extreme flooding will push water past the swallet into this old channel down the Pancake Field to the terminal sinking point at the Water Sinks Cave.

Water Sinks Creek enters the depression from the slopes of Jack Mountain to the west. It flows under Route 609 as two separate streams and then become a single channel that flows across the Pancake Field to the flank of Chestnut Ridge where it becomes a losing stream (Fig. 11.4). It then parallels the ridge for 700 feet to the northeast and turns back to the northwest along the base of Cave Hill. It nearly encircles Cave Hill tumbling over a waterfall and enters a deep sink of the northeast side of the hill and south of Valley View Hill. At this point the stream sinks at the base of a large limestone escarpment 110 feet high (Fig. 11.5). There are several caves located within the Water Sinks Depression. Two of these are Owl Cave and the Water Sinks Cave System (Fig. 11.6). Both of the caves have entrances located close to the Water Sinks Creek.

W.B. White (ed.), *The Caves of Burnsville Cove, Virginia*,
Cave and Karst Systems of the World, DOI 10.1007/978-3-319-14391-0_11,

Fig. 11.1 The Pancake Field looking northeast (*photo* by P. C. Lucas)

Fig. 11.2 The Pancake Field in flood (*photo* by P. C. Lucas)

Fig. 11.3 The Sinking Creek swallet. WBW photo (*photo* by W. B. White)

Fig. 11.4 Water Sinks Creek just upstream from the deep ravine that takes it into the deep part of the depression. The Owl Cave terrace is to the *right*; the Pancake field terrace is at the top of the bank to the *left*. WBW photo (*photo* by W. B. White)

Fig. 11.5 Water Sinks Creek as it flows into the sink against the cliff in the Water Sinks depression (*photo* by P. C. Lucas)

Fig. 11.6 The entrances to the old section of Water Sinks Cave (*photo* by P. C. Lucas)

11.2 Water Sinks Cave: The Old Portion of the System

George Deike, with assistance from members of the Nittany Grotto, surveyed both Owl and Water Sinks Cave in 1958. The caves were then called Siphon Cave No. 1 and Siphon Cave No. 2. The map of Siphon Cave No. 2 (Water Sinks Cave) shows about 1200 feet of cave passage that was known at the time. The map does not show an upper level that has over 500 feet of passages. A resurvey was done in 1990 that included the upper level. This survey totaled 2542 feet of passages with a total vertical relief of 90 feet (Fig. 11.7).

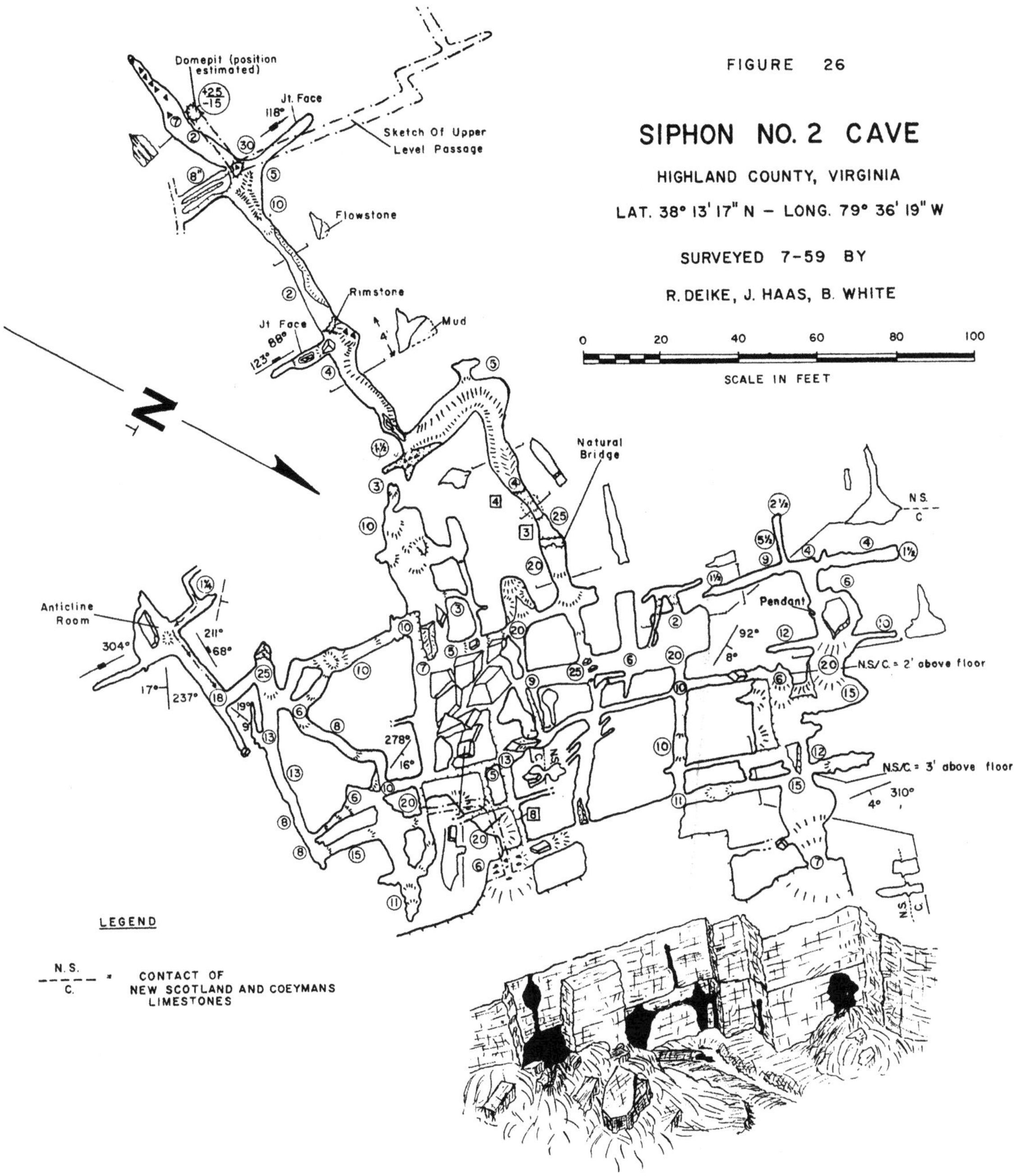

Fig. 11.7 1959 Map of Water Sinks Cave (Siphon #2) from Deike MS thesis (*photo* by P. C. Lucas)

The four entrances to the Old Section of the Water Sinks Cave System are found 20 feet above the terminal sinks of the Water Sinks Creek, in the face of the 110-foot-tall limestone escarpment. It would seem that with four entrances, one of them would provide a straight-forward entry into the cave. Surprisingly, this is not the case. There is some climbing involved either outside, as the entrance is approached, or inside the entrance, negotiating steep mud inclines. The cave is a tight maze tucked inside the cliff. Once inside, the cave passages are a mixture of rounded, tubular phreatic shapes and tall, narrow canyons. The brown walls give evidence that flooding can cover the lower section of the cave. Many intersections are within sight of one of the several entrances (Fig. 11.8). At night, it is surprising to travel some distance through several passages and then reach an intersection where the lights from the house on the hilltop can be seen.

Passages extending to the east reach a muddy, drippy passage with massive flowstone in places. It eventually ends in tight, muddy crawls that pinch out. A small anticline is breached in this area. Impressive fossil displays are present in the ceilings. The westernmost entrance gives access to one of the larger passages in the cave. Unfortunately, the passage becomes filled with sediment after 150 feet. There are several crawlways that branch from this passage, one of which connects to the rest of the cave system.

There is a passage that begins at an intersection in the southwest center of the entrance maze. This tall, narrow canyon extends to the southwest, up a steep sloop of about 50 feet, to a climb up and over an eroded flowstone arch. Soon beyond this point, the passage becomes a muddy crawlway. It eventually reaches the base of a small, 20 foot pit named "The Pit" (Fig. 11.9). This pit can be climbed generally without assistance and opens into the bottom of a larger canyon passage intersection. From here, only one passages escapes being completely choked by sediment. It extends to the northeast where another intersection is reached. Ahead, the floor drops into a short segment of a large passage. From the intersection, by following a flowstone ledge to the right, another crawlway can be found. It leads to the northeast where it becomes nearly choked by mud. A dig at this

Fig. 11.8 The light of day. One of the many Water Sinks entrances (*photo* by P. C. Lucas)

Fig. 11.9 Al Grimm climbing to the upper levels, Water Sinks Cave (*photo* by P. C. Lucas)

Fig. 11.10 High clay banks in the Water Sinks upper levels (*photo* by P. C. Lucas)

location eventually connected the crawl to yet another section of the cave named the "New Section".

The New Section is a series of upper level passages that have both flowstone and impressive banks of sediment nearly 25 feet tall with layers of gray clay and yellow-brown streaks (Fig. 11.10). The New Section reaches the highest point in the cave, 85 feet above the lower entrances, and 30–40 feet higher than any flooding in at least the past 20 years.

Convective air currents are generated within the cave due to the difference in elevation between the uppermost and lowermost entrances. Massive amounts of winter air that flow into the cave reduce the ambient temperature of the cave walls well below the mean annual temperature. During warm weather, cold air flows out of the cave and "pools" in the meadow in the lowest area of the sinkhole, resulting in lower temperatures there than in surround terrain throughout the seasons.

The Old Section of the Water Sinks is developed in the Keyser Limestone. A reef zone is present at the top side of the contact between the Corriganville and New Creek Limestones. There are several areas of speleothems in the cave but most have been discolored by flood waters. Speleothems in the uppermost levels have not been affected.

Few bats have been seen in this cave. Most bats encountered have been pips, although big browns have also been encountered. Raccoons frequent the cave, as evident by plentiful scat and one encounter.

The survey of the Water Sinks Cave System was considered complete after 1990, although we often wondered where all the water went, especially in flood, when a deep temporary lake covered the bottom of the depression. This lake would quickly drain, usually beneath areas of debris washed down during the flooding.

11.3 Water Sinks Cave: The Subway Section

11.3.1 The Subway Passage

The accepted definition of where a horizontal cave entrance begins is the drip line. An arch above the lower four "entrances" of the Old Water Sinks provides a drip line, a place where one can find shelter from rain. Accordingly, the large entrance is 40 feet wide and 25 feet high, the largest in Highland County and in the Burnsville Cove. The Subway entrance is also inside this drip line—an entrance culvert that provides access down to the top of a large passage, the Subway (Fig. 11.11).

Fig. 11.11 Matt Lucas at the top of the new Subway culvert pipe entrance that is chained to the cliff (*photo* by P. C. Lucas)

It is determined from evidence inside the cave that in the past, part of the cliff fell across and blocked the Subway entrance. The collapse was loosely assembled, allowing the stream to flow through but tight enough to filter out large objects such as logs and branches. Accordingly, the cave is relatively clear of flood material that normally fills a stream entrance.

The Subway Passage varies in width from 20 feet to over 50 feet. The lowest ceiling height, at 10 feet, is immediately at the entrance, but quickly rises, averaging 30 feet high (Fig. 11.12). It extends 250 feet south-southwest, past the Really Big Rock, to an intersection with several side passages. It then narrows briefly to 20 feet wide (Fig. 11.13) and continues for 200 feet to the Brass Monkey intersection (Fig. 11.14). At this point, the passage turns to the southeast for 100 feet to the intersection of Sweet Dreams. Throughout this distance, the floor of the passage is composed mostly of broken rock, scrubbed clean from swift flood waters. Much of this rock may have been washed in from the entrance. As the Subway continues past Sweet Dreams, much of its volume has been filled with silt and clay. A steep, sandy slope leads upward for 100 feet and then levels out for another 100 feet to a zone where the shape of the passage loses it identity at the Down Yonder intersection (Fig. 11.15). The Subway is about 1000 feet long and is nicely decorated, especially nearer to the entrance. It is certainly the hallmark of the Water Sinks Cave System (Fig. 11.16).

11.3.2 The Emerald Pool and the Roaring River Passage

A tall canyon passage continues south-southwest from the Really Big Rock—Subway intersection. The passage is choked by breakdown after 50 feet, but a small, narrow fissure extends to the west to intersect another fissure that extends downward into a large pit, dropping into the top of a large passage containing a large stream. The approach to the pit has been enlarged to allow human traverse but still requires some squeezing for larger cavers and a climb of 15 feet down in a narrow fissure. In times of severe flooding, this surely becomes a stream route and the pit becomes a waterfall, although no one has seen the event. The drop from the balcony is 30 feet to a steep boulder slope (Fig. 11.17). At the base of the slope is the Emerald Pool, where the upwelling Roaring River flows. A 20-foot-wide passage leads downstream for 250 feet to a sump (Fig. 11.18). During dry weather, the sump can be entered, albeit a wet entry, for another 50 feet to another sump in a 15-foot-diameter, circular room.

Fig. 11.12 The Subway begins as a large passage. John Sweet and Tommy Sweet are in the foreground, with Scott Olson to the *left* of the Really Big Rock (*photo* by P. C. Lucas)

Fig. 11.13 Nathan Farrar at a formation along the Subway, upstream of the Brass Monkey (*photo* by P. C. Lucas)

Fig. 11.14 Neighbors, Barbara Hall and Jimmy Landrum at the Brass Monkey—Subway intersection (*photo* by P. C. Lucas)

The level of this sump has been determined to be roughly the same as that of the French Lake in Aqua Cave. These two points are 1660 feet apart. Dye tracing has shown that Roaring River flows to Aqua Cave and has approximately the same volume as the Aqua stream.

11.3.3 Sweet Dreams

A clean-washed, wet-weather stream passage named "Sweet Dreams" extends from its Subway intersection for about 350 feet to the north and then to the east. The 10-foot-wide and 15-feet-tall passage follows strata with thin bedding (Fig. 11.19). The exposure of the thin beds along the walls and floor of this passage is quite striking. It begins at the Subway and maintains a steep gradient for most of its length. Eventually, the slope lessens and well-rounded cobbles of limestone and sandstone cover the floor and reduce the walking passage to a hands-and-knees crawl. This crawl ends in a lift tube where flood waters become deep enough that they flow out of this passage *en route* to the Scary Breakdown Room (Fig. 11.20). There are several side

Fig. 11.15 Jeff Uhl, lead tape, surveying up the steep bank of sediment in the Subway (*photo* by P. C. Lucas)

Fig. 11.16 Typical speleothems in the Subway Passage (*photo* by P. C. Lucas)

Fig. 11.17 Emerald Pool Pit, a 30 foot drop down to a rubble slope. Nevin Davis, Mike Futrell, John Sweet, and Mike Ficco (on rope) (*photo* by P. C. Lucas)

passages that branch from Sweet Dreams. One of these intersects the Subway about 100 feet downstream from the Really Big Rock. The passage is a generally small stoopway and pirates flood waters. The other side passages also carry flood water and have cobbles along their routes.

11.3.4 Sideways Passage

At the Really Big Rock, the Sideways Passage branches away from the Subway as a narrow canyon, 3 feet wide and 10 feet tall (Fig. 11.21). It then become lower, but remains a walking passage for several hundred feet, making abrupt turns as it follows the joint pattern of the bedrock. At one right-angle turn, the passage becomes a keyhole just large enough to accommodate the average caver. It then becomes a tall, very narrow canyon. A short section of this canyon has been widened. Eventually, the canyon becomes a wider walking passage, particularly if the caver turns their shoulders sideways. The Sideways Passage leads gently down-slope to a complicated area in both in the horizontal and vertical planes. The lowest point reached is the Scary Breakdown Room. This is where flood waters in this section of the cave flow to. The 40-feet-tall room is filled with washed breakdown, which all have edges rounded from the pounding of flood waters. Directly above is a sink point in the bottom of the Water Sinks, so a lot of the flood waters must flow down into the room from the sink above (Fig. 11.22). Where the flood waters flow beyond the Scary Breakdown Room is unknown, but they don't seem to back up in this passage as they do elsewhere in the cave, leading to suspicion of overflow routes to Aqua Spring.

11.3.5 Double Cross and Slim Pickens Passages

These two passages are upper levels that are separated from the Subway by 50 feet. Both passages have a vertical connection at their southwest ends. The Double Cross' connection is a 40 foot pit that opens

Fig. 11.18 Mike Ficco heading downstream from the Emerald Pool (*photo* by P. C. Lucas)

Fig. 11.19 John Sweet examines thin bedded Tonoloway Limestone in the Sweet Dreams passage (*photo* by P. C. Lucas)

into the ceiling of the Subway at the Brass Monkey intersection. The Slim Pickens' connection is a down-climb at the end of the Down Yonder Passage.

Slim Pickens begins as a generous walking passage from the Sideways Passage. After a few hundred feet, it becomes a crawl that descends into a lift tube with a sand floor. It then continues as a crawl, up a steady grade for another 100 feet and gradually opens to a walking passage for several hundred feet. The passage has several branching side passages that are generally small. It is at the end of one of these that the connection down to the Down Yonder Passage is reached.

The Double Cross Passage starts as a stoopway at the Sideways Passage and gradually becomes larger as it reaches a sculpted area called the "Colonnade". Like Slim Pickens, the passage divides into several passages that divide further before becoming too small. One passage becomes a canyon that opens into the pit down to the Subway. Another small canyon passage that extends northwest has an interesting reverberation from shouts. This passage becomes too tight for a traverse but with cavers on both sides, a light connection was made through to a canyon passage that extends from the Subway above the connection passage to the Emerald Pool.

Fig. 11.20 Crawlway approaching the Sweet Dreams lift tube. Scott Olson and Gregg Clemmer (*photo* by P. C. Lucas)

Fig. 11.21 Al Grimm at the narrow key-hole canyons of the Sideways Passage (*photo* by P. C. Lucas)

Fig. 11.22 Scott Olson at the Slim Pickens intersection with the Scary Breakdown Room (*photo* by P. C. Lucas)

11.3.6 Round Rock Passage

There are several passages that branch northwest from the Subway, just below the Really Big Rock. One of these extends for 200 feet, becoming a belly crawl over flowstone, before getting too tight. A gentle breeze through this crawl could indicate more passage beyond. The Round Rock passage also becomes a crawl and reaches a pinch point where cobbles fill the narrow passage. These cobbles were washed into the passage from beyond this point. An unsuccessful effort to dig past this point was attempted one afternoon. Soda Straws will be handy to push past this restriction.

11.4 Discovery of the Subway Section of Water Sinks Cave

Boy, did I miss my guess. I would have bet good money that the cave was going to be tight, narrow canyons choked full of mud and water with lots of logs and flood debris thrown in for good measure—not a place that was going to be a lot of fun to explore. So it was quite a surprise when I climbed down the newly opened fissure and found myself in a trunk passage that headed back under the hill for as far as I could see. But I'm getting ahead of myself. Let's roll the clock back 20 years.

The Bottom is a very dynamic place. After nearly every flood, we find newly opened drain holes into the subsurface; and we find old drain holes choked from trying to have swallowed something too big. Since buying the property, we've regularly had to burn woody debris that has washed in. Several times a year, following heavy rains, we have seen temporary lakes cover the Bottom, sometimes becoming quite deep. During the big flood of 1996, the lake reached a depth of over 40 feet. In January 1998, after a flood, a significant collapse occurred at the lowest Water Sink Cave entrance, creating a hole down through old flood debris 18 feet deep. Although no opening existed in the bottom of the collapse, it did indicate that a cave existed below this point. It also indicated, at least to me, that the cave was surely full of logs and other stuff that had fallen in with the collapse.

In November, 2006, there was another impressive collapse. This time, Frank Marks came over with his video borehole camera. We attached the camera to a length of PVC pipe, and probed all of the holes in the bottom of the collapse with it, trying to find a way down into the cave. We had no success doing this, but the fact that the collapse kept reoccurring in this area really made the location intriguing. In mid-October 2007, a good, soaking rain that fell over two days caused the creek to rise once again. This time, there was no ponding at the Bottom, even with the strong, steady stream flow. I remember thinking how unusual it was and made a mental note to "check it out."

So on Saturday, October 27th, I took a stroll down into the Bottom to see what new holes might have opened and closed, just as I had done so many times in the past twenty years. Unsurprisingly, I noticed that yet another small fissure had opened up in the wall of the flood debris that lies below the level of the meadow. This one was in the very bottom of a 15 foot subsidence that occurred the year prior. To reach it would have required a climb-down through the flood debris. Not feeling particularly ambitious, and almost having just walked away, I tossed a rock into the hole—nothing. So I tossed another rock into a shadow against the cliff face —it fell further than I expected. I had been down there before but, as always the optimist, I climbed down for a closer look. I noticed a fissure sloping down under the ledge upon which I had been standing. It was the first time I had seen this fissure, and not only did there seem to be a drop there, but there was a strong air current coming out. The fissure was too tight and extended beneath some fearsome looking boulders, but it looked very promising and produced a profound echo when I shouted. Whoa Nelly—it looked good! On October 31st, Frank Marks, Cotton Brown, Roger Baroody and I lowered one of Franks' video borehole cameras down the fissure to see what the video camera would show (Fig. 11.23). About 20 feet down, the video feed suggested a good-sized void, even though the camera kept falling into a narrow crack that limited its field of view. Despite having nothing to use as scale, the void looked large enough to accommodate humans. Inspired, I began making plans and gathering needed materials.

The following weekend presented an opportunity. Scott Olson and his friend, Mike Spencer, were down in the Cove. On Friday, November 2nd, I asked them if they would be interested in helping me on this dig. They agreed, and so we spent Saturday using my new generator and a large handful of soda straws, making our way down through the fissure (Fig. 11.24). Frank Marks joined us in the afternoon to watch all the "goings-on". By 8:00 PM, we had made the fissure wide enough to fit someone and hung a 15 foot cable

Fig. 11.23 Cotton Brown and Phil Lucas prepare to lower video camera down crack while Frank Marks looks on (*photo* by P. C. Lucas)

Fig. 11.24 The soda straw technique used to enlarge the fissure (*photo* by P. C. Lucas)

ladder down into the void below. Concentrating on down-climbing the floppy cable ladder that was too short, I really didn't get a good look until I reached the bottom and turned around. I was astonished to see a large passage that echoed away into the distance and had white speleothems hanging from the ceiling. This was not anything like I had anticipated—not that I was complaining. Not really being prepared to start a survey, we immediately made plans to start the exploration and survey the next morning.

11.5 Exploration of the Subway Section of Water Sinks Cave

11.5.1 Soda Straw Survey

Sunday, November 4, 2007
Scott Olson, Mike Spencer, Frank Marks, and Phil Lucas (reporting)
555 Feet Surveyed
3097 Feet Total

The next morning, when Scott, Mike, and I got together, Frank came over and suggested that we take one of his borehole cameras with us (with 600 feet of cable on a spool) so that he could watch our progress on a surface monitor, virtually exploring the virgin passage, as well as record the exploration. We agreed and Mike strapped the camera to his helmet. The technical aspect of this slowed our trip a bit but we were in the cave before noon (just barely). Naturally, I had my point-and-shoot digital camera and surveying gear. All of this did not make for a rapid trip, but we did manage to survey over 500 feet. We turned around in big passage with plenty of side leads. We named it the Water Sinks Subway.

Basically, we surveyed a large passage that averages 50 feet wide and 20 feet high. It begins at a depth 30 feet below the meadow at the base of the Water Sink Cliff. If it weren't for the meadow—there would be a 40 foot wide entrance below the two canyon entrances visible today. By the point we turned around, the passage had extended under the hill to the southwest and under Owl Cave. It was obvious to us that the cave floods—the entire floor is clean washed, and is composed mostly of rounded breakdown, cobbles, sand and some occasional leaves. The passage has a fairly steep gradient so that we reached a depth of 80 vertical feet below the meadow. In the passage, there is good airflow, some nice speleothems, and several nice side leads.

We came out about 4:00 PM so Mike could hightail it back to Pennsylvania. Besides, we were nearly at the end of our video-cable. Frank had captured about 2 h of footage in swirling black and white images, as Mike's head was nearly always in motion. It made no difference to Frank—he got a lot of virtual virgin caving done while sitting warm and cozy in his vehicle. But next time we want color!

11.5.2 Quick Pix Survey

Wednesday, November 7, 2007
Matt and Phil Lucas (reporting)
461 Feet Surveyed
3559 Feet Total

My son Matt and I descended into the cave and surveyed in the first right hand passage that starts behind the Really Big Rock. At first, it seemed to be just the crack where the big rock had fallen away from the wall, but after 20 feet, it took on the shape of a true solutional passage. It looped around the second big rock and joined the second lead along an area of anastomoses and then sloped uphill to the southwest, where it divided into two parallel hands-and-knees passages, each with an old cobblestone floor. These passages were generally small and floored with sand and cobbles. A crawlway at the end of one passage continues towards the Waterfall Caves but was not surveyed in favor of bigger passage. This is a good direction to find a northern extension of the cave. Another passage startled Matt when he heard his own heart beating. Yet another passage became a belly crawl and headed up-dip to the northwest. After 150 feet, it seemingly ended in a solid wall. But there was a gentle drift of air flowing into the passage from this "dead-end." This required crawling all the way to the blank wall. Sure enough, there was a small fissure hidden in the ceiling that was the source of the air flow. Finally we tied into a station from the previous survey to complete our efforts.

11.5.3 How Sweet Survey

Sunday, November 11, 2007
Gregg Clemmer, Mike Ficco, Scott Olson, John Sweet, and Phil Lucas (reporting)
951 Feet Surveyed
4510 Feet Total

This survey extended the end of the first survey in the downstream direction. Thus far, we have not seen any flowing water in the cave, so when the term downstream is used, it means the direction of flow during flood. The survey began where a side passage that we named Sweet Dreams turns to the northeast and follows the dipping beds for about 500 feet. This appears to be the channel where much of the flood water flows. The walls, ceiling and stream bed are all clean-washed and are following the contact between the Keyser and the Tonoloway Limestones, whose beds are dipping at about 8 degrees (Fig. 11.19). Several smaller side passages branch from Sweet Dreams and follow a parallel down-dip northeast direction. Eventually, the floor of Sweet Dreams becomes covered with sand and cobble and has some leaves scattered along the way. At one point, an interesting collection of pine needles coats the floor and walls. Everyone kept saying "This is not a place to be when it's raining!" But it wasn't raining this day and there was no mud, just clean-washed cobbles and sand.

Finally, we were reduced to a belly-crawl going down-slope to a seemingly dead end. However, there was a gentle breeze blowing through. When pushed, this dead end revealed a lift tube (or dry sump) where the flood waters had pushed up a 14-foot steep slope of sand and cobbles. At the top was a space with one wall of solid bedrock and the other of breakdown. It appears to be the side of a sinkhole and later, when I plotted the data, the location is 180 feet under the present-day south side of the Water Sink's sinkhole. This must have been a monster sinkhole. There are two small canyons leading from the breakdown. One is small, continues for 40 feet, and perhaps "takes" the air, but heads to the northwest (weird direction); the other will require pulling out some rocks. These leads will wait for some future pushing. The two other side/parallel passages also continued down dip—one ended in a muddy gravelly plug (the only place we saw any mud) and the other led to a sandy belly-crawl that briefly opened into two small canyons. Both of these will require either some serious pushing or enlargement.

11.5.4 Down Yonder Survey and Rebar Ladder Installation

Saturday, November 17, 2007
Jeff Uhl, Rick Lambert, and Phil Lucas (reporting)
968 Feet Surveyed
4977 Feet Total

This morning, Jeff Uhl arrived with some rebar that he had bent for rungs. Rick Lambert, Jeff, and I spent a good part of the day drilling holes and pounding in these rebar rungs to make an easy access down the 19 foot entrance drop into the Water Sinks Subway Passage. It worked like a dream. Each rung made ever-increasing in pitch, ringing sounds as they were driven home. If we had had our eyes closed, it would have sounded a lot like a bunch of pitons being hammered home into a good crack. It is a wonderful ladder and so much better than a dangly cable ladder. The entrance is now user-friendly. So, we declared the task complete and strolled down the Subway for a look-see and to find a virgin passage to survey.

Turning right at Sweet Dreams, we stretched the tape up the sand slope heading south by southeast. Well, the sand soon turned into earthy clay that was covered in earthworm casts by the millions. Hum, I think this ought to be named the "Cast-a-way". Ouch! Well, that's its name anyway. No clean washed rock there, but a nice passage nonetheless. It goes right underneath Owl Cave but 120 feet below it and partway into the slope of Chestnut Ridge. However, as it enters Chestnut Ridge, but before it reaches the line of big sinks that start at Castle Rocks, it fills with sediment. At one point, we reached an upward-sloping passage that ended in a collapse of thin, cherty/shaly bedded, rock that might be from the same bed as a 3–4 foot thick bed of shale that I see on the surface near Castle Rocks. We also encountered a small trickle stream coming down a thin passage from the northwest. This might be part of the surface stream that sinks around the Castle Rocks area. This stream joined a reasonable-sized passage that flows down to the northeast. We did not follow this stream. In fact, we turned around at this point and headed home with about 450 feet of survey and leaving five leads.

11.5.5 Brass Monkey Survey

Monday, November 19, 2007
Frank Marks, Nevin Davis, and Phil Lucas (reporting)
407 Feet Surveyed
1.02 Miles Total

The short trip into the Subway today provided just over 400 feet of survey in a rather conspicuous side

Fig. 11.25 The Brass Monkey side lead (*photo* by P. C. Lucas)

Fig. 11.26 A colorful hydrated iron oxide formation (goethite? ferrihydrite?) (*photo* by P. C. Lucas)

lead about 400 feet from the entrance (Fig. 11.25). This passage remained large for 150 feet, until a steep slope took the floor up quickly towards the ceiling. At this point, I could see a climbable 10 foot fissure into a void above. It seemed to me that I could hear a large echo above and beyond. A belay and some form of protection should be used for this pitch. A single bat was seen on the ceiling—the first bat seen in the cave. There was mud in this passage and worm casts were scattered along the mud floor.

Backing up, we took a side passage that we had passed and began to survey again. This nice tube sloped gently up a clean-washed, small cobble slope—another lift tube. The passage then turns southwest again; and has nice, walking dimensions and sculpted walls. Lead tape Marks called back, "There's a nice echo ahead and it leads to a drop." It was somewhat of a surprise that following up a cobblestone slope would lead to a drop, but this was indeed the case. With no way to easily climb down, we turned up another slope into another short passage that ended in some colorful goethite formations (Fig. 11.26). This does not seem to be a simple cave.

11.5.6 Nick of Time Survey

Tuesday, November 27, 2007
Ben Schwartz, Nevin Davis, and Phil Lucas (reporting)
376 Feet Surveyed
1.09 Miles Total

Ben Schwartz came up from Blacksburg (in the nick of time before leaving for Texas) to accompany Nevin Davis and me into the Subway. Ben wanted to collect samples of some mineral crystals that were spotted in the Sweet Dreams Passage and of the goethite formations seen during the Brass Monkey Survey.

Collecting the mineral samples went well. With that completed, we investigated the 20-foot-deep fissure that we left behind last trip after I had declared it to be unwise to downclimb the pit. Accordingly, we brought a ladder and bolt kit to use for this pitch. Naturally, the first thing Ben did was to climb down the fissure, saying that it wasn't all that hard. He said there were passages taking off at the bottom and for us to come on down. I insisted that since we had the ladder and bolt kit with us, we ought to use them; besides I was going to take some pictures anyway (this is always good for 5 min or so—time enough to set a bolt).

Now I need to explain that Ben was carrying a 15 foot cable ladder and I a 10 foot ladder in our cave packs. The latter was in case the drop was deeper than the 15 foot estimate I had noted in the survey book last trip. It is a nuisance to have a ladder come up 5 feet short. Well, guess what, the 15 foot ladder was 5 feet short. Ben said that we didn't need that 5 feet—of course he would say this, after having just climbed back up, gaining 5 vertical feet with each step. As usual, I forgot about Ben's long legs and decided to leave the short ladder in my pack. Needless to say, five chin-ups were necessary to climb back up.

The passages at the bottom were mostly round tubes with muddy floors. We followed the largest one, a stoop way, down the dipping beds. It continued for over 250 feet, gradually getting somewhat bigger and dryer. Eventually we stepped across a number of rimstone dams 2–3 feet high (they had been undermined) and reached the point where the passage turned east along the strike. Here, during a quiet survey moment, I heard a stream!!! Pushing a segment of an old rimstone dam out of the way, we came to an intersection that had large stream cobbles scoured clean. Down to the right was a narrow fissure that had the roar of a stream coming from below and beyond. Though it was tight, Ben managed to continue for another 20 feet or so where he yelled back to us that he was at the top of a 30–40 foot drop into the top of a large passage or room. He explained that the walls seemed to be 40 feet to each side and that the sound of rushing water was coming from a large stream flowing through this passage. Whoa Nelly—this was exactly what we have been hoping to find. Could this be the stream coming from Sinking Creek, Better Forgotten, or Chestnut Ridge? One problem—we weren't quite there yet. Even if we could reach the point that Ben had, rock anchors and a rope would be necessary for the drop. So while Ben was assessing, I picked out the places where soda straws could provide passage enhancement if this proved to be the only way down. But there might be a better way.

With this discovery in mind, we returned to the main entrance passage to a point not far down from the Really Big Rock. There, a nice side lead that obviously takes a lot of water continued on. To enter this side lead, a short climb down is necessary; but with every trip into the cave; folks have peered down into this good looking canyon. It heads in the direction we had just surveyed. So we figured, perhaps this might be a good place to look for an easier access to the stream passage. Sure enough, the stream could be heard coming from a tight fissure on the right wall 40 feet down the passage. We followed the fissure for a short way until we reached the point where our last survey station could be seen across the top of a very narrow canyon. This was directly above the point where Ben had reached earlier. This approach will be a heck of a lot easier than from the other direction.

11.5.7 Hocus-Pocus Survey

Tuesday, December 4, 2007
Al Grimm, John Sweet, and Phil Lucas (reporting)
377 Feet Surveyed
1.16 Miles Total

We took my cordless hammer drill and drilled several holes to enlarge the fissure down to the stream that we had heard last trip. The soda straws worked well, but two of the drill's batteries did not hold a charge. Being out of juice, we headed down to survey a side passage that was missed on the How Sweet Survey. Along the way, Al gave a little whoop and holler about having almost stepped on a snapping turtle. Sure enough, there he was right out in the middle of the trunk passage, as if he had just arrived. We quickly determined that he was no longer living but I don't think that he had been there during our previous trips down this passage. He had not yet begun to stink. I think that perhaps he had entered the cave during the last big rain and had been hanging out in some corner (probably trying to be a bat—a scary one) until recently.

Our side lead began as a nice tube about 6 feet wide and 5 feet tall. In many sections, this tube was either washed clean or had banks of sand and cobbles (Fig. 11.27). We followed the passage in an upstream direction. It branched into several smaller tubes, some of which connected to other side passages and some lead to fill. Eventually, we came to a place where a lot of large sandstone cobbles were covering the floor. Here, I climbed up 6 feet into a small room with breakdown. From the evidence of washing, the fresh air currents, and a slight echo to my scrambling into the room, I sensed we were near something big. I pulled away a couple sticks that were partially blocking my view and looked up through a crack into a black void with booming echoes. This had to be the main passage that we had doubled back upon. My sketch confirmed this looping. I tied a ribbon on a stick in the crack for our return trip back up the main passage to the entrance. Sure enough, walking past the turtle, there was the ribbon in the floor along one wall.

Fig. 11.27 Al Grimm passing below an under-cut flowstone ledge (*photo* by P. C. Lucas)

We could simply bend over to pick it up. This is yet another reminder that despite our careful observations, key connections might still be overlooked.

11.5.8 Touch and Go Survey

Sunday, December 9, 2007
Mike Ficco, Mike Futrell, Nevin Davis, John Sweet, and Phil Lucas (reporting)
387 Feet Surveyed
1.24 Miles Total

Our purpose this day was to gain entry down to the Roaring River (I decided to use this name because I have heard others using it and besides, these two words beginning with R's easily roll off the tongue). With only three straws and the usual rock gardening, we pushed aside the final rubble of rock restrictions! Mike Ficco scooted ahead to the top of the drop and greeted the rest of us with whoops and hollers, producing echoes that said, "big passage!"

The balcony at the top of the decorated dome was a perfect place for rigging. Mike Ficco had the rock anchors drilled and hangers attached before I could arrive with my camera. That is a sweet and fast drill. Soon the rope was attached and the real fun began.

The drop was beautiful. It was decorated with speleothems that rim the top of the 20 foot dome (Fig. 11.28). It measured almost exactly 30 feet down to the landing on a steep scree slope that then dropped another 16 feet to the shore of a wide aquamarine pool with a large stream passage leading away into the distance. This pool was the source of the Roaring River. It has a deep pooled sump that appears to have dive able passage beyond, heading in a northeasterly direction (towards the entrance of the Water Sinks Cave) (Fig. 11.29). Roaring River flows out of the sump and down a large passage to the southwest (towards the Pancake Fields). This was a large stream. I would guess that it had at least half the discharge of the Aqua stream. So, to those who are familiar with this area, it might be assumed that this is a part, perhaps a major part, of the Burnsville Cove drainage. Certainly, more dye tracing is in order. Furthermore, it might be assumed that Sinking Creek (the surface stream), after sinking underground at the head of the Pancake Fields, flows down towards the Water Sinks. The streams of the Chestnut Ridge Cave System that are a part of the Aqua drainage are also last seen flowing in a general northeast direction, as one would expect. So, the Roaring River is flowing in the opposite direction of these assumptions. Obviously, many questions have been raised by this discovery, and there will be a lot of investigations as a result.

Hmmm, I think maybe I have left you hanging as to where the Roaring River Passage goes. This nice, big passage, 15 feet wide and 20 feet high, becomes abruptly lower after 230 feet and becomes a deep pool sump (with a salamander guarding the downstream sump) (Fig. 11.30). It also definitely appears to be dive able.

11.5.9 Durn Tootin Survey

Thursday, December 13, 2007
John Sweet, Phil Lucas (reporting), and Al Grimm
577 Feet Surveyed
1.34 Miles Total

Fig. 11.28 Nevin Davis looks down from the Emerald Pool Pit balcony as Mike Ficco descends into unknown territory (*photo* by P. C. Lucas)

About 200 feet inside of the Subway entrance, where I have taken some pictures of fungus flowers, is a side passage that had been patiently waiting its turn to be explored and surveyed. Today was its day. It was a nicely oval-shaped passage that had soft floor,

Fig. 11.29 The Roaring River upwells from the lower passage seen in the bottom of the Emerald Pool (*photo* by P. C. Lucas)

Fig. 11.30 Mike Ficco on a ledge above the downstream sump (*photo* by P. C. Lucas)

composed of a mixture of fine sand, silt, leaves and a few twigs. I had been wondering where the air was going that pours into the entrance, so that when I entered the passage, the air current was the first thing I observed. No, it was not blowing into this passage, but blowing out—not as strong as the entrance air, but definite air movement nonetheless. So, at this point, the cave's air currents remain very much a mystery.

The passage headed back to the northeast and paralleled the entrance passage. After several jogs to the right and left, the passage took on a classic keyhole shape. The narrowness at this point presented difficulty. Jamming his body into the uppermost space, Al wiggled forward about 10 feet until he dropped down into a wider space below. Thousands of fossils sticking out of the walls made the maneuver easier. This would have been a stopper, had it been a muddy tube.

From there, the passage made another jog right and left before it straightened for 35 feet to an intersection. There were good echoes and a tantalizing air current that was beckoning us onward. However, there was a problem—another tight place just 10 feet short of the intersection. The undulating pinch was about 7–10 inches wide for about 4 feet. These 4 feet would require a couple straws for proper passage enhancement, but it sure looked good ahead. The plot of the survey shows that this intersection is 108 feet southeast of the entrance and 60 feet deeper. This is very close to a swallow hole in the Water Sinks Meadow.

Next, we backed up to another hands and knees crawl off of the Quick Pix survey. This passage has a floor of hand-size, rounded sandstone cobbles, thinly-coated by a film of clay-like silt. Apparently, only a recent flood has caused this mud coating while earlier floods tumbled the cobbles—interesting. We also noticed another slight air current coming in from this passage. The passage continues for over 100 feet to where it turns down a slope into a too-small tube that will require some cobble removal for further penetration.

Finally, we travel down to the Down Yonder intersection and surveyed a side passage at the top of the sediment slope. Earlier, I had wondered if this was actually a passage and not just a wide spot along the main passage. Well, it is a passage that heads to the southwest. It had a slight air current blowing into the cave. However, in 80 feet it became filled with sediment of clay and silt size. We could see ahead to larger passage and felt air in our face, but it will require about 10–15 min of digging to get through.

11.5.10 Howdy Doody Survey

Friday, December 21, 2007
Scott Olson, Gregg Clemmer, Jean Vargas, and Phil Lucas (reporting)
522 Feet Surveyed
1.44 Miles Total

Our plan this day was to survey the side passages off of the Down Yonder Survey located southwest of Sweet Dreams. This we did. In doing so, the cave showed us some mud. This area does not get the flushing from flood currents that the Sweet Dreams area gets, even though they are not that far apart. The passages we surveyed trend to the northeast and were generally 3–4 feet high, 5–6 feet wide, and floored with cobbles and mud banks. Eventually, they became lower until they became too low to follow. Curious circulating air currents seemed to be present in most all of these passages but none were very strong. At one point, the scratch marks of some animal (about the size of an opossum) were seen on a wall. Several beetles and millipedes were noticed as well. All in all, this is not going to be a frequently visited area of the cave.

11.5.11 Lickety Split Survey

Friday, January 4, 2008
Scott Olson, Tommy Shifflett, Phil Lucas (reporting), John Sweet, and Doug Stanley
566 Feet Surveyed
1.55 Miles Total

Tommy, Scott and I came equipped with rope, ladders, hammer drills (two!) and all sorts of paraphernalia to do climbs and passage enlargement. The first was a climb at the end of the Brass Monkey Passage where I thought I might have heard an echo from a canyon up above a steep mud slope. While making an initial assessment, Tommy found a crawlway that led upward into the overhead passage making the climb unnecessary. Unfortunately, it did not continue very far and became too small.

Next, we backtracked to the Durn Tootin Side Passage and lugged our gear down to where we had ended our previous survey with a narrow too-tight canyon. A passage intersection could be seen ahead and there were good echoes. Just before this point, I placed several bolts to make a climb at the top of a keyhole a bit easier. After finishing this task, we were single file in the narrow canyon, going to the final restriction that I intended to make larger. The problem was that Tommy was in the lead and I had forgotten to tell him exactly where it was. So, I was bringing up the rear behind Scott when I realized that they had already passed the restriction and were standing in the intersection.

The intersection sported a choice of equally inviting passages going in opposite directions. We chose right. This was heading to the southeast towards the lower end of the Water Sinks, a promising direction with air. So we surveyed Lickety Split. Well, not

Fig. 11.31 Tommy Shifflett looking at clean-washed breakdown in the Scary Breakdown Room (*photo* by P. C. Lucas)

Fig. 11.32 Tommy Shifflett and Scott Olson in the Scary Breakdown Room intersection (*photo* by P. C. Lucas)

really, because I kept stopping to take some pictures. The passage reminded me of those in the old, upper Water Sinks Cave—tall and somewhat narrow, but generally comfortable walking. We came to several intersections but stayed in a southeast-trending direction. After about 200 feet, we came to a steep-sided room with massive breakdown. The entire breakdown pile had been washed clean and scoured from flood water coming down from above. Some of the breakdown blocks were very large and the whole pile did not have the mud and clay "cement" to help hold it together, as is often the case. Several large chunks in the ceiling seemed to have no visible means of support (Fig. 11.31). Quite frankly, it looked pretty scary. We tip-toed softly across one side and placed a survey station. There was a steep climb in the ceiling but I think it would be madness to attempt it.

To scout a route beyond, Tommy disappeared from view into the floor between breakdown blocks. A few nervous minutes later he returned to say he had found a way down but we would all need to be careful. This was wasted breath as far as I was concerned and we slowly picked our way down surveying cautiously. Soon enough Tommy (lead tape) called back that he was at the top of a cobblestone slope that led down to a small hole blowing air. Scott looked at me and said "Could this be the Sweet Dreams Lift Tube?" It was and so we completed a loop within the cave's passage but confronted yet another mystery—where does the water go? There is obviously a lot of water entering this room during flood conditions into the top of the breakdown (Fig. 11.32). The lift tube carries enough flow during flooding that it pushes cobbles up a 15 foot slope. And the flood waters flow through the passage that we surveyed today into this area. But where does all this water exit? Air currents flow in the same directions as the water and it is probable that it is following the same unknown passage.

Not far from this connection is another tall, narrow canyon leading away from the breakdown in a northwest direction towards the entrance. Being able to get through was not certain, so naturally, Tommy tried it first. As normal, he got through, which led to a lengthy discussion about what he was seeing on the other side. The canyon is approximately 30 feet tall but is only wide enough in some spots to accommodate cavers. Tommy crawled through along the sandy floor, where the canyon was widest. I gave it a try. All went well for a short distance and then, despite frequent exhales and a lot of squirming, my forward progress stopped. Suddenly, I realized I had forgotten to remove my survey notebook and that it had me jammed. Cussing under my breath, I slowly backed out. There was even a discussion as to the worthiness of continuing. I tried again without the notebook and got through successfully, despite a very snug fit for about 20 feet. Beyond was about 150 feet of narrow, clean canyon passage. Through this passage, we made yet another tie-into the previous survey in the Sweet Dreams section. Both ends of these surveys had been sketched by Mike Ficco which accurately described the cave.

11.5.12 Slim Pickings Survey

Thursday, January 10, 2008
John Sweet, Al Grimm, and Phil Lucas (reporting)
824 Feet Surveyed
1.71 Miles Total

Our destination was the Sideways Passage, beyond where we made all of those passage enhancements last week, and then to knock off the side leads. It was a good plan, but these side leads appeared to be bounded by the sink of the huge Water Sinks sinkhole on one side and known cave on the other. It might be slim pickings for a lot of passage.

We first chose the northwest trending passage heading back towards the entrance. We didn't go far before we encountered massive breakdown with clean washed sides. Sure enough, this is the edge of the entrance sink. We could not seem to make much progress in attempting to penetrate this collapse nor did we expect to. So heading back down the Sideways Passage, we continued to survey each short lead, with none going very far. Eventually, we got to the next-to-last intersection. Choosing the left passage to the northeast and crossing a hole in the floor, we came to the scary breakdown room and tied into the previous survey. Then, in the passage leading into the Scary Breakdown, we turned in an upstream/upslope direction and started a survey extension to last week's survey. The walking passage meandered gently upslope out of sight. It started out looking good but how long would this continue? 50 feet, then a 100—it kept going. Lots of small side leads branched from nearly every predictable joint intersection. The floor is a combination of sand, fine sediments and small cobbles. There was a steady drift of cold air flowing down the passage, which kept going in a generally south, southwest direction. Finally, after 150 feet of nice passage, the bedding took a downward turn to the southeast and so the passage height quickly became lower until it became a cobblestone belly crawl going down slope. It was another lift tube.

Groundhog Al slid down and spent several minutes rooting through the cobbles and sand until he passed beneath the belly of the tube. The other side was a nice hands and knees crawl that went again upslope, but more steeply than before. The survey footage accumulated as we made steady progress up the crawl and the passage kept getting bigger. We called it quits in continuing passage 10 feet wide and just as high.

11.5.13 The Double Cross Survey

Tuesday, January 15, 2008
Al Grimm, John Sweet (reporting), and Phil Lucas
992 Feet Surveyed
1.89 Miles Total

It is amazing where one little insignificant-looking lead can take you. The first thing we did was go look around in the Down Yonder area to see if there were any up-trending crevices that had been missed. We saw nothing that looked right so we returned to the little tube leading to the Sideways Section. Before reaching the station where we started off last time, Phil said, "Let's knock off this little passage to the right before going on." Why not indeed? It looked like it would end soon and we could move on to our objective. It got tight but did not end. Al squirmed through and reported an intersection and walking passage. I did not much like the tight spot and there was certainly no room for three people to survey. We determined that it led back into known cave so I went around while Al and Phil surveyed through.

Once through, instead of closing the loop, we set off into new territory. Most of our new passage was pretty nice. It eventually led to a small room with interesting erosional features, including three columns that appeared to be holding up the ceiling (Fig. 11.33). Continuing, we arrived at a "T" intersection, where a fissure went to the right and a smaller passage went up and to the left. I took the lead, setting stations, followed by Al reading instruments and Phil taking notes. I went right, climbing down and under a breakdown block for a short shot, then moved out along the fissure until the floor suddenly ended in inky blackness,

Fig. 11.33 Al Grimm in the Colonnade Room of the Double Cross (*photo* by P. C. Lucas)

Fig. 11.34 Al Grimm, contortionist, negotiates a corner in the upper Double Cross (*photo* by P. C. Lucas)

where I waited for Phil to catch up. "Watch this rock", I said, dropping a fist-sized rock into the void. Silence. A larger rock made enough noise to tell us that it was a fair ways down. Then we push a boulder over the edge, which ricocheted once and then landed with a thud on a muddy floor. We guessed it was around 50 feet deep and then tried using my laser distance-measurer to see if it could pick up a signal off the floor. It turned out to be 48 feet deep. We surveyed out to the edge and looked for good sites to set bolts on the next trip. The void returned a big echo, which led us to wonder if we were looking down a ceiling crevice into the main passage. Al rolled out 30 feet of survey flagging, tied it to a rock and dropped it down the crevice.

We retreated to the junction, where we surveyed to the left and upward, squeezing past some white soda straws into a small room, then on to a low passage with more soda straws. We could have pushed on but not without doing damage. Al probed a clay-floored crawl that might bypass the formations, but it became too small (Fig. 11.34). We decided we would return later with a shovel and try to dig out the bypass rather than break formations. It was getting late anyway and we had not even reached our main objective, station SP38, where we had quit last time.

We retraced our steps, turned right, went through the gravel crawl and arrived at the point we had set out for over 6 h earlier. Here we had two choices. We decided to go right. After about 100 feet, it turned sharply upward. We surveyed to a breakdown plug but left several other leads untested. We had been in the cave nearly 8 h and felt that was enough. Our loose-end-tying trip left more ends loose than it tied.

The last piece we did passed entirely over the Down Yonder section. The potential bypass of the soda-straw passage is now the highest point in the cave and the only part above the survey datum outside of the entrance. On our way out, we walked down the main passage and found a strip of survey flagging hanging from a ceiling crevice right at the entrance to the Brass Monkey section. We had seen that crevice before but there is no way to tell from below whether it connects to anything. Now we know that it does. Surveying down this drop will close a huge loop.

11.5.14 Hem and Haw Survey

Saturday, January 19, 2008
Tony Canike, John Sweet, Scott Olson, Keith Christenson, and Phil Lucas (reporting)
588 Feet Surveyed
2.01 Miles Total

At the Colonnade Room, we split up: the diggers going ahead to the Soda-Straw Room; while the surveyors "tied up loose ends". This time, it worked out as expected, there being no great length to any of the side leads.

The first side lead, surveyed in 11 shots, went for about 160 feet and ended in two breakdown chokes. The second side lead had pretty much the same result. The digging crew returned as we finished up these two passages. They had not broken through but had opened up two crawls somewhat. The hoped-for bypass of the Soda-Straw Crawl seemed not to be going in the correct direction. Scott and Keith headed out to the main passage to take some photos while we, the surveyors, went on to deal with new booty.

Returning to the Soda-Straw Room, we first did a short crawlway to the left that ended in three shots. With no bypass available, it was now time to tackle the Soda-Straw Crawl. The goal was to get through with minimal damage to the formations. I went first and determined that there was indeed passage to survey, so the others followed. A much-too-tight crevice continued straight, while a passage to the left ended in a sediment fill. To the right was a sinuous, tubular passage 3–3.5 feet high by 15–20 inches wide with some air movement—commodious enough just to traverse but it made for tight surveying. To make matters worse, it got smaller as it went.

I led, setting stations, while backing down the tube, since it was very difficult to turn around. Tony followed with the instruments, and Phil brought up the rear taking notes. After just a couple of shots we heard footsteps and voices echoing in the distance. Several loud "Helloooowwwws" brought no reply and soon the sounds faded away, but we were now sure that we were heading toward another ceiling crevice and that we had heard the other crew down in the main passage. After seven shots and about 70 feet gained, the passage still continued with no end in sight, although it still narrowing. The cramped quarters were working on my head, so we opted to quit. Tony also seemed pleased at the thought of more open spaces, while Phil, with the largest girth in the group, would have happily pushed on. The Echo Tube remains for a smaller, or at least less claustrophobic, survey team. A two-man team would be ideal for this passage.

On the way out, we made voice contact with Scott and Keith through the previous ceiling crevice and learned that they had been in the side passage leading to the Roaring River, so our unfinished passage will probably connect to a ceiling crevice in that area. The survey plot shows it headed in exactly that direction. Backtracking through the Colonnade Room to the main junction, we surveyed two more short side leads off of the Slim Pickens passage before leaving the cave. Several of our leads terminated in what has become a straight line of terminations on the southwest edge of the cave, which corresponds to the edge of the hill overlying. These are the highest passages in the cave, being near or above the level of the entrance. The areal extent of the cave is now pretty well defined unless we can make a breakout to the east, towards Chestnut Ridge, or somehow get through or under one of the collapse areas.

11.5.15 Dragon's Diarrhea Survey

Tuesday, January 22, 2008
John Sweet (reporting), Rick Lambert, and Phil Lucas
282 Feet Surveyed
2.06 Miles Total

We set out for the Brass Monkey section with the intent to survey beyond the short ladder drop described in the Nick of Time survey report and, if time permitted, to check a couple of leads below the drop that the NT survey had passed up.

This section of the cave is well known to be wet and muddy and it started off as advertised. Our first survey shot took us from the top of the ladder drop across a floor crevice too narrow to enter to a T-intersection and a slimy down-climb of 10 feet or so. The left lead at the T went steeply up for a short ways until it pinched out; and the right lead sloped gently downward but with the same result. With minimal booty there, we backtracked, rigged the ladder, and descended to a pool of water and a muddy tube passage with a trickle of flowing water.

The first lead was a low crawl to the right that turned back to the known passage in only two shots, so we closed a short loop. I then started up a lead to the left behind some speleothems. It was about a 1.5 feet high, 3 feet wide and sloped slightly upward with a trickle of water on the floor. The solid-looking floor was actually a thin crust that covered a thick layer of orange mud about the consistency of mayonnaise. My elbows, knees, and toes broke through the crust as I crawled but I was able to stay mostly out of the mud. Rick followed with the instruments and Phil brought up the rear with the notebook, each breaking up more of the crust. My turn to sink deeper came as the ceiling lowered to one foot and the slope steepened slightly before emerging into a small room floored with the same goop.

Once in the room, I looked up through what was surely the too-narrow crevice we had looked down a short time ago. The proof of that was a broken stalagmite that had fallen down the crevice as we had crossed and now lay in the mud before us. The crawlway extended only a mercifully short way beyond. This had started out to be, for no particular reason, the Dragon Tail survey but we renamed it more descriptively at this point. Going out was downhill and the crust was fully destroyed so it was a swim in 6 inches of slimy goop. Back in the tube passage, Phil asked if I wanted to see where it went and check one more lead but I was already on my way back to the ladder as I shouted a resounding "NO!" I do want to do that, of course, but not right now!

We ascended the long-enough ladder at the short ladder drop, each burdened with 20 pounds of added goop, and stopped long enough to take one survey shot down the too-tight crevice to close that loop. Phil didn't even open the notebook. We just memorized the three numbers and headed for the entrance.

11.5.16 Hither and Thither

Sunday, January 27, 2008
John Sweet (Reporting) and Phil Lucas
280 Feet Surveyed
2.11 Miles Total

Phil and I decided to check out a few unlikely-looking side leads that had been left on the survey of the Sweet Dreams passage last November. We first checked three small tubes at HS6 and HS7, which took only one survey shot each to mark off the list. The next was a little more complicated, being high on the right wall. It was also a tube, floored with sand and sticks, requiring three shots to a pinch out.

Moving on down, we did a two-shot tube at HS10 and then came to a sand-and-clay-filled passage at HS11. We had brought some digging tools for just this spot, so we set into remove the offending sediment. I dug for a bit, then Phil took over. The passage definitely continues but it will take more digging effort than we were prepared to give it today. The afternoon was winding down and we had one more place to look at, a cross passage near the end of the cobble crawl, just before the lift tube. The left side ended in a sediment plug but the right side went for 50 feet to the ESE before narrowing to only 3 or 4 inches wide. This is an interesting direction, as it could extend the cave into new territory, and there is definite air movement as well as an echo. It would be a difficult dig, requiring movement of the spoils for quite a distance and even then, having no good place to put them.

11.5.17 Knock on Wood Survey

Tuesday, January 29, 2008
Al Grimm, John Sweet, and Phil Lucas (reporting)
431 Feet Surveyed
2.19 Miles Total

There is nothing spectacular to describe about today's surveys. With the exception of about 70 feet of nice walking passage, the rest was crawlways of various sizes and descriptions. At least none of them had dragon diarrhea! In fact, they were all quite pleasant. (Now you gotta know only a caver would say that.)

Well… there is one crawl, a dig really, that is quite interesting to me. It is a real passage about 6 feet wide in some places, maybe bigger. It is hard to say because it is mostly filled with fine gravel, the size of rice cereal with just a bit of silt to hold it together… just barely. It fills the passage to within 6 inches of a nice, arched ceiling. The passage was headed nearly south when we first made its acquaintance and it had a definite flow of cool air (49 degrees) to help dissipate the sweat off of the brow of a fevered digger. It looked like a good prospect to me, and so I put in my share of digging (with the magic green digging tool) until 20 feet had been excavated to a place where the floor seemed to fall away. Well, it didn't fall very far—about 6 inches. There are places that 6 inches would be a lot, but this wasn't one of them. Yet, it was enough to allow me to scoot ahead a few feet to get a better look. Growl, grumble, and snort… the passage turned an abrupt left and continued onward for at least another 10 feet and then out of sight in the same 6 inch high, 6 feet wide gravely crawl with that gentle breeze beckoning onward. The hour was late and rain was in the forecast to start this evening. Besides, we definitely needed a sled to start moving the spoils. So we stopped.

I think this is a good dig. The line plot shows it is generally headed towards the wet weather stream course in Owl Cave where some of the flood waters disappear, flowing toward Chestnut Ridge. This Subway crawlway is headed right for that very spot! It is 60 feet away and 50 feet deeper. The Subway crawl (I think a good name is the Knock on Wood Crawl, KWC) is about 27 feet long and has the same rice-sized gravel throughout its length. How could this gravel have been deposited so uniformly in the crawlway? First, I think the cave back floods from Aqua. When this happens, the KWC serves as an overflow tube (there must be several). The current through KWC is enough to carry finer sediments away but slight enough to sort the fine gravels throughout its length. If this is true, then there is a good possibility that the crawl will intersect with cave passage beyond that connects to Aqua and perhaps Chestnut Ridge. Anybody interested in some digging? It is a pleasant dig.

11.5.18 Good Vibrations Survey

Thursday, May 22, 2008
Jon Lillestolen, Tony Canike, John Sweet, Phil Lucas (reporting), and Al Grimm
172 Feet Surveyed
2.23 Miles Total

The objectives of this trip were: set bolts and permanently rig the drop from the Double Cross down to the Brass Monkey, do some surveying down the Hem and Haw, and perhaps even a little digging off the Colonnade.

John Sweet was coming down with a cold and so waited at the bottom of the Brass Monkey intersection with the drill, bolting equipment, and rope, while the rest of us made our way through the cave to the top of the drop without carrying all of that equipment. Upon reaching the top and dropping a cord, we then pulled up the rope and all the stuff from down below. This worked well. Using Tony's drill, Jon played Spiderman as he set the rigging. Later, all of us used the rope as our exit from this section of the cave—how sweet! Speaking of sweet, it seems that John couldn't bear to be standing around while the bolting was going on, so he traveled up through the cave to join us at the top.

Once the bolting was finished, the passage on the other side of the drop was investigated and surveyed. It didn't go far. After that survey, Jon and I went down the Hem and Haw Passage to complete the going survey left there while the rest of the gang, minus John, went back past the Colonnade Room to dig on a side passage (it didn't go far either). John decided to go all the way back to the bottom of the cave to see if he could tell where the survey Jon and I were doing might connect back in. He suspected this connection because we could hear the voices of Keith Christenson and Scott Olson doing photography in the lower cave on our last trip. We had a pretty good idea where this connection ought to be located (in a ceiling channel), but it was hard to see from below. In fact, it was impossible. So John's curiosity got the best of him and he traveled to this part of the cave and set up a watchful vigil. As soon he heard our voices coming from a fissure from "up above", he started poking around.

The passage where we were surveying got tight. After about 150 feet, it got even tighter. Jon was lead tape (a good trick since we were using a Fat Max) and it was a good thing because he was able to get about 15 or 20 feet farther than I could. The last 10 feet for me was a vertical exhale fissure that had no floor in a few places (and you know how demanding that can be). By this time, we were getting closer and closer to John Sweet, who seemed to be just down the passage a ways. Finally, Jon could see John down an even tighter fissure. Jon said that he might be able to get through but, if so, would not be able to return. I declared that any attempt of forward progress on my part would be insane and that maybe he should stop as well. I then struggled backwards to that one spot where I hoped to be able to turn around. In my old age, I am secure in my belief that I don't have to see every square inch of every cave. Eventually Jon also decided to turn around, but it was harder for him because he is still young and still feels he has to see it all.

11.5.19 Round Rock Dig

Tuesday, June 10, 2008
Al Grimm and Phil Lucas (reporting)

We went to the end of a small cobblestone crawlway in the Quick Pix area of the cave. The passage headed towards the northwest beyond the present day margin of the hill that marks the end of all other passages. The human sized end of this passage was a tiny canyon, 4 inches wide, sloping down with the bedding, filled with round cobbles, and had a steady drift of air blowing up through it—another lift tube. We could see ahead about 10 feet or so and it appeared to be getting bigger. We spent about 2 h digging and reached the point where real progress was being made.

There is a high ceiling in the passage just before the intersection, near the dig. Al, taking a break from digging, wiggled up into the fissure with foot support from me. At one point in his struggling, he stepped on my head, pushing the helmet down onto my nose. Eventually, he reached the top and found about 50 feet of small, phreatic tubes that became choked with speleothems.

We left the Round Rock Dig looking good. There is almost certainly some passage ahead—how much? That remains to be seen.

11.5.20 Slim Pickens Dig 1

Monday, June 21, 2008
Scott Olson, Brad Cooper, Gregg Clemmer, Dave Kohuth, and Phil Lucas (reporting)

This was the first really serious dig attempt at this dig. Good progress (about 20 feet) was made up to and around a corner. What lies ahead will just have to wait for the next digging episode. I will say this—it is a nice dig, being slightly upslope, floored mostly by packed sand and small cobbles, with fresh air, a solid ceiling with a generous width, plenty of space for spoils, and can be reached by short travel time.

11.5.21 Tickle

July 6, 2008
Matt, Tina, and Phil Lucas (reporting)

In June, Scott Olson and I had dug between the Slim Pickens level and the Down Yonder, but failed to make a connection due to the position of large rocks that we could not move. On this day, Tina, Matt, and I, armed with the hammer drill and soda straws, went to attack the dig from the Down Yonder (lower) direction. Climbing up to the very end of the passage, to the breakdown choke, I found what was left of the burnt notebook page from last trip. There was some concern about removing rocks from below a breakdown collapse. But, with the judicious placement of one soda straw and 30 min of digging, we made the connection. We made sure the remaining rocks were stable, and climbed up through to find ourselves in the Slim Pickens passage. This will surely prove to be a useful connection, saving lots of energy and time when traveling to this part of the cave. I named it the "Tickle". Sometimes we get lucky.

11.5.22 Slim Pickens Dig 2

Wednesday, July 20, 2008
Brad Cooper, John Sweet, Jean Vargas, Gregg Clemmer, Mike Kistler, Tommy Shifflett, and Phil Lucas (reporting)

Using the new shortcut, we arrived at the dig site at about 11:30 AM and proceeded to make the dirt fly. Our first order of business was to widen the dig at the turn to allow another person the room to manage buckets and ropes. Then, with that accomplished, buckets full of dirt and sand were pulled down the dig's slope gently. At the end of the day, 200 buckets had been removed and the dig length is now about 40 feet. A possible connection over to the Double Cross Passage was investigated without success.

11.5.23 Setting Transducer and Anchors for Video Camera and Lights

Tuesday, September 9, 2008
Al Grimm and Phil Lucas (reporting)

We went to the Roaring River to estimate the stream flow rate. After making a channel 10 feet long, 3 feet wide, and 4 inches deep, we dropped sections of a foam noodle stick and timed how long they took to float the 10 feet. The average time was 3.5 s, which produces an approximate flow rate of 2.9 cubic feet per second (CFS).

Later that day, we went to the culvert where Aqua Stream flows under the road and determined the flow rate there, using the same methods as in the cave. The result was 3.6 CFS. By these approximate values, it seems that most of the flow from the drainage basin passes through the Emerald Pool on its way to Aqua.

11.5.24 Pushing the Roaring River Sump

Thursday, September 18, 2008
Brad Cooper, Al Grimm, and Phil Lucas (reporting)

We descended to the Roaring River in the Subway Section of the Water Sinks Cave to attempt a push on the downstream sump. We didn't have wet suits, but we did have a rope and face masks! The water level was lower than normal, so there was a little bit of airspace. We tied the rope off and Al went first, wading through the water. By dipping one ear and looking through the scant airspace, he said that he could see the ceiling slightly rising ahead, so he ducked under and came up 8–10 feet down the passage into an area with a more generous ceiling space (Fig. 11.35). He hollered back that the passage was turning slightly to the right and getting bigger. I was most impatient to see this, so I went on through too. It was about 30–50 feet to a circular room. Off to the left of the room, the water was over our heads, but we walked around to the right, only shoulder-deep. The room was 20–30 feet in diameter. On the far side of the room was a steep mud bank. I climbed up until I could see that it got too tight to continue, but it would have been an easy dig. In any case, we had been in the water for a while and were getting cold, so we headed back.

I went out first. At one point, I realized that if I kept using the rope to pull myself along, it would have directed me under the wall. There, where the passage turned, was an undercut wall that the rope slid down to when taunt. Taking a good look at the direction of Brad's light coming through the low airspace, I let go of the rope, ducked down, and pushed through the really low section until I came up on the other side where Brad was waiting. I shouted back to Al. He came through, following the line until he realized that it would take him underwater. He hollered back that the rope wasn't leading him the best way, so I explained to just follow the airspace. The situation could have gotten bad if he had blindly ducked under and followed the rope. But we all got out safely, and we know there's a potential dig there for another time when the water is low.

11.6 The Effects of the April, 2011 Floods

In April, 2011, we had two floods, back to back, following more than 3.5 inches of rain. The first was on the 12th and caused ponding in the Water Sinks Bottom at least 40 feet deep over the Subway entrance (Fig. 11.36). After the meadow was above water again, I went down to check out how this flood had changed things. There was so much debris along the cliff face that the Subway entrance culvert was hidden (Fig. 11.37). The meadow was all mud, a foot deep or so. On April 28, 2011, another flood occurred. This one didn't causing ponding in the Bottom as deep as the first, but it did tear some asphalt off the road running through gorge and washed out culverts under Rt. 609. The day after the flood, on my drive down to the mailbox, I stopped to look at the Water Sinks ponding. At the time, which was past peak flow, the pond was just above the ceiling of the lower Water Sinks Cave entrances. When I returned from checking the mail and driving just up the road to check on the culverts, only 20 min later, the pond in the Bottom had already dropped about 12 feet. What was going on? So I got out and stood there and watched for a while,

Fig. 11.35 Push through the Roaring River Sump (*photo* by P. C. Lucas)

Fig. 11.36 The depth of the flooding in the Bottom after the April 2011 flood. The water level has already dropped a few feet since the peak depth, and is about 41 feet deep at the cliff face (*photo* by P. C. Lucas)

Fig. 11.37 The debris at the base of the cliff face in the Water Sinks Bottom after the April 2011 flood. Oval picture insert of Phil Lucas by the Subway Section entrance, now buried 3–4 feet under the debris (*photo* by P. C. Lucas)

mesmerized by the spectacle. There was a lot of debris on the pond, and I could just watch the pattern of more being brought in and how it assimilated. Then I heard a loud crack, like a rifle or perhaps a tree breaking, but I didn't immediately see anything that might have caused it. I kept looking for the tree broken by the swift current. That was when it caught my eye that all the floating debris was moving towards the cliff face, and as I watched, the whole pond started draining rapidly. It maybe took 5 min for the whole thing to drain—Amazing! As the water drained, I noticed a new, 4-inch crack on the cliff face, right between the two main Water Sinks Cave entrances. That is what I had heard cracking—rock breaking under stress!

A couple days later, I made my way down into the Bottom and waded through the mud to the cliff face. The creek was still flowing with a large volume and there was a large pool at the base of the cliff. The pool was full of floating debris, so much so that it appeared to be solid ground. The Subway culvert was still under water, but I thought, gee, I at least ought to be able to see the lid. Then, all of a sudden, the pool began to drop. For a moment, it seemed that the ground was sinking! I quickly climbed up the bank to get further away. As I watched, the pool dropped 6–8 feet in less than a minute, then stabilized, and then refilled. I watched this cycle two or three times, with it dropping and filling 6–8 feet each round of just a few minutes. There must have been some metering effect below the debris pile. Days later, when the water was even lower, Nathan Farrar and I went back and confirmed my worst assumption—the Subway entrance culvert was gone. All that was left was a chain dangling where it used to be attached to the pipe.

On Sunday, August 7, 2011, six of us descended down into the Water Sinks bottom and proceeded to dig and reopen the Subway entrance. Several days before, Nathan Farrar and I had hung a steel cable down from the top of the cliff with a pulley attached and established a haul system that could lift and lower all the various things that we anticipated would be needed for this project. The first order of business was the removal of all the logs and tree limbs that had accumulated during the April flooding. Then the task turned to excavation. By late afternoon we had dug down to the level of the original culvert position where, once again, we reached an opening into the Subway cave. But where was the culvert? As the last few shovel scoops were made, the sound of plastic being scraped was heard. Sure enough, a rib of a plastic culvert was uncovered. As more and more of the culvert was uncovered, we could see that the culvert had become a crumpled wad. It was collapsed in the same manor that someone might stomp down on a cardboard box to crush it.

Later that evening, Nathan and I entered the cave to see what changes, if any, the flood might have caused. The first thing that I noticed was a lot of new gravel and sand fill along the floor of the Subway passages. In some places there seemed to be several feet of sediment added (Figs. 11.38 and 11.39). The huge rock that was at the base of the rebar rung ladder was gone. It was now a five foot drop below the last rung. I noticed many large rocks were now either missing or

Fig. 11.38 Nathan Farrar examining the debris clogging the way on down to the Emerald Pool, just beyond where the culvert was found (*photo* by P. C. Lucas)

Fig. 11.39 Nathan Farrar examining sediment deposits on the wall of the Sweet Dreams Passage, where it is evident that the passage was filled with sediment to within inches of the ceiling and then eroded back open, over the course of the two April, 2011 floods (*photo* by P. C. Lucas)

rearranged. There were a few more logs washed in but there were no organic debris jams seen. Below the Really Big Rock, we walked over to look down into the passage that leads to the Emerald Pool. And there it was—the culvert! It was the top section, eight feet long, that had been broken away from the four foot piece that we saw crumpled in the bottom of the dig. It was still mostly intact with the remains of the ladder still inside. There was a hole torn in the middle and one end was ragged, but it was still a round culvert. When this top section broke away from the bottom piece, it was forced through a slit about 15 inches wide and 6 feet long. The outside diameter of the culvert is 36 inches so it had to have been greatly deformed to be forced through this slit.

The several weeks following digging open the Subway on August 7th brought a flurry of activity, as we worked to construct a new and improved Subway entrance. We hand-mixed 20 or so bags of quickrete and dumped the mix around the lower, outer-culvert shell. Over the space leading into the cave, yet outside of this outer culvert, we built a wooden roof of double-layer 2 × 4s, attached to 4 × 4s bolted to bedrock and steel-cabled to various large boulders. Any gaps other than the cut doorway in the culvert were then filled with expanding foam. Later in the week, we borrowed Nevin Davis' concrete mixer, and began our most impressive task: mixing nearly five tons of concrete up by our house, where power and water was available and then hauling it, in buckets, down to the Water Sinks Meadow, using Frank Marks' Honda Big Red ATV and my Polaris Ranger ATV. Because the meadow doesn't run all the way to the entrance, a 35 foot-long, 3/8 inch steel-cable zip line was used to zip buckets from the ATVs to the top of the culvert, where they were dumped between the two culvert walls then and reattached to be pulled back up. This made quick work of emptying the buckets. The end result was a 12 foot tall, 48 inch outer-diameter culvert with about 8 inches of concrete between it and the 30 inch interior-diameter culvert.

Using six of the same 3/8 inch steel cables and turnbuckles, the outer culvert is held tight to the cliff face. Also, a multitude of anchor bolts were driven through the outside culvert into the concrete for added strength. A ladder was installed inside the inner culvert and a lid made of a double-layer of 2 × 6s, with a handle on either side and legs to rest on, was attached to the top. Now, a caver can climb up the turnbuckle ladder-of-sorts, stand on the concrete lip, pull open the lid to where it rests on its supports, and climb on down into the cave (Fig. 11.40).

Fig. 11.40 The inner pipe of the entrance and the crack in the limestone pillar (*photo* by P. C. Lucas)

The floor at the bottom of the culvert is sloped concrete, draining into the cave. Just through the cut doors is a large concrete step, and beyond are wooden steps, down to the top of the rebar ladder. The length between the culverts and the rebar ladder has been widened to at least shoulder width. We have also added six more rebar rungs to the ladder down into the Subway Passage.

Still, we had another threat—the cracked fissure from the April floods turned out to be the side of a pinnacle, 20 feet tall by 4 feet wide by 3 feet thick. It was disconcerting that this pinnacle might fall. Even worse, that it may injure someone. So, with pallets guarding the new entrance, the pinnacle was pulled down on Sunday, September 4, 2011. We did so by wrapping a 1-inch diameter nylon rope around the pillar and hauling on the rope with both my ATV and Frank's until it toppled over, causing whoops and hollers from the small audience gathered to watch the spectacle. Because of the pallets cushioning the blow, the Subway entrance culvert was not harmed (Fig. 11.41).

The first test of the new entrance's ability to endure floods is occurring now, September 6, 2011, as Tropical Storm Lee sweeps through. As of this morning, over 2.5 inches of rain had fallen within the past 20 h, causing Water Sinks Creek to swell to a considerable stream. As we observed it at 9 AM this morning, all of the stream was sinking at the cliff face (not ponding!) in newly opened swallets.

Fig. 11.41 Brad Cooper on the completed double pipe and concrete entrance and Nathan Farrar on the rubble formed by pulling down the fractured pillar (*photo* by P. C. Lucas)

11.7 Owl Cave

11.7.1 Description

Owl is a maze cave developed on two levels, having 3004 feet of passages surveyed and developed in the Keyser Limestone (Fig. 11.42). The two main entrances to the cave are located side by side, 15 feet apart, each 10 feet tall (Fig. 11.43). The passages from both entrances join each other in 10 feet, and again at 40 feet, where they become a single passage. Below a section with a log jammed in the ceiling (Fig. 11.44), the passage becomes a hands-and-knees crawlway leading 50 feet gently down to the northeast, where it becomes a belly crawl for a short distance and then slopes up sediment bank (a lift tube) into the Tuxedo Junction Room. Here is a three way intersection. Straight ahead is a tight belly crawl for 20 feet that opens to a walking passage. After advancing up a muddy steep slope for 40 feet, the Talus Room is encountered. The floor of the room continues up a steep slope of broken, loose chert and sandstone blocks that end in a ceiling fissure. On the southwest wall of the Talus Room is an intersection with a narrow canyon (Fig. 11.45) that ascends rapidly to an upper level of walking passages with several stoop ways and a challenging crawlway over a long, shallow pool of water. There are about 500 feet of passages in this upper level. About 70 feet southwest from the Talus Room intersection is a short belly crawl leading to a 15 foot pit that drops down into the lower level of the cave near the first dry siphon. A rope or short cable ladder is handy to negotiate this short drop. The eastern end of the upper level intersects with the Canyon Room opposite of the Talus Cone Entrance. The upper level has more flowstone than the lower level and has sticky mud in places. It is the most challenging section of the cave to explore.

Continuing to the south from Tuxedo Junction is a stoop-way and then a walking passage that averages 10 feet wide and 7 feet tall with a dark humus floor and scattered organic debris from past flooding. This passage continues for several hundred feet and past two side passages that trend southwest. The side passages extend only for about 100 feet before ending in breakdown. Past the second side passage, a 5-foot-tall sediment bank and a low ceiling reduce the walking passage to a 16 inch high belly crawl for 10 feet.

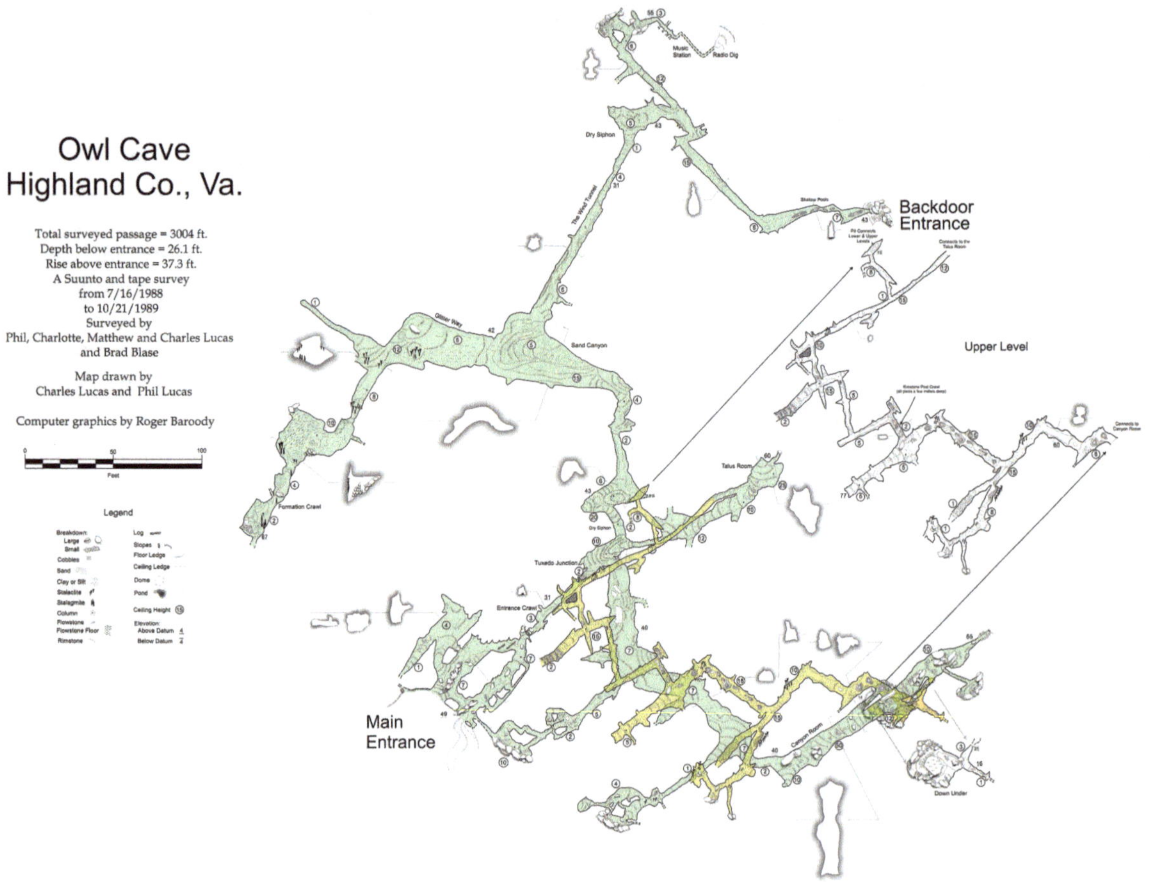

Fig. 11.42 Map of Owl Cave (*photo* by P. C. Lucas)

Fig. 11.43 The Owl Cave entrances as they appeared in June, 2011 (*photo* by P. C. Lucas)

Abruptly, this opens into the tall Canyon Room, trending northeast with a 50 foot ceiling height. This room is 140 feet long and has a steep talus slope of clays and rubble that extends all the way to the ceiling and what once was an entering passage on the south wall. At the top of the talus cone is now an entrance, the Talus Cone Entrance, opened in 2009. The new entrance enters the cave at the bottom of a 30 foot pipe that extends vertically from the bottom of a small sink below a rocky outcrop. The Canyon Room continues past the new entrance intersection and down a stone staircase for 60 feet, where it ends in a breakdown choke. Opposite the new entrance intersection at the top of the scree slope on the north wall of the Canyon Room is an intersection with a passage that is the northeastern end of the upper level maze. At the bottom of the stone staircase is a narrow passage that turns sharply to the southwest and slopes downward and to a 20 foot pit. A rebar rung ladder provides access to the bottom of the pit and the Looking Glass Dig.

To the north of Tuxedo Junction is a passage that slopes down 10 feet to a 16 inch high tube that abruptly turns up a loose cobble slope. The slope has

Fig. 11.44 Log jammed in passage top in owl cave. It has been covered in debris from recent flood. The same log appears in Douglas' book caves of Virginia. Al Grimm is the subject (*photo* by P. C. Lucas)

Fig. 11.45 Al Grimm sliding down a steep canyon floor into Talus Room (*photo* by P. C. Lucas)

large cobbles at the bottom that grade to smaller gravels and then to sands at the top, where the small passage has gradually opened into a much larger room. This is a good example of the three so-called "dry siphons" that exist in this cave. The arrangement is a U-tube. It is not difficult to imagine flood water surging through this U-tube restriction under pressure with enough velocity to flush through all the sediments but grading the material as the current diminishes.

Beyond the room above the "dry siphon", the passage continues as a crawlway for about 40 feet to where it pushes up another "dry siphon" into Sand Canyon. This is a room about 100 feet long with ceiling heights from 6 to 15 feet. A large sediment mound occupies the center of the room. A deep channel has been cut by flood waters moving through the room around the sediment mound and out the Wind Tunnel passage to the northeast. The west end of Sand Canyon turns to the southwest and becomes smaller and lower, finally becoming choked with breakdown. The Wind Tunnel is a walking, then crawling, straight passage for 150 feet. At this point, another "dry siphon" up-slope is encountered. There is much loose material and advancing is "moving up two and back one". Beyond are a larger walking passage and a two-way intersection. To the left, northwest, the passage slopes up and becomes very narrow. After 40 feet, a low area of breakdown is encountered and just beyond is the limit of exploration, the Music Room, into more massive breakdown. On the surface, not far away, on the steep hillside is a crack blowing air between soil-covered boulders. This is called the Radio Dig, where several afternoons have been spent trying to gain a new entrance. It is called the Radio Dig because of the day when we placed a "boombox radio" with its speakers going "full bore" in front of the blowing hole and then entered the cave to the breakdown choked passage until we heard the music—hence the Music Room.

To the right, southeast of the "dry siphon", the walking passage continues for about 200 feet but becomes narrow before reaching the base of a small

Fig. 11.46 Nathan Farrar at the Backdoor Entrance (*photo* by P. C. Lucas)

breakdown choke. At the top is a small hole that can be wiggled through and out—the Back Door Entrance of Owl Cave (Fig. 11.46). This entrance is on the other side of the hill from the main entrances. It is on the hillside 40 feet above the bottom of the large sinkhole that contains the Water Sinks Cave System. During extreme flood events, the Backdoor issues a large stream that tumbles down the hillside to the sink in Fig. 11.47.

11.7.2 The Exploration of Owl Cave

A survey of some parts of the cave was made October 2, 1952 by an unknown party. From that survey, a working map was drawn on October 21, 1952. The name of the cave was listed as "Lockridge Sewer" on the survey notes. This was written in pencil, as was the remainder of the survey notes. However, the name "Syphon" was written in red below the penciled name.

Fig. 11.47 The Backdoor Entrance in flood in 1996 (*photo* by P. C. Lucas)

I believe this was done when the working map was drawn because the name Syphon was used.

On July 14, 1958, Barbara J. Hagen submitted a typed report for Siphon Cave No. 1 with the owners listed as being A. Lee and P.S. Lockridge. Her description reads as follows: "The cave is a siphon for the surrounding water shed and fields. There are two openings next to each other which join about 20 feet inside the cave. The cave consists of an extreme amount of freshly washed-in mud and contains some fresh breakdown. There are quite a few side passages which lead towards the surface and which have considerable wash-in at the beginning of each passage. A pit of about 15–20 feet is at the end of the main passage. This cave is a good place not to be in during wet weather. There are a few old but slow growing stalactites."

Apparently the word Siphon (or Syphon) was used to name what are now known as Owl Cave and the Water Sinks Cave, in which Owl Cave was known as Siphon Cave No. 1 and Water Sinks Cave was known as Siphon Cave No. 2. This was not because the caves had water-they didn't, except for a shallow pool in Owl, in the upper level, backed up by a rimstone dam that is a few inches deep—but because they siphoned water from the surrounding fields and watershed. It is not clear to me what was meant by this siphoning explanation. I assume she meant that the caves were drains for the Water Sinks depression. There is no doubt that during times of flooding, the dry sumps would become actual sumps.

Flooding into Owl Cave is not quite the issue as it once was. A berm has been constructed along Water

Sinks Creek, across the meadow from the main cave entrances. This was done to prevent the creek from overflowing its banks, washing across the meadow, and into the cave. Still, during major floods, the stream overflows the berm, into and through Owl Cave, with considerable portions flowing both to the Backdoor Entrance and to the Looking Glass Dig.

The Backdoor Entrance was discovered in August, 1964 when some Parkersburg Grotto members were visiting both Owl and Water Sinks Caves. They had explored Water Sinks Cave and set out for Owl Cave without knowing that its entrance was on the other side of the hill. One of the party members, L.E. Busch, found a small opening and wiggled into find a narrow, small canyon passage. He yelled to the others who were lagging behind to follow him and set off to explore his find. But his companions did not find this hole or know where he had gone. Busch kept exploring without waiting for them. Eventually, he came out the Owl Cave entrance. By this time, one of the party had gone over the hill to find the main cave entrance and was just arriving when Busch was exiting. Certainly a happenchance discovery/reunion.

In 1958 and 1959, George Deike, with assistance from members of the Nittany Grotto, surveyed both Siphon caves. From these surveys, Deike drew excellent maps, using them, along with complete descriptions of the caves, in his 1960 thesis (Fig. 11.48). After Deike's investigations, visitation to the Water Sinks probably became infrequent. Both

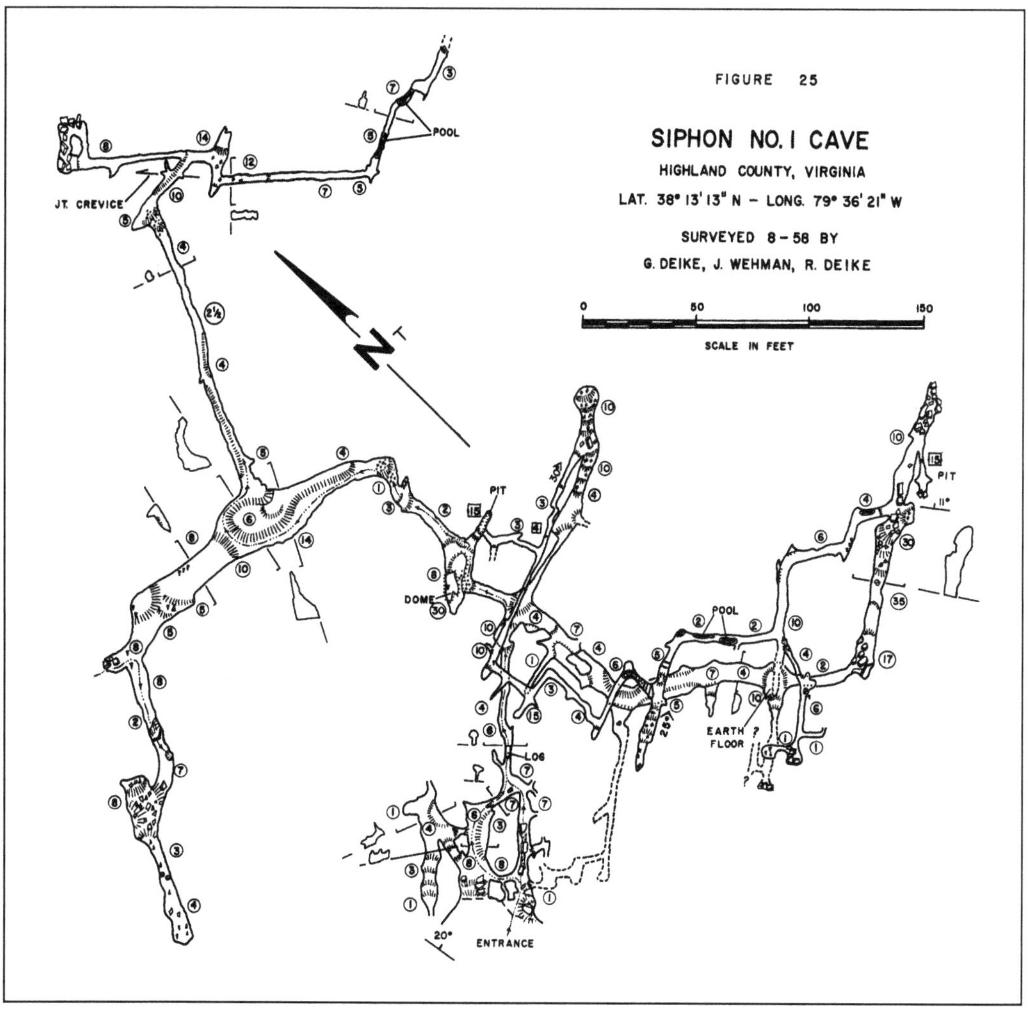

Fig. 11.48 The Deike map of Siphon 1 Cave, now known as Owl Cave (*photo* by P. C. Lucas)

caves were subject to flooding and there was a lot of organic debris washed up against the entrances, especially at the base of the cliff where the Water Sinks Cave entrances are located. This debris created a sort of obstacle course. However, the caves, no doubt, received occasional visits from cavers over the next 30 years, but there are no written records.

The next written record is a deed of sale from February of 1988 when my wife, Charlotte, and I purchased a tract that included the caves. Changes now began to take place more rapidly. The first event involved the names of the caves. The use of the name Siphon was confusing and misleading. The caves were renamed. Siphon Cave No. 1 was renamed Owl Cave when a large Horned Owl was seen roosting in the entrance passage. Also, the twin entrances look like owl eyes when approaching the cave. Siphon Cave No. 2 was changed to Water Sinks Cave—an obvious choice since the USGS had used this name on the topographic maps.

Mother Nature is messy. Charlotte and I foolishly attempt to clean up some of her messes from time to time. The Owl entrances were partially obscured by a scree slope and tangles of limbs and flood debris. So we set about to pull limbs and debris from the cave and made big bon fires. Hiring local contractors we removed the scree slope and constructed a small meadow at the base of a limestone escarpment.

Today, the entrances can be easily seen from across the meadow. The surrounding area has interesting karst features in a pleasant setting, appealing to cavers. In fact, over the years, there have been four wedding ceremonies and one celebration of a life conducted in the meadow.

Owl Cave is an excellent "kids' cave", being mostly horizontal but with enough challenges to keep things interesting. Various youth camp groups and families have enjoyed an afternoon adventure exploring here. The cave has received investigations from those interested in science and conservation such as the Karst Geology Field Trip sponsored by Allegheny Highlands Environmental Council, which visited the cave in 2009. In May 2009, the cave was used by the Navy to train dogs for the Afghanistan War effort.

Besides owls, other animals frequently use the cave. Groundhogs and Alleghany wood rats find the entrance most convenient. Bats find roosting areas further inside, especially in the Sand Canyon. In the spring and summer of 2009, a turkey vulture hatched two chicks in the left-hand entrance area of Owl Cave. The entrance area smelled of rotting meat as the two vulture chicks quickly grew to about the size of a chickens. They made strange noises and fake charges, trying to frighten the cavers away.

11.7.3 Beneath the Talus Cone

About 200 feet inside of the main Owl Cave entrance, the passage divides at a three-way intersection. The right-hand passage is a series of stoop walks and belly crawls that lead to a tall canyon. Spilling into this canyon is a tall cone of scree that enters through a hole in the ceiling. It is a steep slope to climb up and over to continue along the canyon. It is called the Talus Cone. When I first saw it, it seemed clear to me that a connection to the surface must be just beyond the top of this Talus Cone. The canyon passage continues past the Talus Cone but soon reaches breakdown that seals the passage. Just before the collapse is a side passage that cuts back down beneath the Talus Cone to a climbable 20 foot pit. At the bottom are two crawlways that didn't go far before becoming choked with boulders, cobbles, gravels and silt. These sediments were from a fast flowing stream that flowed around the Talus Cone during past flood events. Recently, I have built a berm along the streambed to keep the surface creek (Water Sinks Creek) from overflowing into Owl Cave, so only air currents flow through these crawlways nowadays. (That is, until the April 2011 floods.) The direction these crawls are headed is under Chestnut Ridge and away from any known Owl Cave or Water Sinks Cave passages.

This is a prime location to discover cave between Aqua Cave and the Chestnut Ridge Cave System, potentially connecting them. Owl Cave and the Water Sinks Cave system are tightly contained within Cave Hill, bounded by the large Water Sinks sinkhole. This hill is attached to Chestnut Ridge by a saddle. The Talus Cone Canyon lies adjacent to the saddle and the two crawlways are beneath the saddle, extending east into Chestnut Ridge. Important places need naming. Being a fan of Lewis Carroll, I think "Looking Glass" is an appropriate moniker—a place to pass through into another realm (the cave beyond upstream Aqua Cave) that has eluded our efforts for so long. And so the dig site beneath the Talus Cone is hereby named the "Looking Glass". There are actually two possibilities at the Looking Glass—two crawlways. The right crawl quickly pinches and has less air. The left crawl goes further with more air but no easy place for spoils. I named the right hand crawl "Tweedledum" and the left "Tweedledee". But which way should we dig first? I think Lewis Carroll might have said something like this:

Tweedledum and Tweedledee
Received the queen's directive.
"Find a way to slithy toves
Your choice can be selective."

They whiffled down the talus slope
and came upon a branching
"which way to go" became the show
-a battle most enchanting.

Vorpal blades went snicker-snack
And when the contest ended
Body parts were stuffed in sacks
Their quarrel quite appended.

All this seems a poser be.
Did Tweedledum do Tweedledee
Or was it just conspiracy?

The time has come the Walrus said
To tell another version
The story of the vorpal blades
was simply a diversion.

Tweedledum thought Tweedledee
Had crawled beneath the saddle
Desperate calls for his return
His fright began to rattle

Could shuffling heard around the bend
Be the scaley Jabberwock?
His home invaded by the twin
Now scattered bones among the rock.

'Twas brillig and the slithy toves
Did gyre and gimble in the cave
All mimsy were the borogoves
And the mome raths outgrabe.

If this poem sounds somewhat familiar, it is because I have blended parts of the poem *Jabberwocky* at the end of Chapter I in "Through the Looking Glass", as well as parts of poems in Chapter IV dealing with Tweedledum and Tweedledee and The Walrus and the Carpenter. But most of the poem is my own composition and I hope it would not have embarrassed the Reverend Dobson too greatly.

11.7.4 The Talus Cone Entrance Dig

As I contemplated how to best to accomplish efficient digging at the Looking Glass it became apparent that a new entrance would be most helpful. This entrance would enter the cave at the top of the Talus Cone, which is directly above the Looking Glass dig. It would be a lot of work but once it was completed, only a couple of minutes would be required to reach the Looking Glass dig face, and even more importantly, the transport of tools and equipment needed for the dig would become very easy. This might well make the difference in the success of the dig. So in March of 2008, work began on what was to become the "Talus Cone entrance".

The digging of the Talus Cone entrance turned out to be a very ambitious project but it was eventually completed. Here are some facts, figures, and logistics for the "Talus Cone Entrance" project (Fig. 11.49):

- It began March 15, 2008 and was finished on December 15, 2009.
- A 48 inch, square shaft was dug 30 feet deep.
- Three sides of the shaft are in unconsolidated scree with the forth side being bedrock.
- Accordingly, shoring was needed for the three sides.
- Lots of big rocks were encountered on the way down—all yielded to straws of varying lengths.
- Infrastructure includes: short staircase, platform, ramps, and numerous rock anchors, chain, and rope.

Fig. 11.49 The completed Talus Cone Entrance (*photo* by P. C. Lucas)

- A 2 to 1 advantage pulley system was used to lift 5 gallon buckets of spoils.
- Several 3/8 inch cables supporting large rocks and shoring during excavation.
- Two sections of 36 inch culvert were placed in the shaft as the permanent entryway.
- A 4 to 1 block-an-tackle and a 10 ton come-a-long were used to lower and position culverts.
- A ladder is constructed down the inside of the entrance culverts to allow for permanent access.
- A lid on the culvert restricts the air flow.

Thanks to following people for helping me dig open the Talus Cone entrance: Ed Kehs, Ed Bauer, Al Grimm, Jim Richardson, John Sweet, Tony Canike, Rick Lambert, Tom Blomer, Maret Maxwell, Nathan Farrar, Chris Woodley, Bob Alderson, Frank Marks, Scott Olson, Ben Schwartz, Dave Collings, Russ Carter, Keith Wheeland, and Amanda Martin.

11.8 The Evidence for Potential Connections to Other Caves and Cave Systems

11.8.1 Water

The flood water that flows around the Talus Cone and down into the Looking Glass does not reenter Owl Cave or the Water Sinks Cave. In fact, on the surface during dry weather, the entire Water Sinks Creek sinks before reaching the Water Sinks cliff face and flows directly to Aqua Spring without first flowing through the Water Sink's Emerald Pool, like the Butler Cave and Breathing Cave streams and the surface stream Sinking Creek do. This insurgence of the Water Sinks Creek is located at Castle Rocks, just 500 feet to the south. Castle Rocks is at the end of a line of sinks that presently marks the end of the southeast development of Owl Cave. Apparently at this point, the upper level development of Owl became so large in the geologic past that collapse occurred. During extreme flooding, when the Water Sinks Creek overflows its banks, some of the flood water flows into Castle Rocks and sinks immediately without ponding. Although this sink has not yet been dye traced, undoubtedly, it also flows directly to Aqua Spring. The piracy at the Twisting Sister does the same. If the route of the stream channels beyond the Looking Glass can be followed, then a connection to Aqua Cave is probable.

11.8.2 Air

The winter air flows into the Looking Glass. This air eventually warms and flows out some unknown upper exit. There are no exits on Cave Hill or nearby hilltops that cause snow to melt. Aqua Cave also inhales winter air. The destination of Aqua Cave's air has long been a source of discussion, but I believe it is the Chestnut Ridge Cave System with its high elevation entrances. I think that the Looking Glass air joins the Aqua Cave air in route to Chestnut Ridge. If so, then this offers the possibility of a connection to both Aqua Cave and to the Chestnut Ridge Cave System.

11.8.3 Geology

There is a lineament from the Bullpasture River below Aqua Cave, up Mill Run, and through the Water Sinks feature. This likely represents some kind of structure that is nearly perpendicular to the axis of the Chestnut Ridge anticline. Aqua Cave's passages generally follow this linear feature, as do Owl Cave's, Water Sink Cave's and Helictite Cave's passages. Since dye traces show that the stream from the Emerald Pool in the Water Sinks Cave flows to Aqua Spring, then it is reasonable to conclude that Aqua Cave extends toward the Water Sinks/Owl Cave, beyond what has been explored to date.

There are many of sinks east of Owl Cave. This line of sinks starts at Castle Rocks and extends to the northeast to the top of the hill, or rather what used to be the top of the hill, but is now a large sinkhole. Prior to the discovery of the Subway, my assumption was that all these sinks represented a probable termination or collapse of cave passages in this direction and that finding a way through to Aqua or Chestnut Ridge would be difficult. However, the survey of the Subway reveals several levels exist in the Water Sinks/Owl Cave System. The lowest levels are nearly 200 feet below these sinks above Castle Rocks. Characteristically, these levels are not connected to one another in many places. They seem to be separated by sandy units, perhaps similar to the sandstones in Butler and Breathing Caves. Earlier, it was thought that the Water Sinks Cave and Owl Cave entrances were located at the contact between the New Creek and Licking Creek Limestones. A careful examination and description of the geology is currently underway in this area by JMU professor, John Haynes. The preliminary findings show that much of Owl Cave is developed in the Keyser Limestone and that the lowest level of the Water Sinks Cave is in the Tonoloway Limestone. This would support the distinct development of separate levels. The lowest levels can then easily pass beneath the Castle Rock sinks. As is demonstrated in the known caves, there are probably few if any connections with the upper levels. Ben Schwartz conducted several electrical resistivity measurements in the area of Owl Cave in the summer of 2009. Although he has not yet completed analysis of these measurements, the tentative results indicate the possibility of lower level passages. In short, I believe the geology is good for additional cave extending down along the lineament from Cave Hill to Aqua.

Helictite Cave

12

Philip C. Lucas

Abstract
Helictite Cave was discovered by excavating a small sinkhole above the northeast side of the Water Sinks Depression. There are three main sections: an entrance area maze, the Streamway which is a fragment of master trunk passage, and the updip maze. The cave is formed in the Licking Creek Limestone so there are many chert beds which influence passage morphology. A fault passes through the cave revealed by slickenside surfaces. The cave gets its name from the exceptional displays of helictites found in the entrance area. The cave contains 7.3 miles of surveyed passage and is apparently related to the Emory Spring drainage system.

12.1 Cave Description

Helictite Cave is located 400 feet northeast of Water Sinks on the USGS Williamsville, VA. Quadrangle. It is developed in Licking Creek and New Creek Limestones of the Helderberg Group. There have been just over 7 miles surveyed in a complex maze network of joint controlled passages that extend a lateral distance of 2400 feet to the northeast along a northeast plunging anticline. The cave reaches a depth of 187 feet. At this depth, sediments have filled most of the cave's passages. The cave has two distinctive levels. A large lower level passage developed in the New Creek Limestone extends generally southwest northeast. Apparently, this passage at one time carried a large stream flowing to the northeast. The floor of the passage now has a very gentle gradient and is filled with silts, clays, and small cobbles. The Canyon Maze, an upper level, is a maze complex of joint controlled interconnected canyons that are generally 10–15 feet wide and 20 feet high. These canyons are almost entirely developed within the cherty beds of the Licking Creek Limestone.

The entrance was opened by digging in the bottom of a small steep sided sink on the northeast side of the much larger Water Sinks sinkhole (Fig. 12.1). It is 120 feet vertically above the bottom of the Water Sinks. The cave opens as a small canyon passage that descends in a general northeast direction for several hundred feet where it breaks down through a thick layer of chert and sandy beds of the Licking Creek Limestone and into the more massively bedded New Creek Limestone. Below this point, the cave becomes large and increasingly decorated with areas of speleothems. To the north is a zone of massive breakdown where large areas of slickensides are exposed. Beyond this is a large passage named the Streamway that today only carries a trickle. However, the Streamway appears to have once been the main drain for the Burnsville Cove, but has since been largely filled by sediment. This passage continues 1000 feet to the northeast where the plunging anticline brings the ceiling lower until breakdown is reached in the cherty

P.C. Lucas (✉)
587 Limestone Lane, Williamsville, VA 24487, USA
e-mail: Lucas@VirginiaCaves.org

W.B. White (ed.), *The Caves of Burnsville Cove, Virginia*,
Cave and Karst Systems of the World, DOI 10.1007/978-3-319-14391-0_12,

Fig. 12.1 The steep sinkhole entrance to Helictite Cave. Photo by P.C. Lucas

Licking Creek Limestone. Another 200 feet beyond the breakdown, the passage becomes even lower and finally becomes filled with silts and clays.

The Canyon Maze lies just to the northwest of the Streamway and only two passages connect these two sections. Most of the maze encompasses a rectangular area about 1200 feet by 1400 feet, or about 40 acres. The ceilings of the canyons are generally flat with ripples marks from the time of deposition being prominent in many places. The canyons undulate up and then down along the thickness of the Licking Creek Limestone. The floors are frequently rough and steeply sloping. The walls are often soft with several inches of chert, sand and clays left in place as the carbonates were dissolved. The maze in places is predictable along the joint sets, but in other locations is confusing and has a three-dimensional complexity to confuse all but a slow, plotting survey crew. A small stream flows east along the northern end of the Canyon Maze into Lake Woe-Be-Gone. Upstream, it can be followed for 1800 feet to a room where it flows from beneath breakdown.

A few of the passages within the Canyon Maze are smaller elliptical shaped tunnels. These do not extend very far until they intersect the more typical rectangular-shaped canyon passages. Other distinctive features are large deposits of chert within the Licking Creek Limestone. These chert beds are generally very brittle and weak and have, in several locations, collapsed in large unstable piles. They are black, sharp-edged, and bring an appearance to the passages that look manmade rather than natural. Combined with the layers of chert and sand are outstanding displays of invertebrate Devonian fossils such as brachiopods, corals, gastropods, and crinoids. In places, the fossils that have been separated from the limestone are found in thick layers along the ledges and passage floors.

An interesting aspect of the thick silt deposits is the development of holes being drilled by dripping water. These "drill holes" are a few inches in diameter but can be several feet in depth. One was measured at 6 feet deep. A number of these drill holes are found in the Bullfrog Room, where the water dripping into holes of varying depths produces an interesting variety of notes—the deeper holes create the deeper bass notes. Estimates, based on these notes, place the Grandfather Bullfrog drill hole to be close to 10 feet deep.

Although there are numerous seeps, drips, and trickles throughout the cave, only two tiny streams are found. One of these is last seen in Lake Woe-Be-Gone as mentioned. This stream has been traced (using fluorescein dye) to Emory Spring in the Bullpasture Gorge, 1 mile to the east (Fig. 12.2). The other stream can be seen flowing 200 feet along the Streamway until it sinks into the sands and silt of the streambed in the Bull Frog Room. This has been also traced to Emory Spring. The destination of these streams is interesting because of the close proximity of Helictite Cave to the Water Sinks, which has been traced to Aqua Spring. Apparently, the collapse of the Water Sinks and perhaps the accumulating sediments have effectively sealed any former connections between these two caves. So effective is this seal that when the

Fig. 12.2 A dye trace in Lake Woe-Be-Gone. Photo by P.C. Lucas

Water Sinks is ponding with flood water more than 40 feet deep, none flows into Helictite Cave, which is 50 feet lower in elevation. Because of the area being drained by underground flow routes, it is conjectured that a large, underground stream is flowing beneath or close to Helictite Cave that resurges at Emory Spring.

There are striking displays of helictites, spar, cave pearls, and other speleothems concentrated in several locations. Eighty feet inside the entrance is Helictite Hall, where a dense cluster of translucent helictites extends for over 60 feet along the ceiling (Fig. 12.3). The White Lightning Room has a number of cave pearls. One pearl had a diameter of 1.19 inches and is only 0.009 inch from being a perfect sphere. In the Boot Bag Canyon are long seasonal pools containing large dog toothed spar crystals (Fig. 12.4). Some of these crystals are estimated to be nearly 3 inches long.

Fig. 12.3 Helictites in Helictite Hall. Photo by P.C. Lucas

Fig. 12.4 Scalenohedral calcite in Boot Bag Canyon. Photo by P.C. Lucas

Thick clusters of long soda straws are found in the Attic Room along with delicate rimstone pools, totems, drapery, many helictites, and other outstanding speleothems.

Although the cave was dug open, a lid is normally kept closed to keep the air flow the same as it has been in the recent past to prevent desiccation of those speleothems near the entrance and to prevent destabilization from freeze and thaw processes. Despite the lid being closed, a few Pipistrelle bats (now known as Tri-Colored bats) have been seen roosting in scattered locations throughout the cave. How they enter the cave is unknown. There may be small cracks within the entrance sink that allow them access. In several locations are piles of bats bones, some being covered by calcite deposits. Typical of the Burnsville Cove, there is a sparse number of aquatic fauna. Collections from the tiny streams and drips include two species of amphipod. Terrestrial fauna include millipedes, mites, beetles, and springtails. Tracks of raccoons and other small mammals have been observed in several locations deep within the cave.

There have been investigations of the air currents at the entrance. Generally, with warm outside temperatures, air blows out of the cave's entrance. The opposite is true during cold weather, but with the currents being stronger. However, the cave has a strong breathing phenomenon. Measurements taken with a recording anemometer, along with barometric and temperature recordings, show how the air flow direction and speed change with changing surface conditions. Some of the breathing oscillations take place in a minute or faster and are usually in one prominent direction. At other times, only the flow rate varies and the direction remains constant. There seems to be more than a simple rhythmic change in the flow direction and velocity. With all the data collected, a lot of questions have been raised. One conclusion has been reached—there is a connection(s) at a higher elevation that allows an air flow (convection) to the lower Helictite entrance. A further study of this airflow is planned that will include measurements taken in other caves within the Cove

Helictite Cave has been placed on Virginia's Significant Cave List (Virginia Cave Board) as a very significant cave in the following categories; geologic, hydrologic, aesthetic, atmospheric, paleontological, and in length.

12.2 The Discovery of Helictite Cave

12.2.1 An Entrance Is Hiding

It is a warm, spring day in 1988, soon after Phil and Charlotte Lucas purchased their new property in Highland County, Virginia. Phil is walking along his driveway. On the hillside above him are some outcrops he hasn't noticed before. Scrambling up, he stops at the edge of a fissure 30 feet long and 15 feet wide. About 18 feet below is a rough rocky floor. A ledge on the far side could be hiding a cave passage. He can feel cool air as he slides down into the fissure, but when he bends to check the ledge all he sees is a dirt and rock choke. Shucks! Still, the fissure looks like a good prospect for a dig and Phil makes up his mind that at the first opportunity, a dig it will be.

12.2.2 Bugger Cave

On a hot, sultry July 30, Gregg Clemmer, Tommy Shifflett and Phil escape the heat and start the first dig in the fissure bottom. They choose the downhill side because it pushes more air. This turns out to be a fateful decision. In several hours, they drop 6 feet and the air currents increase.

That summer, the dig continues. A chain and pulley to haul up buckets eventually give way to a boom, so that a block and tackle can raise several buckets at a time. We find air blowing from under the fissure bottom and also from under the fissure wall, 8 feet down. Although digging towards the wall means digging towards the downhill slope, we go that way anyway, as it will make shoring much easier. Soon we have a horizontal tunnel following a pack rat midden. Eventually, the tunnel is 15 feet long. Bats fly in and out.

On September 11, 1988, Phil and Charlie Bill Lucas, Brad Blase and Dave Collings continue the dig. At one point, Charlie Bill sees the headwall quiver. He gives it a kick and the remaining mud and rock topple over into a pit! We are in! The pit proves climbable, though unstable. Surveying yields 300–400 feet of narrow, ugly passage with lots of fossils and breakdown. Through several more digs, we drop to 120 feet below the entrance, but have lost most of the air in breakdown almost directly under the fissure bottom. It was named Bugger Cave but was only an isolated fragment.

12.2.3 Back to Digging

It's August of 1990. Phil hints about starting a new dig on the uphill side of the fissure. He gets a cool response from his friends. Taking the "bull by the horns", Charlotte and Phil start the new dig on August 26, 1990. They choose a joint where winter frost sometimes coats the rocks. Most of the rock is brittle chert that shatters beneath the hammer.

In a subsequent dig, they hit a layer of solid chert about 6 feet down. They move over, following a hint of air. Soon the dig requires shoring and lots of help. Eventually, they reach a more broken surface that really does have air. About 15 feet down, holes start to open up and then close as debris falls back in. Once Ben Schwartz hears rocks falling away and asks for a belay. Sure enough, a larger hole soon opens up below. But any celebration is premature. They chase the hole down 20 feet and over-reach the shoring. The dig looks scary.

They widen the hole. Building a wooden platform over the pit, they start down again, this time with plenty of shoring. The entire project takes place over a period of years when the right weekend puts together the right amount of people who want to dig. All in all, there are at least 15 major digging sessions. To all those diggers, Thank You!

On March 9, 1996, we return and find increased air movement. Something has moved or collapsed below.

After the day's digging, we feel we are close to a breakthrough. Ben asks Phil if he can dig some more the next weekend, a BCCS Pancake Weekend. Phil said yes, but that he will be in West Virginia. Ben teases Phil, "OK, we'll only scoop a couple thousand feet before we turn around."

12.3 Initial Exploration of Helictite Cave

The following log chronicles the exploration of Helictite Cave. In an attempt to capture the history of the cave's exploration, the explorers wrote a recap in a logbook after every trip.

12.3.1 Saturday, March 16, 1996

Ben Schwartz, Cori Giannuzzi, Ed Kehs, and Kim Kehs
91 Feet Surveyed
456 Feet Total (including Bugger)

The last nine buckets of material are removed and Helictite Cave is entered for this first time. But the diggers have a problem: Phil is not around (he's caving in the Culverson Creek Cave System). The diggers survey 8 stations to determine that the cave does indeed "go". The weather is cold and the cave is taking air with a strong current. The roof of the cave 40 feet inside has an array of helictites and more can be seen beyond. Ben calls and gets a message through to Phil. Phil returns to the Water Sinks that night.

12.3.2 Sunday, March 17, 1996

Phil Lucas, Ben Schwartz, Cori Giannuzzi, Nevin Davis, Tommy Shifflett, Jean Hartman, Gregg Clemmer, John Rosenfeld, Bill Royster, Tresh Geiger, Keith Christenson
604 Feet Surveyed
1060 Feet Total

The news has gotten out. We have plenty of help (Fig. 12.5)! We survey down into Helictite Hall. A right turn leads into a narrow, cherty fissure. Phil and Nevin find an upper way through. The room beyond has a 30 foot ceiling (Fig. 12.6) and slopes down for 50 feet to the top of a 50 foot pit. Phil downclimbs the pit. At the bottom, a canyon extends in two ways.

Downstream, the passage fills with sediment after 70 feet. Upstream, the canyon steadily climbs and reaches an intersection of highly fluted passages with ceiling heights of about 50 feet. The air must disappear into the ceiling somewhere. Exactly where is still undetermined.

12.3.3 Tuesday, March 26, 1996

Ben Schwartz, Cori Giannuzzi, and Phil Lucas
394 Feet Surveyed
1670 Feet Total
4 h Trip

We survey each lead going in from the entrance and each one peters-out. Soon, there are only two leads left to check in the cave. The first is a breezy crawl, nearly choked with cherty ledges. Ben goes in with a hammer. Phil is touching up his notes when he hears Cori say, "Phil may have a little trouble getting through here." Phil decides it's time to activate plan B. Going back to Helictite Hall, he climbs up into the last lead in the cave. It has air also. Squirming through a forest of speleothems, he comes to the top of a small pit with helictites and flowstone on the ceiling. Seeing blackness beyond, he squeezes through a small window into walking passage!

He hears Ben hammering down below. Yelling, he asks Ben if he can see any upper passage. The answer is "no." In fact, Ben cannot understand where Phil's voice is coming from. "What's the passage like up there?" Ben asks. "Walking passage that goes!" yells Phil. "Then why am I down here in this tight crawl beating at the walls?" Ben responds.

Phil returns and they all survey the upper passage across the little pit. The character of the cave changes. Breakdown comes in on the left wall and slickensides mark the ceiling. After 50 feet, a slope leads up into the Cruddy Room (Fig. 12.7). The walls, floor and ceiling are broken and dangerous-looking. The rock is nearly all chert. The prospects of finding a way out of this room seem bleak, but Ben does find a hole that needs another look.

On the way out, Phil climbs down the well where he had heard Ben. He looks right into the top of Ben and Cori's crawl. They had been just beyond this point and hadn't seen any overhead passage from their position. This turns out to be a better route through the cave (Fig. 12.8).

Fig. 12.5 April 8, 1996. The second survey party into Helictite cave. Everyone was anxious to go in. Photo by P. C. Lucas

Fig. 12.6 Helictite Hall leads to a narrow canyon that opens into a room leading to a 50 foot pit. Al Grimm in photo. Photo by P.C. Lucas

Fig. 12.7 The Cruddy Room. Photo by P.C. Lucas

12.3.4 Wednesday, March 27, 1996

Ben Schwartz, Cori Giannuzzi, and Phil Lucas
641 Feet Surveyed
2311 Feet Total
5 h Trip

We feel lucky. When we crawl back through the short cut we found last night, Ben finds a huge cave pearl off in a small muddy recess (Fig. 12.9). We christen the little rain well the Cave Pearl Drop. Thanks to Ben's

Fig. 12.8 The crawl passage. Photo by P.C. Lucas

Fig. 12.9 A nest of cave pearls. Photo by P.C. Lucas

sharp eye, the pearl is discovered before it is completely stained with mud. After an unusual method of cleaning (Ben pops it in his mouth, swishes it around a bit and spits out the mud), we are amazed at its size and roundness. It later measures 1.19 inches in diameter and is within 0.009 of an inch of being perfectly spherical.

Back in the Cruddy Room, we cautiously approach the hole that Ben had seen. It is surrounded with a lot of loose rocks. Ben kicks a couple of times and hundreds of pounds of rocks slide down. Then Ben either does a very brave thing or a not-so-smart thing. He slides on through the Cruddy Hole amid the clatter of sliding rock. But soon he returns and says "It goes!" After grooming as much as we can from the Cruddy Hole, we too carefully slide down through and continue with the survey. The cave branches and begins to open up as it follows down the 20° bedding.

Cori is on lead tape and disappears around a corner. Soon we hear strange laughter, a cross between a giggle and a cackle. "I've never heard Cori make a sound like that before!" Ben says. (Ben and Cori were not yet married). And when Ben and Phil catch up to Cori, blackness greets their eyes.

The cave has opened up in a big way! They stand in a large room with a scree slope stretching up behind them and down below. Ceiling heights are 50 feet. Across the scree slope, they come to a bridge between two drops. The right hand drop is a portal looking down into a big room. A 20–30-foot-long white flowstone speleothem hangs from a ceiling nearly 60 feet high (Fig. 12.10). It resembles a streak of lightning in the glow of our carbide lamps, so we call this the White Lightning Room. Ben is so full of enthusiasm that he decides a permanent survey station is needed. He grabs large rocks from the scree slope and builds a 5 foot high rock cairn with a five-sided rock on top. On this top rock he boldly prints "74" on each of the five sides so that it can be seen from any direction. This at least this gives Phil time to catch up with his sketching.

We survey down over large breakdown blocks, crawl under a ledge, and pop into the bottom of the White Lightning Room. We find wonderful pools of calcite crystals, cave pearls, and white flowstone. Some of the pools have upside-down dogtooth spar and some have right-side-up dogtooth spar. Beyond, stands a large white stalagmite (Fig. 12.11). Another "under the wall" crawl leads us to the Sand Room with its "beach" of white sand. Drill holes from dripping water are prevalent in the Sand Room. Some of the drill holes are only an inch wide at the top, but more than 40 inches deep.

A portal from the Sand Room peeks into the Pancake Room (it has flat topped or pancake-like stalagmites). The floor is nearly covered with flowstone and a beautiful drape adorns the opposite wall. What might be best described as a "birdbath" stalagmite stands with other stalagmites too close to the portal (Fig. 12.12). We will wait to survey into this room until a safer way is found.

Instead, we survey down a short passage and discover a nice white flowstone stalactite above a reflecting pool. To the right, we find a floor of red flowstone and then another room with a white cascade flowing from an overhead passage. Phil boosts Ben up

Fig. 12.10 Wall decoration—the "streak of lightning". Photo by P.C. Lucas

Fig. 12.11 The white beauty. Photo by P.C. Lucas

a breakdown block to confirm that there is a passage up there and we leave a sling for the return trip.

12.3.5 Thursday, March 28, 1996

Ben Schwartz, Cori Giannuzzi, Nevin Davis, and Phil Lucas
456 Feet Surveyed
2767 Feet Total
4 h Trip

We climb the large breakdown block using the sling that we had left on our last trip. We begin our survey.

Fig. 12.12 The Birdbath. Photo by P.C. Lucas

At the top of the initial pitch, we have to boost ourselves up into the Attic Room. A very difficult traverse through a forest of speleothems is required to reach the other side (Fig. 12.13). The speleothems in the Attic Room are exquisite (Fig. 12.14). We decide that they should only be visited for photo trips. Unnecessary traffic will take its toll.

Ahead, a slot canyon leads down about 30 feet into a well where a steady shower rains into a pool. Ben pushes a crack at the bottom, but it's too tight. Following the far wall of the Attic Room, we survey into a downward crawl that has a breeze. After about 120 feet, this passage opens up into a room with a finger-sized stream. This stream flows southeast into a small sump with no air space. We wonder if drier weather would make the sump negotiable. On the southwest wall of the room is a crack filled with breakdown. The air must be moving through this crack.

Fig. 12.13 Soda straw stalactites with beginning helictite development. Photo by P.C. Lucas

Fig. 12.14 The Attic Room. Photo by P.C. Lucas

12.4 Exploration of the Streamway

12.4.1 Monday, April 22, 1996

Ben Schwartz, Cori Giannuzzi, and Phil Lucas
1266 Feet Surveyed
4130 Feet Total

Our party splits up to find a way through the breakdown that has stopped our progress to the northeast. Ben heads down a crack. Phil climbs up to a slanting fissure. Ben hollers that he has found his way through and for everyone to join him. Phil yells back that he also has found passage and perhaps that instead they should join him. Ben yells back and says that he really has found something nice that we will all like and that Phil should climb back down. Phil yells back that it had better be good and reluctantly and gingerly climbs back down.

Ben's passage quickly becomes a squeeze down into a crack and then a crawlway. Phil complains that his lead is much better than this. Ben says to just keep coming. The crawl opens into a window. We chimney down into a nice room with a large boulder in the middle. While Ben and Cori set a survey station on top of the boulder, Phil climbs up another fissure. Blackness looms above. The top of the climb opens onto a small ledge overlooking a large passage (20 by 40 feet) that stretches as far as a carbide light will reach. Phil hollers down to Ben that he is in the wrong passage. The really big one is up here. Quickly the other two scramble up and we survey down the "big one". Unfortunately, 100 feet later, the big one reduces to a sandy crawl. But soon, it leads upwards and gets bigger until it is quite respectable again. Finally though, it comes back to the large boulder with the survey station. We had come full circle.

We find a sloping fissure that intersects the ceiling/wall. Climbing up, we recognize slickensides on both the ceiling and floor. This passage appears to be on a fault, with the lower block having dropped several feet to open it up (Fig. 12.15). None of us had seen such a large area of slickensides before.

We survey straight up the slickensides until we reach the fracture where the lower block has broken and fallen. We can see down through the fractures into what looks like larger passage. Excitedly, we survey down the fracture zone. Phil keeps looking to the left to where he thinks the lower passage should be. Finally, around the corner, a walking passage is visible. Phil goes a little further down. Quietly he says, "Uh-oh!" From behind, Ben asks, "What did you say?". Phil says "Uh-OH!!" louder. "What does that mean, Uh-OH?" asks Ben. So Phil responds, "Just come on over here and you'll see." Of course, Ben and Cori scramble to Phil's side and stare into the blackness that stretches beyond. Ben runs down the last of the slope. "Stream Passage! This is stream passage!"

It is stream passage, but with no real stream anymore—only a little rivulet the size of a finger. The stream bed is silt, smooth and flat, with a braided pattern on it that looks like an aerial view of a large river delta (Fig. 12.16). The stream banks are made of very fine silt and are steep or vertical in many places.

Looking upstream, we can see breakdown, but downstream to the northeast appears to go on forever! It's late already, but of course we keep surveying, cursing our 50-foot tape. We take care to make only one set of tracks. Eventually, we leave the stream bed and survey up a mud bank for 50 feet. There, we rejoin the main passage in a large room, over 100 feet across and 40 feet high (Fig. 12.17). Water dripping into "drill" holes of different sizes and depths makes strange sounds, like water dripping into tin buckets. Or like Bullfrogs (Fig. 12.18). We have found the Bullfrog Room.

To the right is a neat "mud city" formed by water dripping onto mud. Each little pillar has a little capstone, much like the pillars in Bryce Canyon. Ahead, the passage still reaches into darkness and we keep surveying, still cursing our 50-foot tape. Finally, the passage character seems to change, so we use this as an excuse and call it quits. It's after 3:00 a.m. Ben and Cori have to go to work in a few hours. The direction of this "paleo stream passage" is to the north-northeast, which takes it right into the hill. Our immediate conclusion is that this passage must have been the original drain for Burnsville Cove.

Fig. 12.15 The Slickenside Room. Photo by P.C. Lucas

Fig. 12.16 The Streamway. Photo by P.C. Lucas

Fig. 12.17 The Bullfrog Room. Photo by P.C. Lucas

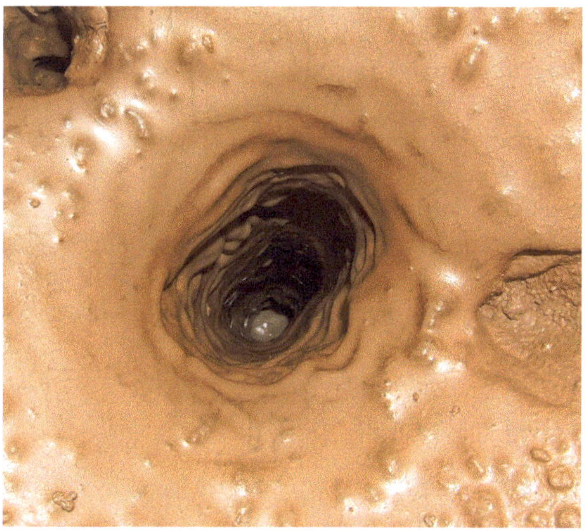

Fig. 12.18 Water drip drill holes in the mud. Photo by P.C. Lucas

12.4.2 Tuesday, April 23, 1996

Ben Schwartz, Cori Giannuzzi, Phil Lucas, and Nevin Davis
721 Feet Surveyed
4972 Feet Total

When the survey party reached the stream passage, Nevin simply said "Well, I'll be damned"! And he turned to me and said "How do you explain this?" Resuming the survey, we found that the passage character was indeed changing. The passage first narrowed considerably, then entered an area of breakdown. Ben finally found a way up through the breakdown and we found ourselves in another Cruddy Room, so we called this one the Cruddy II Room. Working around the northeast parameter of the room, we came to a hole that sloped steeply down into a canyon-like passage that opened up into a smaller version of the stream passage. This passage ended in a silt fill after about 100 feet. There were some short side leads to be surveyed but there appears to be no going passage northeast of the Cruddy II Room.

12.4.3 Wednesday, May 1, 1996

Ben Schwartz, Cori Giannuzzi, and Phil Lucas
631 Feet Surveyed
1.04 Miles Total

We moved down past the Slickensides Fault Room, starting our survey at the animal tracks. These tracks had been spotted during our last trip and are located near where Phil said "Uh-oh". There were two animals that made tracks and we identified the larger set as being a raccoon. The smaller tracks are not as clear as the raccoon's but they could be from a squirrel or rat of some type. We spread flagging tape around the tracks so we will not crawl over them accidentally.

The survey headed toward the passage that they had seen below during the first discovery of the Slickensides Fault Room. The passage leads across large breakdown to a point looking down into a large room. On the right side (northwest) a high upper passage can be seen. In fact, there may be two passages in the ceiling on northwest wall of this room, but both will require some technical assistance to reach. The far end of the room ends in collapse, but in the bottom and beneath the upper leads is a small passage that leads out into the Uh-Oh Passage upstream from our original survey. We tied into this and attempted to survey upstream along the stream level. We were only able to survey approximately 125 feet before we encountered breakdown that we could not penetrate. There are small scallops of the walls in this area. Just before the end was reached, we discovered another fault. This slippage comes up from an about 15° angle. Up-slope and just down from the animal tracks, we also observed a large fold that bent the rock strata into an S shape. We assume that this must be a drag fold associated with the faulting in this area of the cave. Looking up this fold, we can see a passage taking off along the folded bedding in the ceiling. Again, this will require a climbing pole or some type of aid to follow this lead. As the Slickensides Fault Room joins the Uh-Oh passage, there is an area of massive breakdown and there are many large cracks that lead down into spaces below. We decided to check these out and Phil followed a sandy crawl until it was plain that we needed to survey. We went through a series of sandy crawls until we reached a point where darkness can be seen above. Climbing up through breakdown we found ourselves in the large passage that has the slickensides intersecting at the ceiling. This was a lead that we had somehow missed when we surveyed it two trips ago. Apparently, in the past, water entered the area we had just surveyed from into this room and was the reason for all the sand we had been seeing. Coming into this passage from an overlooked lead, humbled us a little and made us wonder what else may have been overlooked.

On the way out, we took a little time and placed a half stick of explosive in the still too-tight bypass to the Cave Pearl Drop. The explosive had to be placed behind a flowstone wall that was making the constriction. We used sand in plastic bags as a mud pack to contain the charge. The electrical blast wire was then unwound back to Helictite Hall, where we connected the wire to the blaster's terminals. The results were good.

12.4.4 Friday, May 24, 1996

Dave Hubbard and Phil Lucas

Dave and Phil spend most of the day picking their way through the cave, carefully scanning all the pools, looking under rocks, and pondering geology. Dave collects springtails, isopods, amphipods, and a flatworm. Dave submits a written proposal that Helictite Cave be included on the very significant listing of the Virginia Significant Cave List.

12.5 Exploration of the Canyon Maze

12.5.1 Sunday, May 26, 1996

Beth Billman, and Charlie Bill, Matt, and Phil Lucas
749 Feet Surveyed
1.31 Miles Total

So far we haven't had any luck getting side leads to "go" off the Streamway. Today, we begin surveying at the next lead upstream from the last survey. Our first passage goes nowhere but does contain some outstanding crinoid stem fossils standing out in bold relief.

The next lead is near the Bullfrog Room. We first come to the bottom of a dome pit about 25 feet high with possible leads at the top. It will be an interesting climb. Next is a side passage to the right that leads to a room going nowhere.

Continuing ahead (north), we slip through one short crawl and then another and come up into walking passage. We survey up a long, steep slope, very similar to all the side leads off the Streamway. But this time, climbing to the top of the slope…there is passage! Soon, a side passage looks into blackness. But the way ahead looks too good! We come to another intersection where collapse is coming in from something to the right. There is also a nice left-hand lead.

Peeking up the scree slope, we see the blackness of a big room. Very excited now, we survey into this large room. We'll call it Chert Hall (Fig. 12.19). The passage continues down to the north and then goes back up another slope, where we come to another intersection. All directions look good! We have run out of time but are happy to have broken out into major cave again.

12.5.2 Sunday, June 2, 1996

Bill Royster, Ben Schwartz, Cori Giannuzzi, and Phil Lucas
1244 Feet Surveyed
1.54 Miles Total

First, we did some mop up in the Bullfrog Room. This is the first time that we have been on the far side of the room and we notice a high dome pit. It's between 50 and 70 feet high and looks to have passage at the top. We find another passage high up the back wall. Access to these leads may be difficult.

As we make our way to Chert Hall, we anticipate crossing northwest over the anticline. The bedding is getting flatter. We survey north in a big passage that ends after only 120 feet. We back up to a lead heading northeast, a nice walking passage that seems to have an intersection every 50 feet. We stay straight until we come to a Y. Each way looks good. We go left, walking and hitting intersections for another 100 feet. We can see that side passages to the right are going into a large passage paralleling us.

Fig. 12.19 Chert Hall. Photo by A.N. Palmer

We start having trouble with the floor. There are holes in it. Not deep, but one finally stops us and forces us to find another way around. No problem. We go back to the big passage we've been paralleling. Then we see speleothems. They start on the floor, but we manage to work around them. Soon they span the floor, but again we can use another passage as a bypass. Finally, we reach a beautiful area with wall to wall rimstone filled with wonderful crystals. Ahead, it's even prettier. We can't continue without doing serious damage. So it's back into a parallel passage. This time we have to back up a good ways and find ourselves in a large breakdown room. Climbing down into a small passage, we are again heading north! Although the walls are washed clean, they have many crystals on them. Soon, we come to a dried up pool with calcite rafts lying askew on the bottom. Ahead is a shimmering area looking like large crystals of brown sugar. Finally we come to water! The pools are 1 to 3 feet deep and have large calcite crystals covering all the rocks and walls—a beautiful sight to behold but frustrating for exploration. The passage is narrow enough to straddle and Phil crosses the pool, but he is stopped after 20 feet at the next intersection, where the passage widens and the flowstone spreads. He can't continue without damaging the speleothems. We'll have to find another way.

12.5.3 Tuesday, July 2, 1996

Ed Richardson and Phil Lucas
1016 Feet Surveyed
1.81 Miles Total
10.75 h Trip

We measure the depths of some of the drill holes in the Bullfrog Room. The deepest is 47 inches, but we can tell that there are even deeper ones by the sounds we hear. Then it's on into the Canyon Maze, where we start surveying in the Big Right Hand Passage. This canyon is typical, with a chert ceiling, crumbly walls and billions of fossils. We name one area Fossil Hall. We need a fossil expert. At another location, fossils are "mounted" on the top of rock pillars, similar to the mud cities in the cave. In yet another spot, we find a striking mud city that has very high pillars. In the southern-most passages, crystal needles line the walls.

The canyon passage keeps dividing and we leave many leads. This area will truly turn out to be a maze. We head up the strike, the passage gradually lowering and then ending in breakdown. Further penetration could be made by moving rocks.

The canyon near the Crystal Pools Passage has a lower level which becomes a separate passage. We look down into it but do not enter. This passage should be surveyed next, as it might lead around the speleothem areas that have stopped us so far.

12.5.4 Sunday, July 14, 1996

Dave Collings, Dave Socky, and Phil Lucas
932 Feet Surveyed
2.06 Miles Total
7 h Trip

We try to go up the Crystal Stream Passage, but the first two attempts end at breakdown chokes. So we climb up near the top of the breakdown and find a way down between blocks and finally a short climb down into a smallish canyon that displays some vadose action. So we continue our survey up this canyon going apparently upstream. It's hard to tell because it has been a long time since any actual water has flowed down in this canyon. There are a few side passages but fewer than normal for this section of the cave. This may be because of the vadose nature of the canyon where most would-be side passages are filled with sediment. Fossils are abundant through this passage.

After a lot of short shots and "up and downs" in this sinuous passage, we suddenly encounter an abrupt 90° right turn. Then, the passage floor becomes sandy and in 36 feet, we reach an intersection of a larger passage. To the left, the passage quickly ends with a rapidly dripping pool surrounded with firm mud and white sand. To the right we go down a good-looking passage that is wide and has a rock covered floor. At the first survey station is an area of breakdown that we later determined to be the source of the water that enters this area. Beyond this is a room that has the peculiar blisters on the ceiling. We have seen this in other sections of the Canyon Maze but this has a most striking appearance so we will call this the Blister Room. At the lower end of the Blister Room, the floor slopes down steeply in a narrow slot where it is necessary to kick rocks ahead of us as we descend. Then the passage climbs back up sharply, but with chert ledges that provide some uncertain purchase. Two survey shots later is a small room where the passage makes another right angle turn to the northeast and goes as a rocky crawl for as far as we can see. This crawl averages 10 feet wide and 2 feet high and descends at an angle of 7°, which is the dip of the plunge of the anticline. The ceiling still has those blister-like bubbles so this will be the Blister Crawl. We surveyed nice, long 51 foot shots and reach the end after 150 feet where the firm mud suddenly becomes flat and therefore reaches nearly to the ceiling. There are a couple of inches above the mud where the passage can be seen to have perhaps reached an intersection with another passage. But the space we are looking through is too small to really see enough to make any firm conclusion. A dig at this point may be useful but more knowledge about other nearby passages would be useful first. At 192 feet, this is the deepest point we have yet reached in the cave. This is about 30 feet lower than the end of the Streamway.

Time to go, so we head back and reach the Chert Hall. We cannot resist the temptation to set several stations down the right hand lead that we passed up on the way in. At first, the passage gradually gets smaller and then becomes a crawl for 30 feet or so. But after the third shot of our survey, we can see big passage ahead. Sure enough, we come to a big intersection where three big passages are heading to the northwest, southwest, and southeast. Damn—what a time to have to stop our survey, but then it is nice to know that we have something to go back to.

12.5.5 Friday, August 2, 1996

Wade Berdeaux and Phil Lucas
441 Feet Surveyed
2.23 Miles Total
8 h Trip

Back in the Canyon Maze, at the Left Hand Formation passage, we put on Moon Suits (suits over our coveralls to keep the speleothems clean). We survey over the speleothems, but our next shot crosses another speleothem area! We lay down a sheet of Tyvex this time. Ahead is yet another massive speleothem. Again we use the Tyvex, surveying to the southwest.

At a speleothem-free (!) intersection, we turn and survey northwest. In one shot, we reach the most massive speleothem yet but are able to creep down one side using some Tyvex. To the left is a 7 foot flowstone waterfall (Fig. 12.20). Using a large sheet of Tyvex, we climb up the flowstone falls, only to reach another intersection with a large pool of water and crystals. We turn right into a passage that has a sandy floor with ripples showing a flow down dip (4°) to the northwest. Like most of our northwest-trending passages, this one is encrusted with needle-like formations. Passing more speleothems, we reach a passage with encrustations and holes in the floor. Ahead is an intersection that looks big. Unfortunately, a 15 foot hole sits just before the intersection and we have no vertical equipment. A short cable ladder would get us down, but climbing up on other side will be difficult.

Backing up to the previous intersection, we find a southwest-trending passage that quickly leads to another pit. Stuck again! Again we back up, and moving past speleothems, find a nice lead that has good air movement. But a white stream of flowstone goes into this hands-and-knees crawl. There is simply no way to continue without extreme damage. This passage will stay virgin!

Fig. 12.20 The Frozen Waterfall. Photo by P.C. Lucas

12.5.6 Friday, October 4, 1996

Dave Socky, Dave Collings, and Phil Lucas
1485 Feet Surveyed
2.52 Miles Total
9 h Trip

We head straight back for the left-hand lead before the Chert Hall that we had surveyed four shots into on our last trip. We have three passages to chose from and, after some discussion, we take the right one. This is heading toward the road and seems to have the best promise to bypass the speleothems. After a steep climb down and then a steep climb back up, we come to an even steeper slope down where climbing can be tough. Across the top, at the ceiling, can be seen what seems to be the passage continuing as a crawl. The ceiling here is the layer of chert called the Bubble Layer. Phil picks his way down to the bottom and finds a very sloppy coating of fossil mud. There is a small fissure that leads down from the bottom but it is tight and full of sloppy mud that wants to slide down in front of you, blocking the way. The only thing intriguing about this is an echo indicating a volume ahead. Phil starts the climb up the other side but soon stops when the slope becomes vertical. This will probably a serious climb where a top anchor is used for the climbdown. It will wait.

So we return to station 436 and start a survey down the next biggest passage to the southwest. This passage looks good! It is 10–20 feet wide and has 15 foot ceilings. Like most passages, it has some sort of intersection about every 50 feet or so. After five shots and a short hands and knees crawl, we come to a large Room 40 wide, 60 feet long, and 30 feet high that we call the Billabong Room because of a tune that a happy Dave Collings has been humming. This room has six passages leading from it and we take the one heading straight across first. This passage goes through a short crawl into a Cruddy Room type passage that continues across large areas of broken chert where the bubble ceiling has collapsed. Some of the "rooms" in this broken stuff have some size but we can't get comfortable being in such an unstable area. At station 499 in the Cruddy Rock Room there is a climb down that we did not do because we assume that we could come to the bottom of it another way. This turns out to be wrong and the lead remains. At station 501, Phil finds a crawl going up a crumbly slope beneath a crumbly ceiling. This crawl would have been dismissed immediately if it were not for the nice flow of air blowing into the crawl. Phil slowly and carefully inches his way up the slope on his belly trying not to touch the ceiling. Finally after about 20 feet, he comes to a flat area where the crawl continues getting slightly larger. Then the passage goes down but the floor is still broken chert and the slope is at such a steep angle that a rock slide can easily be started. Phil finally and very carefully down climbs the slope and continues on down the passage of broken chert. Quickly, the passage becomes collapse again with several possible crawls and climb-ups but Phil calls it quits, figuring that he has pushed his luck enough for one day.

Retreating back to station 494, we branch our survey to the passage that heads down and to the northwest. After two shots, we intersect a passage heading southwest. This passage floor is interesting because of white sand and pools. Phil thought he could see some sort of isopod tracks in one of them. Further down the passage, we encounter drill holes and more white sand. This is a nice looking passage. It has solutional-looking contours. Hopefully it will go for a ways. In three shots (120 feet), it abruptly ends in fill and collapse.

So we go back to the Billabong Room and branch off station 489 into a passage going down and to the northwest. In two shots, we intersect another passage with white sand. This passage also has sparkles on the walls. To the southwest, the passage goes 30 feet and then turns southeast into a very narrow slit full of sparkles. To continue down this passage it will be necessary to dig. To the northeast the passage narrows to a window squeeze that Phil digs open a bit to make his passage easier. Collings, of course, said that this certainly isn't necessary for him and complains about landowner rights while this process takes place. Beyond the window is an intersection where, to the northwest, another window can be seen. This window is much larger, however, and leads to a large passage full of interesting features, including: drill holes, mud formations and fossil beds. We call this area Ludwig Castle passage. To the northeast, we come to an intersection where a pit is encountered. It may be possible to down climb this but it is uncertain if one can climb back up once down. A climb up the southeast leads to a short mud-floored crawl that ends in 50 feet. Returning to the Ludwig intersection, we survey to the southwest. The passage narrows and then reforms into a confusing area of vertical canyons with upper levels. Phil climbs up some of these but can make no real progress. We retreat to the window intersection and survey through the passage to the southeast. This passage turns back up into the Billabong Room and we tie into station 488.

12.5.7 Monday, October 14, 1996

Matt and Phil Lucas
452 Feet Surveyed
2.61 Miles Total
8 h Trip

We traveled into the canyon maze to station 436 where the last trip had begun and started our survey into the crawl to the southeast. The passage was following a straight line but, almost immediately, we encountered holes in the floor that required climbing down and back up, time after time. In 140 feet, we came to a reasonably-sized room where our passage going straight ahead seemed to end, but to make sure it will require someone to climb up about 30 feet. There is a passage going southwest but it is a tight hole that will require some digging or rock removal. To the northeast is a down-sloping passage that levels out and comes to a crawl. This crawl has sticky gooey clay in it and after about 10 feet, the clay is mounded up into a hump that will require some digging to make further progress. At this point, an echo of a larger space can be heard ahead and there is a current of air flowing through this point. Definitely, we need to return to this lead.

We then backed up and traveled to the Billabong Room to station 487. The lead that Dave Collings had located last time did not look all that good to Phil. Before starting the survey, he asked Matt to check it out to see if it really went. Matt crawled on in and soon came to walking passage. So it's "off to the races" again! Matt's passage heads southeast for about 100 feet and then turns southwest in a sandy crawl that quickly opens up into a nice size passage that runs straight for nearly 200 feet where it ends in sediment fill. There are two promising leads near the end of this passage. One lead heading west will need a hammer but the lead to the southeast needs surveying! This lead looked to be blocked by a large rock. Matt pulled at this rock and it fell towards us revealing a passage beyond. Something to return to!

12.6 Breakout!

12.6.1 Sunday, November 3, 1996

Ben and Cori Schwartz, Nevin Davis, and Phil Lucas
1601 Feet Surveyed
2.91 Miles Total

We are going to drop a pit off Chert Hall. The rock is solid and we quickly place a bolt. The 20 foot ladder doesn't reach (6 feet short) but we can step off to get down. The pit is dead bottom with fluted pendants that look like daggers. Dagger Pit (Fig. 12.21) seems to have a passage at the top, but that's another 15 feet above the top of our cable ladder.

There also seems to be a passage on the same level as the top of the ladder, but around a corner to the right. To reach it will require long legs, a belay, and a bit of nerve. Nevin looks at the lead and says that it doesn't look like it goes. He probably says that just to goad Ben, who immediately ties on a belay and scampers across.

Fig. 12.21 The daggers of Dagger Pit. Photo by P.C. Lucas

Ben is gone a few minutes, then returns, saying to Phil that he is going to like what is ahead. But how to get the rest of us across? Ben does some mud excavation and works his way down the pit to set a bolt for a 15 foot cable ladder. This time, the end of the cable ladder is only 3 feet off the floor. The two cable ladders are only 6–8 feet apart. We must descend one and climb the other. By the time we reach the top we are all slimed.

After a few stations, the ceiling takes on that flat blister surface. The passage gets bigger! This area is joint controlled and every 30 feet or so there are side passages. After 100 feet, the passage opens further. We call this the Big Bird Room because of the white calcite splatters on the floor. They look like the droppings of huge birds! Around the corner are more speleothems, but nothing that we can't avoid (Fig. 12.22).

There are side passages everywhere! Soon, we come to the bottom of a clean-washed, 25 foot dome. There may be passage off of the top but we have better leads. We back up one station and head south, climbing into a room where breakdown forms a balcony. We see an unusual grouping of white to translucent stalactites attached to the ceiling by very thin stems (Fig. 12.23). They all line up on one side of the room for us, so we call this the Convention Room. Down below in the room are flowstone "flows." This area is absolutely beautiful (Fig. 12.24).

Fig. 12.22 Calcite pool crystals. Photo by P.C. Lucas

Another 100 feet down the passage there are pools with large wedge-shaped crystals growing along their edges. With the many side passages, it's difficult to choose which way we want to go. At one point, we come to a division we call The Triplex because of the three equally good looking passages to choose from.

After several more intersections, we pass a side passage with a sandstone cobble stream bed. We continue in our passage and come to a 10–15 foot pit that will require a rope. Backing up to the cobble passage, we head northeast for nearly 250 feet before we decide to turn left to avoid a climb. Our lead turns into a hand and knees crawl that passes a smaller side passage with nice air. We continue down our passage until we reach a large intersection. There, we decide to leave the ladders and come back to rig a traverse across the top of Dagger Pit.

12.6.2 Sunday, November 24, 1996

Russ Carter, Dave Collings, and Phil Lucas
818 Feet Surveyed
3.07 Miles Total
10.75 h Trip

We head to the Big Bird Room and take the right hand lead. This area has patches of "bird poo" but nothing

Fig. 12.23 The Convention Room. Photo by P.C. Lucas

Fig. 12.24 Passage cross-section at the Convention Room. Photo by P.C. Lucas

we cannot pass. Soon, we come to branching leads and we chose the more promising ones as we survey. At station 617, we come to a major intersection and decide to take the right-hand passage to the northeast. This is a straight passage for a while with dogtooth spar found in a little side passage to the left. Abruptly, as usual, the passage ends its northeast trend and we turn left to the northwest in a passage that slopes down on a hard mud slope that has clusters of aragonite crystals growing from small rocks at the bottom. Here, the passage intersects a rain well that has a beautiful flowstone-lined pool beneath it. We call this the Hot Tub because of its appearance (Fig. 12.25). Although, because the rain well was very active this day, we cannot see the bottom very clearly because of the agitations. However, we could tell that the bottom must be composed of large softball sizes clumps of crystals, but we couldn't tell if they were dogtooth spar or what. The rain well had speleothems all up its sides and the top seemed to have leads.

There was a low crawl heading to the southwest and a head-high tube leading to the northeast with a clump of aragonite crystals in the middle of it, making it difficult to check out. We would have let this go except for a nice echo beyond, which is always hard to ignore. Phil struggled up into this tube and manages to pass the aragonite crystal. Beyond, the tube opens into a tall fissure with a small tube leading down, filled with calcite rafts. But at the top is what seems to be a passage above a chert layer. Climbing up, Phil sees that a crawl does indeed continue and in just 10 feet it opens into the top of a large passage intersection. Best of all, he can see a sheet of Tyvex that he had placed on an earlier trip into the formation section. Finally, we are able to make a connection! Dave and Phil bring the survey through and tie to station 427. Unfortunately, this connection doesn't provide any bypass to the area beyond the speleothems, as it connects to the speleothem passage in between the line of speleothems.

We once again back up, this time to station 617, where we head northeast into a smaller passage that heads down and turns left into a hand and knees crawl over rocks. Russ looks through and sees blackness (now how can you see blackness?) so we survey on through. Sure enough, in one shot, we emerge into a large intersection area where six passages branch. We turn northeast again and follow a passage that branches and turns until, after about 250 feet, it comes to a pit 15–20 feet deep and just a little too wide to cross. It is frustrating because we can see another intersection just beyond it. Climbing down would be frustrating because the other side is vertical and looks to be composed of mud. Later, when we plot this, we see that

Fig. 12.25 The Hot Tub. Photo by P.C. Lucas

had we been able to cross the pit, we would have come to the passage near the end of the last survey, which is where we had originally planned to go. Instead, we back up again to station 617 and survey the large passage to the southwest. Two shots bring us to a speleothem climb we call the Cream and Butterscotch. The main passage ascends up the speleothems where a large passage can be seen, but messing up the speleothems is out of the question and besides, we probably can find another way around without too much trouble. A crawl with air is found on the way back out at station 641. This will probably be a way around but the hour has grown late. We survey an obvious loop around and tie to station 617 and call it a day.

12.6.3 Tuesday, December 10, 1996

Gregg Clemmer, Tommy Shifflett, and Phil Lucas
1166 Feet Surveyed
3.29 Miles Total

We head to the Middle Finger leads across the Dagger Pit. We decide to take a nice-looking lead at station 582 that heads in a direction away from the center of the Canyon Maze. Immediately, we start encountering intersections at every survey shot. Most intersections are only 30–35 feet apart which makes the survey footage go a little slower than otherwise. We survey generally straight for three stations to where some speleothems are encountered and take a left passage for one shot to another intersection. Tom notices air coming from the left but after a couple shots, we come to a ceiling passage not easily reached. Heading in the other direction, we come to a breakdown area, which we follow to the southeast until unbroken passage continues first to the northeast and then southeast. Here, the passage slopes down and aragonite crystals coat the walls of a small passage to the left. Heading to the southwest, the aragonite crystals become more numerous, coating cobbles, stones on the floor, and the walls of a short crawl. We decide to call this the Frosty Way. Following upstream(?) in the Frosty Way breakdown, we encountered two leads that continue as a crawl. Backing up, we survey up the slope of a larger passage and then on the right wall, up a somewhat tricky climb into a canyon passage with two levels. The lower level ends in a tacky mud fill but the upper passage takes us to a northeast trending passage that steadily becomes larger with rooms that contain large breakdown blocks.

At a connection between these rooms is where Gregg "saves Phil's Life". Phil is standing on soft material near the edge of a climbdown and is sketching when suddenly the floor starts to collapse beneath him. Ever alert, Gregg Clemmer quickly grabs Phil's hand and sleeve to hold him up. As Phil feels the floor give way beneath him, he jams his feet against the wall. So there they are, frozen for a bit in time. Gregg is tightly clutching Phil's hand while Phil is braced above a 4 foot deep collapsed floor, balancing the sketch book in one hand and looking with some dismay at the muddy glob that has a death grip on what had once been his clean sketch hand. Phil slowly looks up at Gregg and says "Gregg, I really appreciate your saving my life. Don't misunderstand, I really do appreciate your really quick reflexes and natural instinct to grab me. But look at the hand that you grabbed! My sketch hand is a mud ball! Aarrg!"

From this point on, we survey down nice size passage, arbitrarily choosing which way to go. Tommy says, "What is holding this hill up anyway?" to which Gregg replies, "There must be enough columns left between the passages, like a room and pillar coal mine." An interesting mud sculpture at station 673 divides a horizontal northeast passage from a near pit into a northeast trending passage. At station 674 is a side passage that opens into a really nice, big, straight passage that is a least 150 feet long where flowstone can be seen in the distance. We can't climb down, however, because of the width. Near station 683, we see some "iron" fragments attached to the wall. At station 685, Gregg crawls down a low lead for 30 feet and sees void through a broken rock plug that can be easily removed. Our time "runs out" at station 691 where some small breakdown blocks a forward direction and we take an uneventful journey back to the entrance where we emerge at 9:30 p.m. in 20° temperatures.

12.6.4 Thursday, January 16, 1997

Matt, Charlie Bill, and Phil Lucas
960 Feet Surveyed
3.53 Miles Total
13.5 h Trip

Dad and sons are on a trip into Helictite for some serious surveying. We have a cable ladder and drop the pit at station 595. A wall pendant makes a handy anchor. Matt climbs down. The pit is vertical for only 15 feet. At the bottom is a three way intersection. We continue straight, following dry cobblestone passage downstream. We encounter delicate frost work. The camera comes out and we spend some time trying to do justice to these incredible speleothems.

There are leads everywhere. We continue "downstream" in a hands and knees passage that we call the Sparkle Way because of the glittering ceiling and walls. The mud floor is deeply cracked. We survey a side passage at station 714 that trends northwest. The passage looks good but comes to a mud-filled halt in 30 feet. At the end are translucent speleothems with an evaporated appearance. They look old and ice-weathered. The stalagmites have soda straw tops that no longer touch the ceiling and have irregular ends that look "melted".

Our cobblestone passage becomes too low and will have to be dug open. It looks like it will take about 2 h, but there is a steady breeze flowing out. We take a parallel passage that has intersections with upper level passages every 30 feet or so. We attempt to climb up into the upper levels at station 720, only to find that fill has sealed off all leads. So we turn back to a crawlway that heads first northwest and then northeast. This passage is a belly crawl in places but is sand-floored. It finally gets too tight. There is also a slight air current here. We leave a side passage requiring a small amount of digging that could bypass the constriction.

Turning around, we head up a sandy crawl to the northwest and pop out into a canyon. Heading northwest and then southwest, we again reach a constriction that will have to be enlarged. There are some very pretty speleothems and crystals. Backing up, we take an upward sloping passage. This passage is heading southwest and keeps getting bigger as it passes by several intersections. Finally, we come to a sizable passage/room. There is an area of collapse and breakdown, but we see passages leading both left and right around it. Proof that these passages continue will have to come on the next trip.

12.6.5 Saturday, January 25, 1997

Nevin Davis, Ben Schwartz, and Phil Lucas
1760 Feet Surveyed
3.86 Miles Total
10.5 h Trip

Picking up a lead at station 606, the passage almost immediately gets challenging. Holes in the floor must be climbed down or chimneyed across. After several stations, we arrive at the top of a canyon about 20 feet high. It is wide enough that only long legs will reach. Ben is in the lead and Phil straddles after, but when Nevin reaches the Straddle Way, he doth protest. After an attempt at the high route, he climbs down to the bottom (no easy feat in itself) and then climbs back up the other side.

The Straddle Way stops at the Jump. The Jump is a four way intersection with another hole in the floor. Ben steps out and peeks around to the left (northwest—the direction he wants to go) and sees nice passage taking off. Straight ahead looks good too! But getting across and down from this point may not be easy.

Ben steps back to tape the survey and Phil steps out to see. Looking across, Phil sees a nice sloping sand bank on the opposite wall of the hole in the floor. Deciding that returning is an obstacle that can be resolved later, he jumps across into the soft sand…a perfect landing! Ben jumps with the same result. When Nevin reaches the intersection, he looks down and across. He lifts his head. He says "Come on guys. How did you do this?" We tell him that we jumped. We feel returning will be only a small problem. Nevin declares that he will not jump. So we watch while the old master maneuvers and kicks various footholds in the loose sand walls to no avail. Finally, he manages to lower himself a body length when his luck, or more exactly his crumbly footholds, let go and he reluctantly makes a rather puny leap to the sand.

Following a short discussion on the potential methods of climbing back, we continue our survey on across the intersection. In a couple of shots, we come to a three way intersection with speleothems. There is a hole in the floor of the lead trending southeast. Recognition flashes in Phil's mind. We are on the other side of the pit that stopped Wade Berdeaux and Phil on 8/2/96. Another connection is made! Best of all, an alternate route is established around the prettiest speleothem passage.

Continuing northeast, the passage becomes larger and we climb over breakdown. There are a number of good looking leads to the southeast. Several shots further we come to a pretty formation rain well with three crawls continuing. All are either wet, muddy, or low. We're mapping big passage today, so we go back up the main passage to the first lead to the southeast. Phil asks Ben (forgetting that southeast is a direction Ben doesn't want to go) to check it out. Ben soon hollers back that he has come to a room that he can turn around in and is doing so.

This doesn't sound particularly inviting, so we fall back to the next southeast lead. This time, Phil pushes it. Ben says that he believes it will just go into what he was already in. Phil pushes on a little further anyway. Suddenly, he comes into a large room with beautiful speleothems. He hollers back to Ben and says, "This couldn't be the same place, because I'm in a large room with lots of leads." Ben says sheepishly, "Yes it is the same room." Phil says, "No you're wrong. This is a big, pretty room worth surveying and not the small room that you were in. Come on down here and see." Ben does and says, "Yes this is it." Phil says, "Ben how can this be? Why would you turn around in a big room like this?" Ben says, "Well, this is going in the wrong direction." The moral of the story is…watch Ben carefully when you're trying to push him in a direction he doesn't want to go.

We do in fact survey the room and a northeast-trending passage leading from it. This passage has clean washed walls and a cobblestone floor. At an intersection, a water crawl appears to be the other end of the water crawl from the rain well at the end of the main passage. We continue on up a cobblestone crawl into a low canyon with sparkles. Our passage becomes a narrow canyon that leads to a speleothem room. Once again, we sense recognition. This is the room in the Second Formation Passage where speleothems had stopped the survey early last year. Not only have we made a connection, but by placing a couple of rocks, we can exit from the cave this way. This will be a much shorter route to this section of the Canyon Maze.

We backtrack to The Jump and take the northwest lead (the direction that Ben wants to go). In two shots, we come to a large room with a scree slope coming down from the northwest. The scree is hard packed and leads up into a room that has an almost perfectly domed ceiling. Ben and Nevin call this the Stope Room (a mining term).

Ben circles around the bottom of the scree slope and finds a small lead. We continue the survey. After nearly 500 feet, we come to a T-intersection with a smaller passage. There are standing pools of water. Turning left across a pool and piles of sand, we break up into a room and set our last station. The passage continues and we observe some very strong breathing air currents. Nevin believes this may indicate a nearby entrance, but we are a long way from any known possibilities.

12.6.6 Sunday, February 9, 1997

Dave Hubbard and Phil Lucas
527 Feet Surveyed
3.96 Miles Total
10.5 h Trip

Dave drove in this morning and was a little late because of snow still on the roads and slow traffic so it was noon before we get into the cave. We slowly make our way down through the Canyon Maze to the speleothem canyons. Here, we spend some time taking photos of the Dog Tooth Spar Pool, having put on our Tyvex boot covers. Dave's camera has a telephoto lens and we try to get some close up shots of the crystals. We have brought in some clear plastic to cover the Crystal Pools in the next speleothem passage over where we will be crossing. After installing the plastic, we cross the pool and take some more photos of The Scissors, which are two long calcite crystals that look somewhat scissor blades.

Eventually, we head on through the Crystal Pool and into the Little Stream Canyon where we start really looking at fossils. Dave has a keen eye for spotting unusual fossils among the thousands that are exposed every few feet along some sections of the cave passage. An especially large Brachiopod is found at the beginning of the Wrong Way Passage. We set it aside for the return trip when we intended pack it carefully and take it out of the cave for identification.

We travel on to the end of the last survey. It was nearly 6:30 p.m. before we begin our survey. At our second station, we take a sharp turn to the left (northwest) and encounter a tight canyon that appears to be too tight at the bottom. There is extremely wet and sloppy mud covering the walls and many of the ledges crumble as we try to climb up to the top. We finally crawl through a small hole in the ceiling of the canyon and then climb down 10 feet or so to a passage heading northwest. This is a walking passage and soon, to our surprise, we come to an intersection with a small stream flowing from the northwest. We decide to survey upstream because it is the larger passage.

The passage continues mostly straight for nearly 400 feet with a fine silt floor that was so soft that we left deep footprints. A nice looking side passage to the southwest was intersected 200 feet upstream but we continued on straight. The passage opened up in a room/ intersection with two upper-level passages about 20 feet up. At this intersection, the passage turns right (northeast) and follows the plunging structure that has a dip of about 10°. We set one more station and stop because of the late hour and wet coveralls starting to cause a chill. Ahead, we can see a short crawl will be necessary to reach another intersection about 40 feet ahead.

12.6.7 Thursday, February 27, 1997

Matt, Charlie Bill, and Phil Lucas
773 Feet Surveyed
4.11 Miles Total
9.5 h Trip

We head to the stream passage found on the 2/9/97 survey. At station 806, we survey in a side passage with white sand floors and frosted cobblestones. It is a delight to survey, but after 400 feet, we come to a dry sump where fill comes to within inches of the ceiling. There is air moving through, but about 10 feet will have to be dug out. The passage is heading southwest up the Cove. It is probably the other end of a passage we surveyed until it became too low on 1/16/97.

We return to the stream passage and survey downstream, following the little trickle channel in the wide, flat silt floor. The ceiling quickly comes down and further penetration will require some slurpy digging. There is no air movement and it seems pretty hopeless.

We continue on upstream past the point where Dave Hubbard and Phil had stopped on the last trip. Soon, we can see ahead to an intersection, but the way is low, wet and very muddy. With just enough space to hold our chests above the paste, the rest of our bodies are getting thoroughly slimed. Soon, things turn from slime to ultra slime. The passage is very low ahead, and Phil is able to slide through only by squeezing out a path through toothpaste-like goo (Fig. 12.26). After another couple of turns, the passage becomes too tight.

Fig. 12.26 Upstream at this point is 7 inches high. The bottom is very soft mud, so penetration through this passage was no problem. Photo by P.C. Lucas

We retreat a short distance to the Slime Room. Charlie Bill and Matt need to change carbide, so Phil climbs a fissure that leads up steeply to the northwest. The bottom of the fissure is narrow but it gets wider higher up. The climb proves a bit difficult. Soon, Phil sees that the fissure is getting bigger and retreats. Surveying again, we reach the top of the fissure, a room at least 70 feet high, 60 feet long and 40 feet wide. This "Tall Room" is drippy and slopes steeply over breakdown to the northwest. At the far end of the room a high canyon leads away into darkness. We attempt to climb to the bottom of the canyon, but the last 10 feet bell out.

12.6.8 Saturday, March 1, 1997

Nevin Davis, Cori Schwartz, and Phil Lucas
1235 Feet Surveyed
4.34 Miles Total
12 h Trip

The three of us head back to the lead at the bottom the Tall Room. It's a long haul back to this section. Nevin says for the third time that Helictite is not a wimp cave anymore. Reaching the canyon, we set a bolt and drop a ladder down. The drop is only 12 feet. Next time we'll leave an etrier as a permanent rig.

The passage below is a stream passage (with no stream). The walls are clean, smooth, and coated with crystals. We have descended into the New Creek Limestone, which gives a very different kind of passage shape. There are rounded corners and scalloping on the walls and ceiling.

We survey downstream first in a wide stoop-walking passage floored with sand and firm mud. Soon we pick up a little stream. After about 300 feet, the passage becomes too small to follow without lying in the water and hammering at the walls. A side lead goes the opposite direction but ends in collapse after 80 feet.

So we head back upstream. The passage stays wide and about 3–4 feet high with a mud floor that becomes increasingly soft and sticky. At one point, a large rock has fallen from the ceiling. It nearly blocks the passage but we are able to scoot around it. Some higher, larger passages intersect in places on the right wall, but we keep to the stream passage, getting some long survey shots while crawling on soft mud.

Soon, the mud gets softer and has a film of water on it. The mud is red-brown. We call the passage the Chocolate Way. We encounter two pools about 1.5 feet deep. A stream about 6 inches wide and 0.5 inches deep flows from one pool to the next then disappears. Now, we are not only muddy but wet. But the passage keeps going, making abrupt turns right and left, overall trending northwest.

At station 868, a passage heads straight, but is partially blocked by breakdown that will need a little poking to get by. Finally, we get to an area where there is no mud. Instead, a tiny canyon has incised a foot or two into the floor. Ahead, we can see breakdown blocking the passage, but with blackness above it. We climb up into a decent-size room. Cori also crawls into it while checking a side lead.

We see what looks like the same stream passage continuing on the other side of the room, but collapse again blocks our way. Cori crawls through a crack but reappears behind us. She says there are some other leads but we have accomplished a decent survey and decide to call it quits at this point, leaving several good leads return to.

12.6.9 Saturday, March 22, 1997

Ben and Cori Schwartz, Nevin Davis, Jeff Maddox, and Phil Lucas
1414 Feet Surveyed
4.61 Miles Total
13.74 h Trip

We travel to station 594, which is just short of the 20 foot pit that Charlie Bill, Matt and Phil had dropped during the January 16, 1997 survey. We hope that by branching off and down a side passage at this point we can avoid the pit. We know our plan succeeds after a couple shots and see that our cobblestone passage in one or two more shots will tie in below the pit.

But we are intrigued by a side passage that heads to the northwest and southeast. Ben “declares” that this is the right direction and we branch away from our tie-in point. After only a couple shots, we intersect a large overhead canyon room. We find that the easiest way to approach it is to go a short distance northwest and come up into a larger crack. When we move to the right, we see a survey station in the middle of a cobblestone intersection from the previous survey and we tie in.

Now heading up into the canyon room, we reach a southeast trending-passage that leads to another room that seems to have leads in all directions. We want to head generally to the southwest, which will take us down the hill along the Pancake Fields and toward Burnsville. Only one lead looks good but it is a high slot in the wall of the room. Ben manages to quickly climb up and is gone awhile before returning. He reports that where he stopped, it is big enough to turn around in, and that we should survey up that way. Ben says that before we climb up, he needs to stabilize the loose rocks in the fissure that he has just climbed and for us to move back. For the next several seconds, large rocks come tumbling down, some the size of refrigerators. Finally, one rock becomes jammed halfway down so Phil climbs up to help dislodge it. However, the both of them cannot move it so the slope is deemed safe for pedestrians and we all climb up the fissure, leaving Fallen Rock Room below. At the top of the scree slope is the edge of what seems to be a higher room from where the rock has fallen. To try to push up higher, but it is too dangerous so we skirt around on the right into a northwest trending passage. This passage is 12 feet wide and is up in the chert layer, therefore it is full of broken rock. The passage turns left, then right, and eventually heads back to the northeast where we come to a station 743 marked on the wall. This is the end of the previous survey into this area. Also, this is the survey station that we originally intended to start our survey from and we have approached it from the direction we intended to survey!

We try to push leads from this point to the northwest but nothing wants to go, so we “back up” to the previous survey, station 738. Here, Phil climbs up into an upper-level passage that starts out as a crawl across crystals

Fig. 12.27 Ice-like speleothems. Photo by P.C. Lucas

Fig. 12.28 The Lucas family in Helictite Cave. Photo by P.C. Lucas

pools and soda straws. Soon, he yells, "I've got room enough to turn around, so come on up and let's survey." The passage leads to an intersection, and taking the left, to the northwest, we see an amazing passage coated with aragonite crystals. Beyond this, the passage turns right. But in that intersection is a beautiful area of helictites so clear that they look like ice (Fig. 12.27).

Carefully crawling on further, we reach yet another intersection and take the left turn. Here is a slope down into a sizable area with beautiful flowstone covering one wall and part of the floor. Heading to the southeast, we follow a canyon passage that soon becomes a small crawl that eventually intersects another canyon passage. Ben and Phil explore this passage to an end in both directions. The cave does not seem to want to go in the southwest direction at this point in the hill. We probably need to be further beneath the hill.

Heading back to the station 917 intersection, we survey to the southeast into what, at first, seems to be a grand junction of passage intersections. After checking all the passages, only one seems to have any great promise but it will require some special equipment (boot covers). The floor is covered with flowstone and white rimstone dams for as far as we could see (40–50 feet) down the passage.

12.6.10 Saturday, March 29, 1997

Beth Billman, and Charlie Bill, Charlotte, and Phil Lucas
403 Feet Surveyed
4.68 Miles Total
6 h Trip

With Charlotte on the crew, all of Phil's family have participated in the survey of Helictite Cave (Fig. 12.28). Arriving at Chert Hall, we survey southeast from the end of the passage. To our surprise, it ends in just two shots. Another dies after only 80 feet. We then survey a short crawl that leads to a room with a steeply sloping breakdown floor. Now we're getting somewhere! Then, three shots down the slope, we once again come to a dead end. There are lots of fossils and some neat mud drill holes, but not much else.

We retreat to the Streamway to the first passage heading east. This ends after 25 feet in a predictable mud plug. Phil climbs into a canyon heading east. This lasts for 100 feet before ending in a mud plug. We decide that we have been cursed enough for one trip and head out.

12.6.11 Saturday, April 5, 1997

Ben and Cori Schwartz, and Phil Lucas
559 Feet Surveyed
4.79 Miles Total
9.75 h Trip

Our destination is the high canyon beyond the Tall Room. It takes 2.5 h to get there from the entrance, primarily because of Phil's glasses getting steamed up and having to stop and cool down. On the way, a chert ledge breaks from under Ben's foot and he bruises his arm against a sharp edge as a result. Maybe this will not be one of those great trips that we've become accustomed to having.

Reaching the bottom of the Tall Room, Phil pulls out the etrier he has made to use in place of a cable ladder. The etrier however is about 5 feet too short but we decide that we can get back up anyway and slip on down. Our survey starts near the bottom of this climb at station 845 and we continue in a middle level passage that heads in a northwest direction with several connections to the stream passage. Too soon, it ends in a series of dripping domes. These domes could have passages in their tops but, short a climbing pole, there is no way to climb them.

We return to station 957 and find some high leads going up into the tall canyon that we could see from the Tall Room. Ben climbs up through a small hole and soon finds a larger way that we can more easily climb up through, so Phil joins him. The canyon looks good but we quickly reach a climb down. From here, we can see some distance ahead, however, we can also see that some climbing will be necessary to negotiate this passage. Finally, we reach a constriction where the way through will require a 4 foot dig.

We had left Cori behind so we retrace our steps and begin a survey of this canyon. Coming to the 4 foot crawl, we each take turns digging. Although the floor of the dig seems firm and rather dry, we are able to peel layers away, revealing soft gooey mud underneath. Ben is able to tear chunks from the rotten walls. He laughs, holding up his fist, and says "In this cave, you've always got your hammers with you." Soon, we have a nice opening to crawl through. On the other side is another enlargement of the canyon passage that slopes down with the dip of the bedding, which was an approximate 10° angle dipping to the northwest. At the end of the room, the canyon ended in fill mixed with collapse. There is a small passage that went to the west for about 20 feet before ending.

Now we have a decision to make—where to go from here. One obvious choice would be to continue to the end of the Chocolate Way and survey the leads there. But we are really slimed from pushing the canyon and decide to return to station 789 where a good lead takes off, that looked dry when we surveyed past it several trips ago. Unfortunately, this lead did not go anywhere except to rejoin the main canyon passage in two survey shots. Making the second shot, Phil leans across an intersection over a small pit 10–15 feet deep, to brace his hands on the opposite wall. The walls are close together here and his pack straps catch on a projection and stops him just a few inches from the opposite wall. Phil had leaned well beyond his balance point and could do nothing except hang there with his hands outstretched should the pack straps suddenly let go. "Help" he says, "I'm in a rather precarious position and I need someone to pull me back." Ben and Cori laugh at his predicament and finally Cori pulls him back to an upright position. To add insult to injury, once Phil is on the other side of the intersection, the dirt ledge he is standing on suddenly slides off and drops Phil down 6 feet before he can stop.

12.6.12 Saturday, April 15, 1997

Judy Davis, Tom Roehr, and Phil Lucas
300 Feet Surveyed
4.85 Miles Total
9.75 h Trip

We head down to the Streamway to two canyons that extend west that have not yet been entered. This is Tom and Judy's first trip through the Crack Up, through the Slickensides Room, and down to the Streamway and both of them accomplish these maneuvers with no difficulty in one hour's travel time.

The first challenge is to approach these canyons without disturbing the patina of the sand bank that we have been so careful to preserve. Phil finds a rather circuitous route following gravel bars and features in the banks that will help conceal their journey. Upon entering the first canyon, he determines that a survey is warranted because of the passage that he can see ahead. Soon, both Tom and Judy are with him and they begin the survey from station 142. The canyon immediately takes an upward trend at about a 30° slope and soon we are climbing up and over a restriction in the passage. At this point, we're in the passage two survey shots (30 feet) and Phil finds himself beneath yet another drop that requires a 15-foot climb-up. He can feel a definite flow of air coming down the drop and can see the passage continuing onward and upward. After climbing up, Phil belays the others up only to find that the passage ends in a small, low, mud-filled room in just 30 feet. Phil is mystified as to where the air that he felt earlier could have been coming from. It must be one of those circulating air currents that can be so baffling. They all retreat down to the bottom and prepare to survey the adjacent passage. Once again, they are faced with the dilemma of creeping around the edge of the passage to minimize any signs of passage. This passage went up at a fairly steep angle for about 50 feet and ends in a clay plug like the other passage. One interesting thing about both of these passages is that they both have smooth walls as if water had been flowing through them, unlike the rest of the canyon maze.

From here, we travel around the Bullfrog Room to the point below the large overhead passage that can be seen from below. Here is a fluted chimney where Phil is able to climb 30 feet up into the passage. Unfortunately, the passage has no floor. Instead, there is a series of rain wells, one after another, with no ledges to climb around on. Realizing that belays would be required, if not direct aid, Phil retreats to the bottom. This is strike three, but the survey crew is determined. So we travel up to the passage that leads into the canyon maze but turn left and advance to a window that drops away into the Bullfrog Room. Around to the right is a lead that requires steps being cut into the steep mud slope which Phil does with his green digging tool. Once the passage is reached, Phil goes up a slope where the mud-covered walls change to bare rock walls and narrow to a keyhole-shaped squeeze. Beyond, the passage continues but Phil is unable to get through this point. Thinner people will be required.

We now travel up to the Chert Hall intersection and take the left-hand lead to the Billabong Room and on to the side passage with the Fallen Door lead. Here, we start a survey from station 550. The passage goes about 50 feet across really rotten rock but then goes down through some breakdown where a hands and knees crawl can be seen heading northeast. The crawl goes 30 feet to an apparent dead end. However the "end" is the bottom of a scree slope and at the top, there is a small space that we quickly enlarge. We are now able to climb up into a large passage heading northeast as far as we can see. The floor immediately drops into a semi-pit which Phil climbs down and then back up the opposite side, which is a near vertical mud wall about 30 feet high. At the top, Phil finds a thin partition (2 inch) and on the opposite side there is a pit that drops 30–40 feet to the floor of the passage that continues to the northeast and seems to be a stream passage. With no rope or ladders, we retreat once again, but this time all the way back to the entrance.

12.6.13 Saturday, May 17, 1997

Dave Hubbard, Bill Morrow, Ben Schwartz, Rick Lambert, and Phil Lucas
7.5 h Trip

The baits that Phil left a week ago have attracted many springtails and other critters. Unfortunately, sometime during collecting, Dave accidentally loses his wedding ring when removing his gloves. Although the search is diligent, the ring is not found. Now the cave has another category of significance. Gold can be found in Helictite.

Ben, Rick and Phil spend a lot of time taking slides in the Attic room, Bullfrog Room and the Formation Passages. On the way out, Ben slides down a small hole in the Cruddy Room that leads to a passage that he had seen on the way in. The walls of this hole are very crumbly and as Phil gently touches a rock at the top to determine the stability of the area, two large rocks suddenly roll down the slope toward Ben. Fortunately, Ben is not hurt but as we join him below, we are extremely cautious to avoid other rock falls. The new passage goes for about 30 feet and may continue further with some hammering. Phil finds another way out into the passage at the Cave Pearl Drop.

12.6.14 Monday, May 26, 1997

Keith Christenson and Phil Lucas
543 Feet Surveyed
4.99 Miles Total

We head straight for the Dagger Pit. After crossing to the other side, we start our survey at the first lead. This passage starts as a steep climb up to the bubble ceiling of the main passage, to a walking and reasonably good-looking lead. We set three stations and shoot a splay over to what we presume to be the top of the other side of Dagger Pit. The passage continues ahead and seems to open into another intersection. Here, the rock becomes very rotten and crumbly. Even the large rocks on the floor cannot be trusted to have a solid center. The fossils are everywhere and making a decision where to place our feet is tedious. At station 1005 is a side lead that goes down a short pit clogged with a couple large, but crumbly-looking, rocks in the bottom. Also to the right is a hole going down a drippy fossil-lined slope. We survey on down the passage, which takes a short offset to the right and then becomes low over some flowstone drips on breakdown. Beyond this short crawl it opens into a room with large slab breakdown but with a lead at station 1011 that will require some digging (10 min?). We can see 15 feet into this lead where it seems to open up again. This area needs to be visited again. We back up to station 1007 and survey down the drippy hole on the west side. Here, we get really slimed as we slide down and then through a muddy passage that ends in rainwell dead end. The only redeeming aspect of this passage is the outstanding display of crinoids that are sticking out of the wall in the rainwell. Returning to the main passage, we continue to check side leads. The next one is a hole going down, dirty and tight. It can wait. The next is plugged. The next is a tight crawl that connects to the Bird Poo Room. We only make a voice connection through this to avoid the digging required.

At station 560, Phil crawls into a small lead, moving some rocks and gets in about 20 feet before encountering more rocks in the way that he can see through. Suspecting that this is a passage accessible from another direction, he retreats back to Keith at station 560. To prove this theory, we go through the Bird Poo Room and survey down a canyon passage to the northeast. At the next intersection, sure enough, is the other end of the crawl. Heading northwest we survey one shot to a restriction that will require a 5 min dig to some passage seen beyond.

12.6.15 Wednesday, June 4, 1997

Jay Paul and Phil Lucas
243 Feet Surveyed
5.04 Miles Total

We travel down to the White Lightning Room where more pictures are taken. Phil has a roll of a new Kodak film that supposedly is self-adjusting for various light conditions. Then, we travel on to the Canyon Maze and across Dagger Pit. From here, we go to station 643 where a crawl promised to lead to the Convention Room bypassing the tight crawl that stopped Russ Carter. Some small amount of digging is necessary to get started but it quickly opens up. In just 30 feet, we come to a room that has beautiful speleothems that require some more photos. We decide to call this room the Chapel because many of the speleothems look like large candles. Another crawl follows this room and again, it opens quickly into a larger passage that has flowstone at the beginning and large breakdown at the end, where there is the intersection of the passage that leads to the Convention Room. Off the middle of this passage is a side passage that goes 20 feet to the right and intersects another passage that we discover is the top of the Butterscotch Climb. We tie these surveys together.

From here, we travel back to station 643 and then to station 635 where a nice looking passage heads northeast. We survey down this passage but, unfortunately, after one shot the passage stops in a mud choke. The top of this passage can be seen continuing on for 50 feet plus but we can see no way to climb up into it. A 15 foot climbing pole will be required or perhaps a ledge can be excavated and a tricky traverse made. Better yet, approaching it from another direction is a strong possibility. At station 634, we dig open another lead but this opens into a room that ends in a mud choke in 30 feet.

12.6.16 Saturday, June 7, 1997

Ed Saugstad, Dave Collings, and Charlie Bill and Phil Lucas
1218 Feet Surveyed
5.27 Miles Total
10.5 h Trip

We arrive at the beginning of the stream slot canyons on the far side of the crystal pool and begin to trace the route of the paleo-stream. A few feet down the slot, we discover a place where the stream was pirated from the canyon. A short dig uncovers a tight crawl leading away, but more digging and hammering will be needed and we move on.

Our next stop is a short loop which needs mapping, but we find more than we expected. The passages are mostly low glittery crawls with multiple intersections. Several hundred feet later, we find ourselves on the other side of the Dog Tooth Spar Pool that had stopped a previous survey. Dave and Charlie Bill crawl down a parallel passage and discover an even better display of spar in a dry pool. Phil, Dave and Ed survey the side lead while Charlie Bill digs open a lead branching off at the pool. The new passage is a low crawl that we follow for 75 feet or so to a floor covered with calcite rafts.

The next lead takes us to a dome with an upper level shooting off to the southeast and a low crawl that connects to station 1038. Another lead branches from station 756 as a low sandy crawl. This crawl brings us to the bottom of the pit which had stopped Phil and Wade Berdeaux on an earlier trip.

We push on to our primary objective, the stream lead at station 797. We believe this lead is our best chance to find a way to continue northeast. It starts as a slot about 10 feet high, but is divided into upper and lower routes by ledges. Phil and Charlie Bill climb up and over to a 10 foot drop to deep water. Phil discovers the rottenness of the ledges during a dramatic plummet to the water. Charlie Bill follows in a more understated way. Ed and Dave survey the lower route.

This is wetsuit country. The passage branches, with the left lead going downstream to a low airspace. We try the drier stuff first. Ahead, the survey leads through a narrow slot to a deep water sump. A tight, slimy slot goes up 25 feet and ends.

Oh well… we move on to the low airspace. Phil scouts ahead and shouts back to bring the survey through. "I found walking passage that keeps going!" Soon, we stop the survey at a junction. A low, but promising, sand-floored crawl continues to the northeast. A walking passage heads northwest and brings a small trickle creek. We call the wet area we've just mapped Lake Woe-Be-Gone.

12.6.17 Friday, June 13, 1997

Mike and Andrea Futrell, and Phil Lucas
847 Feet Surveyed
5.23 Miles Total
9.75 h Trip

We first travel to the Formation Canyons and gather up 3 boot covers. From here, we cross Dagger Pit and approach the Convention Room via the Chapel. Donning the boot covers that Andrea calls "Pooper Stompers", we begin our survey and cross the speleothems that have been blocking the large side lead heading southeast. After we survey three stations, we find ourselves at the bottom of the pit that Shifflett, Clemmer and Phil discovered at station 674. This pit is unclimbable from below as well, so we branch off into a lower side lead and in five shots, join into the previous survey at station 673.

We travel northeast to station 676 and climb about 15 feet up into a high lead that looks good. However, after only a few stations, we encounter a pit that has an undercut mud wall. It would even be difficult to descend with a ladder. Mike placed 1084 with large letters on the ceiling so it might be seen from below coming from some other direction. The next series of side leads were either one shot wonders or other encounters with pits. At one point, we come to a room that has a dome ceiling from the rock collapse. We call it Stope Two Room. Getting somewhat discouraged, we quit this area and travel to the Triplex Canyon leads. These two fine-looking leads quickly end in collapse with a short, frosty lower passage that hooks back into station 647.

12.6.18 Saturday, July 12, 1997

Ben and Cori Schwartz, and Phil Lucas
251 Feet Surveyed
5.47 Miles Total
7.5 h Trip

We continue past the right turn to Lake Woe-Be-Gone and head to station 801. A dig here might bypass the lake. We dig for an hour in pasty clay and finally give up. Phil goes on through Lake Woe-Be-Gone to attempt a voice connection. He yells for Ben and Cori. They can faintly hear him through the dig, but getting through will take a long time and a lot of effort, so Ben and Cori get wet and join Phil on the other side of Lake Woe-Be-Gone.

First, we map the side passage where the little side stream enters from Chocolate Way. It only goes 70 feet. Then, we push into the low (but dry!) crawl continuing beyond Lake Woe-Be-Gone. It soon opens into hands and knees and then walking passage! We stretch the tape out! With clean solid walls and a firm, cracked-mud floor, it really looks good. Then suddenly, the lead turns and dumps down into a sticky mud sump. The passage obviously goes but digging it out will be tough. A 15 foot high rain well just beyond the sump has no passage at the top. What a disappointment. We feel that if we had gone another 200 feet to the northeast, we'd have gotten beyond the sediment fill that seems to limit this end of the cave.

On the way back out, Ben and Phil push a passage for 25 feet to another mud plug. Both mud plugs are much lower than Lake Woe-Be-Gone and prove that the lake is perched. Soaked through and through, we head out of the cave for an early exit.

12.6.19 Saturday, August 23, 1997

Bob Hoke, Tom Spina, and Phil Lucas
497 Feet Surveyed
5.57 Miles Total
10.33 h Trip

Phil and Tom are both a little fragile on this trip because of previous injuries, but Bob is healthy and will end up doing all the climbing. This is Bob's first visit to Helictite so we take a quick tour to the Pancake Room before heading into the Canyon Maze, across the Dagger Pit, and through the Convention Room. We start at station 606 and survey a number of side leads, all of which either dead end or loop back into the main passage. Nothing of any consequence is discovered and this turns out to be a rather uninspiring survey trip as far as new discoveries go. However, it is one of our most successful trips in reducing the number of leads in the cave.

12.6.20 Tuesday, September 16, 1997

Myke Coughlin, Cori Schwartz, and Phil Lucas
293 Feet Surveyed
5.62 Miles Total
8 h Trip

We travel to the Attic Room where Cori slips down the passage to check on the sump (it is still full). We then proceed down to the Streamway to station 155 where a high lead on the left is waiting. Soon, all of us are clambering up a steep mud slope. At the top is a confusing tangle of intersecting fissures in breakdown and bedding plane enlargements. We survey first northeast and find sediment filled passages. Surveying to the southwest leads to a crack between breakdown blocks with slight drift of air. Myke and Cori hammer away. Myke slips through and finds shallow pools of water and passage getting bigger. Cori slips through and yells "I found station 350!" It's the end of a Canyon Maze passage that Ed Richardson and Phil Lucas surveyed. This is the second connection from the Streamway into the Canyon Maze and creates a large loop.

Cori and Myke survey this connection, yelling the measurements to Phil who is hopelessly too big to wiggle through the squeeze. After the survey is finished, Myke explores another crawl and finds himself in the Streamway behind Phil. A third connection! This time the passage is large enough for Phil to get through.

12.6.21 Saturday, October 11, 1997

Cori and Ben Schwartz, and Phil Lucas
744 Feet Surveyed
5.76 Miles Total
11.5 h Trip

We pack our boot covers and head beyond the Ice Palace to the Rimstone Way (Fig. 12.29). We were turned back there last time by a rimstone-covered floor. The boot covers work well and we survey down a beautiful passage with lots of delicate rimstone. We pass a nice side lead after about 30 feet, but it has a high gravel bank that will slide onto some speleothems if we attempt to enter it.

Continuing on, the Rimstone Way reaches an intersection with leads going northeast and southwest. A high lead continues ahead. Northeast looks good for 50 feet, but then "ends" in a high lead above flowstone. A right lead continues for 20 feet to another intersection. In one direction, we hit the Rimstone Way at the lead with the gravel bank. To the northeast, we again encounter flowstone and crystal pools. There is no way on.

Back at the end of the Rimstone Way, we survey southwest. The passage turns right, then reaches an end in mud after 50 feet. Just before the end are some of the most exquisite "ice speleothems" yet seen in the cave. Many are shaped like upside down Dairy Queen cones made of clear ice (Fig. 12.30). They hang from thin soda straws. We name this area "Christmas in October."

Back past the Ice Palace, we survey a lead heading southwest. It begins as a steep climb down, then passes through a series of drippy rooms with white sand banks. After about 120 feet, the passage turns left into a sand crawl. At an intersection 50 feet further, we follow a breeze to the southwest, but find a breakdown collapse with no apparent way through. Heading northeast from the intersection we fare no better, hitting mud fill after 50 feet.

12.6.22 Saturday, October 18, 1997

Earl and Charles Thierry, and Phil Lucas
2 h Trip

This was a treat to take Earl Thierry caving. We took a leisurely stroll down to the Helictite Hall and turned around at the Cave Pearl Drop.

12.6.23 Saturday, November 22, 1997

Ted Andrus, Rick Lambert, and Phil Lucas
775 Feet Surveyed
6.06 Miles Total
10.25 h Trip

The goal today is to reach 6 miles! We head to station 882 at the aragonite crawl in the Canyon Maze. Our plan is to follow a loop around the Fallen Rock Room and back to the Crystal Crawl. We soon find ourselves in a tight maze section with lots of aragonite crystals in many places. We quickly survey over 700 feet and are only a third of the way around our planned loop. We reach the 6 mile mark and realize that the Canyon Maze will contain a lot more passage.

12.6.24 Saturday, January 3, 1998

Ben Schwartz, Paul Stevens, and Phil Lucas
719 Feet Surveyed
6.19 Miles Total
9 h Trip

We are returning to the same area that we surveyed the last trip—the Fallen Rock Room. As Paul and I climb up into the Fallen Rock Climb, Ben decides to check out an adjacent climb up to the southeast. I call to Ben that it has been checked out, but he insists that he will double check it anyway. When Paul and I are safely up

Fig. 12.29 The Ice Palace. Photo by P.C. Lucas

Fig. 12.30 Christmas in October. Photo by P.C. Lucas

in the top of the climb, we wait for Ben. Where is he? Finally, we hear him call that he has found (actually dug open) a passage. So, Paul and I climb back down and prepare to make the climb to where Ben is waiting. Paul goes first and sets the first station. As Phil is waiting, a strange gurgling sound comes from his intestines. Suddenly he realizes that a Quick Potty stop is necessary- real quick! All his plans to prepare for a burrito bag are to no avail as he barely has time to get the coveralls down. Alas, the best plans of men and mice are often foiled but at least his pants were not soiled.

The new passage that Ben has dug open looks good and continues for nearly 200 feet before coming to an abrupt end. At one point, Ben digs a while at a little stope scree slope but it doesn't amount to anything. So, we climb back down into the Fallen Rock Room and back up toward the direction we first climbed up.

From here, we start picking off side passages on the right. Most of these double back to the already surveyed passage or tie into the canyon maze to earlier surveys. Finally, we tie back into station 743, the same station that we had coincidental tied into on the March 1997 survey. By this time, we have become cold (except for Phil, of course) and decide that a return to the surface in time for a spaghetti supper would not be a bad idea. And that's what we did.

12.6.25 Saturday, January 24, 1998

Nevin Davis, Bob Thren, Tony Preston, and Phil Lucas
610 Feet Surveyed
6.31 Miles Total
8 h Trip

Our destination is the Convention Room and the leads beyond the formation-crossing passage. After some sight-seeing with Bob and Tony, we head that way, stopping by the Bullfrog Room to hear the croaking. Even though we have had a lot of wet weather, the dripping is slow and the big Bullfrog is slow. Arriving at the Convention Room, we walk cross the plastic sheets covering the formations and then survey the first lead to the left that we come to. This is a passage with sparkles on the wall and a chert bridge at one point 5 feet long and only a couple inches thick. This passage ties into a known passage at station 570. This will be convenient for future trips to this area, providing a bypass around the Convention Room.

We next travel 200 or so feet to the next lead at station 1080 where a southwest-trending passage branches. This leads alternately to high areas and then to low crawls. At one point, there are several small tube-like pits in the Corriganville or New Creek Limestone. Narrowness defeated our efforts to slip down these interesting-looking tubes, but they obviously are taking water down to a lower level in the cave.

Finally, we tie back into a previous surveyed passage at station 669 near the Devils Den. This seems to be a good point to call it quits, so we do.

12.6.26 Saturday, April 25, 1998

Bob and Chris Alderson, Barry Horner, Tom Spina, and Phil Lucas
414 Feet Surveyed
6.39 Miles Total
9 h Trip

We head for the place that Wade Berdeaux and Phil had stopped, at the end of survey 472. This is at the bottom of a 20 foot pit. Once again, we call on Bob to make a climb and this time Barry and Phil provide footholds by supporting his feet for as high as they can reach. Bob struggles up to the top and then ties an elaborate attachment point of webbing and hangs the cable ladder down to us. We then all climb the short, but awkward, pit. Heading down the passage toward the entrance, we start a survey at the first side passage, at station 469. This passage climbs up to a short crawl and opens at the top of a 20 foot tall room. The way down is another 20 foot pit with a wall of mud and sand. Here, the second cable ladder we had brought was used. By the time we were all down, much of the wall had fallen also. Going back could be interesting!

At the bottom, we survey out a crawl that encounters an intersection where we turn right. Surveying through a series of "hands and knee" crawls, we come to another 3-way intersection. Straight ahead looks the tightest and naturally Bob (our advance scout) goes for it. Soon, he calls from the other side that we might as well come on through. We ask why. He says that the passage opens up and we should come that way.

"That way" is tight! Phil goes first with one arm ahead and one behind. It is necessary to turn sideways halfway through. Halfway through, he gets stuck. Wiggling backwards, he restarts with both arms forward. This works but Chris who is slightly claustrophobic has seen enough to become concerned. Barry comes through next in the same manner as Phil with no trouble. Now, Chris decides that she will wait there for the rest of the party to finish the survey and return. Phil, realizing that the big cave passage ahead may be known passage, pushes on ahead. Sure enough, he finds that the passage ahead is "The Jump" where he and Ben had jumped down from the high left hand passage wondering if they could return.

Phil reports that this is the best way out of the cave. Chris decides that if the crawl is made bigger, she will come through. Luckily, Phil has his green digging tool and soon the crawl is enlarged sufficiently. With a yell, Chris comes roaring through. By the time we tie into station 753, it is time to head out. A fine trip, even though just 414 feet were surveyed. We accomplished several climbs and made some important closures. The sling was left at the short ladder drop for future trips.

12.7 And Future Trips There Were!

Throughout the rest of 1998, there were at least monthly trips, mostly mop-up and photography trips—no major discoveries. Also, in 1998, work began on the Super Sweet Dig, at the north end of the Streamway. Great airflow and easy digging prompted 11 digs between September, 1998 and November, 2001. The January, 2001 NSS News ran an extended article on Helictite Cave, with many supplemental photographs. On August 18, 2002, the cave passed the 7 mile mark, and was considered completely surveyed. On March 11, 2009, a finalized map of Helictite Cave, drawn by Phil Lucas and digitized by Roger Baroody, was published. In November of 2009, Nathan Farrar and Chris Woodley organized teams to systematically check the remaining leads. After seven trips, the cave was only extended 700 feet, most of which was found in La Fin du Monde, past the previous end of the Chocolate Way. Clearly, the surveyors of the late 90s had been thorough. Helictite Cave currently sits at 7.19 miles in length. However, this figure may soon jump to over 12 miles, if a connection with the newly discovered Wishing Well Cave is made, quite possibly via the Super Sweet Dig.

Probing the Wishing Well

13

Nathan Farrar and Philip C. Lucas

Abstract
Wishing Well Cave was discovered by an intensive excavation operation that followed a wisp of air movement from a rubble pile in a shallow sinkhole. Following two years of excavation, the cave was entered and exploration began in 2010. The cave consists of two major passages in the upper portion of the Licking Creek Limestone. Most of the cave is beneath the sandstone caprock and is northeast of the main Burnsville Cove drainage. The cave appears to be a segment of the paleodrainage to Emory Spring. The cave contains extensive displays of speleothems including helictitees and ice-clear stalactites.

13.1 Quite the Dig

It is just as well that we do not know all the struggles, trials, and tribulations that will eventually face us as we begin what is seemingly a simple project. If we did, we may never begin. Such is the case with the Wishing Well Dig/Cave.

It all began for Phil Lucas (although he did not know it at the time) in the late 1950s when he was a teenager. Along with some companions, he scampered down into a shallow sinkhole to look at a small hole blowing air. Apparently, the sink and the blowing hole were unremarkable, because none of the residents that Phil has since interviewed, who were around at that time, can remember anything more than a shallow sinkhole being there.

So this is an account of the efforts to dig open what has become known as the Wishing Well. It is located along the Burnsville Road at the northeast end of the Burnsville Cove. Storm water runoff along the road would flow down into this sink. Where this water resurged was not known. Air blows out from the sink during warmer temperatures and vice versa in cold weather. This convection air current establishes this as a low "entrance". The sink was used as a dump site during a period in the mid-1900s. When the county dirt road was improved and paved in the late 1950s, the sink may have been partially covered with fill material but storm runoff flowing down into the natural drain kept washing it open.

Phil revisited the sink in 1989 and found a moderate breeze still flowing from a small hole on the side of the sink. In the 1990s, a new owner constructed a house on the property and, in the process, filled in the sink. Again, storm water kept washing it back open, despite the owner's repeated efforts to refill it.

When the property changed hands in 2007, the new owners were most interested in having a cave on their property and took a keen interest in the air that still managed to flow up through cracks and small holes that had washed open. When Phil met his new neighbors, he learned of their interest and volunteered

N. Farrar (✉)
10308 Annaberg Court, Burke, VA 22015, USA
e-mail: ncf4w@virginia.edu

P.C. Lucas
587 Limestone Lane, Williamsville, VA 24487, USA

W.B. White (ed.), *The Caves of Burnsville Cove, Virginia*,
Cave and Karst Systems of the World, DOI 10.1007/978-3-319-14391-0_13,

his assistance. Frank Marks, another neighbor and caver, also volunteered his assistance and together they embarked on a project that became ambitious and quite challenging.

The location of the Wishing Well is interesting because it falls within the Emory Spring drainage. The primary watershed for Emory Spring must be the southwest flank of Jack Mountain, 1.5 miles to the northwest, where limestone beds are exposed. A broad syncline extends from the crest of Jack Mountain to the Bullpasture Mountain. Although the valley floor is underlain with shales and sandstones, where the limestone beds are exposed on the slope of the mountain, there is a karst landscape with numerous sinkholes and sinking streams. The air blowing from the Wishing Well Cave during warm weather is driven by a simple convection current, in which the cold cave air, being denser, essentially discharges from the lower cave entrance. For this convection to occur, a higher entrance that receives inflowing air must exist. The upper entrance must be located high on the slope of Jack Mountain, in one of the many sinkholes that have developed at that elevation. This upper entrance is likely choked with sandstone boulders that have eroded from the sandstone exposures higher up on the mountain. Even so, there must be enough cracks in the rubble for air to pass through and thereby allow for a strong convection current—strong enough that it manages to force the air up through the choked sinkhole below, filled with dirt, rocks, and an accumulation of many years of trash.

A good air current is a strong motivator for cave diggers. We know that the air is coming from a cave—it is just a matter of time (and luck) before we dig down to it. But how far down is the unknown factor in the equation. In this setting, we knew that the limestone beds were overlain by sandstone, as it is exposed in numerous ledges along both sides of Route 609 (Burnsville Road) that runs right adjacent to the Wishing Well. Generally, the Oriskany sandstone is about 60 feet thick but it can vary. Underlying the Oriskany Sandstone is the Licking Creek Limestone, and below that are a number of other limestone and sandstone beds, adding up to a total thickness of about 400–600 feet or thicker. So there is plenty of room for cave development. We figured that if a cave could be reached by digging down through this sinkhole, then there is a good chance that it will lead to a confluence with a larger stream flowing from Jack Mountain—the Emory Spring stream. If this larger stream passage could be followed in the upstream direction, then it may lead to the upper entrance on the slope of the mountain a mile and a half away, creating a major cave system.

Missy Douglas, the new land owner, began the digging in the summer of 2007, determined to find out if there was a cave below. She just took a shovel in her hand and started digging. Soon, she was able to engage the help from some of the young camp councilors from Camp Alkulana. As they dug down, they uncovered old bottles and a lot of miscellaneous items. It became clear that this was the site of an old dump that had been covered over. But the air kept blowing up through cracks and crevices. On September 1, 2007, Scott Olson, Gregg Clemmer, Tony Canike, and Phil Lucas spent an afternoon digging. They made some progress in making the hole deeper but encountered only dirt and trash from the old dump. There was no sign of any bedrock. It became more and more obvious that a backhoe would be the best option for digging at this point.

On October 2, 2007, we employed Steve Burns, a local excavator, to use his backhoe to uncover the cave entrance. Missy and Jimmy were away for weekend but they were comfortable with Phil directing the digging of a hole in their front yard; and dig a hole Steve did. Soon, there was quite a pile of dirt and old trash, including old fence wire and rocks that were probably gathered from the surrounding fields and thrown into the old sinkhole. Still, there was no bedrock to be seen. The air kept blowing but was now coming up between layers of the old dump. Phil attached a burning incense stick to the end of a fishing pole and held it down into the excavated hole to determine where the air was flowing. Instead of blowing up from below, it was now coming from the sides of the excavation. The hole, instead of getting deeper, got longer as they moved horizontally, chasing the air flow. At the end of the second day of digging, a very large area had been excavated. Steve was having trouble finding places to pile the dirt and we still could not determine exactly where the air was coming from and had not discovered bedrock. The Landrum's yard looked like a huge bomb crater or maybe an open pit mine (Fig. 13.1).

We were running out of time. Steve had to go to another job and colder temperatures were on their way. The air would only flow up through the cave, wherever it was, when it was warm. The only way to

Fig. 13.1 The Landrum's yard after the first round of excavations. *Photo* by P. C. Lucas

Fig. 13.2 Missy Landrum begins the second round of digging. *Photo* by P. C. Lucas

determine where to dig was to "follow the air". Trying to dig in the winter, when the ground is frozen, is not practical. Phil simply could not leave a huge hole in the yard that would be there all winter. So after he made a number of measurements, he made a timely decision—he told Steve to fill the whole thing back in. Phil knew that Missy and Jimmy would be disappointed not to have a new, uncovered cave when they returned home. They were. In fact, they would have been willing to accept a huge hole in their yard if it would lead to a cave. But what was done was done and the winter closed down the operation.

Next spring, in 2008, Missy and Jimmy noticed that a wet weather stream flowed off the hillside down into a small hole that had opened near the edge of the filled-in excavation. They could hear water running down below. When Phil checked out the hole, he thought he felt some air movement. When things dried out, on April 28, 2008, Missy began digging again (Fig. 13.2)—she was determined to find that cave! Phil joined in on May 1 and soon, they had a hole deep enough to require some shoring. It was a good reason to get some honest exercise and so Phil would come over and dig awhile from day to day. When possible, others would join in the fun and some weeks later, they had excavated a 10-foot-deep hole against a large sandstone rock (Fig. 13.3). This hole (and rock) was just to one side of the former excavation. They optimistically hoped that this large rock was bedrock and that we could dig down beside it and find the way to the cave below. But the more they dug, the more it became plain that the rock was not bedrock, but just a huge "floater".

Then one day after a serious rain, runoff caused one wall of the hole to collapse, despite the shoring. Having enough of that, Phil again made arrangements for the contractor Steve Burns to dig a big hole in the Landrum's front yard (Fig. 13.4). This time, on June 26, 2008, he dug a deep hole beside the big rock and then constructed a ramp for access. Using lumber made available by Jeff Uhl, Phil constructed a wagon track and staircase combination down to the bottom of the dig. The hauling system used a wagon, a long rope, two pulleys, and a four-wheel-drive ATV for pulling spoils up the ramp (Fig. 13.5). It was a great haul system and soon, the diggers had a large pile of material excavated. They were finding more and more air and eventually, dug down to a layer of field stones with little dirt. The air was flowing between these rocks and the more we uncovered, the stronger the air current got.

The subject of the air flow was always a good topic of conversation for the diggers. Phil began to wonder if there might be a connection to Helictite Cave. On September 29, 2008, Frank Marks, Nevin Davis, and

Fig. 13.3 The hole is enlarged. *Photo* by P. C. Lucas

Fig. 13.4 The excavated hole and the big rock. *Photo* by P. C. Lucas

Phil Lucas conducted an experiment to determine if there was such a connection. They placed a tarp over a slope of the excavation where most of the air was blowing out. This slope was composed mostly of field stones that probably had been thrown in the original sinkhole. A hole was cut in the tarp, the same size as

Fig. 13.5 The haul system. *Photo* by P. C. Lucas

the mechanical anemometer (Fig. 13.6). This anemometer was connected to a data logger. Then a fan was placed on top of Helictite Cave's culvert entrance. The fan would reverse in a time sequence, blowing air into and out of the cave. The results showed that there was a measureable change in the air flow at the excavation, indicative of a subterranean air connection between Helictite Cave and the dig site.

As they continued the digging, they reached a dilemma. The air was coming from the somewhat-loose fieldstone slope and heading under the north wall of the dig. Soon, Phil declared the situation to be getting too risky and in need of either some extensive shoring or a larger hole. They elected for the latter, which called for some more machinery. The backhoe could not dig any deeper. It was time for a tracked

Fig. 13.6 Tarp installed at the Landrum dig site. *Photo* by P. C. Lucas

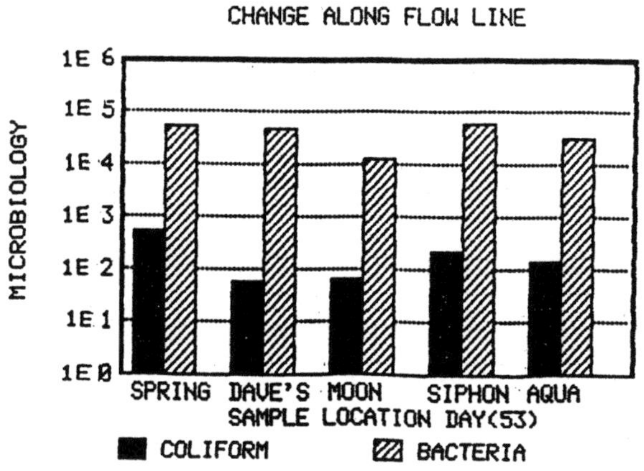

Fig. 15.1 Bacterial counts at various locations in Butler Cave. Data from Chess (1987)

medium and placed in an incubator within 12 h of collection. The laboratory was about a 6 h drive from Burnsville Cove. Dan and whoever he could sucker into being his field assistant drove down to the Homestead the day before the sample run. Up in the morning, they would race through the surface sites collecting their samples, then into the cave for the underground samples, then drive like hell back to Penn State to begin processing the samples. With luck all of the samples would be in the incubator within 12 h and before the bars closed.

The thesis was successful (Chess 1987). The results were presented at the Sault Saint Marie NSS Convention in 1987 but have not been written up for publication. The results from one day's sampling run are shown in Fig. 15.1. The "spring" sample was water taken from the spring at the Homestead. There are samples from Dave's Gallery, from the small infeeder at the Moon Room, and from the Sinking Creek Sump. These results may be compared with the bacterial counts at Aqua Spring. As a practical matter, there are no differences. The bacterial contamination is essentially constant throughout the system. There is no die-off.

15.4.3 Meteorology

By Meteorology (or cave weather) is meant the temperature, relative humidity, and barometric pressure within the cave and the variations of these parameters with time. Both short term variations (minutes to hours to days) and long term variations (seasonal effects) are of interest. Fred Wefer began meteorological observations in Butler Cave in 1984. His first account (Wefer 1984) discusses the methodology and gives some preliminary results. Results from a second year of data collection are given in another lengthy article (Wefer 1985). The next report appeared three years later (Wefer 1988). It was becoming apparent that there were significant variations in temperature and relative humidity but they were quite subtle and required careful measurements and careful calibration of the equipment. At this stage of the investigation 330 data sets had been collected. The results on the entrance series measurements were analyzed and presented at the 1989 NSS Convention in Sewanee Tennessee. This analysis appears in the *BCCS Newsletter* (Wefer 1989). Analysis of the data collected along the trunk channel was presented at the Appalachian Karst Symposium in Radford, Virginia, in 1991 which is the only formal publication of the meteorological data (Wefer 1991). This presentation also appears in the *BCCS Newsletter* (Wefer 1990). The last report was an account of temperature and relative humidity deep within the system (Sand Canyon) (Wefer 1992).

Overall, an incredible quantity of data was accumulated. The 1992 report claims 680 temperature and relative humidity measurements. Many of the details are reported in the sequence of *BCCS Newsletter articles* but a scientific digestion of the data for more formal publication has not been previously undertaken. Chapter 22 summarizes the overall results.

15.4.4 Burnsville II?

Given the new cave discoveries and opportunities for new geological, mineralogical, and possibly biological research, it seemed a good idea to compile a new set of papers for a second Burnsville Cove issue of the *NSS Bulletin*. This project was launched at the 1991 annual meeting with Fred Wefer and Will White assigned the job of organizing and editing the document. An impressive list of 17 possible papers was compiled [*BCCS Newsletter* **17**, 58 (1991)]. A few of these topics such as the sediment study and the meteorological study were already well underway. Most of them, however, were nothing more than titles to which some hapless individual's names were attached. There was a great deal of data gathering, interpretation, and writing that would have to be done. That was part of the reason that the project never really got off the ground, but the main reason was that new discoveries from Chestnut Ridge were coming thick and fast. The proposed compilation seemed premature and so the entire project was put on hold. The project was taken off hold in anticipation of the 50/40 celebration and its reincarnation is the present volume.

15.4.5 Paleoclimate

A hot topic during the past twenty years has been the use of cave deposits as records for climatic changes during the Pleistocene and Holocene. Speleothems, particularly cylindrical (broom-handle) stalagmites are the main data source. Think of these speleothems as drill cores obtained without the necessity of drilling—or anything to drill into. Stalagmites have a microstratigraphy that can span from 40,000 to 100,000 years for a six-foot stalagmite. Samples taken along the axis of the stalagmite can be dated by U/Th isotope methods to extract a time line for the growth of the entire stalagmite. Growth may or may not have been uniform, there may be breaks where growth ceased for extended periods of time, and the top of the stalagmite may have ceased growing a long time ago. Other measurements on the growth layers then give information on the climate above the cave at the time that the layer was deposited. Paleotemperatures have been obtained from the variation in oxygen isotope ratios along the stalagmite and type of surface vegetation from variation in carbon isotope ratios. Most speleothems are luminescent under UV light due to the incorporation of humic substances in the calcite. There is often a banding in the luminescence that can be correlated with changes in precipitation on the land surface.

Sometime, somehow, and by persons unknown, a broom-handle stalagmite was broken in the Huntley's Section of Butler. Segments of this stalagmite, each about 16 inches long, were found neatly stacked on a fill bank above the Natural Bridge. The top segment was removed, sliced longitudinally, and used for luminescence measurements as part of Bryan Crowell's M.S. thesis (Fig. 15.2). By mounting the stalagmite slab onto an automated stage and focusing a laser to a spot size of a few micrometers, it was possible to program a computer to collect the luminescence brightness at one spot, then advance the stage by some pre-programmed amount—typically 5–10 μm—collect another brightness datum, then move to the next and so on for thousands of points. The result was an oscillatory curve showing the rise and fall of luminescence intensity along the growth bands of the stalagmite. Further measurements with an electron microprobe showed that strontium and magnesium concentrations in the calcite of the stalagmite also had an oscillatory pattern. Unfortunately, we were unable to arrange for U/Th dating of the stalagmite so the Butler data are floating in time although the measurements were a good demonstration of technique. These and data from other caves were packaged into Bryan's M.S. thesis (Crowell 2001).

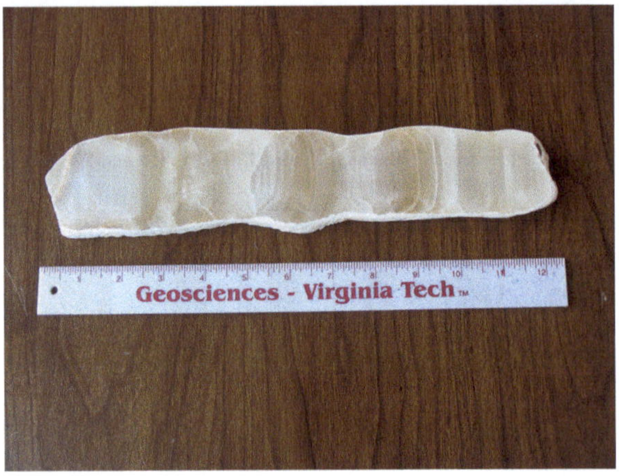

Fig. 15.2 Polished slab from Huntley's Cave stalagmite. The top of the stalagmite is to the *right*. The scale is in inches

Although there is now a substantial literature on luminescence banding and isotope profiles of speleothems, studies of the distribution of trace elements are sparse. A small segment of the stalagmite was mounted in an electron microprobe and rastered for Mg and Sr (Fig. 15.3). These trace elements were uniformly distributed except for narrow bands where the elements were depleted.

15.4.6 Moonmilk

Moonmilk is a soft, white, pasty material often with the appearance of cottage cheese. It occurs in Butler Cave as small blobs associated with aragonite in the Crystal Passage and may occur in the Chestnut Ridge caves as

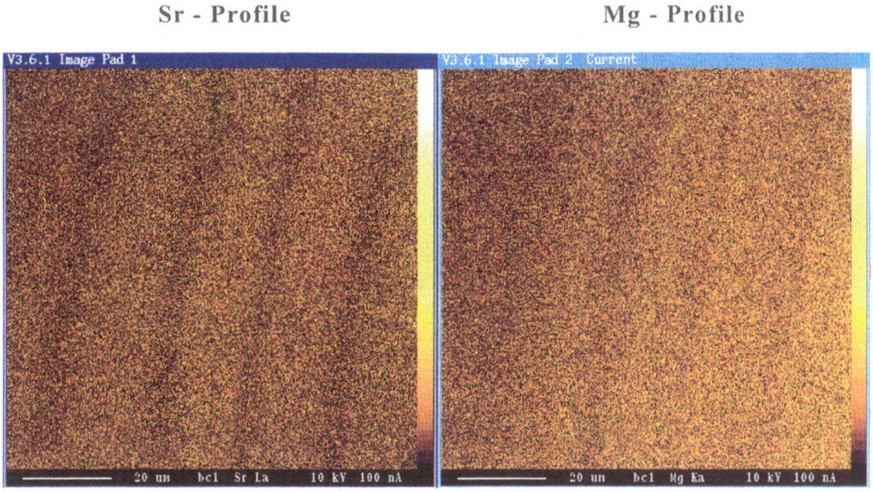

Fig. 15.3 Electron microprobe map of the distribution of Sr and Mg in the Huntley's Cave stalagmite. The scale bar is 20 μm which is also the approximate spacing between bands

well. The moonmilk mineral in Butler is hydromagnesite, $Mg_5(CO_3)_4(OH)_2 \cdot 4H_2O$. Because of its fine-grained pasty appearance there has been an ongoing argument as to whether or not moonmilk is precipitated through the action of microorganisms. Thermodynamic calculations (White 1997) show that hydromagnesite is a stable phase under cave conditions of temperature, water vapor, and CO_2 partial pressure. However, microbiologists have found bacterial remnants in moonmilk. Are the bugs essential to the precipitation of moonmilk or are they just hanging out in a cozy habitat?

A nifty test to distinguish between these possibilities was devised for Tiffani Heil's B.S. thesis in Geosciences at Penn State [*BCCS Newsletter* **20**, 20 (1994–1995)] (Heil 1995). She obtained sterile bottles with filter cap lids. The filters would permit the passage of water vapor and CO_2 but not any airborne particulates. To provide a sterile source material, magnesium ribbon was burned and the ash allowed to fall directly into the sterile bottles. One bottle was capped with the filter cap; the other bottle was left uncapped. Pairs of bottles were placed in various caves one of which was Butler Cave. Bottles were placed at the Rabbit Hole, at the Natural Bridge and at the Moon Room. The bottles remained in the caves for three months, were then removed, and the contents examined by X-ray diffraction, scanning electron microscopy, and infrared spectroscopy.

The ash from burnt magnesium ribbon was MgO which appeared in the SEM images as perfectly formed little cubes, a few micrometers on a side (Fig. 15.4). This highly reactive MgO combined with water vapor and CO_2 from the air to form a platy material quite different in appearance from the unreacted MgO (Figs. 15.5 and 15.6). The material was non-crystalline and could not be identified by X-ray diffraction but appears to be a precursor to hydromagnesite. The material in the open bottle seems to be more completely reacted suggesting that access to the open atmosphere, and thus, presumably, microorganisms, enhanced the rate of reaction. What is needed is a new set of experiments with the bottles placed in the cave for at least a year.

Fig. 15.4 SEM Image of the ash from freshly-burned magnesium ribbon

Fig. 15.5 SEM image of Natural Bridge sample after three months in a closed bottle

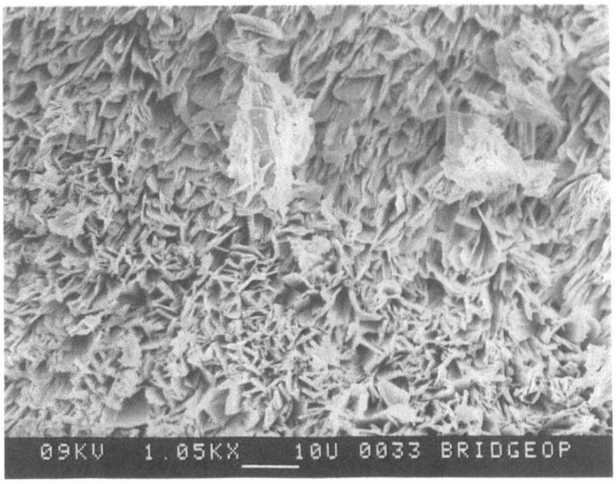

Fig. 15.6 SEM image of Natural Bridge sample after three months in an open bottle. Note that the scales of these images are roughly equal

15.4.7 Sulfur Isotopes in Gypsum

Gypsum is the second most common cave mineral and occurs in many dry caves in the Appalachians including Butler and the Chestnut Ridge System. Dr. Christopher Swezey of the U.S. Geological Survey has been investigating the origin of cave gypsum by measuring the concentrations of the isotopes of sulfur in $CaSO_4 \cdot 2H_2O$. With BCCS permission, he removed four samples from Butler Cave and determined the enrichment/depletion of ^{34}S with respect to the common isotope ^{32}S. The Butler samples showed depletion in the range of 1–5 parts per thousand indicating that the Butler gypsum was derived from the oxidation of pyrite. Had the cave gypsum been the result of dissolution, transport, and recrystallization of primary gypsum beds, the ^{34}S isotope would have been enriched by roughly 15 parts per thousand (Swezey and Piatak 2003).

15.5 Current Research: Ongoing at the Compilation of this Volume

15.5.1 Tracing Cave Connections by Means of Induced Air Currents

Phil Lucas, Frank Marks and Nevin Davis have made a surprising discovery by injecting air into one cave entrance and measuring the response at another entrance (Lucas et al. 2009). They demonstrate that mechanically changing the airflow at one entrance will rapidly affect the rate of airflow at other entrances as measured by sensitive anemometers. There is a simple mechanical technique to enhance the sensitivity of the airflow measurements.

The method involves directing a fan, with a 480 W motor, in and out of a cave entrance in a periodic fashion. The in and out reversals of the fan are recorded on a datalogger. At the same time an anemometer, recording to a datalogger, is placed in another remote entrance location. The anemometer signal and the fan reversals are recorded over a period of time. This is repeated between entrances with known connections to each other and between those with no known connections. The result of the measurements shows that the induced airflow causes a measurable change in the rate of the airflow in delayed synchronism with the fan reversals at the connected remote entrances and between some cave entrances where there is no known connection. These initial results are included in Chap. 22. Although the technique demonstrates an air connection between caves, it does not guarantee that the connection is humanly accessible or that the connection can be actually found.

15.5.2 Recording Floods at the Water Sinks Depression

The newly-discovered Water Sinks Cave floods rapidly after large rainstorms. The question is how rapidly, and what is going on inside the flood overflow passages. At the lowest level of the cave, a large stream up-wells from a deep sump and flows nearly 300 feet before entering another sump. This stream is the major portion of the sub-surface Burnsville Cove drainage. Nearly all of the cave's lower passages are washed clean from fast flowing flood waters. Leaves stuck to the ceiling near the entrance demonstrate the depth of flooding (150 feet).

Phil Lucas and Frank Marks placed video cameras in the cave, remotely controlled from the surface (Lucas 2009). A video recording has been accomplished that shows a flood event as it occurs inside the cave. The first camera was attached to the cliff face above the entrance (a vertical culvert pipe) to the new section of the cave. The surface creek is seen first starting to flow, and then overflowing its banks against the cliff reaching a depth of over 15 feet. Finally the flood level rises above the camera. The second camera was positioned at the bottom of the 20-foot entrance drop. The video shows the flood water starting to cascade down the drop as a small waterfall and then becoming a raging torrent 40-feet wide. A third camera recorded a view of a rocky floor with the flood coming in a series of rapids around the rocks.

15.5.3 The Geomorphic and Hydrologic Evolution of the Burnsville Cove Drainage System

The long-standing question, going back to George Deike's thesis on Breathing Cave is: What is the sequence of events recorded in these caves? What are the ages of the events and how do they relate to the evolution of the surface landscape? Recent investigations by Benjamin Schwartz and Dan Doctor suggest a much more complicated history than had previously been considered. The evolutionary scheme was presented at the 15th International Congress of Speleology (Schwartz and Doctor 2009). The abstract follows with a more complete development of the ideas in Chap. 24.

"The Burnsville Cove in Bath County, Virginia, hosts an extensive karst system (>100 km of mapped passages) developed under a complex synclinal valley in the Silurian-Devonian Keyser Limestone of the Helderberg Group. Observable patterns of cave development are the result of hydrogeologic controls on paleo-conduit formation, landscape evolution, and several episodes of conduit modification as the subterranean drainage system adjusted to changes in surface drainage and morphology. Currently, most of the mapped caves contain vadose free-surface streams that drain north-northeast toward the Bullpasture River, which flows south into the James River. However, many passages are dominated by solutional and sedimentary features indicative of former phreatic conditions and substantially different paleo-drainage characteristics. Widespread evidence of deep phreatic development under low-velocity conditions exists throughout the Burnsville Cove in cave passages between 0 and >230 m above modern spring elevations. Mineralogical and geological evidence of hydrothermal fluid migration and volcanism in the area has recently been recognized and may have contributed to the formation of proto-conduits which influenced early stages of karstification."

"Observations of surface topography and cave passage morphologies indicate at least two stages of phreatic passage development in the Burnsville Cove. The first was low-velocity deep circulation and dissolution that predates all modern landscape controls and karst drainage features. Prior to the development of the current surface drainage pattern, groundwater (and surface water) may have flowed to the south toward what is now Dry Run Gorge. This early stage of development was likely disrupted by the incision of the Bullpasture River Gorge, which pirated the river to the east near what is now Williamsville. After the initial piracy and lowering of the water table, the second stage of phreatic development was related to a stable period during incision of the Bullpasture River Gorge. Incision redirected drainage toward a paleo-spring in the incipient Bullpasture River Gorge and initiated development of a large, nearly horizontal phreatic trunk. Most recently, renewed and rapid incision of the Gorge lowered the water table again, rearranging drainage patterns and dissecting what may originally have been one larger drainage basin into the four smaller basins which exist today."

15.6 Reviews and Descriptions

Burnsville Cove has come to be recognized as one of the most significant karst areas in the United States. With a total cave length of 70 miles, Burnsville Cove cannot compete (so far!) with the Mammoth Cave area, or with the Carlsbad—Lechuguilla and related caves, or with the 100-mile maze caves of the Black Hills. But it makes up for length in geological complexity and thus of scientific interest. The Burnsville Cove caves appeared as an entry in the massive volume *Speleogenesis*, published by the NSS (White 2000) and as an article in the *Encyclopedia of Caves* (Clemmer 2005).

15.7 Conclusions

The chapters that follow give a summary of the current status of research in Burnsville Cove. Although there has been impressive progress, there is a great deal of additional science that can be done. Some would be suitable for thesis work for interested graduate students. Some of the research could be undertaken by interested cavers. This volume celebrates 50 years of progress but it should not be taken as a historic document but rather as a roadmap to the future.

References

Chess, D.L. 1983. Thesis proposal for an M.S. in environmental pollution control. *BCCS Newsletter* 9: 26–32.

Chess, D.L. 1987. Comparisons of microbiology and selected anions for surface and subsurface stream waters for the Aqua Spring watershed in Burnsville Cove, Virginia. M.S. thesis in Environmental Pollution Control, The Pennsylvania State University, 116 pp.

Chess, D.L., C.A. Chess, I.D. Sasowsky, V.A. Schmidt, and W. B. White. 2010. Clastic sediments in the Butler Cave—sinking creek system, Virginia, USA. *Acta Carsologica* 39: 55–70.

Clemmer, G.S. 2005. Caves of Burnsville Cove, Virginia. In *Encyclopedia of Caves*, ed. D.C. Culver, and W.B. White, 60–71. Amsterdam: Elsevier Science.

Cournoyer, D.N. 1954. The speleo-barometer. *NSS Bulletin* 16: 91–93.

Crowell, B.E. 2001. *Luminescence of speleothems as in indicator of paleoclimate: correlation with contemporary climate records*. M.S. in Environmental Pollution Control: The Pennsylvania State University.

Davis, N.W., and J.W. Hess. 1982. Hydrogeology of the drainage system, Burnsville Cove, Virginia. *NSS Bulletin* 44: 78–83.

Deike, G.H. III. 1960a. Cave development in Burnsville Cove, west-central Virginia, with special reference to Breathing Cave. M.A. thesis in Geology, University of Missouri, 155 pp.

Deike, G.H. III. 1960b. Origin and geologic relations of Breathing Cave, Virginia. *NSS Bulletin* 22: 30–42.

Faust, B. 1947. An unusual phenomenon. *NSS Bulletin* 9: 52–54.

Harmon, R.S., and J.W. Hess. 1982. Ground water geochemistry of the Burnsville Cove area, Virginia. *NSS Bulletin* 44: 84–89.

Heil, T.L. 1995. Biological involvement in the formation of moonmilk. B.S. thesis in Geosciences, The Pennsylvania State University. 37 pp.

Holsinger, J.R. 1982. A preliminary report on the cave fauna of Burnsville Cove, Virginia. *NSS Bulletin* 44: 98–101.

Lucas, P.C. 2009. Recording a flood event inside the water sinks cave. Proceedings of the 15th International Congress of Speleology 3: 1593–1596.

Lucas, P.C., F. Marks, Jr. and N.W. Davis. 2009. A method for detecting cave connections by induced air flow. Proceedings of the 15th International Congress of Speleology 3: 2100–2105.

Schmidt, V. 1958. The ins and outs of Breathing Cave. *Speleodigest* 2-107–2-109.

Schwartz, B.F. and D.H. Doctor. 2009. Geomorphic and hydrogeologic evolution of karst in the Burnsville Cove, Virginia, U.S.A.: new evidence and perspectives. Proceedings of the 15th International Congress of Speleology 2: 984–990.

Swezey, C. and N. Piatak. 2003. Gypsum in the Butler Cave—Sinking Creek Cave System, Bath County, Virginia. *BCCS Newsletter* 28:15–22. Also see GSA abstract (2002).

Wefer, F.L. 1984. Cave meteorology. *BCCS Newsletter* 10: 33–49.

Wefer, F.L. 1985. More on cave meteorology. *BCCS Newsletter* 11: 25–56.

Wefer, F.L. 1988. Still more on cave meteorology. *BCCS Newsletter* 14: 76–84.

Wefer, F.L. 1989. Yet still more on cave meteorology. *BCCS Newsletter* 15: 18–38.

Wefer, F.L. 1990. Even yet still more on cave meteorology. *BCCS Newsletter* 16: 20–41.

Wefer, F.L. 1991. Meteorology of the Butler Cave-Sinking Creek System. In *Appalachian Karst*, ed. E.H. Kastning, and K.M. Kastning, 65–74. Huntsville: National Speleological Society.

Wefer, F.L. 1992. Meteorological conditions far inside a large cave system. *BCCS Newsletter* 18: 36–41.

White, W.B. 1982. Mineralogy of the Butler Cave—Sinking Creek System. *NSS Bulletin* 44: 90–97.

White, W.B., and J.W. Hess. 1982. Geomorphology of Burnsville Cove and the geology of the Butler Cave—Sinking Creek System. *NSS Bulletin* 44: 67–77.

White, W.B. 1997. Thermodynamic equilibrium, kinetics, activation barriers, and reaction mechanisms for chemical reactions in Karst terrains. *Environmental Geology* 30: 46–58.

White, W.B. 2000. Cave development in Burnsville Cove, Virginia, U.S.A. In *Speleogenesis: Evolution of Karst Aquifers*, Klimchouk, A.B., Ford, D.C., Palmer, A.N., and Dreybrodt, W., eds., National Speleological Society, Huntsville, 362–366 pp.

The Geology of Burnsville Cove, Bath and Highland Counties, Virginia

16

Christopher S. Swezey, John T. Haynes, Richard A. Lambert, William B. White, Philip C. Lucas, and Christopher P. Garrity

Abstract

Burnsville Cove is a karst region in Bath and Highland Counties of Virginia. A new geologic map of the area reveals various units of limestone, sandstone, and siliciclastic mudstone (shale) of Silurian through Devonian age, as well as structural features such as northeast-trending anticlines and synclines, minor thrust faults, and prominent joints. Quaternary features include erosional (strath) terraces and accumulations of mud, sand, and gravel. The caves of Burnsville Cove are located within predominantly carbonate strata above the Silurian Williamsport Sandstone and below the Devonian Oriskany Sandstone. Most of the caves are located within the Silurian Tonoloway Limestone, rather than the Silurian-Devonian Keyser Limestone as reported previously.

16.1 Introduction

This chapter describes the geology of Burnsville Cove, a karst region that forms the southwestward extension of the valley occupied by the Bullpasture River in Bath and Highland Counties of western Virginia (Figs. 16.1 and 16.2). Burnsville Cove is approximately 6 miles long and 2–3 miles wide. The Cove is bounded on the west by Jack Mountain, on the east by Tower Hill Mountain, on the south by Warm Springs Mountain, and on the north by the Bullpasture River (Fig. 16.2). Chestnut Ridge is a north-trending ridge within Burnsville Cove. The strata exposed in Burnsville Cove consist of limestone, sandstone, and siliciclastic mudstone (shale) of Silurian and Devonian age, and these strata provide the template upon which the numerous caves have formed.

16.2 Historical Background

The geological features of Bath and Highland Counties were first described during the nineteenth century. William B. Rogers, the State Geologist of the

C.S. Swezey (✉)
US Geological Survey, 12201 Sunrise Valley Drive, MS 926A, Reston, VA 20192, USA
e-mail: cswezey@usgs.gov

J.T. Haynes
Department of Geology and Environmental Science, James Madison University, Harrisonburg, VA 22807, USA

R.A. Lambert
PO Box 151, Monterey, VA 24465, USA

P.C. Lucas
587 Limestone Lane, Williamsville, VA 24487, USA

W.B. White
Department of Geosciences, The Pennsylvania State University, University Park, PA 16802, USA

C.P. Garrity
US Geological Survey, 12201 Sunrise Valley Drive, MS 956, Reston, VA 20192, USA

W.B. White (ed.), *The Caves of Burnsville Cove, Virginia*,
Cave and Karst Systems of the World, DOI 10.1007/978-3-319-14391-0_16,

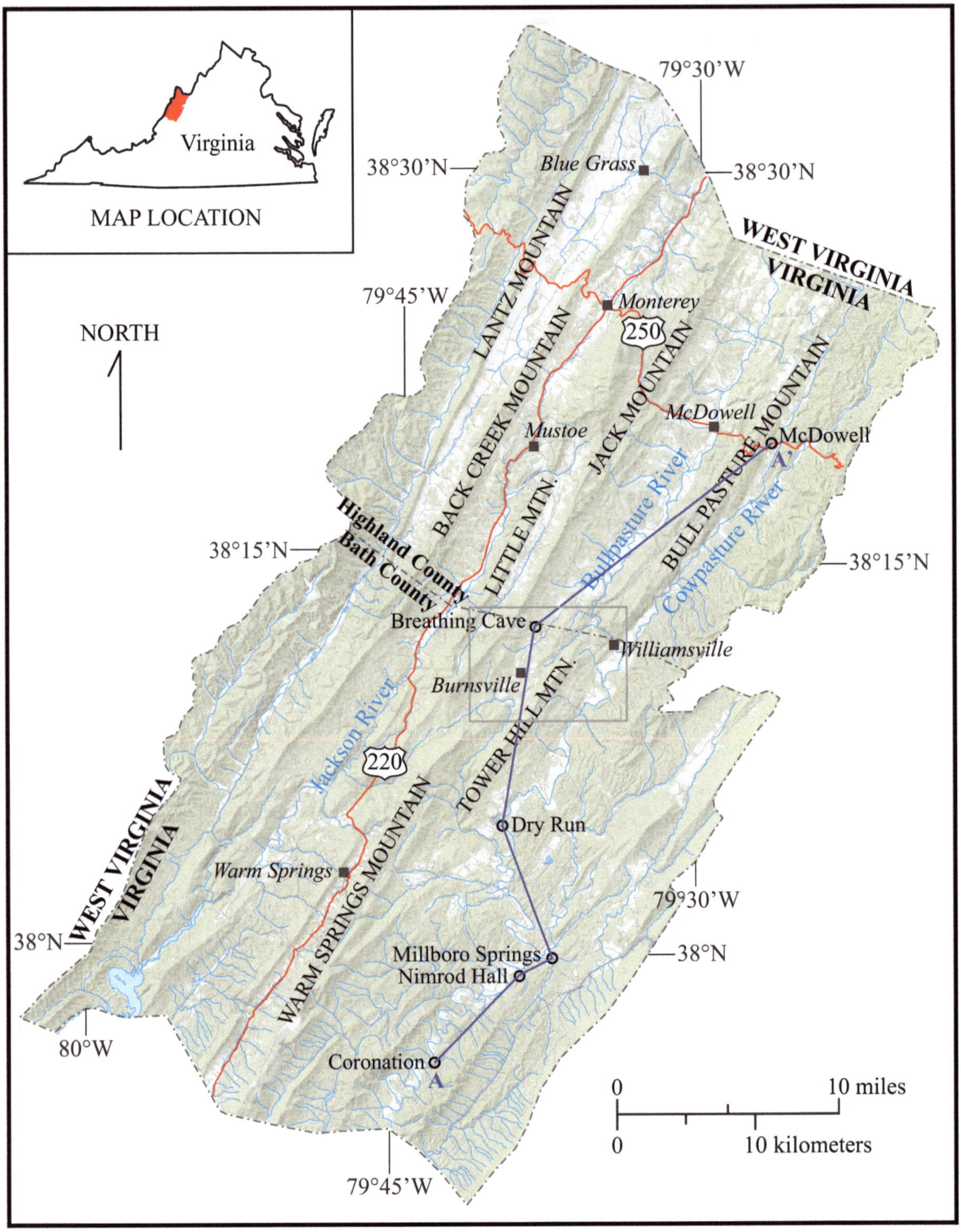

Fig. 16.1 Map of Highland County and Bath County, Virginia. Towns and other communities are denoted by *black squares* and names in *italics*. *Rectangle* in *middle* of figure outlines the area shown in Fig. 16.2. *Red lines* denote major roads. *Purple line* denotes location of cross section A–A′ (Fig. 16.17), and *open circles* denote outcrops used in the construction of the cross section. *Blue* denotes hydrological features *MTN.* MOUNTAIN

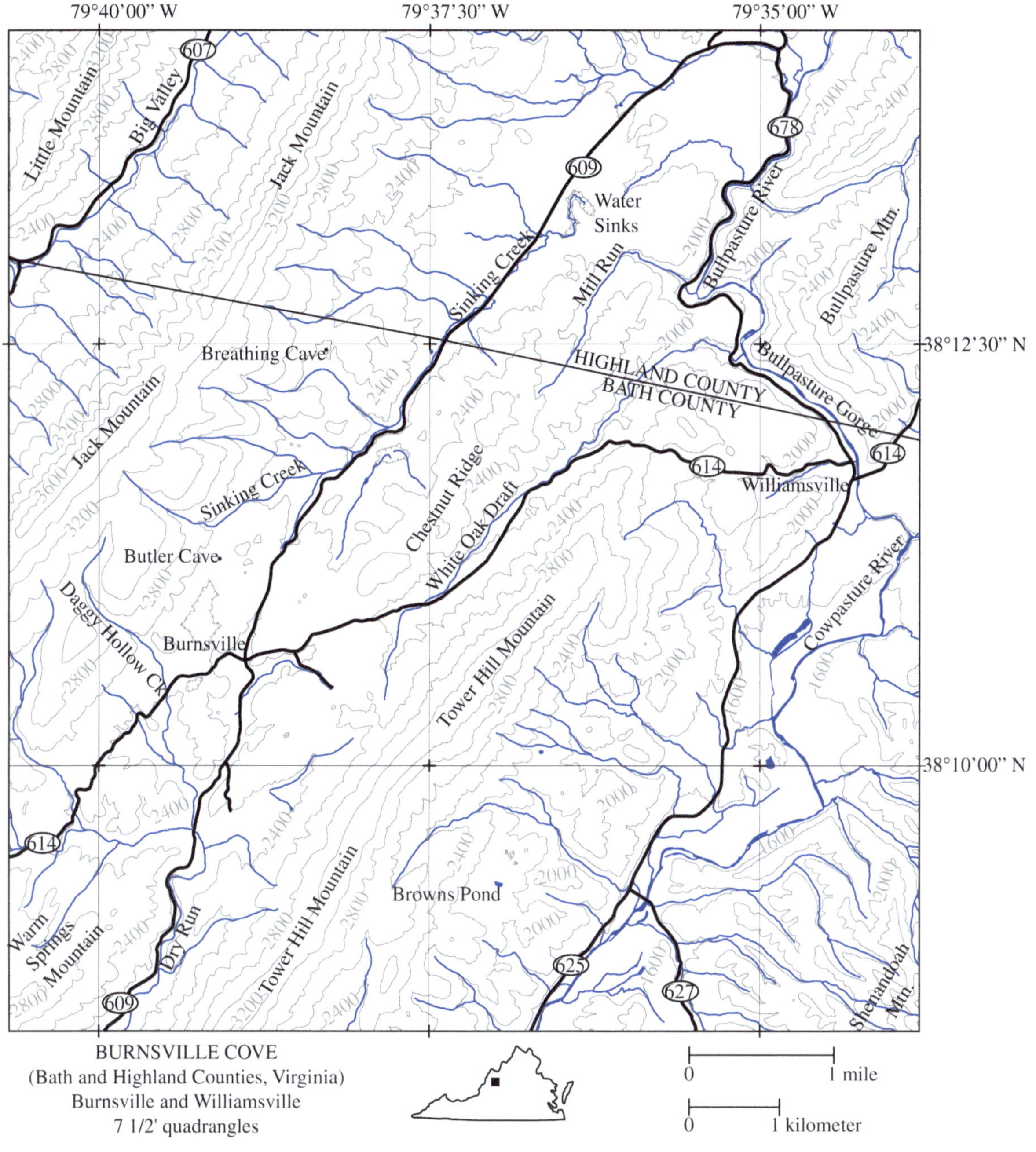

Fig. 16.2 Topographic map of Burnsville Cove, Bath and Highland Counties, Virginia. Data derived from U.S. Geological Survey 7.5 min topographic maps of the Burnsville quadrangle (1982) and the Williamsville quadrangle (1999). *Blue* denotes hydrological features. Contour elevations are given in feet above mean sea level

Geological Survey of Virginia (1835–1842) and a Professor of Natural Philosophy at the University of Virginia (1835–1853), provided initial descriptions of the strata and structural features of the region (Rogers 1843). After the American Civil War, John L. Campbell of Washington and Lee University provided more detailed descriptions of the geology of Bath County (Campbell 1879). Similar detailed descriptions of the geology of Highland County were later published by Campbell (1885), the Superintendent of the

Public Schools of Highland County, and by Hotchkiss (1885), a school teacher in Staunton and cartographer for Thomas "Stonewall" Jackson during the American Civil War. The first truly comprehensive geological map of Bath and Highland Counties, however, was published by Darton (1899), a geologist with the U.S. Geological Survey.

Specific geological features of Burnsville Cove were first described and mapped systematically during the 1940s and 1950s. The initial mapping effort was by Ramsey (1950), who completed a Master's thesis at the University of Virginia on the geology of the southern part of the Bolar Anticline (also known as the Warm Springs Anticline). Ramsey's work was soon followed by studies by George H. Deike III and other members of the Nittany Grotto at The Pennsylvania State University (Deike 1959, 1960a, b). Deike's work has been used for many later interpretations of the geologic setting of specific caves in Burnsville Cove (e.g., Hess and Davis 1969; Davis and Hess 1982; Harmon and Hess 1982; White and Hess 1982; Schwartz and Doctor 2009).

Subsequent studies related to Burnsville Cove were published by the Virginia Division of Mineral Resources (VDMR). In 1962, the VDMR published a report by Bick (1962) on the geology of the Williamsville quadrangle. The area of this report encompasses 235 square miles in Bath and Highland Counties (including Burnsville Cove), as well as some western parts of Rockbridge County (immediately east of Bath County). At the time that he conducted this work, Bick was a professor of geology at Washington and Lee University. In 1965, the VDMR published a report by Kozak (1965) on the geology of the Millboro Quadrangle. Although this quadrangle is located south of Burnsville Cove, the report includes parts of southern Bath County. At the time that he conducted this work, Kozak also was a professor of geology at Washington and Lee University. In 2001, Eugene K. Rader and Gerald P. Wilkes of the VDMR published a 1:100,000 scale geological map of the Virginia portion of the Staunton 30 × 60 min quadrangle (Rader and Wilkes 2001). This map encompasses most of Bath and Highland Counties, as well as Augusta County (immediately east of Highland County). As with the map by Bick (1962), the map by Rader and Wilkes (2001) shows the general stratigraphic and structural features of Burnsville Cove, but not in great detail.

In 2009, John T. Haynes and Steven J. Whitmeyer (and their students at James Madison University), along with Richard A. Lambert and Philip C. Lucas, began to publish results of various detailed geology research projects in Bath and Highland Counties. Their work has focused on deformation in Devonian shale (Haynes et al. 2009; Caro et al. 2010; Haynes and Whitmeyer 2011), petrographic studies of Devonian shale (Haynes et al. 2010b), and studies of Silurian and Devonian stratigraphy in the vicinity of Burnsville Cove (Walker et al. 2010; Haynes et al. 2010a, 2011).

16.3 Stratigraphy

The geologic map (Fig. 16.3) and accompanying stratigraphic columns (Figs. 16.4 and 16.5) show a variety of lithologies that range in age from Ordovician through Devonian. The following section, however, describes only the strata that outcrop in Burnsville Cove. These strata consist primarily of limestone, sandstone, and siliciclastic mudstone (shale) of Silurian to Devonian age. Chert is also a component in some of the Silurian to Devonian strata. In addition to these older stratigraphic units, Quaternary gravel, sand, and mud (not shown in Fig. 16.3) are found throughout Burnsville Cove, both in caves and on the surface.

The stratigraphic nomenclature applied to these rocks of Burnsville Cove has varied greatly over the years, but most of the caves in Burnsville Cove have formed within the Silurian Tonoloway Limestone and the Silurian-Devonian Helderberg Group. During the middle to late 20th century, the names applied to these rocks were taken from the publications of Swartz (1930), Butts (1940), and Woodward (1941, 1943), each of whom used many stratigraphic names derived from localities in New York. During the late 1960s through early 1970s, Zeddie P. Bowen (a professor at the University of Rochester) and James W. Head (a graduate student at Brown University) published regional studies of the Helderberg Group (Bowen 1967; Head 1972, 1974), and they claimed that it was inappropriate to use some of the New York nomenclature south of central Pennsylvania. Head's generalized stratigraphic section for Bath and Highland Counties is shown in Fig. 16.6. During the 1980s, with continued geologic investigations of the caves of Burnsville Cove, White and Hess (1982) produced a stratigraphic column (Fig. 16.7) that draws on the nomenclature of Swartz (1930), Butts (1940), Deike (1960a, b), and Head (1972, 1974). This stratigraphic column has been widely copied and it appears in a number of publications. Since 2000, studies by the authors of this chapter have resulted in

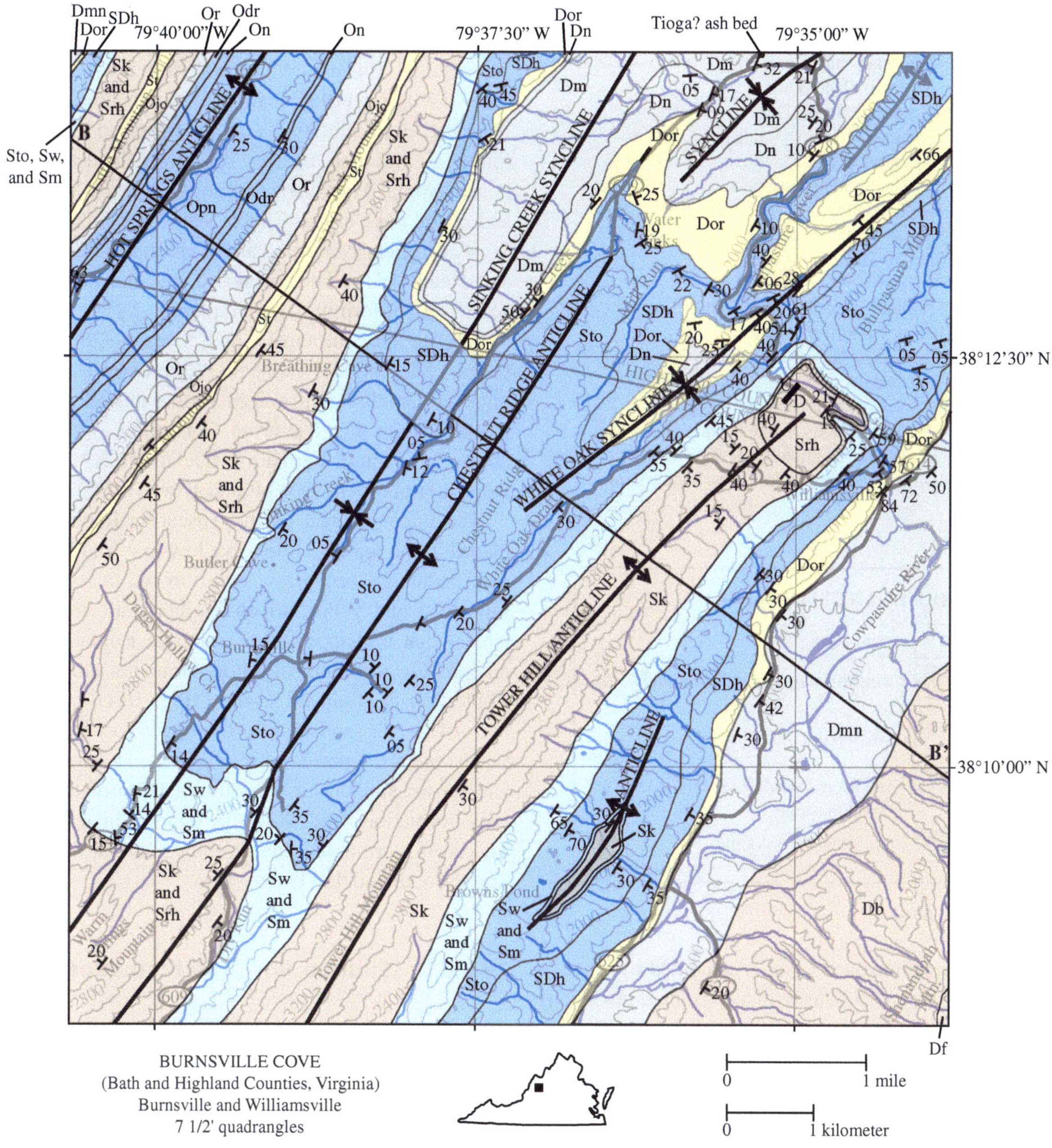

Fig. 16.3 Geologic map of Burnsville Cove, Bath and Highland Counties, Virginia. Map base is same as Fig. 16.2. *Sources of geological data* Deike (1960a, b), Bick (1962), and field work by the authors. Cross section B–B′ is shown in Fig. 16.33

additional understanding of the strata in Burnsville Cove. The following sections of this chapter describe these strata in the vicinity of Burnsville Cove according to this most recent work (Fig. 16.8). Additional details regarding stratigraphic nomenclature, type sections, and the contacts between stratigraphic units may be found in Appendix A of this volume.

16.3.1 Silurian Tuscarora Formation

Description: The Tuscarora Formation is a 40–120 foot-thick unit of white to yellow-gray sandstone that forms very hard and resistant ridges. The sandstone is a quartzarenite (according to the classifications of McBride 1963; Folk et al. 1970) composed primarily

Df = Devonian Foreknobs Formation
Db = Devonian Brallier Formation
Dmn = Devonian shale:
Millboro Shale (Dm)
Needmore Shale (Dn)
Dor = Devonian Oriskany Sandstone
SDh = Silurian-Devonian Helderberg Group
Sto = Silurian Tonoloway Limestone
and Silurian Wills Creek Formation
Sw = Silurian Williamsport Sandstone
Sm = Silurian McKenzie Formation
Sk = Silurian Keefer Formation
Srh = Silurian Rose Hill Formation
St = Silurian Tuscarora Formation
Ojo = Ordovician Juniata Formation (Oj)
and Ordovician Oswego Sandstone (Oo)
Or = Ordovician Reedsville Shale
Odr = Ordovician Dolly Ridge Formation
On = Ordovician Nealmont Limestone
Opn = Ordovician pre-Nealmont carbonate strata

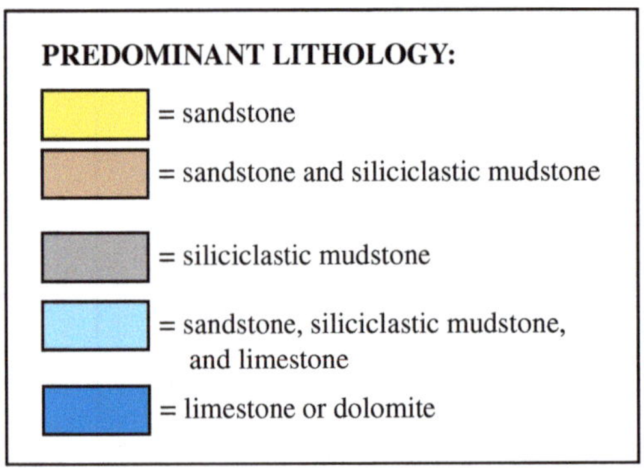

Fig. 16.3 continued

of monocrystalline quartz grains cemented by silica as quartz overgrowths. Lenses of conglomerate (mostly quartz pebbles) are present at a few places. The unit contains beds that are 1–5 feet-thick, and some display prominent cross-bedding. Good exposures of the Tuscarora Formation are found on the crest of Jack Mountain along U.S. Route 250 in Highland County (Fig. 16.1).

Contacts: The lower contact of the Tuscarora Formation is placed at the top of a unit of red sandstone and siliciclastic mudstone (mapped as the Juniata Formation), just below a unit of white to yellow-gray sandstone that is mapped as the Tuscarora Formation. Rader and Gathright (1984) and Diecchio (1985) stated that the Juniata Formation is capped by an unconformity at some locations in Bath and Highland Counties. This unconformity is likely to be a regional unconformity named the Cherokee Unconformity, which extends throughout the Appalachian, Michigan, and Illinois Basins (Wheeler 1963; Swezey 2002, 2009). Wheeler (1963) named this unconformity the "Taconic discontinuity," and he indicated that it is found at the base of the Tuscarora Sandstone (in Pennsylvania, Maryland, and northern West Virginia) and at the base of the Clinch Sandstone (in southern West Virginia, Virginia, and Tennessee). Dennison and Head (1975) later proposed that the name "Taconic disconformity" be replaced by the name "Cherokee Discontinuity," and Swezey (2002) proposed changing the name to "Cherokee Unconformity." In southwestern Virginia, this unconformity is found within the Tuscarora Formation, which lies above the Juniata Formation (Dennison et al. 1992; Dorsch et al. 1994; Dorsch and Driese 1995).

The upper contact of the Tuscarora Formation is placed at the top of white to yellow-gray sandstone at most localities, just below a unit of red to red-brown sandstone and yellow to olive siliciclastic mudstone (shale) that is mapped as the Silurian Rose Hill Formation. On the flank of Little Mountain along U.S. Route 250 in Highland County (Fig. 16.1), however, beds of quartz sandstone in the uppermost Tuscarora Sandstone are reported to interfinger with beds of red and brown siliciclastic mudstone (shale) of the Rose Hill Formation, and thus the upper contact of the Tuscarora Formation may be transitional at this location (Woodward 1941).

16.3.2 Silurian Rose Hill Formation

Description: The Rose Hill Formation is a 300–800 foot-thick unit of red sandstone (containing some cobble-size to pebble-size intraclasts of siliciclastic mudstone) and yellow to olive siliciclastic mudstone (shale). The sandstone is a quartzarenite to litharenite, and the sand grains are cemented by hematite. Fossils include marine ostracods, crinoids, trilobites, and brachiopods, as well as abundant *Cruziana* trace fossils. Good exposures of the Rose Hill Formation are found in Bullpasture Gorge as ledges in the river and as bluffs along the river bank (Fig. 16.3).

Contacts: The lower contact of the Rose Hill Formation is the upper contact of the Tuscarora Formation. The upper contact of the Rose Hill Formation is placed at the top of a unit of red to red-brown

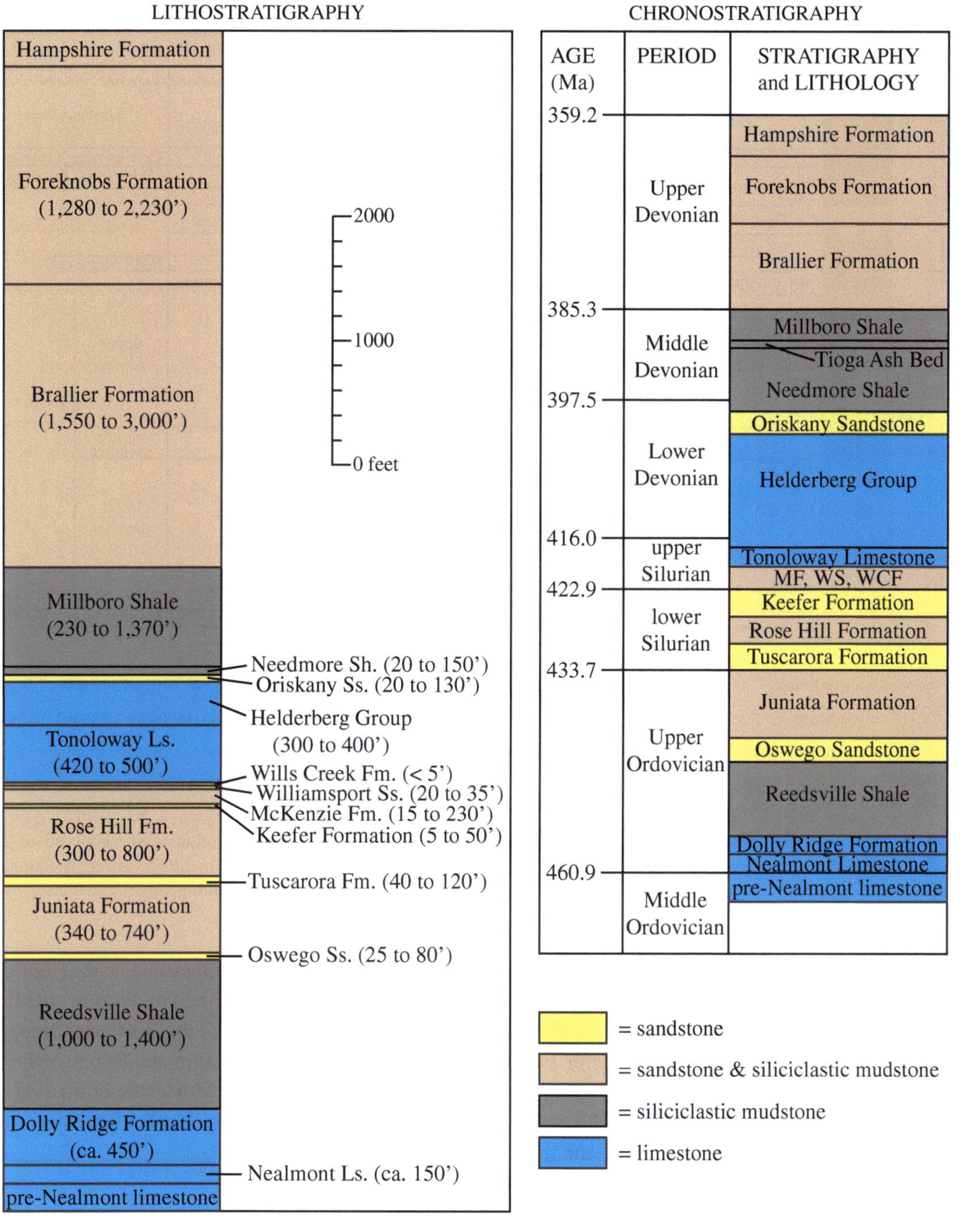

Fig. 16.4 Lithostratigraphic and chronostratigraphic charts for strata in the vicinity of Burnsville Cove, Bath and Highland Counties, Virginia. Time scale is from Gradstein et al. (2004). *Fm.* Formation; *Ls.* Limestone; *Sh.* Shale; *Ss.* Sandstone; *MF* McKenzie Formation; *WCF* Wills Creek Formation; *WS* Williamsport Sandstone

sandstone or yellow to olive siliciclastic mudstone (shale), just below a unit of white to yellow-gray sandstone that is mapped as the Silurian Keefer Formation.

16.3.3 Silurian Keefer Formation

Description: The Keefer Formation is a 5–50 foot-thick unit of white to yellow-gray sandstone.

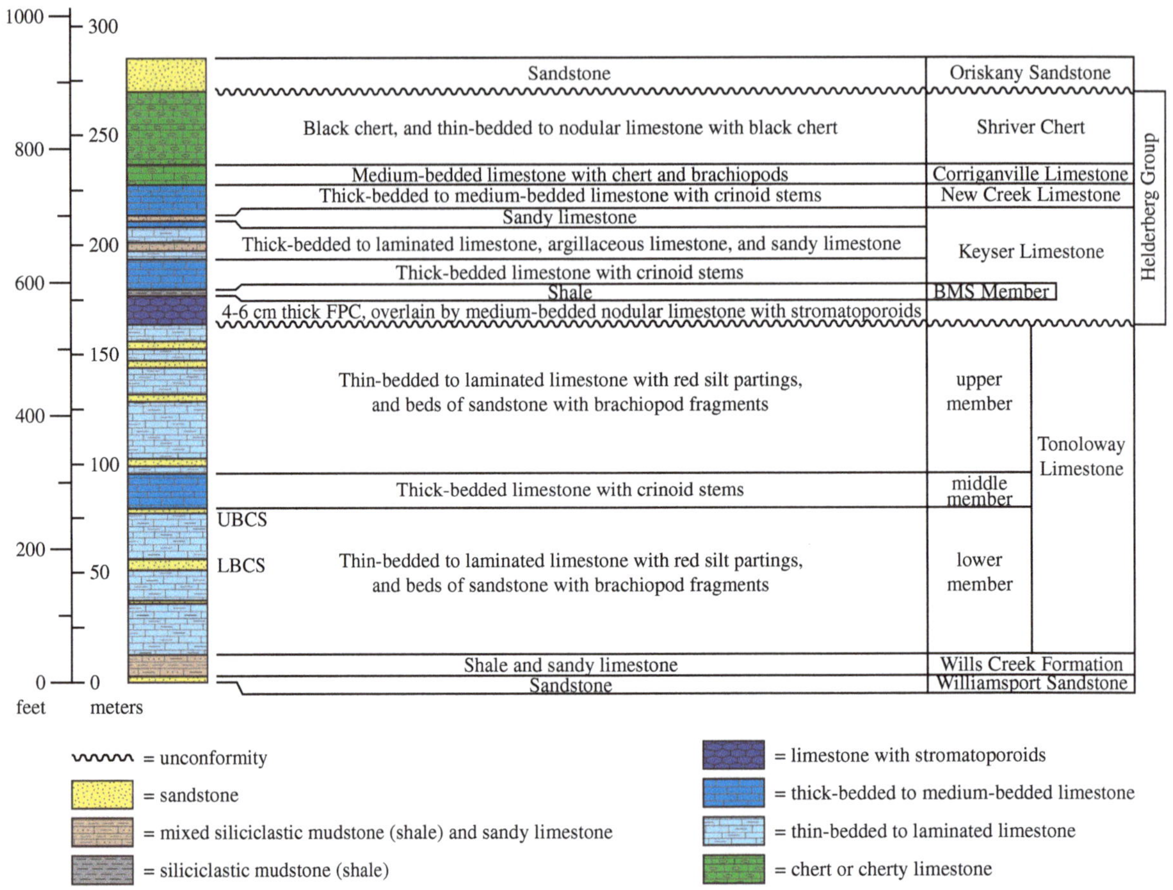

Fig. 16.5 Silurian-Devonian strata exposed along U.S. Route 250 on east side of Bullpasture Mountain from recent work by J.T. Haynes and colleagues. *BMS* Big Mountain Shale; *FPC* flat-pebble conglomerate (conglomerate composed of flat pebbles of carbonate mudstone); *LBCS* lower Breathing Cave sandstone; *UBCS* upper Breathing Cave sandstone

The sandstone is a quartzarenite composed primarily of monocrystalline quartz grains cemented by silica as quartz overgrowths. Fossils include vertical burrows (*Skolithos*) and rare brachiopods. The Keefer Formation is 30 feet thick in Bullpasture Gorge along State Road 678, and 8 feet thick at Lower Gap (Haynes et al. 2011). At Blue Grass, the sandstone of the Keefer Formation is absent, and a bed of hematitic and chamositic oolites (cemented by ferroan dolomite) is present at approximately the same position.

Contacts: The lower contact of the Keefer Formation is the upper contact of the Rose Hill Formation. The upper contact of the Keefer Formation is placed at the top of a unit of white to yellow-gray sandstone, either just below the first distinct bed of limonitic and oolitic limestone (mapped as the McKenzie Formation) or just below a thin bed of calcite-cemented shale that in places contains siltstone, limestone, and dolomite. This thin bed of shale was identified by Woodward (1941) as a separate formation named the Rochester Shale, and it was later mapped by Diecchio and Dennison (1996) as the lowermost bed of the McKenzie Formation. Deike (1960a) suggested that the Keefer Formation may be capped by an unconformity.

16.3.4 Silurian McKenzie Formation

Description: The McKenzie Formation is a 15–230 foot-thick unit of gray shale, white to yellow-gray sandstone, and yellow-orange to black sandy and oolitic limestone and gray limestone. The sandstone is a quartzarenite, and the limestone ranges from carbonate mudstone to sandy and oolitic packstone to grainstone

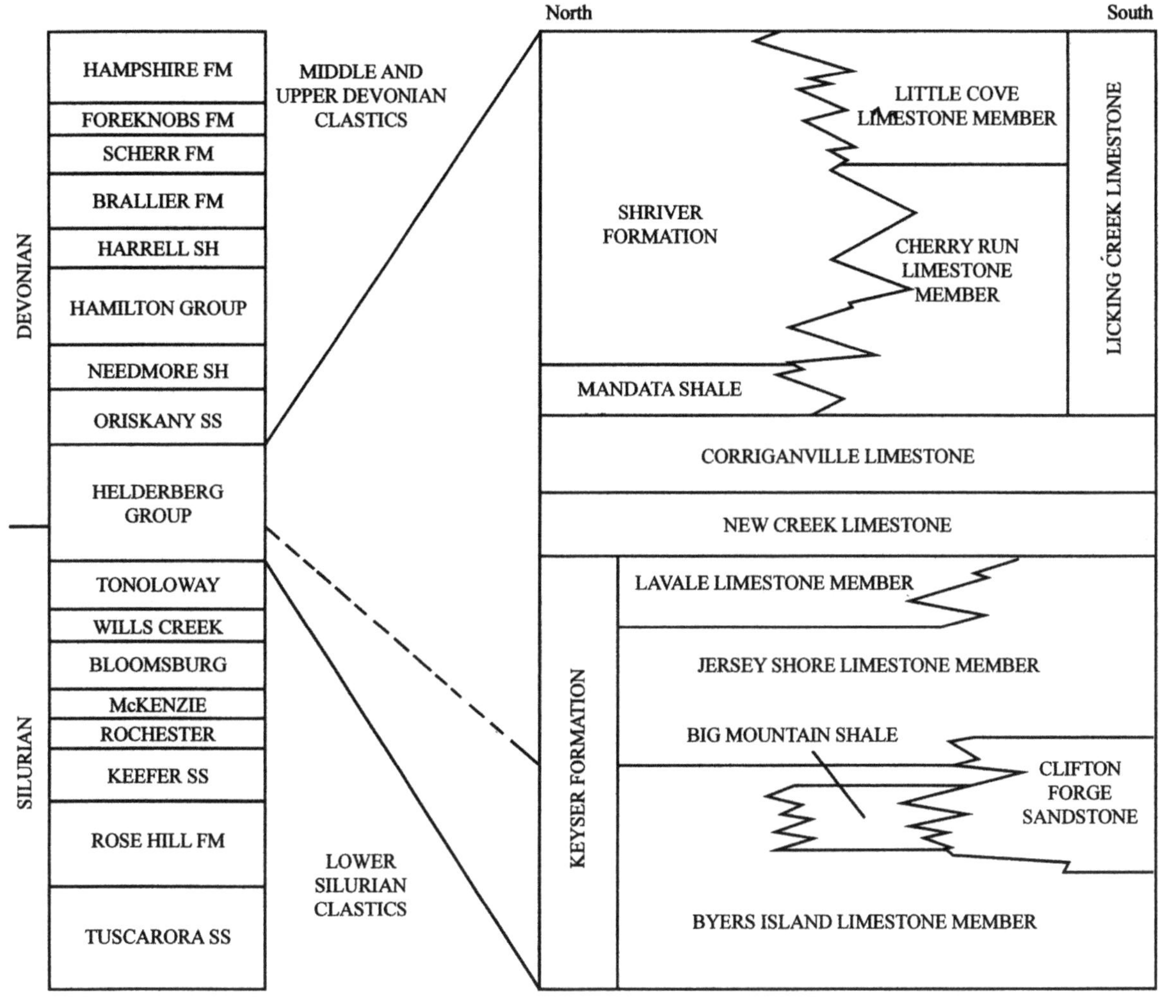

Fig. 16.6 Stratigraphic column showing the nomenclature of Head (1972). *FM* Formation; *SH* Shale; *SS* Sandstone

(according to the classification of Dunham 1962). Sedimentary structures within the McKenzie Formation include cross-bedding and laminations. Haynes et al. (2011) report that the McKenzie Formation is 170 feet thick in Bullpasture Gorge (Fig. 16.3).

Contacts: The lower contact of the McKenzie Formation is the upper contact of the Keefer Formation. At most localities, the upper contact of the McKenzie Formation is placed at the top of a unit of dark gray siliciclastic mudstone (shale) and limestone, just below a bed of brown sandstone (with ripple marks) that is mapped as the Williamsport Sandstone. Along U.S. Route 250 just east of McDowell, Woodward (1941) reported that the top of the McKenzie Formation is gradational with the base of the overlying Williamsport Sandstone. The authors of this paper, however, have observed that the contact is sharp and not gradational in Bullpasture Gorge along State Road 678 and by the river, where dark gray carbonate mudstone of the uppermost McKenzie Formation is overlain by sandstone of the lowermost Williamsport Sandstone.

16.3.5 Silurian Williamsport Sandstone

Description: The Williamsport Sandstone is a 20–35 foot-thick unit of sandstone with minor amounts of glauconite (Fig. 16.9). The sandstone is a quartzarenite composed primarily of monocrystalline quartz grains cemented by silica as quartz overgrowths. The unit contains some cross-bedding and prominent ripple structures. Fossils include pyritized ostracode shells

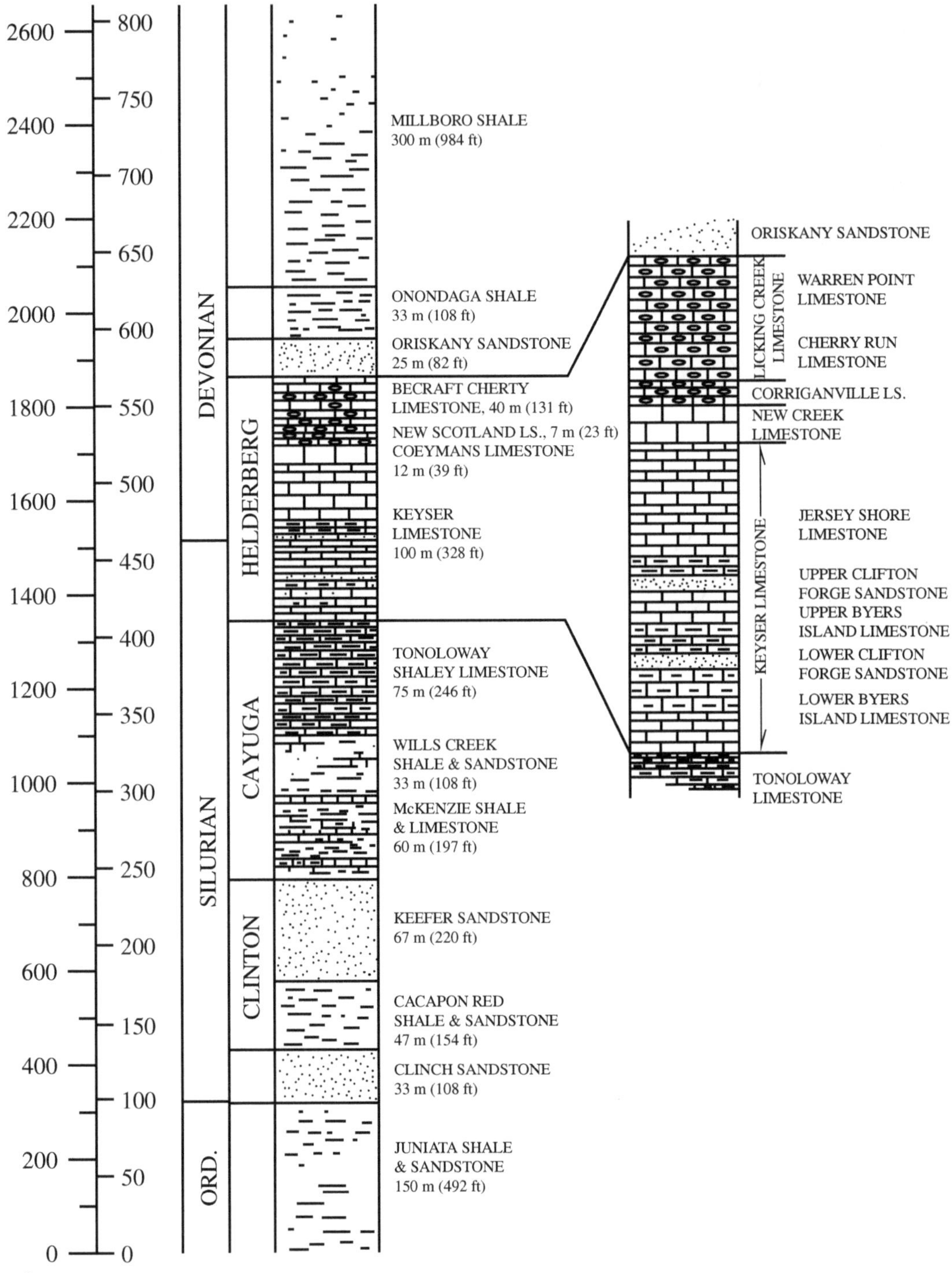

Fig. 16.7 Stratigraphic column from Burnsville Cove, as published by White and Hess (1982) LS. Limestone

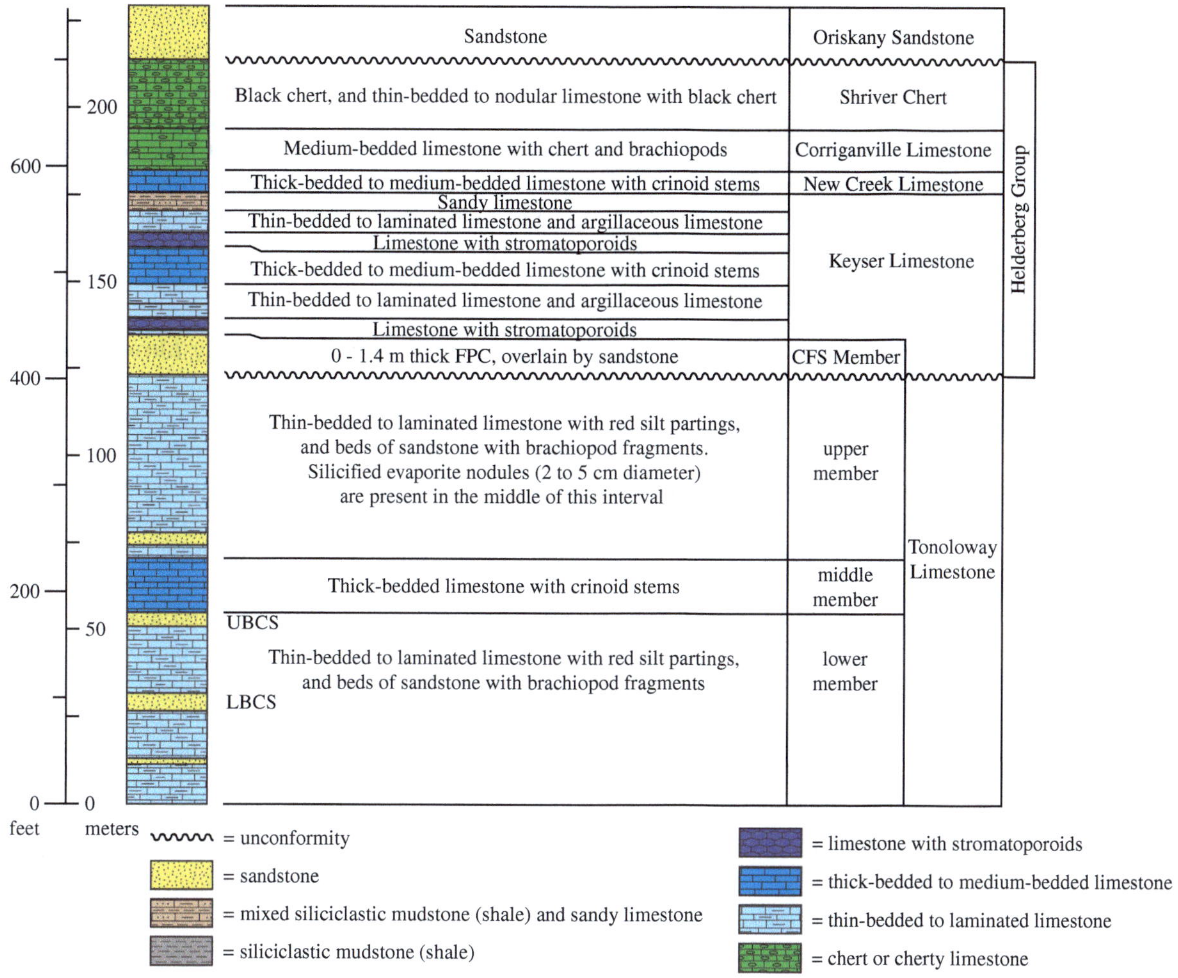

Fig. 16.8 Composite stratigraphic column from Water Sinks, Chestnut Ridge, and Burnsville Cove from recent work by John Haynes and colleagues. *CFS* Clifton Forge Sandstone; *LBCS* lower Breathing Cave sandstone; *UBCS* upper Breathing Cave sandstone

and horizontal trace fossils. The Williamsport Sandstone is 13 feet thick on the east side of Bullpasture Mountain along U.S. Route 250 (Fig. 16.5) and 27 feet thick in Bullpasture Gorge (Haynes et al. 2011).

Contacts: The lower contact of the Williamsport Sandstone is the upper contact of the Keefer Formation. The upper contact of the Williamsport Sandstone is placed at the top of a unit of brown to yellow-brown sandstone, just below a unit of brown to pea-green calcite-cemented sandstone, sandy limestone, or siliciclastic mudstone (shale) that is mapped as the Wills Creek Formation. Good exposures of the Williamsport Sandstone are found in Bullpasture Gorge and on State Road 614 just north of Williamsville.

16.3.6 Silurian Wills Creek Formation

Description: The Wills Creek Formation is a 0–212 foot-thick unit of primarily brown to green shale, sandstone, sandy limestone, and limestone. Ripple structures, laminations, desiccation cracks, and molds of evaporite crystals are present in some of the beds. Other beds contain conglomerate composed of flat pebbles of carbonate mudstone. Fossils are not abundant, but include leperditian ostracods, stromatolites, and brachiopods. The Wills Creek Formation is 36 feet thick on the east side of Bullpasture Mountain along U.S. Route 250 (Fig. 16.5) and <5 feet thick in Bullpasture Gorge. On Fig. 16.3, the unit is mapped with

Fig. 16.9 The Silurian Williamsport Sandstone in Bullpasture Gorge. Photo by J.T. Haynes

Fig. 16.10 Laminated limestone in the upper member of the Silurian Tonoloway Limestone in the Water Sinks Cave System. "Water Sinks" location is shown on Fig. 16.2. This passage is known as the "Sweet Dreams Passage" and is named after John Sweet, who is shown in this photograph. Photo by C.S. Swezey

the overlying Tonoloway Limestone. Good exposures of the Wills Creek Formation are present on U.S. Route 250 and State Road 613, both on the southeastern flank of Bullpasture Mountain.

Contacts: The lower contact of the Wills Creek Formation is the upper contact of the Williamsport Sandstone. The upper contact of the Wills Creek Formation is placed at the top of a unit of yellow-brown calcite-cemented sandstone or calcite-cemented shale, just below a thick unit of gray to blue thin-bedded to laminated limestone (carbonate mudstone) that is mapped as the Tonoloway Limestone. Woodward (1941), however, stated that the contact of the Wills Creek Formation with the overlying Tonoloway Limestone is a transitional contact, although the Tonoloway Limestone is more uniformly limy and slabby than the underlying Wills Creek Formation. A good exposure of the Wills Creek-Tonoloway contact can be found along U.S. Route 250 on the east flank of Bullpasture Mountain east of McDowell.

16.3.7 Silurian Tonoloway Limestone

Description: The Tonoloway Limestone is a 420–500 foot-thick unit of gray to blue limestone and dolomitic limestone, with some prominent beds of fine-grained to medium-grained sandstone and siliciclastic mudstone. The limestone and dolomitic limestone are primarily thin-bedded to laminated carbonate mudstone, with some beds of bioclastic and oolitic packstone to grainstone (Fig. 16.10). The sandstone is composed of quartz sand and fossil fragments, and a clay matrix is present at some places. A distinctive bed of white to gray bioclastic chert occurs at the base of one of the sandstone beds (upper Breathing Cave sandstone), and blocks derived from this sandstone and associated chert are present on the crest of Chestnut Ridge and the southeastern flank of Jack Mountain and Bullpasture Mountain (Fig. 16.3). The Tonoloway Limestone is 495 feet thick along U.S. Route 250 on the east flank of Bullpasture Mountain (Fig. 16.5) and 450 feet thick in Bullpasture Gorge.

Fig. 16.11 Red silt partings in black carbonate mudstone in the lower member of Silurian Tonoloway Limestone, Grand Canyon passage of Breathing Cave (location shown on Fig. 16.2). Photo by J.T. Haynes

Fig. 16.12 Encrusting stromatoporoids and *Favosites* coral in the middle member of the Silurian Tonoloway Limestone on State Road 614, east of Burnsville. Photo by J.T. Haynes

Fig. 16.13 Silicified evaporite nodules in the upper member of the Silurian Tonoloway Limestone, Sweet Dreams Passage in the Water Sinks Cave System. "Water Sinks" location is shown on Fig. 16.2. Photo by C.S. Swezey

In studies of this unit from Pocahontas County (West Virginia) to Allegany County (Maryland), Woodward (1941) and Bell and Smosna (1999) divided the Tonoloway Limestone into the following three members: (1) a 94–212 foot-thick lower member of gray to black limestone (carbonate mudstone), argillaceous limestone, dolomite, and dolomitic sandstone. This member is characterized by abundant thin beds and laminations (some of which are notably pink to red at some locations; Fig. 16.11), mud cracks, and silicified evaporite nodules. Fossils include ostracods, gastropods, stromatolites, and *Tentaculites*; (2) a 64–120 foot-thick middle member of thick-bedded gray to black limestone (packstone to grainstone), knobby irregularly bedded blue limestone, gray limestone (carbonate mudstone), and scattered beds of gray calcite-cemented sandstone. Fossils include corals, brachiopods, and crinoid stems. At some localities, the lower third of the middle Tonoloway Limestone contains a prominent zone of stromatoporoids (Fig. 16.12). Denkler and Harris (1988) identified and described a new index species of conodont (*Homeognathodus peniculus*) from the middle member of the Tonoloway Limestone, and their holotype specimen was obtained from this stromatoporoid-bearing unit of the middle Tonoloway Limestone on U.S. Route 250 on the east flank of Bullpasture Mountain; and (3) a 182–220 foot-thick upper member of limestone (carbonate mudstone), argillaceous limestone, dolomitic sandstone, and dolomite characterized by abundant thin beds and laminations, mud cracks, and silicified evaporite nodules (Fig. 16.13). Fossils include rare ostracods and *Tentaculites*.

In Highland County, a good exposure of the Tonoloway Limestone is present on U.S. Route 250 approximately 3 miles east of the town of McDowell (Fig. 16.14). At this outcrop, the lower member of the Tonoloway Limestone is a black to gray to tan thin-bedded to laminated limestone (carbonate mudstone) with two prominent beds of sandstone that contain both quartz grains as well as skeletal remains of bryozoans, brachiopods, crinoids, and trilobites. On U.S. Route 250 and on State Road 614 east of Burnsville, the middle Tonoloway Limestone is a massive to thick-bedded limestone (packstone to grainstone) with a highly diverse and abundant fauna dominated by crinoids, stromatoporoids, and corals (*Favosites*, *Halysites*). At the outcrop on U.S. Route 250, the upper Tonoloway Limestone is a gray to tan thin-bedded to laminated limestone (carbonate mudstone) that includes some thin beds of conglomerate (composed of flat pebbles of carbonate packstone to grainstone), dolomitic fine sandstone, dolomitic siltstone, and dolomitic limestone (carbonate mudstone). Other features of this upper member include thin beds, laminations, mud cracks, silicified evaporite nodules and breccias, and stromatolite fossils. In stream passages in many caves of Burnsville Cove, potholes have formed in the thin-bedded and laminated limestone (carbonate mudstone) of the Tonoloway Limestone (Fig. 16.15).

Fig. 16.14 Contact between the Silurian Tonoloway Limestone and the overlying Silurian-Devonian Keyser Limestone along U.S. Route 250, east side of Bullpasture Mountain. Photo by J.T. Haynes

Fig. 16.15 Potholes in the Silurian Tonoloway Limestone in Butler Cave (location shown in Fig. 16.2). *Red* 3.5 inch long pocket knife for scale. Photo by C.S. Swezey

Contacts: Some of the lithologic units in the Tonoloway Limestone are similar to those of the overlying Keyser Limestone, and thus stratigraphic differentiation of these two formations can be difficult. Woodward (1941) stated that the Tonoloway Limestone has distinct thin beds and laminations, and is platier and less fossiliferous than the overlying knobby Keyser Limestone. Woodward (1941) specifically identified the upper contact of the Tonoloway Limestone in Highland County approximately 1 mile east of Crabbottom (later renamed Blue Grass) at the top of a bed of calcite-cemented shale, just beneath beds of stiff and platy limestone (mapped here as the Keyser Limestone). Butts (1933) noted that the Tonoloway Limestone is generally marked by the ostracode *Leperditia alta*, whereas the overlying Keyser Limestone is marked by the brachiopod *Chonetes jerseyensis* and the coral *Cladopora rectilineata*. The authors of this paper noted that in the vicinity of Burnsville Cove the limestone beds of the lower and upper members of the Tonoloway Limestone are predominantly thin-bedded carbonate mudstone, whereas the limestone beds of the Keyser Limestone are predominantly crinoid grainstone, nodular packstone and wackestone, and coral and stromatoporoid boundstone.

The lower contact of the Tonoloway Limestone at some locations is the upper contact of the Wills Creek Formation. A good exposure of the Wills Creek-Tonoloway contact can be found along U.S. Route 250 on the east flank of Bullpasture Mountain east of McDowell (Fig. 16.1). In Bullpasture Gorge, however, the Wills Creek Formation is thin or absent, and the Tonoloway Limestone rests directly on the older Williamsport Sandstone.

The upper contact of the Tonoloway Limestone at some locations is placed at the top of a unit of gray laminated limestone, just below a unit of thick-bedded to medium-bedded massive to nodular limestone that is mapped as the Keyser Limestone. In the Water Sinks Cave System, however, the uppermost bed of the Tonoloway Limestone is a 10–12 inch thick carbonate boundstone with prominent "cabbage-head" stromatolites. This bed is capped by an unconformity, above which lies a conglomerate (composed of flat pebbles of carbonate mudstone) and then a cross-bedded sandstone (composed of quartz sand and fossil fragments) that are mapped collectively as the Silurian Clifton Forge Sandstone Member of the Keyser Limestone. Dorobek and Read (1986) identified this unconformity in southwest Virginia, but it has only recently been recognized in Bath and Highland Counties.

In his study of Breathing Cave, Deike (1960a) placed the upper contact of the Tonoloway Limestone at the top of a unit of dark laminated limestone (containing a few beds of pink calcite-cemented mudstone), just beneath a unit of thin-bedded limestone with beds of gray shale and argillaceous limestone. In Breathing Cave, this contact is located 12 feet below a sandstone bed that is mapped as the lower Breathing Cave sandstone (described below). In contrast, White and Hess (1982) placed the upper contact of the Tonoloway Limestone at the top of a unit of predominantly shale 98 feet below the lower Breathing Cave sandstone. Recent unpublished work by John Haynes, Rick Lambert, and Phil Lucas, however, suggests that both the "upper Breathing Cave sandstone" and "lower Breathing Cave sandstone" are located within the Tonoloway Limestone rather than within the Keyser Limestone (based on correlations with outcrops on U.S. Route 250). The "upper Breathing Cave sandstone" and associated basal chert forms the ceiling in much of Breathing Cave and many other caves in Burnsville Cove. Petrographic descriptions of the upper Breathing Cave sandstone and the lower Breathing Cave sandstone are provided in Fig. 16.16.

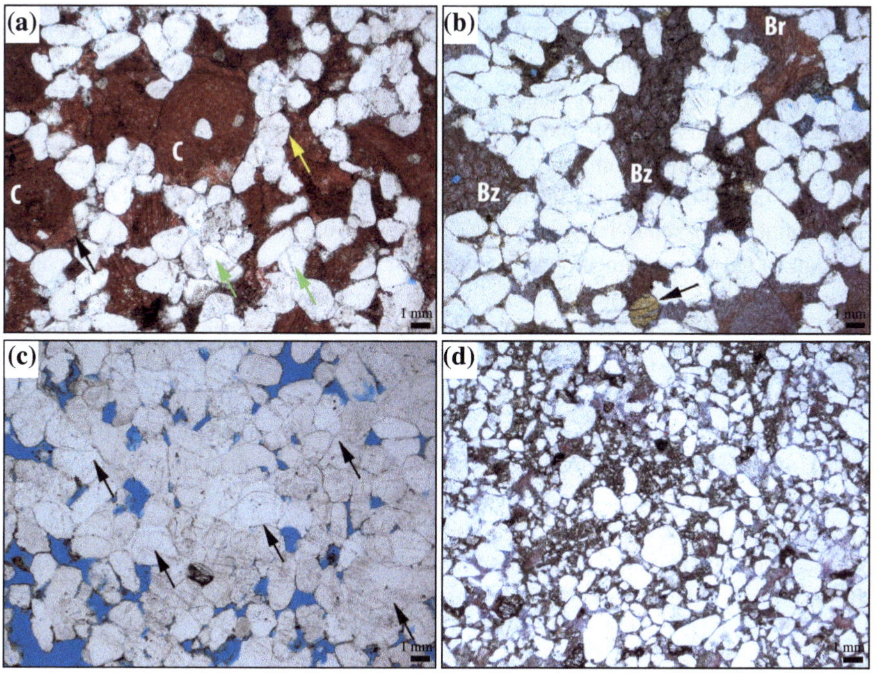

Fig. 16.16 Photographs of thin sections (shown in plane-polarized light) of the upper Breathing Cave sandstone and lower Breathing Cave sandstone. *Scale bar* in *lower right corner* of each image is 0.1 mm. All photos by J.T. Haynes

Photograph A, Fig. 16.16: Upper Breathing Cave sandstone from the entrance to Breathing Cave. Composition from thin section analysis is 81 % framework grains (F), <1 % matrix (M); 14 % cement (C); and 5 % porosity (P), which is denoted as $F_{81}M_{tr}C_{14}P_5$ (based on 300 point counts, 1 thin section; tr = trace amount). The framework grains consist predominantly of monocrystalline quartz (white) and fossil fragments of calcium carbonate (red). Nearly all of the non-carbonate framework grains are subrounded, medium to fine sand (0.350–0.177 mm) composed of monocrystalline quartz, with trace tourmaline and collophane. Most of the carbonate grains are crinoid fragments (Cr), with some bryozoan, brachiopod, and trilobite fragments. There are only trace amounts of non-calcareous silty clay matrix. The predominant cements are quartz overgrowths (81 % of total cement) around the quartz grains, ferroan and non-ferroan calcite overgrowths (19 % of total cement) around the carbonate framework grains, and minor ferroan dolomite (<1 % of total cement), primarily in remnant interparticle pore spaces between quartz grains. "Dust lines" are visible between the quartz framework grains and the quartz cement overgrowths. Porosity (blue) is almost entirely secondary, from the dissolution of carbonate grains and cements.

Photograph B, Fig. 16.16: Upper Breathing Cave sandstone from outcrop near the old (Nicholson) entrance to Butler Cave. Composition is $F_{77}M_{tr}C_{21}P_2$ (300 point counts each, 2 thin sections). The framework grains consist predominantly of monocrystalline quartz (white) and carbonate shell fragments (lilac and red). Most of the grains are subrounded, medium to fine sand (0.350–0.177 mm) composed of monocrystalline quartz. The carbonate grains include crinoid fragments, lilac-stained bryozoans fragments (Bz) of ferroan calcite, and a red-stained brachiopod fragment (Br) of non-ferroan calcite. A well-rounded detrital tourmaline grain (T) is present at the lower center of the image. As with the sample in photograph A, there is only a trace of matrix, and it is comprised of non-calcareous silty clay to predominantly clay. The cement consists of quartz overgrowths (43 % of total cement) around the quartz framework grains, ferroan and non-ferroan calcite overgrowths (49 % of total cement) around the carbonate framework grains, and minor ferroan dolomite (8 % of total cement), primarily in remnant interparticle pore spaces between quartz grains. Some of the calcite cement is ferroan (iron-rich) calcite. Porosity is almost entirely secondary, and there is slightly more porosity in the sample shown in (B) than in (A).

Photograph C, Fig. 16.16: Upper Breathing Cave sandstone from outcrop along U.S. Route 250 on the east slope of Bullpasture Mountain. Composition is $F_{65}M_{tr}C_{18}P_{17}$ (300 point counts, 1 thin section). The framework grains consist almost exclusively of rounded to subrounded medium to fine sand (0.350–0.177 mm) composed almost exclusively of white monocrystalline quartz. The cement consists exclusively of quartz overgrowths (100 % of total cement) around the quartz grains. These quartz overgrowths bind the rock and prevent it from disintegrating into loose sand when the carbonate minerals are dissolved. The abundant porosity (17 % of the total sample) results from preservation of secondary porosity that developed as the original carbonate grains and associated carbonate cements were dissolved.

Photograph D, Fig. 16.16: Lower Breathing Cave sandstone from outcrop at the new (SOFA) entrance to Butler Cave. Composition is $F_{63}M_{29}C_8P_{tr}$ (300 point counts each, 2 thin sections). The framework grains consist predominantly of monocrystalline quartz (white) with minor feldspar and carbonate fossil fragments (lilac and red). Most of the quartz grains are subrounded to subangular fine to very fine sand (0.2–0.1 mm) although grains sizes range from medium sand (0.3 mm) to silt (<0.062 mm). The carbonate grains include a few small brachiopod and crinoid fragments, but most carbonate material is not recognizable as to origin because the particles are too small. The presence of appreciable matrix is another important difference between this sandstone and the upper Breathing Cave sandstone. The matrix is a silty clay of mostly illite, mixed-layer illite/smectite (I/S), and chlorite, and it varies in color from pale green to dark green to dark yellowish brown to brownish black. The I/S may be derived from altered volcanic ash (tephra) that was mixed with the original mud. The cement consists of the matrix itself together with minor ferroan dolomite (56 % of the non-matrix cement) and ferroan calcite (44 % of the non-matrix cement. The presence of appreciable matrix evidently prevented the development and growth of quartz cement overgrowths on the quartz framework grains. There is <1 % porosity in the sample.

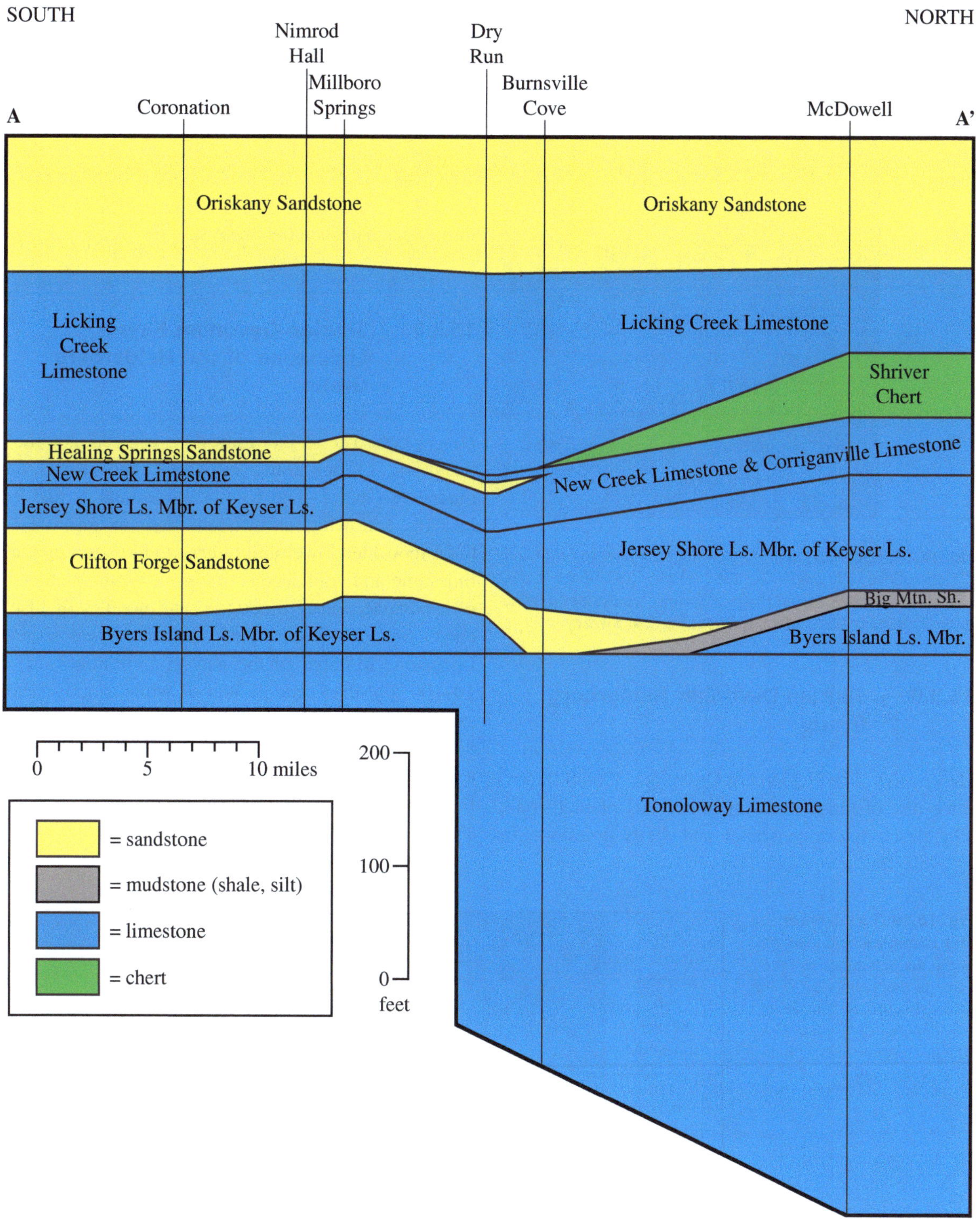

Fig. 16.17 South-to-north cross section through Silurian and Devonian strata in Bath and Highland Counties. Stratigraphic nomenclature follows the terms used in this paper, and not necessarily the terms used in the original published descriptions of the outcrops. See Fig. 16.1 for outcrop locations. *Sources of data for measured sections* Coronation (Woodward 1943); Dry Run (Swartz 1930); McDowell (Woodward 1941, 1943); Millboro Springs (Dorobek and Read 1986); Nimrod Hall (Dorobek and Read 1986); Ls. Limestone; Mbr. Member

Fig. 16.18 Phil Lucas at the Upper (dug) entrance to Owl Cave ("Talus Cone entrance") in the vicinity of Water Sinks (Fig. 16.2). *HSS* Devonian Healing Springs Sandstone Member of the Corriganville Limestone. Photo by C.S. Swezey

16.3.8 Silurian-Devonian Helderberg Group

Description: The Helderberg Group is a 300–400 foot-thick unit of limestone with some beds of sandstone, siliciclastic mudstone (shale), and cherty limestone. In Bullpasture Gorge, the unit is 335 feet thick and is divided into the following five formations (from base to top): (1) Silurian-Devonian Keyser Limestone; (2) Devonian New Creek Limestone; (3) Devonian Corriganville Limestone; (4) Devonian Licking Creek Limestone; and (5) Devonian Shriver Chert. Each of these formations is described below and shown in Fig. 16.17. *Contacts:* See descriptions of individual formations and members below.

16.3.9 Silurian-Devonian Keyser Limestone of the Helderberg Group

Description: The Keyser Limestone is a 100–200 foot-thick unit of limestone, sandstone, and siliciclastic mudstone (shale). The Keyser Limestone is divided into the following units (from base to top): (1) a 0–75 foot-thick lower unit of gray to blue-gray to pink limestone and argillaceous limestone, with some beds of sandstone, sandy limestone, and nodules of black chert (mapped collectively as the Byers Island Limestone Member of the Keyser Limestone); (2) a 10–100 foot-thick middle unit of white to gray sandstone (mapped as the Clifton Forge Sandstone Member of the Keyser Limestone) that grades north into a 10–30 foot-thick unit of green shale (mapped as the Big Mountain Shale Member of the Keyser Limestone); (3) a 25–150 foot-thick upper unit of gray to

Fig. 16.19 Rick Lambert, Phil Lucas, and John Haynes at the main entrance to Owl Cave in the vicinity of Water Sinks (Fig. 16.2). This cave was denoted as "Siphon Cave #1" in Deike (1960a). Photo by C.S. Swezey

Fig. 16.20 South-southwest side of Water Sinks, showing five of the entrances to Old Water Sinks Cave. Location shown on Fig. 16.2. This cave was denoted as "Siphon Cave #2" by Deike (1960a). Photo by P.C. Lucas

blue-gray to pink limestone and argillaceous limestone, with some beds of sandy limestone (mapped collectively as the Jersey Shore Limestone Member of the Keyser Limestone); and (4) a 3–30 foot-thick unit of laminated gray sandy limestone to calcite-cemented sandstone (mapped as the LaVale Limestone Member of the Keyser Limestone). These members of the Keyser Limestone are described in more detail below. Good outcrops of the Keyser Limestone may be found in Bullpasture Gorge and in the vicinity of Water Sinks (Figs. 16.18, 16.19, and 16.20). The Keyser Limestone is 140 feet thick at Water Sinks.

Contacts: The lower contact of the Keyser Limestone varies from place to place. In the vicinity of Burnsville Cove, the lower member of the Keyser Limestone is missing, and the lower contact of the Keyser Limestone is placed at an unconformity at the base of either a unit of sandstone (composed of quartz sand and fossil fragments) that is mapped as the Silurian Clifton Forge Sandstone Member of the Keyser Limestone (Fig. 16.21) or at the base of a unit of thin-bedded to laminated gray limestone that is mapped as the Tonoloway Limestone. This contact can be seen in the main passage of the Water Sinks Cave System and on the north bank of Bullpasture Gorge. To the north and south of Burnsville Cove, however, the lower member of the Keyser Limestone is present (e.g., north on U.S. Route 250 along Bullpasture Mountain; south on State Highway 42 by Nimrod Hall, south of Millboro Springs), and the lower contact of the Keyser Limestone is placed at the base of a unit of medium-bedded nodular limestone, just above a unit of gray laminated limestone that is mapped as the Tonoloway Limestone.

The upper contact of the Keyser Limestone is placed at the top of a 3–30 foot-thick unit of dark gray, thin-bedded calcite-cemented sandstone to sandy limestone, just below a unit of gray crinoid grainstone that is mapped as the New Creek Limestone of the Helderberg Group. Woodward (1943) stated that the

Fig. 16.21 Rick Lambert pointing at the Silurian Clifton Forge Sandstone (showing prominent cross-bedding) at the Sink of Sinking Creek. Photo by J.T. Haynes

contact between the Keyser Limestone and the overlying Coeymans Limestone [New Creek Limestone] may be disconformable.

16.3.10 Byers Island Limestone Member of the Keyser Limestone of the Helderberg Group

Description: The Byers Island Limestone Member of the Keyser Limestone is a 0–75 foot-thick unit of gray to blue-gray to pink limestone and argillaceous limestone, with some beds of sandstone, sandy limestone, and nodules of black chert. The limestone consists of skeletal grainstone, skeletal-pellet packstone, nodular skeletal-pellet wackestone, carbonate mudstone, and stromatoporoid boundstone and mud mounds. The skeletal grainstone consists of pink to gray very coarse-grained beds that contain pelmatozoan debris (primarily crinoid), and lesser amounts of brachiopods, bryozoans, corals, and well-rounded quartz grains. This pink to gray limestone resembles the younger New Creek Limestone, as noted by Woodward (1943). The skeletal-pellet packstone consists of gray, medium-grained to fine-grained beds that contain poorly sorted and abraded skeletal debris (crinoids, brachiopods, ramose bryozoans, and rare trilobites), pellets, and scattered whole fossils (brachiopods, corals, and rare stromatoporoids). Beds are irregular, locally argillaceous, and range from thick to thin, with structureless intervals, intervals of low-angle cross-bedding, and intervals of gently undulatory to planar-parallel laminations. The skeletal-pellet wackestone and carbonate mudstone consist of gray nodular beds with a few beds of skeletal packstone. The wackestone skeletal grains are poorly sorted and have diverse fossil assemblages. In places, beds of skeletal wackestone and carbonate mudstone are capped by unconformities, above which lie beds of intraclasts. Hardgrounds are also common. The argillaceous limestone consists of skeletal sand lags, pellet packstone, bryozoan boundstone, and limy shale. The Byers Island Limestone Member is absent at Water Sinks, but is 60 feet thick along State Highway 42 near Coronation (Bath County), and 40 feet thick on U.S. Route 250 east of McDowell on the southeastern flank of Bullpasture Mountain.

Contacts: In Bullpasture Gorge and in the Water Sinks area, the Byers Island Limestone Member has been removed by erosion, and the Clifton Forge Sandstone Member of the Keyser Limestone rests on an unconformity on the Tonoloway Limestone. Where the Byers Island Limestone Member is present north of Burnsville Cove, the lower contact of the unit is the lower contact of the Keyser Limestone. The upper contact of the Byers Island Limestone Member is placed at the top of a unit of nodular limestone, just below a unit of white to gray sandstone (mapped as the Clifton Forge Sandstone Member) or a unit of olive-gray shale (mapped as the Big Mountain Shale Member). In northern Highland County, the Big Mountain Shale Member occurs near the top of a faunal zone characterized primarily by the brachiopod *Chonetes jerseyensis* (Woodward 1943).

16.3.11 Clifton Forge Sandstone Member of the Keyser Limestone of the Helderberg Group

Description: The Clifton Forge Sandstone Member of the Keyser Limestone is a 10–100 foot-thick unit of white to gray sandstone and sandy limestone. The sandstone consists of moderately sorted to very well sorted, very coarse-grained to fine-grained quartz sand (primarily monocrystalline quartz) with abundant carbonate skeletal grains and peloids, yellow grains of limonite and silt, and argillaceous lenses and thin beds. Quartz and carbonate skeletal grains are commonly segregated into alternating laminations. Skeletal grains include abundant crinoids, and less abundant bryozoans,

Fig. 16.22 John Haynes looking at conglomerate composed of flat pebbles of carbonate mudstone within the basal part of the Silurian Clifton Forge Sandstone in the main passage of the Water Sinks Cave System. Cross-bedding within the Clifton Forge Sandstone is visible above the measuring tape. The flat-bedded conglomerate rests on an unconformity that caps the Tonoloway Limestone (visible in *lower part* of *photograph*). "Water Sinks" location is shown in Fig. 16.2. Photo by P.C. Lucas

pelecypods, favositid corals, stromatoporoids, brachiopods, and trilobites. Cross-bedding is prominent in some beds (Fig. 16.21). The sandstone is cemented by calcite and minor amounts of ferroan dolomite. The Clifton Forge Sandstone Member is 33 feet thick in the Water Sinks Cave System. In the main passage of this cave, a 1–5 inch thick bed of conglomerate (a "flat-pebble" conglomerate composed of clasts of carbonate mudstone) is present at the base of the Clifton Forge Sandstone, and this conglomerate rests on an unconformity on the Tonoloway Limestone (Figs. 16.22 and 16.23). Towards the north (at the approximate latitude of 38° 16′ to 38° 19′ North), the Clifton Forge Sandstone Member grades into green calcite-cemented shale that is mapped as the Big Mountain Shale Member of the Keyser Limestone (Fig. 16.17).

Fig. 16.23 Conglomerate composed of flat pebbles of carbonate mudstone within the basal part of the Silurian Clifton Forge Sandstone in the main passage of the Water Sinks Cave System. "Water Sinks" location is shown in Fig. 16.2. Photo by C.S. Swezey

North of the town of Clifton Forge in Alleghany County (Virginia), the Clifton Forge Sandstone Member is reported to split into two or three sandy beds (Swartz 1930; Woodward 1943), with the lower of these sandy beds merging towards the north with the Big Mountain Shale Member of the Keyser Limestone, and the upper of these sandy beds extending north into Highland County where it presumably pinches out. Deike (1960a, b) considered the 12 feet thick sandstone beds that form the floor and ceiling of Breathing Cave to be the same two sandstone beds that Swartz (1930) and Woodward (1943) identified as tongues of the Clifton Forge Sandstone Member, and he informally named these beds the "lower Breathing Cave sandstone" and the "upper Breathing Cave sandstone." White and Hess (1982) later applied the name lower Clifton Forge Sandstone to the lower bed and the name upper Clifton Forge Sandstone to the upper bed in Breathing Cave. Recent unpublished work by John Haynes, Rick Lambert, and Phil Lucas, however, suggests that the "upper Breathing Cave sandstone" and "lower Breathing Cave sandstone" are both located within the lower member of the Tonoloway Limestone, and that they are not tongues of the Clifton Forge Sandstone Member of the Keyser Limestone (Fig. 16.8).

Contacts: At some locations, the lower contact of the Clifton Forge Sandstone Member is the upper contact of the Byers Island Limestone Member.

At other locations where the Byers Island Limestone Member is absent (e.g., Bullpasture Gorge and Water Sinks), the lower contact of the Clifton Forge Sandstone Member is the upper contact of the Tonoloway Limestone. The upper contact of the Clifton Forge Sandstone Member is placed at the top of a unit of white to gray sandstone, just below a unit of gray to blue-gray nodular limestone (with abundant corals and stromatoporoids and some beds of sandy limestone) that is mapped as the Jersey Shore Limestone Member of the Keyser Limestone.

16.3.12 Big Mountain Shale Member of the Keyser Limestone of the Helderberg Group

Description: The Big Mountain Shale Member of the Keyser Limestone is absent in Burnsville Cove, but farther north it is a 10–30 feet thick unit of olive-gray shale, with some thin beds of sandstone and limestone (Fig. 16.17). In places, this unit has knobby and irregular beds. Burrows are also present, as well as rare brachiopods, ramose bryozoans, and crinoid fragments. The Big Mountain Shale Member is 16 feet thick along U.S. Route 250, and the unit grades south into the Clifton Forge Sandstone Member (Fig. 16.17).

Contacts: The lower contact of the Big Mountain Shale Member is the upper contact of the Byers Island Limestone Member. In northern Highland County, the Big Mountain Shale Member occurs near the top of a faunal zone characterized primarily by the brachiopod *Chonetes jerseyensis* (Woodward 1943). The upper contact of the Big Mountain Shale Member is placed at the top of a unit of dark thin-bedded siliciclastic mudstone (shale) with thin beds of limestone, just below a unit of gray to blue-gray nodular to massive limestone (with abundant corals and stromatoporoids and some beds of sandy limestone) that is mapped as the Jersey Shore Limestone Member of the Keyser Limestone.

16.3.13 Jersey Shore Limestone Member of the Keyser Limestone of the Helderberg Group

Description: The Jersey Shore Limestone Member of the Keyser Limestone is a 25–150 foot-thick unit of gray to blue-gray to pink limestone and argillaceous

Fig. 16.24 Stromatoporoids in the Silurian Jersey Shore Limestone Member of the Keyser Limestone at main entrance to Owl Cave in the vicinity of Water Sinks (Fig. 16.2). Pocket knife is 3.5 inches long. Photo by C.S. Swezey

limestone, with some beds of sandy limestone. The limestone consists of skeletal grainstone, skeletal-pellet packstone, nodular skeletal-pellet wackestone, carbonate mudstone, and stromatoporoid-coral boundstone (Fig. 16.24) and mud mounds. The skeletal grainstone consists of pink to gray very coarse-grained beds that contain pelmatozoan debris (primarily crinoid), and lesser amounts of brachiopods, bryozoans, corals, and well-rounded quartz grains. As noted by Woodward (1943, p. 76), this pink limestone resembles the younger New Creek Limestone. The skeletal-pellet packstone consists of gray, medium-grained to fine-grained beds that contain poorly sorted and abraded skeletal debris (crinoids, brachiopods, ramose bryozoans, and rare trilobites), pellets, and scattered whole fossils (brachiopods, corals, and rare stromatoporoids). Beds are irregular, locally argillaceous, and range from thick to thin, with structureless intervals, intervals of low-angle cross-bedding, and intervals of gently undulatory to planar-parallel laminations. The skeletal-pellet wackestone and carbonate mudstone consist of gray nodular beds with a few beds of skeletal packstone. The wackestone beds are poorly sorted and have diverse fossil assemblages. In places, beds of skeletal wackestone and carbonate mudstone are capped by unconformities, above which lie beds of intraclasts. Hardgrounds are also common. Good exposures of the Jersey Shore Limestone Member may be found at the entrances to Old Water Sinks Cave and the main entrance to Owl Cave (Figs. 16.20 and 16.21). The unit is 110 feet thick in the vicinity of Water Sinks.

At most localities, the Jersey Shore Limestone Member contains 2 or more beds of gray to pink crinoid limestone, each 10–20 feet thick. In many places, the lower of these crinoid beds overlies the Big Mountain Shale Member and contains the brachiopod *Merista typa*. In addition to the crinoid beds, bioherms and biostromes of corals and stromatoporoids (Fig. 16.24) occur within the Jersey Shore Limestone Member. At the community of Mustoe (Fig. 16.1), one such bioherm is particularly well exposed and is interpreted as an ancient reef (Smosna and Warshauer 1979; Smosna 1984; Stock and Holmes 1986; Stock 1996; Dennison et al. 1997). This bioherm is within the upper Jersey Shore Member of the Keyser Limestone, and is overlain by the LaVale Limestone Member of the Keyser Limestone (Stock 1996; Dennison et al. 1997). Furthermore, the bioherm at Mustoe is located at approximately the same latitude as the facies change from the Clifton Forge Sandstone Member to the Big Mountain Shale Member, and thus it is possible that both the bioherm location and the facies change are associated with a down-to-the-north tectonic hinge (Dennison 1985; Dennison et al. 1997). Within the Jersey Shore Limestone Member, Woodward (1943) described two bioherms (composed of corals, bryozoans, and stromatoporoids), approximately 50 feet apart stratigraphically, that he followed across Pendleton County and into Highland County. These two bioherms are present at Water Sinks, where the lower one is 6 feet thick, and the upper one is 15 feet thick (and 50–55 feet above the lower bioherm).

Contacts: In Bath County and in southern Highland County, the lower contact of the Jersey Shore Limestone Member is the upper contact of the Clifton Forge Sandstone Member. In northern Highland County, where the Clifton Forge Sandstone Member grades north into the Big Mountain Shale Member, the lower contact of the Jersey Shore Limestone Member is the upper contact of the Big Mountain Shale Member. The upper contact of the Jersey Shore Limestone Member is placed at the top of a unit of gray to blue-gray nodular limestone with abundant corals and stromatoporoids and some beds of sandy limestone. In some places, this unit of nodular limestone lies just below a unit of gray sandy limestone to calcite-cemented sandstone that is mapped as the LaVale Limestone Member of the Keyser Limestone of the Helderberg Group (Fig. 16.20). In other places, this unit of nodular limestone is capped by an unconformity, above which lies a unit of gray crinoid grainstone that is mapped as the New Creek Limestone of the Helderberg Group.

16.3.14 LaVale Limestone Member of the Keyser Limestone of the Helderberg Group

Description: The LaVale Limestone Member of the Keyser Limestone is a 3–30 foot-thick unit of gray sandy limestone to calcite-cemented sandstone. The limestone is a carbonate mudstone. The unit displays prominent laminations, and usually drapes the underlying nodular beds of the Jersey Shore Limestone Member. Good exposures of the LaVale Limestone Member may be found on the hillside above the main entrance of the Water Sinks Cave System (Fig. 16.20) and at the waterfall in the Water Sinks Cave System.

Contacts: The lower contact of the LaVale Limestone Member is the upper contact of the Jersey Shore Limestone Member. The upper contact of the LaVale Limestone Member is placed at the top of a unit of gray sandy limestone to calcite-cemented sandstone, just below a unit of crinoid grainstone that is mapped as the New Creek Limestone of the Helderberg Group.

16.3.15 Devonian New Creek Limestone of the Helderberg Group

Description: The New Creek Limestone is an 8–60 foot-thick unit of gray to pink limestone. The limestone is a grainstone that consists of coarse-grained

Fig. 16.25 Geology students from James Madison University standing on outcrop of the Devonian New Creek Limestone and the overlying Devonian Corriganville Limestone above the Talus Cone entrance to Owl Cave in the vicinity of Water Sinks (Fig. 16.2). Photo by J.T. Haynes

beds of pelmatozoan debris (primarily crinoids) and lesser amounts of brachiopods, bryozoans, corals, and well-rounded quartz sand. The unit displays prominent cross-bedding in places. The New Creek Limestone is 36 feet thick along U.S. Route 250 on the east flank of Bullpasture Mountain (Fig. 16.5), and 23 feet thick on the hillside above the main entrance to Owl Cave (in the vicinity of Water Sinks). Good exposures of the New Creek Limestone may also be found at the upper entrance ("Talus Cone entrance") to Owl Cave (Fig. 16.25) and in Bullpasture Gorge.

Contacts: The lower contact of the New Creek Limestone is the upper contact of the LaVale Limestone Member of the Keyser Limestone. The upper contact of the New Creek Limestone is placed at the top of a unit of gray limestone that is a crinoid grainstone. In some places, this crinoid grainstone lies just below a gray limestone and cherty limestone (brachiopod packstone to wackestone) that is mapped as the Devonian Corriganville Limestone of the Helderberg Group. In other places, the crinoid grainstone lies just below a unit of gray sandstone that is mapped as the Healing Springs Sandstone Member of the Corriganville Limestone of the Helderberg Group.

16.3.16 Devonian Corriganville Limestone of the Helderberg Group

Description: The Corriganville Limestone is a 5–30 foot-thick unit of gray limestone and cherty limestone. In places, the unit contains a basal bed of gray sandstone that is mapped as the Healing Springs Sandstone Member of the Corriganville Limestone. The cherty limestone displays irregular beds, and contains nodules and ribbons of white and gray chert. Both the limestone and cherty limestone are packstones to wackestones. The packstones are

Fig. 16.26 Devonian Healing Springs Sandstone Member of the Corriganville Limestone at the upper entrance ("Talus Cone entrance") to Owl Cave in the vicinity of Water Sinks (Fig. 16.2). Scale is 3.5 inch long pocket knife. Photo by C.S. Swezey

medium-grained to fine-grained, and locally argillaceous. Fossils consist primarily of brachiopods (including *Macropleura* sp.) and corals. The Corriganville Limestone is 40 feet thick along U.S. Route 250 on the east flank of Bullpasture Mountain (Fig. 16.5), and 25 feet thick at Water Sinks and in Bullpasture Gorge. Good exposures of the Corriganville Limestone may also be found above Owl Cave (Fig. 16.18), and approximately 1 mile southeast of Water Sinks at the junction of the Bullpasture River and Mill Run (Fig. 16.2).

Contacts: The lower contact of the Corriganville Limestone is the upper contact of the New Creek Limestone. The upper contact of the Corriganville Limestone is placed at the top of a unit of gray limestone and cherty limestone (brachiopod packstone to wackestone). Butts (1940) stated that the New Scotland [Corriganville] Limestone is capped by an unconformity, but this unconformity is not obvious in the vicinity of Burnsville Cove. In some places (e.g., Water Sinks and Bullpasture Gorge), the Corriganville Limestone is overlain by a unit of chert, cherty carbonate mudstone, and calcite-cemented siliciclastic mudstone (shale) that is mapped as the Shriver Chert of the Helderberg Group. In other places (e.g., south of Water Sinks), the Corriganville Limestone is overlain by a unit of gray limestone (with abundant nodules and ribbons of black chert) that is mapped as the Devonian Licking Creek Limestone of the Helderberg Group.

16.3.17 Devonian Healing Springs Sandstone Member of the Corriganville Limestone of the Helderberg Group

Description: The Healing Springs Sandstone Member of the Corriganville Limestone is a 0–20 foot-thick unit of gray sandstone. The sandstone contains quartz grains and crinoid fossils, and is cemented by calcite. The northernmost known outcrop is at the upper entrance ("Talus Cone entrance") to Owl Cave in the vicinity of Water Sinks (Fig. 16.26). At this location, the Healing Springs Sandstone is 3 inches thick, and it thickens towards the south.

Contacts: The lower contact of the Healing Springs Sandstone Member is the upper contact of the New Creek Limestone. The upper contact of the Healing Springs Sandstone Member is placed at the top of a unit of sandstone, just below a unit of gray limestone and cherty limestone (brachiopod packstone to wackestone) that is mapped as the upper Corriganville Limestone of the Helderberg Group. Dennison et al. (1992) indicated that the Healing Springs Sandstone Member is capped by an unconformity at some localities.

16.3.18 Devonian Licking Creek Limestone of the Helderberg Group

Description: The Licking Creek Limestone is a 85–200 foot-thick unit of gray cherty limestone, argillaceous limestone, and sandy limestone. The cherty limestone and argillaceous limestone are wackestones to carbonate mudstones. Fossils include brachiopods, rugose and colonial corals, and sponge spicules. Beds are irregular to nodular, and the chert consists of black nodules. The sandy limestone ranges from packstone to grainstone, and contains pellets and skeletal grains (crinoids, brachiopods, corals, ramose bryozoans, and rare trilobites). Beds are irregular and range from thick to thin, with structureless intervals, intervals of low-angle cross-bedding, and intervals of gently undulatory to planar-parallel laminations. In places, the beds are sandy and (or) argillaceous. The Licking Creek Limestone is 120 feet thick in Bullpasture Gorge. The Licking Creek Limestone grades north into a unit of bedded chert, cherty mudstone, and

Fig. 16.27 Phil Lucas beneath outcrop of the Devonian Shriver Chert at the entrance to Helictite Cave in the vicinity of Water Sinks (Fig. 16.2). Photo by C.S. Swezey

Fig. 16.28 George Deike and granddaughter Wendy Kelly on outcrop of Devonian Oriskany Sandstone in the vicinity of Water Sinks. Note the presence of joints in the sandstone. Photo by C.S. Swezey

calcite-cemented mudstone (mapped collectively as the Shriver Chert of the Helderberg Group).

Contacts: The lower contact of the Licking Creek Limestone is the upper contact of the Corriganville Limestone. The upper contact of the Licking Creek Limestone is placed at the top of a unit of gray limestone (containing quartz sand grains in some places, and abundant nodules and ribbons of chert in other places), just below a unit of brown to yellow sandstone that is mapped as the Devonian Oriskany Sandstone. In some places, the Licking Creek Limestone is capped by an unconformity, above which lies the Oriskany Sandstone (Butts 1940; Dennison 1985; Dennison et al. 1992). This unconformity may be the Wallbridge Unconformity, which extends throughout the Appalachian, Michigan, and Illinois Basins (Wheeler 1963; Swezey 2002, 2009). However, Dennison and Head (1975) and Dennison (1985) contend that the Wallbridge Unconformity is the unconformity that caps the Oriskany Sandstone rather than the unconformity beneath the Oriskany Sandstone.

16.3.19 Devonian Shriver Chert of the Helderberg Group

Description: The Shriver Chert is a 5–66 foot-thick unit of gray to black chert, cherty limestone, and siliciclastic mudstone (shale). Beds range from massive to nodular to irregular with thin laminations. The Shriver Chert is 66 feet thick along U.S. Route 250 on the east flank of Bullpasture Mountain (Fig. 16.5). The southernmost known outcrop of the Shriver Chert is at the entrance to Helictite Cave in the Water Sinks area (Fig. 16.27), and farther south the Shriver Chert grades into a unit of gray limestone (with abundant nodules and ribbons of black chert) that is mapped as the Devonian Licking Creek Limestone of the Helderberg Group.

Contacts: The lower contact of the Shriver Chert is the upper contact of the Corriganville Limestone. The upper contact of the Shriver Chert is placed at the top of a unit of bedded chert, cherty carbonate mudstone, and calcite-cemented siliciclastic mudstone (shale), just below either gray skeletal packstone to grainstone that is mapped as the Licking Creek Limestone or just below a unit of brown to yellow sandstone that is mapped as the Devonian Oriskany Sandstone. In some places, the Shriver Chert is capped by an unconformity, above which lies the Oriskany Sandstone (Butts 1940; Dennison 1985; Dennison et al. 1992).

16.3.20 Devonian Oriskany Sandstone

Description: The Oriskany Sandstone is a 20–130 foot-thick unit of white to yellow to red-brown sandstone (Fig. 16.28). The sandstone is a quartzarenite composed of coarse to very fine quartz sand.

Fossils include common molds of the large brachiopod *Spirifer arenosus*. In most places the sand is cemented by calcite, although in a few places the cement is lacking and the unit is a friable mass of loose sand. The Oriskany Sandstone is 80–100 feet thick along U. S. Route 250 on the east flank of Bullpasture Mountain (Fig. 16.5).

Contacts: South of Helictite Cave and Water Sinks (Fig. 16.2), the lower contact of the Oriskany Sandstone is the upper contact of the Licking Creek Limestone. North of Helictite Cave and Water Sinks, the lower contact of the Oriskany Sandstone is the upper contact of the Shriver Chert. An unconformity may be present between the Oriskany Sandstone and the underlying strata (Butts 1940; Dennison 1985; Dennison et al. 1992).

The upper contact of the Oriskany Sandstone is placed at the top of a unit of white to yellow to red-brown sandstone, just below a unit of gray to black siliciclastic mudstone (shale) that is mapped as the Devonian Needmore Shale. In some places, the Oriskany Sandstone is capped by an unconformity, above which lies the Needmore Shale (Butts 1940; Deike 1960a; Dennison and Head 1975; Rader 1984; Dennison 1985; Dennison et al. 1992; Rader and Wilkes 2001). This unconformity may be the regionally extensive Wallbridge Unconformity (Dennison and Head 1975; Dennison 1985).

16.3.21 Devonian Needmore Shale

Description: The Needmore Shale is a 20–150 foot-thick unit of green to gray to black siliciclastic mudstone (shale), with some beds of gray limestone. Nodules of barite have been described from the Needmore Shale in Bath County (Clark and Mosier 1989), and several petrographic and geochemical studies have been published on the Needmore Shale in Highland County (Woodard et al. 1997; Combs and Sethi 2001, 2002; Sethi et al. 2002).

Contacts: The lower contact of the Needmore Shale is the upper contact of the Oriskany Sandstone. In some places, an unconformity is present between the Needmore Shale and the underlying Oriskany Sandstone (Butts 1940; Deike 1960a; Dennison and Head 1975; Rader 1984; Dennison 1985; Dennison et al. 1992; Rader and Wilkes 2001). The upper contact of the Needmore Shale is placed at the top of a unit of gray siliciclastic mudstone (shale), just below a unit of black siliciclastic mudstone (shale) that is mapped as the Millboro Shale. In some places, the basal part of the Millboro Shale contains distinct beds of sand-sized ash, separated by beds of silty tuff and tuffaceous shale (mapped collectively as the Tioga Ash Bed).

16.3.22 Devonian Tioga Ash Bed

Description: The Tioga Ash Bed is a 3–20 foot-thick unit of volcanic ash, silty tuff, and tuffaceous shale. The unit is 3.2 feet thick at a roadside shale quarry 2.0 miles south of Williamsville (Dennison and Textoris 1971). Using uranium-lead dating techniques, Roden et al. (1990) obtained an absolute age of 390 ± 0.5 Ma (million years) from the Tioga Ash Bed in Pennsylvania.

The Tioga Ash Bed is comprised of material that is relatively unstable, and many landslides have occurred along this unit. One such landslide took place during 1998 in Highland County along U.S. Route 250 on the east side of Bullpasture Mountain (Whisonant et al. 1998). The site of this landslide is now covered with riprap.

Contacts: The lower contact of the Tioga Ash Bed is placed at the base of the lowest bed of volcanic ash, just above a unit of gray siliciclastic mudstone (shale) that is mapped as the Needmore Shale. The upper contact of the Tioga Ash Bed is placed at the top of the highest ash bed, just below a unit of black siliciclastic mudstone (shale) that is mapped as the Millboro Shale.

16.3.23 Devonian Millboro Shale

Description: The Millboro Shale is a 230–1370 foot-thick unit of black shale that displays prominent joints (Fig. 16.29). The unit contains abundant carbonate concretions that range up to 3 feet in diameter and 1 feet in thickness, as well as nodules of barite (Nuelle and Shelton 1986; Clark and Mosier 1989). In addition, the upper part of the Millboro Shale in Bath and Highland Counties contains several discontinuous beds of limestone nodules that are correlated with the Tully Limestone (described below). Several petrographic and geochemical studies have been done on the Millboro Shale in Highland County (Woodard et al. 1997; Combs and Sethi 2001, 2002; Sethi et al.

Fig. 16.29 Bette White at outcrop of the Devonian Millboro Shale along the west side of U.S. Route 678, approximately 2.5–3 miles south of U.S. Route 250. Photo by W.B. White

2002). Good exposures of the Millboro Shale may be found along the west side of Route 678 between U.S. Route 250 and Bullpasture Gorge.

Contacts: The lower contact of the Millboro Shale is placed at the base of a unit of black siliciclastic mudstone (shale), just above beds of sand-sized volcanic ash (separated by beds of silty tuff and tuffaceous shale) that are mapped collectively as the Tioga Ash Bed. Where the Tioga Ash Bed is absent, the lower contact of the Millboro Shale is the upper contact of the Needmore Shale. In Bath and Highland Counties, the upper contact of the Millboro Shale is placed at the top of the black siliciclastic mudstone (shale), just below a thick sequence of brown and green siliciclastic mudstone (shale) that is mapped as the Brallier Formation. North of Highland County, strata that are equivalent to the Millboro Shale are overlain by a bed of limestone that is mapped as the Tully Limestone. In some places, the Millboro Shale is overlain by an unconformity (Avary and Dennison 1980; Hasson and Dennison 1988; Dennison and Schwietering 1995). This unconformity is a regional unconformity named the Acadian Unconformity, which extends throughout the Appalachian, Michigan, and Illinois Basins (Wheeler 1963; Swezey 2002, 2009).

16.3.24 Devonian Tully Limestone

Description: The Tully Limestone is a bed of limestone that is clearly identifiable north of Highland County. In Bath and Highland Counties, however, this bed of limestone is a discontinuous 1–8 foot-thick bed of limestone nodules that are not present at every outcrop (Avary and Dennison 1980; Hasson and Dennison 1988). In some places, the limestone nodules are located within the upper part of the Millboro Shale. In other places the limestone nodules are located at the top of the Millboro Shale and the nodules are capped by an unconformity. The Tully Limestone is not shown as a separate unit in Fig. 16.3, but nodules of the Tully Limestone are exposed in outcrops between McDowell and Burnsville (Fig. 16.1).

Contacts: Where it is possible to map this unit, the lower contact of the Tully Limestone is the upper contact of the Millboro Shale. The upper contact of the Tully Limestone is placed at the top of the bed of limestone nodules. In some places, the limestone nodules are overlain by black siliciclastic mudstone (shale) that is mapped as the Millboro Shale. In other places, the limestone nodules are capped by an unconformity (de Witt et al. 1993; Dennison and

Fig. 16.30 Quaternary gravel and sand along the east bank of Sinking Creek near Owl Cave in the vicinity of Water Sinks (Fig. 16.2). Hammer shown for scale. Photo by C.S. Swezey

Schwietering 1995), above which lies a thick unit of gray to green to yellow-brown siliciclastic mudstone and sandstone that is mapped as the Devonian Brallier Formation. This unconformity above the Millboro Shale is the Acadian Unconformity (Wheeler 1963; Swezey 2002, 2009).

16.3.25 Quaternary Sediments and Terraces

Although not shown on the geologic map (Fig. 16.3), Quaternary terraces and sediments are present in the valleys of the rivers and creeks of Bath and Highland Counties. Some of these terraces are composed of Quaternary gravel and sand (Fig. 16.30), whereas other terraces are erosional features that are cut into Devonian shale (i.e., the terraces are strath terraces, and not accumulations of Quaternary sediments). Where the rivers have eroded through Devonian shale and into underlying Silurian-Devonian limestone (e.g., Burnsville Cove), the waters flow underground and a karst landscape has developed.

Some Quaternary sediments are present in Bath and Highland Counties as soil, regolith, and colluvium. These sediments may be remobilized during episodes of heavy rainfall. During November 1985, for example, three days of heavy rainfall in western Highland County caused numerous slope movements, most of which involved colluvium on the Ordovician Reedsville Shale (Jacobson et al. 1989). During this same rainfall, several large debris-slide avalanches were triggered in sandstone regolith high on ridges. Most of these avalanche sites were located on slopes that dip 30°–35° and are parallel to the bedding.

Browns Pond (Fig. 16.2) is one location where a Quaternary sedimentary record has accumulated (Kneller and Peteet 1993, 1999). At this site, several cores have been recovered and radiocarbon dates have obtained (Fig. 16.31). In addition to sediment analyses, pollen samples recovered from these cores have been used to reconstruct changes in the species and abundance of surrounding flora, and to infer past changes in temperature and precipitation. The overall interpretations from these studies by Kneller and Peteet (1993, 1999) are summarized as follows [dates below are given in uncorrected radiocarbon years before present (BP), where 0 BP is equivalent to 1950 A.D.]:

4870 to 0 years BP:	No interpretations given.
7500 to 4870 years BP:	Fluctuations in water depth at Browns Pond.
7500 years BP:	Brief episode of colder but moist climate.
8410 to 7500 years BP:	Pond was present. Fluctuations in pond water depth.
10,050 to 8410 years BP:	Browns Pond was wet ground or marsh land (water depth lower than at present). Deciduous forest with evidence of fire and opening of forest canopy. Average temperature increased during the period. Average precipitation was approximately equal to present values.
11,000 to 10,050 years BP:	Water level in Browns Pond decreased during the period. Deciduous forest. Average

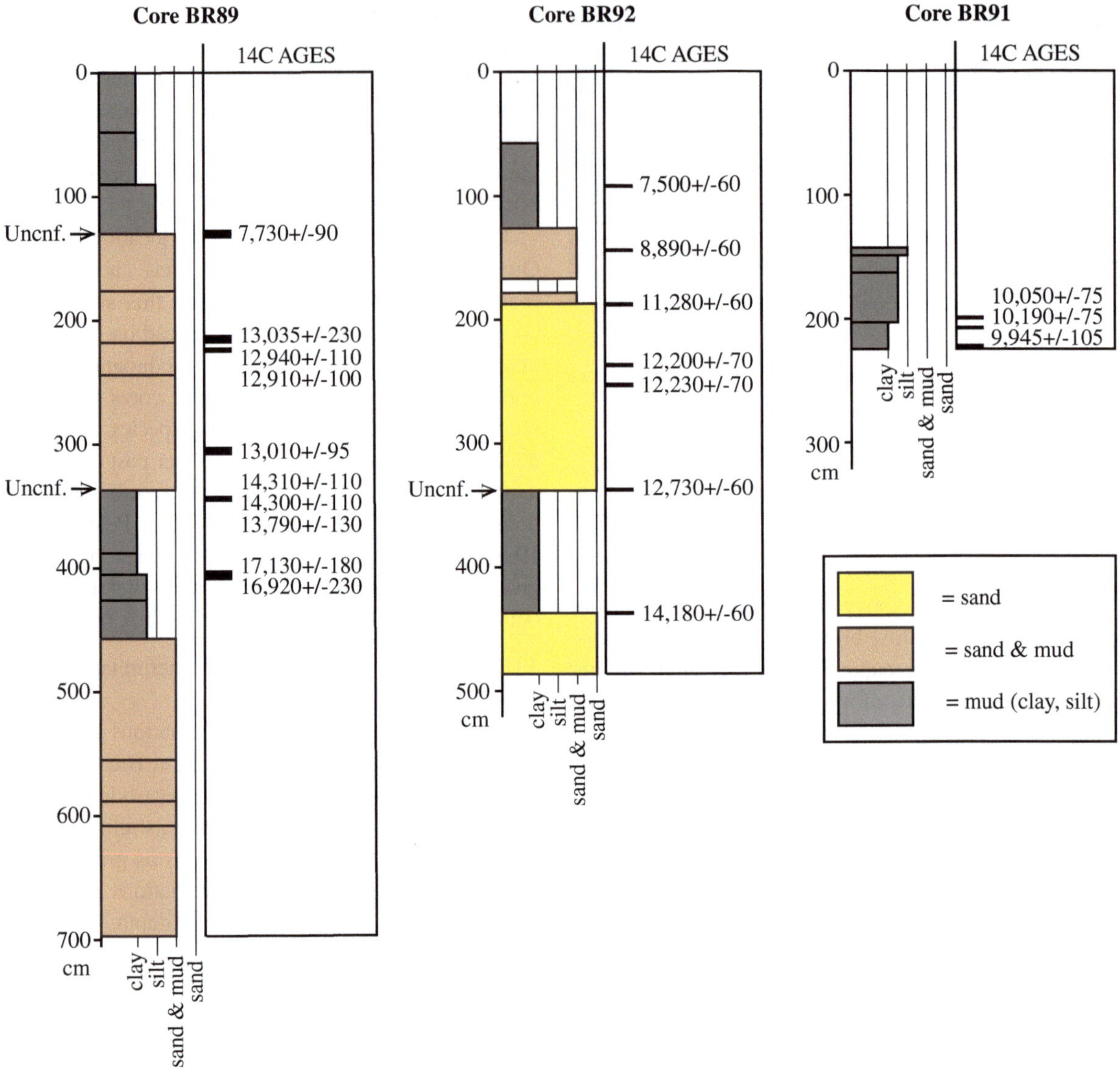

Fig. 16.31 Cores taken from Browns Pond, Bath County, Virginia (compiled from data presented in Kneller and Peteet 1993, 1999). Core BR91 is 4 m east of Core BR92, and Core BR92 is 18 m east of Core BR89. 14C *AGES* Ages in radiocarbon years before present (BP), where 0 BP is equivalent to 1950 A.D. The ages are given with 1 sigma standard deviation of age uncertainty. *Uncnf.* unconformity

	temperature increased during the period.
11,000 years BP:	Disappearance of conifer trees.
11,280 to 11,000 years BP:	Water level in Browns Pond decreased during the period. Percentage of deciduous trees increased. Average temperature increased during the period.
12,200 to 11,280 years BP:	Water level in Browns Pond deeper than at present. Percentage of deciduous trees increased during the

	period. Warmer average temperatures but still cooler than at present. Climate more moist than at present.
12,260 to 12,200 years BP:	Water level in Browns Pond shallower than at present. Colder average temperatures than before.
12,730 to 12,260 years BP:	Water level in Browns Pond deeper than at present. Percentage of deciduous trees increased during the period. Warmer average temperatures but still cooler than at present. Climate more moist than at present.
14,180 to 12,730 years BP:	Water level in Browns Pond deeper than at present. Dense montane spruce-fir forest. Average temperatures cooler than at present.
17,300 to 14,180 years BP:	Average temperatures cooler than at present.
18,000 to 17,300 years BP:	Boreal forest. Average temperatures cooler than at present.

16.4 Structural Geology

The structural features of Bath and Highland Counties include anticlines, synclines, faults, and joints. Most of the anticlines and synclines trend northeast, and are asymmetric with beds dipping more steeply on the western flanks (Fig. 16.32). Thrust faults that dip to the southeast are present on the northwestern flanks of many of the major anticlines. Minor thrust faults are common, and joints are pervasive. Most joints trend northwest, perpendicular to the trend of the anticlines and synclines.

Most of the structural features of Bath and Highland Counties developed when the North American tectonic plate collided with the African and European plates to form one giant continent called Pangaea. This episode of tectonic plate collision is called the Alleghanian Orogeny, and it is thought to have occurred approximately 325–270 million years ago (e.g., Hatcher 1989; Ryder et al. 2008). During this event, rocks along the eastern margin of North America were thrust from east to west and folded into the anticlines and synclines that are visible today. Much later, an episode of igneous activity occurred in Highland County approximately 50–35 million years ago (Fullagar and Bottino 1969; Southworth et al. 1993; Tso et al. 2004), but no igneous bodies of this age have yet been identified within Burnsville Cove.

16.4.1 Anticlines and Synclines

The main structural features of Bath and Highland Counties are anticlines and synclines that trend approximately N40E (Figs. 16.3 and 16.33). Most of these features are asymmetrical, and strata generally dip more steeply on the western flanks. For example, in most places the beds on the southeastern sides of the anticlines dip to the southeast at angles of 30°–35°, whereas the beds on the northwestern sides of the anticlines dip more steeply to the northwest and in places range from vertical to slightly overturned. One exception is the Tower Hill Anticline exposed in Bullpasture Gorge (Fig. 16.3), where the structure is more symmetrical.

Burnsville Cove is located on the east flank of a major northeast-trending anticline with an axis that lies between Little Mountain and Jack Mountain (Fig. 16.3). This anticline is part of the greater Hot Springs Anticline of Butts (1933, 1940). Several minor anticlines and synclines are superimposed on the east flank of this major anticline, including: (1) Sinking Creek Syncline or Burnsville Cove Syncline; (2) Chestnut Ridge Anticline or Bullpasture Mountain Anticline; (3) White Oak Syncline; and (4) Tower Hill Anticline. These minor anticlines and synclines in the vicinity of Burnsville Cove plunge northeast. Most of Burnsville Cove lies along Sinking Creek Syncline, which plunges northeast at approximately 46 m/km (Deike 1960b). Breathing Cave is located on the west flank of this syncline, and the main trunk channel of Butler Cave follows the axis of this syncline.

Fig. 16.32 Near-vertical beds on the western flank of Bullpasture Mountain near western end of Bullpasture Gorge. Photo by W.B. White

Many caves are also located along both flanks of Chestnut Ridge Anticline and several caves follow the axis of White Oak Syncline. Tower Hill Anticline forms the eastern border of Burnsville Cove. This structure is not a classic anticline, but an approximately symmetric box-fold.

16.4.2 Faults

Major thrust faults have been identified on the west flanks of some anticlines in Bath and Highland Counties, but no such faults have been identified within Burnsville Cove. Bick (1962) showed a major thrust fault along the west side of the Tower Hill Anticline (west side of Bullpasture Gorge), but the authors of this paper have not been able to verify the presence of the thrust fault. Instead, the authors of this paper identified only a relatively minor fault in the Rose Hill Formation along State Road 678 in Bullpasture Gorge (Fig. 16.34) and a kink fold within the lower member of the Tonoloway Limestone in Bullpasture Gorge along the north bank of the river. Many relatively minor thrust faults have been identified throughout Burnsville Cove and some of these faults are well exposed in caves. Most of these minor faults have fault planes that dip to the southeast and show displacement ranging from a few inches to 5 feet.

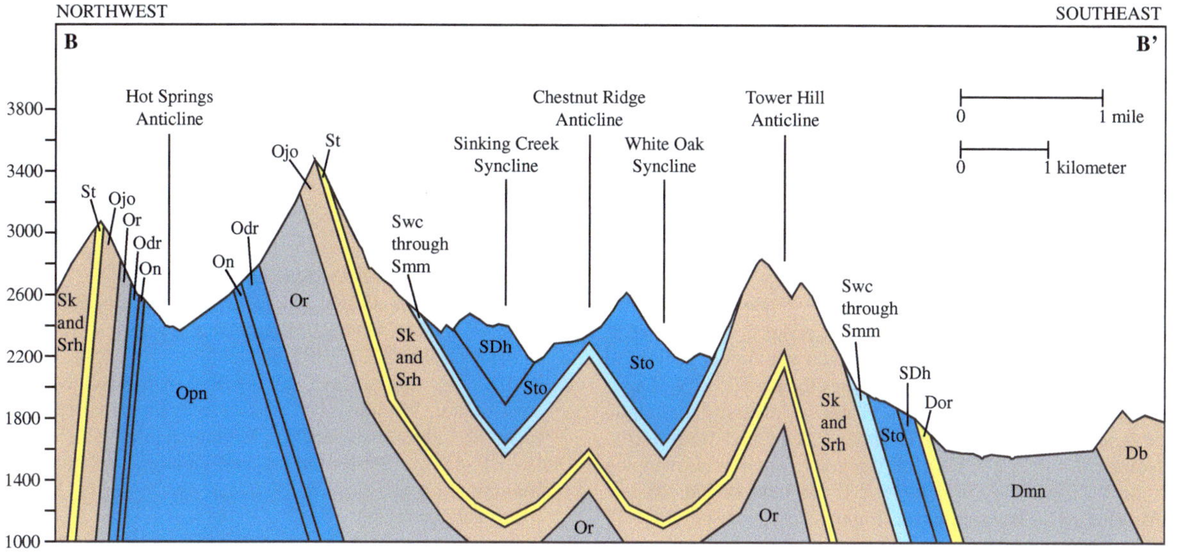

Fig. 16.33 Geologic cross section B–B′ across Burnsville Cove. Cross section location, color code for lithologies, and abbreviations of strata names are shown on Fig. 16.3. Elevation is given in feet above sea level

Fig. 16.34 Selene Deike (daughter of George Deike) at outcrop showing fault in the Silurian Rose Hill Formation by State Road 678 in Bullpasture Gorge. Photo by C.S. Swezey

Fig. 16.35 Small folds and thrust faults along the 90 Ugh Crawl in Butler Cave (Fig. 16.2). Selene Deike is looking approximately east. Photo by C.S. Swezey

In Butler Cave, for example, numerous such thrust faults are visible along a southwest-trending passage named the 90 Ugh Crawl (Fig. 16.35). One unusual thrust fault (the Cathedral Passage fault) is visible in Breathing Cave (Deike 1960a). This fault dips 14° northwest (opposite the direction of most thrust faults in the Appalachian region) and has displacement of approximately 30 feet.

16.4.3 Joints

Joints are prominent throughout Bath and Highland Counties, and most are oriented perpendicular to bedding planes. Such joints are particularly well-exposed in outcrops of the Devonian Millboro Shale along the west side of State Route 678 approximately 2.5–3 miles south of U.S. Route 250 (Fig. 16.29).

According to Deike (1960b), the most prominent joint trend in Burnsville Cove is N50W. Part of Mill Run follows this joint trend for approximately 1 mile (Fig. 16.2). The second most prominent joint trend in Burnsville Cove is N40E, and a third prominent joint trend is N60E. Joints are well exposed in some caves of Burnsville Cove, and many of the cave passages are developed along the N50W joint trend.

Acknowledgments Many thanks are extended to Gregg S. Clemmer for encouragement with this project. Christopher S. Swezey thanks Nadine M. Piatak for help with initial geological studies in the area, and John T. Haynes thanks Nevin W. Davis and numerous students of James Madison University for help in the field. In addition, John T. Haynes benefitted from a U.S. Geological Survey (USGS) EdMap grant for bedrock mapping in the Williamsville Quadrangle (NE quadrant of Fig. 16.2 of this paper). This manuscript was improved by reviews by USGS geologists Nadine M. Piatak and John E. Repetski.

References

Avary, K.L., and J.M. Dennison. 1980. Back Creek Siltstone Member of Devonian Brallier Formation in Virginia and West Virginia. *Southeastern Geology* 21: 121–153.

Bell, S.C., and R. Smosna. 1999. Regional facies analysis and carbonate ramp development in the Tonoloway Limestone (U. Silurian; Central Appalachians). *Southeastern Geology* 38(4): 259–278.

Bick, K.F. 1962. Geology of the Williamsville quadrangle. *Virginia Division of Mineral Resources Report of Investigations* 2: 40 pp.

Bowen, Z.P. 1967. Brachiopoda of the Keyser Limestone (Silurian—Devonian) of Maryland and adjacent areas. *Geological Society of America Memoir* 102: 103 pp.

Butts, C. 1933. Geologic map of the Appalachian Valley in Virginia with explanatory text. *Virginia Geological Survey Bulletin* 42: 56 pp.

Butts, C. 1940. Geology of the Appalachian Valley in Virginia. *Virginia Geological Survey Bulletin* pt. I, 52: 568 pp.

Campbell, J.L. 1879. Geology of Virginia: Continuation of section across the Appalachian chain. *The American Journal of Science and Arts* 18: 119–128.

Campbell, J.K. 1885. Highland County Virginia. *The Virginias* 6(8): 117–119.

Caro, N., K. Walsh, S.M. Walker, S.J. Whitmeyer, and J.T. Haynes. 2010. Analyses of deformed zones in the Millboro Shale, Highland County, VA: Implications for Alleghanian tectonics [Abstract]. *Geological Society of America, Abstracts with Programs* 42(1): 91.

Clark, S.H.B., and E. Mosier. 1989. Barite nodules in Devonian shale and mudstone of western Virginia. *U.S. Geological Survey Bulletin* 1880: 30 pp.

Combs, L.L., and P.S. Sethi. 2001. Petrology, macro- and micro-structures, and scanning electron microscopy of the Devonian Millboro and Needmore Shales, Highland County, Virginia [Abstract]. *Virginia Journal of Science* 52(2): 112.

Combs, L.L., and P.S. Sethi. 2002. Assessment of the petrologic, sedimentologic, and engineering characteristics of the Devonian Millboro and Needmore Shales, Highland County, Virginia [Abstract]. *Geological Society of America, Abstracts with Programs* 34(2): 46.

Darton, N.H. 1899. Monterey Folio, Virginia-West Virginia. *U.S. Geological Survey Geologic Atlas of the United States, Folio* 61: 9 pp.

Davis, N.W., and J.W. Hess. 1982. Hydrogeology of the drainage system, Burnsville Cove, Virginia. *National Speleological Society (NSS) Bulletin* 44(3): 78–83.

Deike III, G.H. 1959. Origin and geologic relations of Breathing Cave, Virginia [Abstract]. *Geological Society of America Bulletin* pt. 2, 70(12): 1803.

Deike III, G.H. 1960a. Cave development in Burnsville Cove, west-central Virginia, with special reference to Breathing Cave. M.A. thesis, University of Missouri (Columbia), 155 pp.

Deike III, G.H. 1960b. Origin and geologic relations of Breathing Cave, Virginia. *National Speleological Society (NSS) Bulletin* 22(1): 30–42.

Denkler, K.E., and A.G. Harris. 1988. *Homeognathodus peniculus* (Conodonta), a new earliest Pridolian index species, and the Ludlovian-Pridolian boundary in the central Appalachian Basin. *U.S. Geological Survey Bulletin* 1837-C, C1–C8.

Dennison, J.M. 1985. Devonian sea level curve for Virginia and West Virginia compared with eustatic curve of Johnson, Klapper, and Sandberg (1985) and relations to hydrocarbons. *Appalachian Basin Industrial Associates, Program-Fall Meeting*, 9: 57–81.

Dennison, J.M., and J.W. Head. 1975. Sea level variations interpreted from the Appalachian basin Silurian and Devonian. *American Journal of Science* 275(10): 1089–1120.

Dennison, J.M., and J.F. Schwietering. 1995. Considerations on an unconformity within Devonian shales of the southern Appalachian basin [Abstract]. *Geological Society of America, Abstracts with Programs* 27(2): 48.

Dennison, J.M. and D.A. Textoris. 1971. Appendix I—stop 8; Tioga Bentonite, 2.0 miles southwest of Williamsville, Virginia. In *Field trip to the Igneous Rocks of Augusta, Rockingham, Highland, and Bath Counties, Virginia*, eds. R. W. Johnson, Jr., C. Milton, and J.M. Dennison. *Virginia Division of Mineral Resources Information Circular*, 16, pp. 47–50.

Dennison, J.M., R.K. Bambach, S.L. Dorobek, J.K. Filer, and J. A. Shell. 1992. Silurian and Devonian unconformities in southwestern Virginia. In *Geologic Field Guides to North Carolina and Vicinity*, eds. J.M. Dennison, and K.G. Stewart. *University of North Carolina-Chapel Hill, Department of Geology Geologic Guidebook*. 1: 79–105,

Dennison, J.M., F.M. Linn, and D.A. Textoris. 1997. Late Silurian buildups related to syntectonic sedimentation across flexure in Virginia and West Virginia [Abstract]. *Geological Society of America, Abstracts with Programs* 29(3): 13.

de Witt, W., Jr., J.B. Roen, and L.G. Wallace. 1993. Stratigraphy of Devonian black shales and associated rocks in the Appalachian basin. *U.S. Geological Survey Bulletin* 1909: B1–B47.

Diecchio, R.J. 1985. Post-Martinsburg Ordovician stratigraphy of Virginia and West Virginia. *Virginia Division of Mineral Resources Publication* 57: 77 pp.

Diecchio, R.J., and J.M. Dennison. 1996. Silurian stratigraphy of central and northern Virginia and adjacent West Virginia. *University of Tennessee at Knoxville Studies in Geology* 26: 107–127.

Dorobek, S.L., and J.F. Read. 1986. Sedimentology and basin evolution of the Siluro-Devonian Helderberg Group, central Appalachians. *Journal of Sedimentary Petrology* 56(5): 601–613.

Dorsch, J., and S.G. Driese. 1995. The Taconic foredeep as sediment sink and sediment exporter: Implications for the origin of the white quartzarenite blanket (Upper Ordovician-Lower Silurian) of the central and southern Appalachians. *American Journal of Science* 295(2): 201–243.

Dorsch, J., R.K. Bambach, and S.G. Driese. 1994. Basin-rebound origin for the "Tuscarora Unconformity" in southwestern Virginia and its bearing on the nature of the Taconic Orogeny. *American Journal of Science* 294(2): 237–255.

Dunham, R.J. 1962. Classification of Carbonate Rocks according to depositional texture. In *Classification of Carbonate Rocks*, ed. W.E. Ham *American Association of Petroleum Geologists (AAPG) Memoir* 1: 108–121.

Folk, R.L., P.B. Andrews, and D.W. Lewis. 1970. Detrital sedimentary rock classification and nomenclature for use in New Zealand. *New Zealand Journal of Geology and Geophysics* 13(4): 937–968.

Fullagar, P.D., and M.L. Bottino. 1969. Tertiary felsic intrusions in the Valley and Ridge province, Virginia. *Geological Society of America Bulletin* 80(9): 1853–1858.

Gradstein, F., J. Ogg, and A. Smith. 2004. *A Geologic Time Scale*. Cambridge: Cambridge University Press. 2004, 589 pp

Harmon, R.S., and J.W Hess. 1982. Ground water geochemistry of the Burnsville Cove area, Virginia. *National Speleological Society (NSS) Bulletin* 44(3): 84–89.

Hasson, K.O., and J.M. Dennison. 1988. Devonian shale lithostratigraphy, central Appalachians, U.S.A. In *Devonian of the World*, eds. N.J. McMillan, A.F. Embry, and D. J. Glass. *Canadian Society of Petroleum Geologists Memoir* 14(2):157–177

Hatcher, R.D., Jr. 1989. Tectonic synthesis of the U.S. Appalachians. In *The Appalachian-Ouachita Orogen in the United States*, eds. R.D. Hatcher, Jr., W.A. Thomas, and G. W. Viele. *The Geology of North America*, v. F-2: 511–535.

Haynes, J.T., and S.J. Whitmeyer. 2011. Contrast in deformation styles within the Silurian-Devonian sequence, Valley and Ridge Province, Highland County, Virginia [Abstract]. *Geological Society of America, Abstracts with Programs* 43 (1): 141.

Haynes, J.T., A.K. Hoge, C.M. Morris, and S.J. Whitmeyer. 2011. Stratigraphy, facies changes, and petrology of sandstones in the Silurian sequence of the Valley and Ridge Province in Highland and Bath counties, Virginia [Abstract]. *Geological Society of America, Abstracts with Programs* 43 (1): 135.

Haynes, J.T., S.E. Porter, P.C. Lucas, R.A. Lambert, T. Rose, S. A. Leslie, and S.J. Whitmeyer. 2010a. Reservoir potential of calcareous sandstones in a carbonate and evaporitic tidal flat sequence: Silurian Tonoloway Formation, Highland County, Virginia [Abstract]. *American Association of Petroleum Geologists (AAPG), 39th Annual Eastern Section Meeting (Kalamazoo, MI), Program with Abstracts*, p. 42.

Haynes, J.T., T.R. Rose, K.E. Goggin, S.J. Whitmeyer, and S. M. Walker. 2010b. Petrography and micropore systems of mudcracks, concretions, nodules, and ash from Devonian shales in Highland and Bath Counties, Virginia [Abstract], *American Association of Petroleum Geologists (AAPG) Annual Convention*, Abstracts Volume 19:42.

Haynes, J.T., S.M. Walker, S.J. Whitmeyer, and K.E. Goggin. 2009. What came first: Folding, faulting, and fracturing in the Devonian Needmore and Marcellus Shales of Highland County, Virginia [Abstract]. *American Association of Petroleum Geologists (AAPG), 38th Annual Eastern Section Meeting (Evansville, IN), Program with Abstracts*, 39–40.

Head, J.W. 1972. Upper Silurian—Lower Devonian stratigraphy and nomenclature in the central Appalachians. In *Stratigraphy, Sedimentology, and Structure of Silurian and Devonian Rocks along the Allegheny Front in Bedford County, Pennsylvania, Allegany County, Maryland, and Mineral and Grant Counties, West Virginia*, eds. J.M. Dennison, W. de Witt, Jr., K.O. Hasson, D.M. Hoskins, and J.W. Head III, field trip leaders. *Guidebook for the 37th Annual Field Conference of Pennsylvania geologists*: 96–103

Head, J.W. 1974. Correlation and paleogeography of upper part of Helderberg Group (Lower Devonian) of central Appalachians. *American Association of Petroleum Geologists (AAPG) Bulletin* 58(2): 247–259.

Hess, J.W., Jr., and N.W. Davis. 1969. Butler Cave—Sinking Creek System. *5th International Congress of Speleology (Stuttgart, 1969)*, v. 2 (Speleogenesis Section), p. S7/1-7.

Hotchkiss, J. 1885. The geology of Highland County. *The Virginias* 6(8): 121.

Jacobson, R.B., E.D. Cron, and J.P. McGeehin. 1989. Slope movements triggered by heavy rainfall, November 3–5, 1985, in Virginia and West Virginia, U.S.A. In *Landslide Processes of the Eastern United States and Puerto Rico*, eds. A.P. Schultz and R.W. Jibson. *Geological society of America special paper*, 236: 1–13.

Kneller, M., and D. Peteet. 1993. Late-Quaternary climate in the Ridge and Valley of Virginia, U.S.A.: Changes in vegetation and depositional environment. *Quaternary Science Reviews* 12(8): 613–628.

Kneller, M., and D. Peteet. 1999. Late-glacial to early Holocene climate changes from a central Appalachian pollen and macrofossil record. *Quaternary Research* 51(2): 133–147.

Kozak, S.J. 1965. Geology of the Millboro quadrangle, Virginia. *Virginia Division of Mineral Resources Report of Investigations* 8: 19 pp.

McBride, E.F. 1963. A classification of common sandstones. *Journal of Sedimentary Petrology* 33(3): 664–669.

Nuelle, L.M., and K.L. Shelton. 1986. Geologic and geochemical evidence of possible bedded barite deposits in Devonian rocks of the Valley and Ridge Province, Appalachian Mountains. *Economic Geology* 81(6): 1408–1430.

Rader, E.K. 1984. Stratigraphy of the western anticlines—A summary. In *Stratigraphy and Structure in the Thermal Springs Area of the Western Anticlines*, eds. E.K. Rader and T.M. Gathright II. *Virginia geologic field conference guidebook* 16: 1–12.

Rader, E.K., and T.M. Gathright II. 1984. Road log. In *Stratigraphy and Structure in the Thermal Springs Area of the Western Anticlines*, eds. E.K. Rader and T.M. Gathright II. *Virginia geologic field conference guidebook* 16: 23–55.

Rader, E.K., and G.P. Wilkes. 2001. Geologic map of the Virginia portion of the Staunton 30 × 60 minute quadrangle. *Virginia Division of Mineral Resources Publication* 163 (1:100,000 scale), 1 sheet.

Ramsey, E.W. 1950. Geology of the southern portion of Bolar Anticline, Highland County, Virginia. M.S. thesis, The University of Virginia (Charlottesville), 114 pp.

Roden, M.K., R.R. Parrish, and D.S. Miller. 1990. The absolute age of the Eifelian Tioga Ash Bed, Pennsylvania. *Journal of Geology* 98(2): 282–285.

Rogers, W.B. 1843. On the connection of thermal springs in Virginia, with anticlinal axes and faults. In *Reports of the First, Second, and Third Meetings of the Association of American Geologists and Naturalists, Proceedings and Transactions*, pp. 323–347.

Ryder, R.T., C.S. Swezey, R.D. Crangle, Jr., and M.T. Trippi. 2008. Geologic cross section *E-E'* through the central Appalachian basin from the Findlay Arch, Wood County, Ohio, to the Valley and Ridge Province, Pendleton County, West Virginia. *U.S. Geological Survey Scientific Investigations Series Map* SIM-2985 (2 sheets with 48-page text).

Schwartz, B.F., and D.H. Doctor. 2009. Geomorphic and hydrogeologic evolution of karst in the Burnsville Cove, Virginia, USA: New evidence and perspectives. *15th International Congress of Speleology, Proceedings*, vol. 2, pp. 984–990.

Sethi, P.S., L.L. Combs, and M.J. Woodard. 2002. Mineralogy, petrology, scanning electron microscopy and backscattered microscopy of Middle Devonian shales from Highland County, southwestern Virginia [Abstract]. *Virginia Journal of Science* 53(2): 94.

Smosna, R.A. 1984. Diagenesis of a stromatoporoid patch reef. *Journal of Sedimentary Petrology* 54(3): 1000–1011.

Smosna, R.A., and S.M. Warshauer. 1979. A very early Devonian patch reef and its ecological setting. *Journal of Paleontology* 53(1): 142–152.

Southworth, C.S., K.J. Gray, and J.F. Sutter. 1993. Middle Eocene intrusive igneous rocks of the Central Appalachian Valley and Ridge Province; Setting, chemistry, and implications for crustal structure. *U.S. Geological Survey Bulletin* 1839: J1–J24.

Stock, C.W. 1996. Stop 5. Mustoe. In *Silurian Stratigraphy of Central and Northern Virginia and Adjacent West Virginia*, eds. R.J. Diecchio and J.M. Dennison., *University of Tennessee at Knoxville Studies in Geology* 26: 119–121

Stock, C.W., and A.E. Holmes. 1986. Upper Silurian/Lower Devonian stromatoporoidea from the Keyser Formation at Mustoe, Highland County, west-central Virginia. *Journal of Paleontology* 60(3): 555–580.

Swartz, F.M. 1930. The Helderberg Group of parts of West Virginia and Virginia. *U.S. Geological Survey Professional Paper* 158-C: 27–75.

Swezey, C.S. 2002. Regional stratigraphy and petroleum systems of the Appalachian Basin, North America. *U.S. Geological Survey Geologic Investigations Series Map* I-2768 (1 sheet).

Swezey, C.S. 2009. Regional stratigraphy and petroleum systems of the Illinois Basin, U.S.A. *U.S. Geological Survey Scientific Investigations Series Map* SIM-3068 (1 sheet).

Tso, J.L., R.R. McDowell, K.L. Avary, D.L. Matchen, and G. P. Wilkes. 2004. Middle Eocene igneous rocks in the Valley and Ridge of Virginia and West Virginia. *U.S Geological Survey Circular* 1264: 137–161.

Walker, S.M., J.T. Haynes, P.C. Lucas, R. Lambert, and S. J. Whitmeyer. 2010. Correlation of sandstones and flat-pebble conglomerates in carbonates of the Tonoloway and Keyser Formations (Silurian-Devonian), Highland and Bath Counties, Virginia, and a revised interpretation of the principal cave-forming stratigraphic horizons [Abstract]. *Geological Society of America, Abstracts with Programs* 42(1): 108.

Wheeler, H.E. 1963. Post-Sauk and pre-Absaroka Paleozoic stratigraphic patterns in North America. *American Association of Petroleum Geologists (AAPG) Bulletin* 47(8): 1497–1526.

Whisonant, R.C., M.J. Woodard, J.B. Shelton, C.F. Watts, and P.S. Sethi. 1998. Stability analysis of a landslide involving the Tioga Ash Beds along U.S. Highway 250, Highland County, VA [Abstract]. *Geological Society of America, Abstracts with Programs* 30(7): 252.

White, W.B., and J.W. Hess. 1982. Geomorphology of Burnsville Cove and the Geology of the Butler Cave—Sinking Creek System. *National Speleological Society (NSS) Bulletin* 44(3): 67–77.

Woodard, M.J., P.S. Sethi, and C. Watts. 1997. Geochemistry and highway rock stability of Devonian shales, Bullpasture Mountain, Highland County, Virginia, USA [Abstract]. *Geological Society of America, Abstracts with Programs* 29(6): 447.

Woodward, H.P. 1941. Silurian System of West Virginia. *West Virginia Geological Survey Report,* 14, 326 pp.

Woodward, H.P. 1943. Devonian System of West Virginia. *West Virginia Geological Survey Report,* 15 655 pp.

Hydrogeology of Burnsville Cove

17

Nevin W. Davis

Abstract

The drainage from Burnsville Cove is mostly underground until it reaches the surface at the four springs along the Bullpasture River. Drainage patterns, the interconnections of observed surface and underground streams, and the boundaries of the individual spring drainage basins were determined by an extensive set of tracer experiments with fluorescent dyes. This chapter is an update of the paper with the same title in the Burnsville Cove Symposium issue of the NSS Bulletin, Davis and Hess (Nat Speleol Soc Bull 44:78–83, 1982). The last dye trace covered in that paper was done 21 Dec 1974 even though the revised manuscript was accepted 13 Jan 1977 but was not published until July 1982. Since that time over 35.7 miles of virgin cave passage have been mapped, much of it under Chestnut Ridge. These new data along with more dye tracing have further elucidated the recharge areas for the four major springs in the Bullpasture Gorge. It has also raised new questions covered in the conclusions section of this chapter. During this study, a total of 27 individual sink-to-spring dye tracings were done. These enabled the determination of the spring recharge boundaries, which led to interesting observations of the interrelations between the basins and the spring flow characteristics. Included are descriptions of change in the flow regimes under flood and base flow conditions. Also eight internal traces to the Butler Cave—Sinking Creek System and one internal trace in Barberry Cave were conducted. Interesting observations from these traces are also discussed.

17.1 Introduction

The problem of the hydrogeology of Burnsville Cove, Virginia, has attracted the attention of speleologists for many years. The first important study of the geology and hydrology of the area was done by Deike (1960a, b), with particular emphasis on Breathing Cave. His study illuminated many topics concerning cave development and hydrology in the cove, but left many others untouched. Of particular interest was the relationship between the springs in the Bullpasture River Gorge and the sinking streams on the limestone uplands of the cove. The first stream tracing was done in December, 1960, by Holsinger (1961) and showed a connection between the Butler Cave—Sinking Creek System (E) and Aqua Spring (Fig. 17.1). The question of what happens to the other sinking streams, how they relate to the geology and the springs, and how the springs behave under various flow conditions was not answered until the present study, which began in 1967.

N.W. Davis (✉)
12512 Dry Run Road, Williamsville, VA 24487, USA
e-mail: nwdavis@mgwnet.com

W.B. White (ed.), *The Caves of Burnsville Cove, Virginia*,
Cave and Karst Systems of the World, DOI 10.1007/978-3-319-14391-0_17,

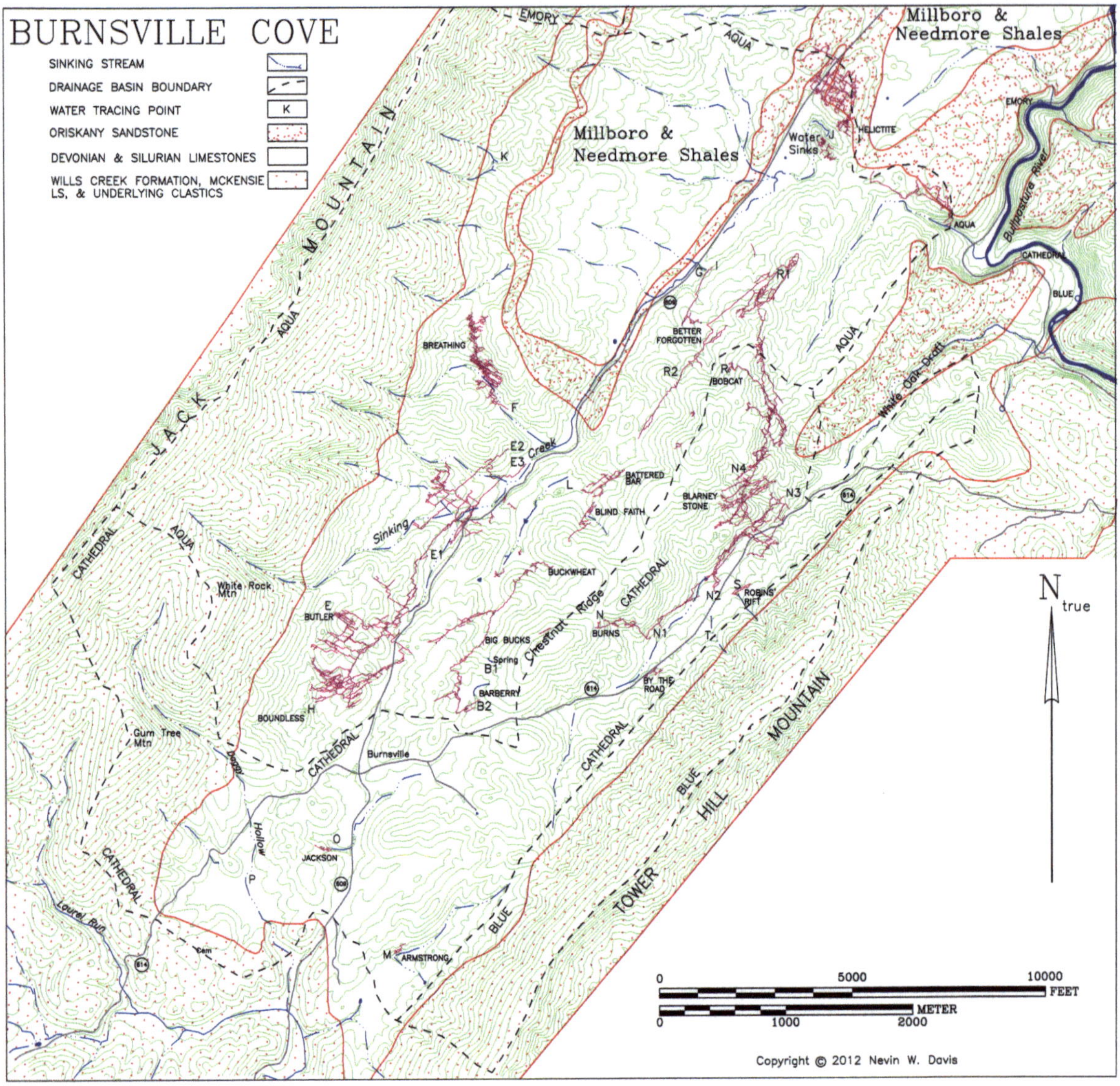

Fig. 17.1 The caves and spring drainage basins of Burnsville Cove

Carbonate aquifers and associated cave systems have been classified according to their flow types by White (1969) into three categories: diffuse flow, free flow, and confined flow. Each of these flow types is controlled by the hydrogeologic setting. The Butler Cave—Sinking Creek System in Burnsville Cove is an example of a confined flow aquifer. Actually, only part of the drainage system is confined today; the rest is free flow. The confined flow situation is caused by an impervious bed that is folded in such a manner as to force ground water to flow at depth under hydrostatic pressure.

For detailed discussion of the geology and geomorphology of the cove and associated cave systems, see Deike (1960a, b) and other chapters in this book. In general, the carbonate aquifer consists of the Tonoloway and Keyser limestones, with several interbedded sandstones playing an important confining role.

Drainage from the clastic rock slopes of Jack and Tower Hill mountains (Fig. 17.1) encounters the limestone at elevations of 1960–2460 ft where it usually sinks. The subterranean drainage from the Aqua and Emory Spring drainage basins first flows down the dip of the enclosing limestones, where it is confined to particular beds. At some point, usually near the synclinal axis, the water flow assumes a direction parallel to the strike and then crosses the structural grain to emerge at the two springs in the Bullpasture River Gorge at elevations 1770 and 1764 ft respectively. Discharge from the Cathedral and Blue Spring drainage basins also appears to be influenced by structural and stratigraphic controls from their sources to the springs on the Bullpasture River.

17.2 Climatic Setting

Burnsville Cove receives an average of 41.4 in. (1051 mm) of precipitation per year. A maximum occurs during the early summer and a minimum occurs in late fall. The mean temperature is 51.2 °F (10.7 °C) with a mean July temperature of 70.3 °F (21.3 °C) and a mean January temperature of 32.6 °F (0.3 °C). A summary of the monthly climatic conditions is given in Table 17.1. Using the Thornthwaite method (Thornthwaite 1948), a potential evapotranspiration of 25.0 in. (635 mm) was calculated for the area. Using the above values for precipitation and evapotranspiration, the runoff can be calculated by subtracting the evapotranspiration from the precipitation. In this way, a mean annual runoff of 16.4 in. (417 mm) was calculated. The mean annual runoff for the Bullpasture River basin above Williamsville, Virginia is 15.3 in. (389 mm) based on 11 years of records (1961–1971). These two runoff values are in good agreement, considering that the 1960s were a period of below average precipitation.

17.3 General Hydrology

Twenty-seven individual sink-to-spring tracings, eight tracings within the Butler Cave—Sinking Creek System, and one trace in Barberry Cave were conducted over a 35 year period, using the tracing procedure described in Appendix A. The results of these tracings are shown in Tables 17.3, 17.4, 17.5, 17.6, 17.7 and 17.8 in Appendix B. The transit time for the dye is expressed, in most cases, as a time less than *t* number of days, since in most cases the dye had already passed when the detectors were recovered.

In two cases, Sneaky Creek and Breathing Cave, detectors were changed every 12 h and the actual time of the dye pulse peak was observed. It should be noted that during a 35 year period, the stream discharge varied considerably and because the dye transit time is a function of discharge, the transit time also varied considerably. Tracings were attempted at medium and low flows to avoid dye loss due to adsorption on suspended clay and silt and to prevent large variations in the transit time.

Spring discharge measurements were made with a current meter. All of the stream discharge measurements indicated in Tables 17.3, 17.4, 17.5, 17.6, 17.7 and 17.8 were estimated by measuring the channel cross-section at a place that was more or less uniform and timing the flow of water through a measured length of the channel.

The individual dye tracings, along with topographic and geologic considerations, were used to delineate the drainage basins of the four springs along the Bullpasture River. Drainage divides on the clastic rocks are simply the surface drainage divides. Divides on the carbonate rocks were not as obvious. Of particular interest was the drainage divide between Aqua and Cathedral Springs. Chestnut Ridge, the anticlinal ridge that bisects Burnsville Cove, should be a divide between these two springs. If no confined conditions exist under the ridge, clastic rocks underlying the carbonate rocks should act as an impervious barrier to

Table 17.1 Summary of mean monthly temperature and precipitation for Burnsville Cove area, Virginia

	°C	°F	mm	in.
January	0.3	32.6	76.5	3.01
February	0.9	33.6	72.1	2.84
March	4.5	40.1	99.8	3.93
April	10.5	50.9	79.2	3.12
May	15.6	60.1	93.2	3.67
June	19.6	67.2	105.9	4.17
July	21.3	70.3	106.2	4.18
August	20.5	68.9	118.1	4.65
September	17.1	62.7	81.8	3.22
October	11.4	52.6	74.7	2.94
November	5.3	41.6	70.4	2.77
December	0.8	33.4	73.2	2.88
Year	10.7	51.2	1051.0	41.38

Data averaged over the period 1961–1971

water crossing the regional strike. Dye tests in Better Forgotten Cave (I), Bobcat Cave (Chestnut Ridge Blowing) (R), Woodzell Sink (L), and Burns Chestnut Ridge Cave (N) placed the divide between Better Forgotten and Bobcat Cave, and between Burns Chestnut Ridge and Woodzell Sink. A check of the dip and strike near Bobcat Cave shows that, even though it is on the northwest side of Chestnut Ridge, it is near the crest of the anticline. It turns out that the crest of the Chestnut Ridge anticline is perhaps 600 ft to the northwest of the summit of the ridge.

Chestnut Ridge ends near Burnsville and is replaced by a rolling area of large dolines. Two of the largest dolines, adjacent to one another, contain streams that drain to separate springs. The Boundless Cave stream (H) drains into the Butler Cave—Sinking Creek System (E) and emerges at Aqua Spring, while the Jackson Cave stream (O) resurges at Cathedral Spring.

In other places in the cove, water bypasses the sandstone in much the same way as it bypasses Taggard Shale in the Greenbrier Limestone karst area of Pocahontas County, West Virginia (Werner 1972). The water from the upper carbonate rocks flow on the upper surface of the impervious shale layer to a nearby hillside spring, crosses the impervious layer by surface flow, and sinks into the carbonate rocks below the impervious layer. This is especially visible near the entrance to Armstrong Cave (M), where water emerges at the top of a sandstone layer and runs down the hill to a sink and presumably joins the stream in Armstrong Cave. Another example of this flow pattern occurs near the entrance to the Butler Cave—Sinking Creek System (E). Water draining from the sandstone flows across and down an access road, causing a mudhole. From here, it flows into the Burnsville Sink (H) and enters the Butler Cave—Sinking Creek System.

The major stream draining into Water Sinks (J) heads on the clastic rocks of Jack Mountain and crosses a band of carbonate rocks before sinking at Water Sinks. This is the only stream observed to cross the carbonate rocks on the flanks of Jack Mountain under normal flow conditions. There are several possible explanations. One is that perhaps the very active sinking streams to the north and south of this stream have diverted water away from this area in the past, retarding development of underground drainage. Another is that the streams entering Water Sinks carry a heavy load of clastic material so that the karstic drainage (if it exists) is clogged.

17.4 Individual Drainage Basins and Springs

There are four springs in the Bullpasture Gorge that drain the limestone highlands of Burnsville Cove, the eastern flank of Jack Mountain north of the cove and the western flank of Tower Hill Mountain. The springs and their drainage basins are listed in Table 17.2.

17.4.1 Cathedral Spring and Blue Spring Drainage Basins

The boundaries of the basins are shown on Fig. 17.1. The combined basins stretch from the crest of Jack Mountain on the west to the crest of Tower Hill Mountain on the east and include the valley of White Oak Draft, east of Chestnut Ridge. Both Cathedral and Blue Springs are included in one discussion because their drainage basins are closely interrelated. Blue Spring emerges from a submerged solution passage which descends steeply along a trend of 300° to a depth of 48 ft. At the bottom on the northwest, water wells up through sand and one-inch diameter pebble boils (Ron Simmons 1999 personal communication. See Chap. 4). The middle Keyser Limestone at the spring dips 75° northwest. After a storm on 27–28 May 1973, Blue Spring was observed discharging orange-brown water at a rate of 32 cfs. Precipitation during the storm ranged from 2 to 3 in. and fell from approximately 5 P.M. on the 27th to 7 A.M. on the 28th. The greatest discharge was not from the main submerged opening, but rather from many boils in the adjacent spring pool. Water of the same color was also squirting out of an opening in the road 3 ft above the spring pool. River water at the time was of a different muddy color and could easily be distinguished from the spring water. Evidently, the spring outlet is partially blocked, and, during flood, water in the submerged conduit develops a considerable head.

Cathedral Spring issues from a pile of large talus blocks at the base of a 90 foot cliff whose upper exposed parts are Licking Creek Limestone/Chert dipping 12° southeast. The spring is probably at or

Table 17.2 Drainage areas and discharges of Bullpasture River Springs

Basin	Spring Horizon[c]	Altitude[c]	Drainage area	Low flow discharge		Discharge per unit area		Colinform[d] count
				28 Oct 72	25 Aug 73	28 Oct 72	25 Aug73	MPN/100 ml
Blue Spring	Keyser limestone	520 m	3.7 km^2	52 l/s	60 l/s	14.1 l/s/km^2	16.2 l/s/km^2	11
		1704 ft	1.4 $miles^2$	1.84 cfs	2.12 cfs	1.31 cfs/$miles^2$	1.51 cfs/$miles^2$	
Cathedral Spring	Corriganville —New Creek contact	522 m	11.1 km	131 l/s	170 l/s	11.8 l/s/km^2	15.3 l/s/km^2	33
		1713 ft	4.3 $miles^2$	4.63 cfs	6.00 cfs	1.08 cfs/$miles^2$	1.40 cfs/$miles^2$	
Aqua Spring	Keyser limestone	540 m	21.1 km^2	273 l/s	138 l/s	12.9 l/s/km^2	6.5 l/s/km^2	79
		1770 ft	8.1 $miles^2$	9.64 cfs	4.87 cfs	1.19 cfs/$miles^2$	0.60 cfs/$miles^2$	
Emory Spring	Corriganville —Licking Creek contact	538 m	15.8 km^2	163 l/s	109 l/s	10.3 l/s/km^2	6.9 l/s/km^2	8
		1764 ft	6.1 $miles^2$	5.76 cfs	3.85 cfs	0.94 cfs/$miles^2$	0.63 cfs/$miles^2$	
Bullpasture River below springs			284.9 km^2 109.9 $miles^2$	3370[a] l/s		11.8 l/s/km^2		
				119.0[a] cfs		1.08 cfs/$miles^2$		
			284.9 km^2 109.9 $miles^2$	1458[b] l/s		5.1 l/s/km^2		
				51.5[b] cfs		0.47 cfs/$miles^2$		

[a] Mean discharge over 5 years August 1960–September 1965
[b] Mean discharge for October over 5 years 1960–1964
[c] Deike (1960a, b)
[d] Most probable number of coliforms per 100 ml of water

near the New Creek-Corriganville Limestone contact. There is a tight, water-floored passage in the talus blocks about 3 ft above the river level leading to a very tight submerged solution passage. Enlargement of this passage by Simmons (1990) on several trips in 1989 led to a series of narrow submerged canyons paralleling the Bullpasture River until in 200 ft a large conduit was reached at 24 ft depth. This passage continues for an additional 750 ft to a depth of 150 ft where it appears to drop vertically (see Chap. 4). Under storm and flood conditions, this spring becomes muddy and remains cloudy long after the river has cleared. Although the normal flow is 4.6 cfs, storm flows of more than 150 cfs have been observed. Under high flow conditions, water emerges from the cliff as far as 150 ft upstream from the main resurgence, which is in the general vicinity of the end of the large, blocked off, main conduit.

The vertical to near-vertical limestone beds on the northwest flank of Tower Hill Mountain capture the drainage from the clastic rocks of the mountainside. Most water flows among the boulders just out of view and slowly disappears upon leaving the clastics. The limestone is covered with clastic colluvium that sharply restricts infiltration into the limestone. In one dye tracing of a surface stream (T) which splits and sinks in two places 100 ft apart, dye emerged from both Blue and Cathedral Springs. On the other hand, during wet weather, dye placed in an ephemeral stream at the bottom of Robins Rift Cave (S) emerged only at Cathedral Spring. Flowing water occurs in Robins Rift Cave only in times of high runoff. The conclusion that can be drawn from the dye tracing is that under low runoff conditions Blue Spring drains the Tower Hill mountainside. At higher flows some of the runoff reaches the Cathedral Spring basin. The spring response to high runoff conditions further shows this to be the case. Cathedral Spring has more variation between flood and low flows than Blue Spring; further, bacterial counts show that Blue Spring is less polluted than Cathedral or Aqua springs, because there are no dwellings or pastures on the mountainside above Blue Spring. The drainage divide between the springs is shown on the map to be just downslope of the Tonoloway—Keyser limestone outcrop line. Table 17.4 shows what appears to be anomalous behavior for the dye trace in Chestnut Ridge Blowing Cave (R) as compared to Robins Rift

Cave (S). Although Chestnut Ridge Blowing Cave has the shortest straight-line traverse distance, it had nearly the longest transit time in the Cathedral Spring drainage basin. Robins Rift Cave, 3300 ft further from the spring had the shortest transit time. The trace of Robins Rift Cave was done under relatively high water conditions, and dye transit time is directly related to flow rate. The trace of Bobcat Cave (Chestnut Ridge Blowing) was performed under very low flow conditions. The little stream that was dyed flows slowly from pool to pool, in a southeasterly direction, following a small passage down the dip of the anticline for 1500 ft to where it passes under a large paleo-trunk passage. It continues downward for a total drop of 632 ft to a sump at 1765 ft msl only 52 ft above the spring. In the Boondocks section of the Chestnut Ridge Cave System another stream carrying what is probably 80 % of Cathedral Spring, sumps at 1734 ft elevation only 21 ft above the spring. No wonder that detectors from Cathedral Spring still tested positive 14 and 39 days after the first positive indication, showing the dilution and stretching of the dye pulse.

There are no surface streams of any significant size on the clastic rocks of Tower Hill Mountain except during flood conditions. The average width of the drainage area from the ridge crest to the first limestone outcrop is only 1800 ft whereas on Jack Mountain this distance is 3900 ft. Under flood conditions, some of the water falling on the Cathedral Spring drainage basin escapes to the south. This water is from the 1.5 miles^2 drainage basin of Daggy Hollow Run (P) and the small stream to the southwest. Under normal flow, all water in these streams sinks into the carbonate rocks and emerges at Cathedral Spring. Overflow during flood conditions enters Dry Run and flows south to the Cowpasture River.

17.4.2 Aqua Spring Drainage Basin

Aqua Cave Spring flows from an underwater opening 2 ft high and 9 ft wide at the base of a 30 foot cliff of Keyser limestone, which dips 22° southeast. The spring is 50 ft above the river. Diving in the spring led to the discovery of Aqua Cave. The stream that feeds the spring can be followed 1800 ft through a vadose passage, which averages 30 ft wide and high, to a sump. The stream cannot be followed through this sump because of breakdown. At a point about 1000 ft from the entrance, a left turn and climb to a higher level brings one to French Lake and an overflow sump 29 ft above the cave entrance. This sump has been penetrated 1800 ft to a depth of 256 ft by Ron Simmons (personal communication). What's remarkable about this, other than the Herculean effort by Ron Simmons, is that the deep point is about 170 ft below the Bullpasture River, the supposed base level! At the French Lake sump the Clifton Forge sandstone dips about 7° to the northwest toward the axis of one of the many minor folds. The passage is constrained to follow the sandstone downward. After any storm Aqua Spring rapidly becomes muddy. After a large storm, flow measurement at the culvert under route 678 by B.F. Schwartz on September 28, 2004, gave a spring discharge of 400 cfs.

The boundaries of the Aqua Spring Drainage Basin are shown in Fig. 17.1. These boundaries are essentially the same as those of the surface Sinking Creek, with the inclusion of the Burnsville Sink (H) and the Mill Run valley. Unlike the other drainage basins studied, this one has but one outlet. Except for evapotranspiration losses, no water leaves the basin except through Aqua Spring under all flow conditions. Even when the sinks on the flanks of Jack Mountain are unable to carry flood flows, the overflow streams sink at either Water Sinks (J) or the Sink of Sinking Creek (surface) (G). In at least three cases, Sink of Sinking Creek (surface) (G), Water Sinks (J), and Woodzell Sink (L), the sinking streams go underground in formations stratigraphically higher than those containing Breathing Cave (F) and the Butler Cave—Sinking Creek System (E). The fact that all of these streams emerge at the same spring indicates that the water is able to penetrate the upper Breathing Cave Sandstone in places.

On 23 and 24 October 1970, a tracing experiment was run to determine which, if any, of the major streams in the Butler Cave—Sinking Creek System traveled through the stream passage in Better Forgotten Cave (I). To this end, different tracers were placed in the three major streams in the Butler Cave—Sinking Creek System: 11 kg of NaCl in Sneaky Creek, 160 g of fluorescein in Sinking Creek, and 250 g of Rhodamine B in Slippery Creek. A careful measurement of the stream flows was made at the same time. 24 h later the stream in Better Forgotten was tested for the presence of tracers. None were found. Experience had indicated that under the existing gradient, the dye pulse should have passed in 24 h and the duration of the pulse of detectable dye should be at least 10 h. The above results are inconclusive. Convincing evidence that the Better Forgotten stream is independent of the Butler Cave—Sinking Creek System water is that the stream discharge at Better Forgotten was smaller than that of any of the streams leaving the Butler Cave—Sinking Creek System. The tracing from the stream at the bottom of Better Forgotten Cave (I) to Aqua Spring is of interest because of the low gradient. The change in elevation from Better Forgotten stream sump to the Siphon Room Sump is 33 ft. This gives a gradient of about 30 ft/mile, compared to the 177 ft/mile gradient of the vadose portion of Sinking Creek in the Butler Cave—Sinking Creek Cave System. It is likely that most of the passage from Better Forgotten to Aqua is completely flooded. There is evidence in the form of obliterated footprints that the Better Forgotten stream passage floods to a depth greater than 40 ft. This flooding could come from two causes: If there is a constriction in flow downstream of the sump in the cave, flood waters entering the cave could simply pool behind it. Another explanation is that since the Better Forgotten sump is only 33 ft above the upstream sump in Aqua, flood waters from all the major streams feeding Aqua Spring back flood into Better Forgotten Cave. In either case, the cave acts as a reservoir and releases water to Aqua Spring as the other inputs begin to recede.

Fig. 17.2 Emory Spring discharging into a muddy Bullpasture River during the Hurricane Francis storm of September, 2004. Photo by Philip C. Lucas

17.4.3 Emory Spring Drainage Basin

Emory Spring issues from the base of a cliff of Licking Creek Limestone at its contact with the Corriganville Limestone. State route 678 runs along the base of this cliff at the river's edge, and the spring opening is buried by the road fill. One hundred fifty ft northeast of the spring, the Licking Creek dips 11° northwest toward the axis of one of the minor synclines on the Chestnut Ridge anticline.

The response of Emory Spring to a flood pulse can best be illustrated by observations made during a storm on 8–9 July 1970, when it rained 4 in. in 19 h. Between 17 and 41 h after the rain stopped, Emory Spring was observed to be gushing clear water. Afterward, the spring became muddy with a very high flow rate. Ten days after the rain stopped, Emory spring had resumed normal flow but the water was still murky. Similar storm responses have been observed at other times (Figs. 17.2 and 17.3).

Fig. 17.3 Emory Spring 14 h later. Photo by Philip C. Lucas

For the input flood pulse to be carried rapidly to the spring with the resultant discharge of clear water, it is surmised that the spring must have an extensive series of flooded passages containing a large quantity of water. When the flood water increases the head on the input side of the system, the response is rapidly transmitted to the spring. When the total storage of the flooded spring conduit is exhausted, the muddy input water emerges from the spring. Even under flood conditions the complete expulsion of all the water in the system required more than 60 h. A quick calculation indicates a flooded volume of between 350,000 and 3,500,000 ft^3.

Observations of Aqua and Emory springs after a storm on 5–6 October 1972, which yielded 5 in., indicated a similar response. Less than 12 h after the rain started, Aqua was gushing chocolate-colored water and Emory was spouting clear water. In the 1970 storm, the river crested 5 ft above normal about one hour after the rain stopped. 40 h later, the river had dropped to within 8 in. of normal level but was muddy with water from springs. Obviously, Aqua Spring has either a shorter series of submerged passages than Emory or has an input close to the spring. Probably both are true because the Water Sinks input is only 3800 ft from Aqua Spring. Since Emory Spring is at river level, it is more influenced by rising and falling river levels.

Bacterial counts show Emory spring to be the least polluted of the four springs, indicating that the recharge area for this spring is similar in character to that of Blue Spring. The drainage basin for Emory Spring is shown on the map (Fig. 17.4). Located north of the Aqua Spring basin, it extends from the crest of Jack Mountain down the southeast flank to the Oriskany Sandstone outcrop. Streams originating on the Clinton Formation sink at the limestone contact under normal flow conditions. Under flood conditions,

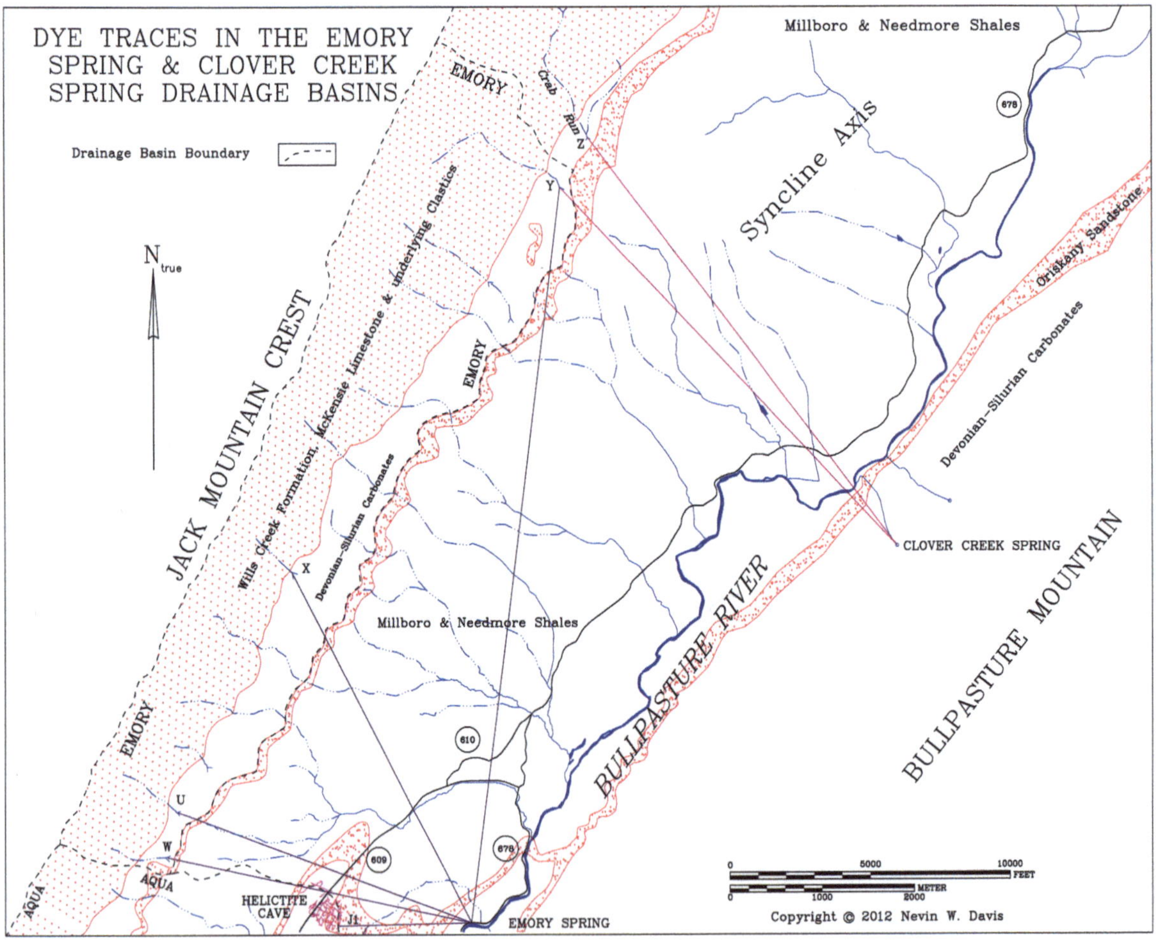

Fig. 17.4 The Emory Spring drainage basin

the sinks overflow and the water crosses the Millboro Shale and empties directly into the Bullpasture River. Five dye tracings were conducted in the Emory Spring basin and one trace in the Clover Creek Spring basin. These tracings were sufficient to accurately determine the boundaries of the basin. The northern boundary was determined by traces Y and Z in adjacent sinking streams. What is remarkable is that trace Y emerged at both Emory and Clover Creek Springs despite the fact that Clover Creek Spring is 185 ft higher than Emory Spring and Emory Spring is nearly 5 miles distant. To reach either of these springs the water must be confined beneath the Millboro Shale and the Oriskany Sandstone. Most likely the separation of the water into two streams occurred close to the tracing point as happened in dye trace T in the Blue and Cathedral Springs drainage basins.

There is another small undelineated catchment area which drains into Emory Spring. This is Helictite Cave with the trace marked as J1 on Fig. 17.4. The cave is in the Licking Creek Limestone, the uppermost unit of the Helderberg limestone series. What is curious about this catchment area is that it is perched high on the north side of Water Sinks which at the present time drains into Aqua Spring. It is very possible that in the past, before erosion and corrosion cut down into the Corriganville and Keyser limestones, Helictite Cave received the Burnsville Cove surface drainage and channeled the water to Emory Spring.

17.5 Internal Drainage of Individual Caves

17.5.1 Butler Cave-Sinking Creek System

Some interesting observations have resulted from dye tracing and visiting the Butler Cave—Sinking Creek Cave System (Fig. 17.5) at various water levels. Difficulty Creek, which collects water from the sink at the entrance of the cave, disappears into a small passage at (5). This water crosses under the main stream passage and emerges at (10), the Sinking Creek resurgence, about 630 ft from where it disappears. Water following the Complaint Section Canyon (11) drops down a pit and disappears into a hole too small to explore. This water evidently follows the same pattern as Difficulty Creek, even though the dry passage continues to its intersection with the main stream passage near Sand Canyon (12). Water from near Penn State Lake (3) and from the sumps beyond Natural Bridge (4) both emerge at the Sinking Creek Resurgence (10). These streams completely fill the passages they follow. The water during base flow conditions follows a new lower route around the main stream passage from Penn State Lake to the Sinking Creek Resurgence.

Under flood conditions, the small bypass passages cannot transmit all the flow. The water in Penn State Lake rises about 10 ft and a stream flows from it to the sump beyond Natural Bridge (4). When the small passages in this area can no longer handle the flow, water backs up and floods the passages to a depth of about 10 ft. Water then flows out of these passages, under the Natural Bridge, and down the main stream passage past Sand Canyon (12) to join the waters from the Sinking Creek Resurgence.

Downstream, Sinking Creek disappears into a passage too small to explore (1). It reappears in several side passages further downstream and finally disappears near the Dry Sumps (13), only to reappear in Marlboro Country as Stream (7). This passage to Marlboro Country is too small to explore, but is large enough to carry all the water under most flood conditions. This passage is probably much more recent than the large passage through the Dry Sumps and down Sneaky Creek (14). Probably, the paleo-flow direction passed through the Dry Sumps and down the French Passage (14) as it does now during large floods.

Sneaky Creek (14) flows to Rat's Doom Siphon (15) and on to Aqua Spring. The cross-section of the passage near Rat's Doom Siphon is small compared to that near Last Hope Siphon (2). Likely, the paleo-flow of Sneaky Creek went through Dave's Lake (9) and on to Last Hope Siphon (2). High flood water even now may do the same. The dye tracings from the Hanging Rock Room (8) to Dave's Lake Area (9) and eventually to Last Hope Siphon (2) show that the present low-water streams do not flow preferentially through the large passages. They cut beneath passages and through walls seemingly at random. It would appear that the cave is still undergoing of active stream downcutting and solution below the present vadose levels.

Water leaves the explored passages in the Butler Cave—Sinking Creek System through five sumps. Two of these, Last Hope (2) and Rat's Doom Siphons (15), are the most accessible and are well known to most explorers of the cave. These sumps are above the Lower Breathing Cave Sandstone, which separates

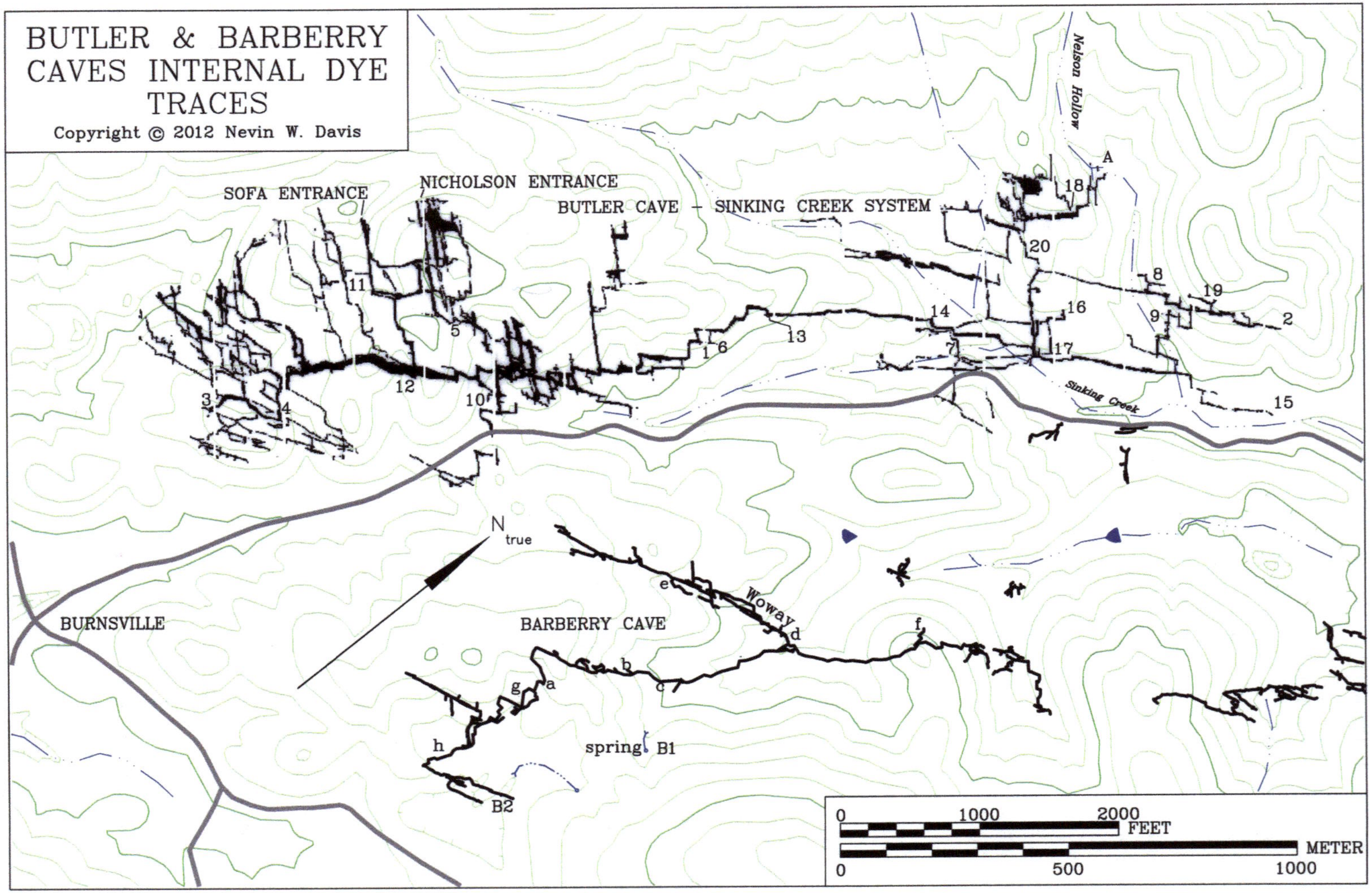

Fig. 17.5 Butler and Barberry Caves—internal traces

Marlboro Country from the overlying French Passage (14). Last Hope Siphon occurs where the Upper Breathing Cave Sandstone ceiling dips below the water level. In Marlboro Country, Sinking Creek is joined by a stream of equal size (16) and by another, smaller stream, to form the largest stream in the cave. This exits the cave through a final sump (17) and flows to Aqua Spring.

17.5.2 Barberry Cave

An enhanced line map of Barberry Cave is shown in Fig. 17.5. The entire cave, all 3.36 miles, is located in the lower member of the Tonoloway Limestone. The Williamsport sandstone forms the floor of the cave just downstream of the Big Bucks entrance, downstream from (c), and in one place in the Woway. The Lower Breathing Cave Sandstone is penetrated in two locations where upper levels exist. One location is in the Woway and the other is an upper level in the upstream portion of the cave, between (g) and (h) where a side passage heads west from the main passage. Even during dry conditions, a small stream, approximately 0.2 cfs, flows the length of the cave from (h) to (f). During wetter times a stream enters an entrance at (B2) and exits at (f), the downstream sump. This sump has been dived by Australian diver Tim Payne for 100 ft to a cobble plug with the cobbles being held in place by the water flow. Another smaller stream enters from the Woway at (d) and joins the main flow. The stream exiting the cave was dye traced to Aqua Spring.

The small stream from a surface spring (B1) sinks within 200 ft into the lower member of the Tonoloway and enters Barberry cave at multiple locations through the Lower Breathing Cave Sandstone (and shale). In the dye trace of this spring, no dye was present in the stream from (h) to (g). A stream entering through Rimstone Pools at (a) contained dye. Dye was captured from the Woway stream at (d), from a ceiling dribble at (b), and from a small stream entering from the east at (c). The Woway stream probably received the dye from a ceiling waterfall through the Lower Breathing Cave Sandstone at (e). It is obvious that the small sinking stream encountered a distributary system above the Lower Breathing Cave Sandstone and entered the cave at multiple locations. The same occurs at other locations in the Cove! The stream sinking in Nelson Hollow (A) on Fig. 17.5 is also distributed to at least two locations in the Butler Cave—Sinking Creek System. The dye exits the Dynamite Section through the Frothing Slosh (20) and dye was also captured in the stream exiting Last Hope Shaft (19). Dye was not present in the water dribbling out of the Hanging Rock Room (8).

17.5.3 Chestnut Ridge Cave System

The Chestnut Ridge Cave System consists of Burns Chestnut Ridge, Blarney Stone, and Bobcat Caves and contains 21.07 miles of passage at this writing. When the dye traces of the small entrance streams in Burns and Bobcat were done, approximately 1000 ft of passage were known in each. Both streams exit the entrance sections of the cave through tight passageways and descend the southeast limb of the Chestnut Ridge anticline following the dip. Blarney Stone does much the same. The Burns entrance stream sumps just before the continuing passage encounters a large stream passage nearly on grade (N1) (Fig. 17.1). The stream parallels Chestnut Ridge for 1300 ft till it plunges down a couple of waterfalls and then shortly sumps. The size and location of the stream indicates that it must drain the Burnsville area. At (N2) on Fig. 17.1, a stream emerges from a sump in the Burns section of the system that carries an estimated 80 % of the flow of Cathedral Spring. It can be followed downstream until it gets too low to follow. The stream reemerges in the Boulder Dash section of Blarney Stone Cave and can be followed intermittently to a sump at (N3). Here the "Cathedral River" is only 21 ft above the spring. This is the last place it is seen in the cave.

The entrance passage of Blarney Stone Cave descends Chestnut Ridge in the northeast direction as do four other nearly parallel passages. None of these passages intersect the large paleo-trunk passage on grade. They have cut below it by an increasing amount from south to north. Where Rolling Rock Canyon (N4) cuts across the trunk, the stream has cut 95 ft below the floor of the trunk. Clearly the large trunk is much older than the canyon passages. The Rolling Rock Canyon sump must emerge at Cathedral Spring if for no other reason than it is at the elevation of the French Lake sump in Aqua.

Fifteen hundred ft north of where the Bobcat Cave entrance passage intersects the paleo-trunk at the

Camp Room, the Aqua—Cathedral Spring drainage boundary is drawn. There has been only one dye trace done to date in the north end of the cave. This is at the sump (R1) of the stream draining the Burnsville Turnpike (R2), the 622 Sump. There are other small streams at this end of the cave which should be traced, but all should go to Aqua Spring. Another dye trace was done on this side of the divide. Dye was injected in the bottom of Woodzell Sink at the lowest point in the sink, not where the stream is indicated to sink on Fig. 17.1, and this dye was observed in the stream in the Burnsville Turnpike (R2).

17.6 Conclusions

The carbonate aquifer of Burnsville Cove is developed in folded and jointed Tonoloway and Keyser limestones, with interbedded sandstones acting as confining layers. Surface drainage from the clastic rock slopes of Jack and Tower Hill Mountains generally sinks upon encountering the carbonate rocks. Subterranean drainage in general follows structural and stratigraphic controls from its source to springs along the Bullpasture River. Drainage basins of four major springs along the Bullpasture River (Blue, Cathedral, Aqua Cave, and Emory springs) have been delineated by dye-tracing experiments and characterized by observations of spring discharge behavior. Results of these measurements are summarized in Table 17.2. Two low-flow measurements and their resulting discharges per unit area are given because of the different results under different flow rates. They apparently indicate two different sets of flow conditions or basin characteristics in response to past precipitation events. On 28 October 1972, the discharges per unit area were essentially the same for all 4 drainage basins, but on 25 August 1973, they were different, those for Aqua and Emory springs being significantly lower than those for Blue and Cathedral springs.

Three major factors control the discharge pattern of the springs: (1) precipitation amount and distribution, (2) surface basin characteristics, and (3) aquifer characteristics. The interaction of these factors may vary between basins, causing the springs to behave differently, and explains why two different sets of values were obtained for the springs along the Bullpasture River. Precipitation amounts and distribution within time and space were different before the two sets of measurements. Previous to the 28 October, 1972 measurements, it rained 0.6 in. after a relatively dry three-week period. Prior to the 25 August, 1973 measurements, it rained 0.25 inch after a relatively moist previous month. Surface drainage basins vary in percentage of carbonate and non carbonate rock, which will affect the rates and volumes of recharge. An individual carbonate basin can vary in size and point of recharge to the aquifer.

Surface and subsurface divides do not necessarily correspond. A change in the point of sinking of a surface stream may change the underground basin that it recharges, thus affecting the drainage area of the spring and its apparent discharge per unit area. The surface basins of Burnsville Cove do vary in the above ways, which helps to explain the variations observed between springs. Carbonate aquifer characteristics, such as length and openness, flow path, and storage, will affect the rate and volume of water movement within it. Differences between springs thus might be explained by the amount of water in storage and the openness of the flow path. It has been shown that the Burnsville Cove springs do behave differently under high flow conditions, indicating variations in flow path and storage characteristics between basins.

The carbonate basins studied represent 17 % of the Bullpasture River drainage basin. Large portions of the basin are composed of clastic rocks, most notably the Millboro shale. Flood overflow from subbasins, such as that of Emory Spring, drains directly to the river. It is not surprising, therefore, that the river exhibits flood characteristics intermediate between that of totally carbonate and totally clastic basins. The mean-annual flood peak discharge per square mile is 31.5 cfs/miles^2. This value is higher than those for limestone basins reported by White and Reich (1970) of 8.9–19.7 cfs/miles^2 but lower than their value of 40.6 cfs/miles^2 for the Jordan Creek carbonate basin. The Jordan Creek basin has a low percentage of carbonate to clastic rock, as does the Bullpasture River basin.

Appendix A: Dye tracing techniques

To determine the drainage basins in the Burnsville Cove, a method of tracing the sinking water to the springs in the gorge was needed. Because of the expected long transit times and the fact that the area was visited only an average of once every two weeks, it was necessary to have some method of introducing a tracer at one time and recovering it at the suspected resurgences at any convenient later time and continue until the tracer was detected. The Dunn (1957) method allows this field procedure and was used with Zotter's (1961) dye receptor variation. Refinements of the techniques used are briefly as follows:

1. Place receptors consisting of about 20 g of activated coconut charcoal enclosed loosely in pieces of nylon stockings, secured with nylon cord to an anchor, in the main flow of all suspected resurgences.
2. Inject a tracer into the source to be investigated. Dyes suitable for use as tracers include fluorescein and rhodamine WT. The less expensive fluorescein was used in most of the work reported here.
3. After allowing sufficient time, collect all receptors from the resurgences and replace them with fresh ones.
4. Place 10 g of the exposed charcoal in a test tube.
5. Release the dye with a 5 % solution of KOH in ethyl alcohol, using only enough to cover the charcoal by 5 mm.
6. Use an intense blue-white light, not an ultraviolet lamp or the sun, to look for the dye. A penlight flashlight with fresh cells will work. Observe the green fluorescence at a right angle to the beam. For very weak tests wait 24 h before concluding that the trace is negative.
7. If the test is positive, repeat steps 3 through 6 until all springs are negative before resuming the tracing program. If the test is negative, repeat steps 3 through 6 until confident the dye has not and will not appear in any of the springs before resuming the tracing program.

For small dye concentrations (less than the easily visible 0.1 ppm) and constant flow rate, the amount of dye captured is directly proportional to the amount of dye that passes through the receptor. Dilution of the dye in the flow system is not important as long as no dye is lost in the system and providing the receptors maintain their full dye-adsorbing capacity. The cove drainage is mainly conduit flow, and apparently very little dye was lost to adsorption on sediments. In the tracing program in the cove, as little as 10 g and as much as 160 g of dye were used for a single trace. An amount of 80 g (one sea dye marker) was determined to be sufficient to trace path lengths as long as 5 miles and dilutions from a 0.001 cfs stream to a 10 cfs resurgence. In one dye tracing, 160 g of dye was used to dye an estimated 28,000,000 ft^3 of water. This was the amount of water that issued from the spring while the receptors still indicated a positive test. As far as is known, dye was never present in visible concentrations in any of the springs.

Appendix B: Tabulation of Individual Dye Traces

See Tables 17.3, 17.4, 17.5, 17.6, 17.7 and 17.8.

Table 17.3 The Aqua Spring Basin

Source name	Map identification (Fig. 17.1)	Date dm/year	Horizon	Discharge at time of trace	Low flow	Dye transit time, t (days)	Dye horizontal distance	Dye vertical distance	Comments
Sinking Creek (sump downstream from Moon Room)	E1	22/03/1969	Tonoloway (below lower sandstone)	35.4 l/s 1.25 cfs	7.4 l/s 0.25 cfs	t < 13	4850 m 15,910 ft	92 m 303 ft	To Aqua Spring
Slippery Creek (Last Hope Siphon)	E2	5/4/1969	Tonoloway (below upper sandstone)	14.2 l/s 0.5 cfs	0.8 l/s 0.03 cfs	t < 13	3880 m 12,730 ft	59 m 194 ft	To Aqua Spring
Sneaky Creek (Rat's Doom)	E3	14/06/1969	Tonoloway (below lower sandstone)	28.3 l/s 1.0 cfs	2 l/s 0.07 cfs	t = 8	3814 m 12,511 ft	40 m 130 ft	To Aqua Spring
Breathing Cave (Pseudopsyphon)	F	1/6/1969	Tonoloway between sandstones	0.6 l/s 0.02 cfs		t = 9	3700 m 12,136 ft	60 m 197 ft	To Aqua Spring
Sink of Sinking Creek (surface)	G	23/08/1969	Tonoloway(top upper member)	14.2 l/s 0.5 cfs		t < 7	2013 m 6604 ft	82 m 270 ft	To Aqua Spring
Boundless Cave	H	13/09/1969	Tonoloway (probably upper member)	0.3 l/s 0.01 cfs		t < 13	6372 m 20,900 ft	150 m 490 ft	To Aqua Spring
Better Forgotten Cave	I	28/11/1969	Tonoloway	11.3 l/s 0.4 cfs	0.6 l/s 0.02 cfs	t << 21	1888 m 6604 ft	16 m 54 ft	To Aqua Spring
Water Sinks (Siphon #2 Cave)	J	20/03/1970	Tonoloway (probably upper member)	14.2 l/s 0.5 cfs		t < 2	1320 m 4329 ft	60 m 200 ft	To Aqua Spring
Fourth Stream North of Breathing Cave	K	27/04/1972	New Creek	2.8 l/s 0.1 cfs		1 < t < 3.8	3625 m 11,890 ft	204 m 670 ft	To Aqua Spring
Woodzell Sink	L	15/10/1972	Tonoloway (upper member)	0.1 l/s 0.004 cfs		t < 7	3718 m 12,191 ft	104 m 340 ft	To Aqua Spring
Spring at south end of Springhouse farm	B1	10/6/1978	Tonoloway (above upper sandstone)	0.3 l/s 0.01 cfs		t < 15	4984 m 16,349 ft	210 m 690 ft	To Aqua Spring
Original Barberry Cave Entrance	B2	19/02/1982	Tonoloway (middle member)	0.1 l/s 0.004 cfs		t < 5	5259 m 17,248 ft	204 m 670 ft	Aqua discharge 50 cfs
Black Canyon—Bobcat Cave	R1	26/02/1986	Tonoloway (probably lower member)	4.3 l/s 0.15 cfs		t < 7	1383 m 4537 ft	23 m 75 ft	To Aqua Spring
Woodzell Sink	L	19/04/1986	Tonoloway (probably upper member)	0.1 l/s 0.004 cfs		t < 1	957 m 3140 ft	34 m 111 ft	Visual trace to Burnsville turnpike

Table 17.4 Individual dye traces: Cathedral Spring Basin

Source name	Map identification (Fig. 17.1)	Date d/m/year	Horizon	Discharge at time of trace	Dye transit time, t (days)	Dye trace distance	Elevation above spring
Armstrong Cave	M	5/4/1970	Tonoloway (below upper sandstone)	2.8 l/s 0.1 cfs	t ≪ 27	7450 m 24,436 ft	175 m 574 ft
Burns Chestnut Ridge Cave	N	3/1/1970	Tonoloway (below upper sandstone)	1.4 l/s 0.05 cfs	t < 14	4410 m 14,465 ft	209 m 686 ft
Jackson Cave	O	28/06/1969	Tonoloway	2.8 l/s 0.1 cfs	4 < t < 17	7171 m 23,520 ft	185 m 607 ft
Daggy Hollow Run	P	26/03/1971		5.7 l/s 0.2 cfs	t < 15	7628 m 25,021 ft	197 m 647 ft
Chestnut Ridge Blowing (Bobcat) Cave	R	28/08/1970	Tonoloway (below upper sandstone)	0.03 l/s 0.001 cfs	15 < t < 22	2445 m 8018 ft	207 m 680 ft
Robins Rift Cave	S	19/04/1972	Tonoloway (upper sandstone)	2.8 l/s 0.1 cfs	0.8 < t < 1.9	3471 m 11,387 ft	67 m 221 ft
Below Robins water supply	T	31/07/1971		0.3 l/s 0.01 cfs	1 < t < 4	3900 m 12,792 ft	175 m 574 ft

Table 17.5 Individual Dye Traces: The Blue Spring Drainage Basin

Source name	Map identification (Fig. 17.1)	Date d/m/year	Horizon	Discharge at time of trace	Dye transit time, t	Dye trace distance	Elevation above spring
Below Robins water supply	T	31/07/1971	?	0.3 l/s 0.01 cfs	1 < t < 4	3900 m 12,792 ft	175 m 574 ft

Table 17.6 Individual dye traces—Emory Spring and Clover Creek Spring Basins

Source name	Map identification (Fig. 17.4)	Date d/m/year	Discharge at time of trace	Dye transit time, t (days)	Dye trace distance	Elevation above spring	Resurgence name
Cycle sink	U	20/04/1969	5.7 l/s	t < 13	3454 m	175 m	Emory Spring
			0.2 cfs		11,329 ft	574 ft	
Sink 1/4 miles south of Cycle Sink	W	10/5/1970	8.5 l/s	t < 18	3424 m	175 m	Emory Spring
			0.3 cfs		11,232 ft	574 ft	
N38° 15′ 29″	X	22/09/1974	2.8 l/s	7 < t < 26	4250 m	230 m	Emory Spring
W79° 36′ 34″			0.1 cfs		13,939 ft	755 ft	
Stream south of Crab Run N38° 17′ 43″	Y	21/12/1974	8.5 l/s	t < 21	7940 m	194 m	Emory Spring
W79° 34′ 32″			0.3 cfs		26,043 ft	635 ft	
Stream south of Crab Run N38° 17′ 43″	Y	21/12/1974	8.5 l/s	t < 21	5292 m	137 m	Clover Creek Spring 7.0 cfs
W79° 34′ 32″			0.3 cfs		17,358 ft	450 ft	
Crab Run	Z	3/8/1974	5.7 l/s	7 < t < 20	5504 m	131 m	Clover Creek Spring 6.9 cfs
			0.2 cfs		18,052 ft	430 ft	
Helictite Cave (Phil Lucas trace)	J1	2/3/2004	?	t < 2	1579 m	34 m	Emory Spring
					5180 ft	112 ft	

Table 17.7 Individual dye traces—Buller Cave Sinking Creek internal drainage

Source name	Map ident. (Fig. 17.5)	Date d/m/year	Discharge at time of trace	Dye transit time, t (h)	Source to resurgence straight line distance	Resurgence name	Map identification (Fig. 17.5)	Discharge at time of trace
Pool beyond Penn State Lake	3	19/12/1970	1.1 l/s 0.04 cfs	$8 < t < 24$	614 m 2015 ft	Sinking Creek Resurgence	10	?
Sump beyond Natural Bridge	4	27/06/1970	2.8 l/s 0.1 cfs	$t < 12$	451 m 1480 ft	Sinking Creek Resurgence	10	14.2 l/s 0.5 cfs
Difficulty Creek	5	4/7/1970	1.1 l/s 0.04 cfs	$t < 22$	189 m 620 ft	Sinking Creek Resurgence all leads bet.	10	?
Sinking Creek Sump	1	18/07/1970	36.8 l/s 1.3 cfs	$t < 1$	186 m 610 ft	Sinking Creek Sump and Dry Sumps	13–6	?
Sinking Creek Sump	1	22/03/1969	36.8 l/s 1.3 cfs	$t = 8$	546 m 1790 ft	Marlboro Country Stream #1	7	26.8 l/s 1.0 cfs
Hanging Rock Room	8	22/03/1969	2.8 l/s 0.1 cfs	$t = 0.5$	69 m 227 ft	Dave's Lake Area	9	2.8 l/s 0.1 cfs
Dave's Lake Area	9	5/4/1969	2.8 l/s 0.1 cfs	$t = 1$	268 m 880 ft	Last Hope Siphon	2	2.8 l/s 0.1 cfs
Nelson Hollow	A	30/06/1976	2.8 l/s 0.1 cfs	$t \ll 69$	390 m 1280 ft	Last Hope Shaft and Frothing Slosh	19–20	?

Table 17.8 Other dye traces

Source name	Map Ident. (Fig. 17.5)	Date d/m/year	Discharge at a time of trace	Dye transit time, t (days)	Source to resurgence straight line distance	Resurgence	Map identification (Fig. 17.5)
Sinking Spring on south boundary of Springhouse Farm	B1	15/01/1997	0.25 l/s 0.01 cfs	t < 4	269 m 882 ft	Flowstone stream upstream Barberry	a
				t < 4	185 m 607 ft	Stream from lower sandstone near Big Bucks	b
				t < 4	172 m 563 ft	Stream entering from east near Big Bucks	c
				t < 4	371 m 1218 ft	Woway stream	d

References

Davis, N.W., and J.W. Hess. 1982. Hydrogeology of the drainage system, Burnsville Cove, Virginia. *National Speleological Society Bulletin* 44: 78–83.

Deike, G.H. III. 1960a. Cave development in Burnsville Cove, West Central Virginia, with special reference to Breathing Cave. M.S. thesis, University of Missouri, 155 p.

Deike, G.H. III. 1960b. Origin and geologic relations of Breathing Cave, Virginia. *National Speleological Society Bulletin* 22: 30–42.

Dunn, J.R. 1957. Stream tracing. *MAR Bulletin* 2: 7.

Holsinger, J.R. 1961. Fluorescein studies in the Butler Cave—sinking creek system. *D. C. Speleograph* 7(3): 12–13.

Simmons, R. 1990. In search of the cathedral system. *BCCS Newsletter* 16: 1–8.

Thornthwaite, C.W. 1948. An approach toward a rational classification of climate. *Geographical Review* 38: 55–94.

Werner, E. 1972. Development of solution features, Cloverlick Valley, Pocahontas County. *West Virginia Speleological Survey Bulletin* 2: 1–53.

White, E.L., and B.M. Reich. 1970. Behavior of annual floods in limestone basins in Pennsylvania. *Journal of Hydrology* 10: 193–198.

White, W.B. 1969. Conceptual models for limestone aquifers. *Ground Water* 7(3): 15–21.

Zotter, H. 1961. Feminine versions of a "Dunn Bug". *Netherworld News* 9: 146.

Geology of Breathing Cave

18

George H. Deike III

Abstract

Breathing Cave, on the northwest side of Burnsville Cove, is a rectangular maze of strongly joint-controlled passages in a limited stratigraphic zone. The cave is confined to a 77-foot section of shaley limestone between two sandstone beds and is little affected by numerous minor folds and faults. It follows the dipping flanks of a syncline through a vertical range of 340 feet. Breathing Cave, which is far from the ground water outlets in the Bullpasture Gorge, exhibits deep bedding-controlled development but it contains some evidence of horizontal enlargement cause by former stable positions of the water table.

[**Editor's Note**: This chapter was constructed from the author's MS thesis and from the account of the thesis published in the NSS Bulletin. Although these documents were published in 1960, they remain the best available information on Breathing Cave. New information on the paleomagnetism of the Breathing sediments was added. Breathing Cave was re-mapped in detail in the early 2000s. The map is to be included in an extensive monograph that has not been published as of October, 2013.]

18.1 Introduction

Breathing Cave (aka Burnsville Saltpetre Cave) is one of the few caves in the Cove with a large natural entrance and so has been known probably since the first settlers arrived in the Cove in the 18th Century. It was mined for saltpetre in the late 1700s and early 1800s and again during the Civil War (Douglas 1964). The crawlway leading into the deeper parts of the cave often exhibited an oscillatory air current which was the origin of the name "Breathing Cave" (Cournoyer 1954). The cave was first visited by Nittany Grotto (including the author) in 1954 and an extensive project to survey the cave was undertaken. See Chap. 2 for more details of this early endeavor. The present investigation was carried out in 1957–1958. It formed the basis for an MS thesis (Deike 1960a) of which a summary was published in 1960 (Deike 1960b).

18.2 Cave Description

The entrance to Breathing Cave (Fig. 18.1) is in a sinkhole on the ridge separating Red Hollow from Nelson Run lying to the northeast on the lower slopes of Jack Mountain. The cave is a complex maze (Fig. 18.2) with the entrance located along the west side of the maze. The limestone beds dip east at

With a contribution by Victor A. Schmidt, Ira D. Sasowsky, and Gary D. Storrick.

G.H. Deike III (✉)
PO Box 10, Greenbank, WV 24944, USA
e-mail: George@shalimarfarm.com

W.B. White (ed.), *The Caves of Burnsville Cove, Virginia*,
Cave and Karst Systems of the World, DOI 10.1007/978-3-319-14391-0_18,

Fig. 18.1 The entrance to Breathing Cave. Photo by Philip C. Lucas

12–20° and the cave follows the bedding so that the eastern side of the maze is 150–200 feet lower than the western side. The passages can be divided into two groups. Main passages follow the regional dip joints at about N 50°W. although the local dip is about due east. The cross passages follow other joints, the most common ones being at N 60°E.

The entrance passage follows a N 50°W regional dip joint and slopes steeply down a rubble pile. At the inner end it is filled to the roof with cobble fill, but a crawl leads to the southwest into the Main Section of the cave, which lies south of the entrance. Shortly before the end of the entrance passage another crawl leads northeast to the Historic Section, north of the entrance.

The Historic Section of the cave has been known for many years and was apparently the section of the cave that was mined for saltpetre. Most passages in this section end in breakdown on the west and are filled to their roofs down dip to the east. The crawl to this section leads through a small maze which, to the east, is filled with cobbles. The first landmark is a large chamber, Big Room, whose extension to the northwest ends in breakdown. Another maze of small walking passage, some with bedrock floors, leads to the Chert and Pit Rooms, which lie en-echelon along the N 50°W joint trend. The Chert Room has a wet breakdown floor sloping east. The Pit Room is a large passage with a false floor of large collapse bocks. The passages leading east from this room join and pinch out against fill at Janet's Crawl. There are extensive dry, level-fill floors near the Pit Room. One crawl leading north from the southeast end of the room leads to the 42-foot ladder drop into George's Gorge. Other maze connections to the gorge cannot be traversed without very difficult climbing.

To the east, the gorge and its branch gradually lower until the roof nearly meets the fill floor. The gorge to the northwest is filled with a precipitous pile of breakdown 60 feet high. Near the top of the pile, two large tunnels with fill floors lead northeast to a crawl into the August Section, discovered in August, 1957. The section contains dry fill floors, and is a simple group of intersecting passages most of which end in fill. The Gumband Section is the furthest northwest part of the cave and is a maze with breakdown and fill floors. There is standing water in some of its lowest parts.

The crawl from the cave entrance to the Main Section leads to a wide passage developed en-echelon along N 50°W joints called the Camp Room. The passage has a dry fill floor. A small maze of dry bedrock and fill-floored passages to the south contains an old wooden ladder at Ladder Corner which leads to a tight crawl. The next prominent feature is Sand Alley, a dry silt-floored room developed along a joint at N 60°E.

Beyond Sand Alley, the passages become wetter and about 150 feet south is a 15-foot climb down into the first of the Three Parallel Passages. At the western end of these passages is a lead down into the Lower Levels, a series of high, narrow, wet fissures which lie beneath the Three Passages and end at a siphon 80 feet northeast of a shaft called Rain Well. The southern end of the western of the Three Passages is filled with breakdown, and an obscure route down through this rubble is a short-cut into the Junction Room.

The usual route southeast from the Three Parallel Passages leads to a high passage with speleothems, where it is necessary to squeeze past the Nutcracker. This main passage continues southeast as the Cathedral Passage, but a detour around a high fill bank must be taken through the Junction Room. Several routes lead upward into the Upper Levels above the Junction Room.

East of this area is a maze of passages called the Rain Well Area. The Rain Well can be reached by

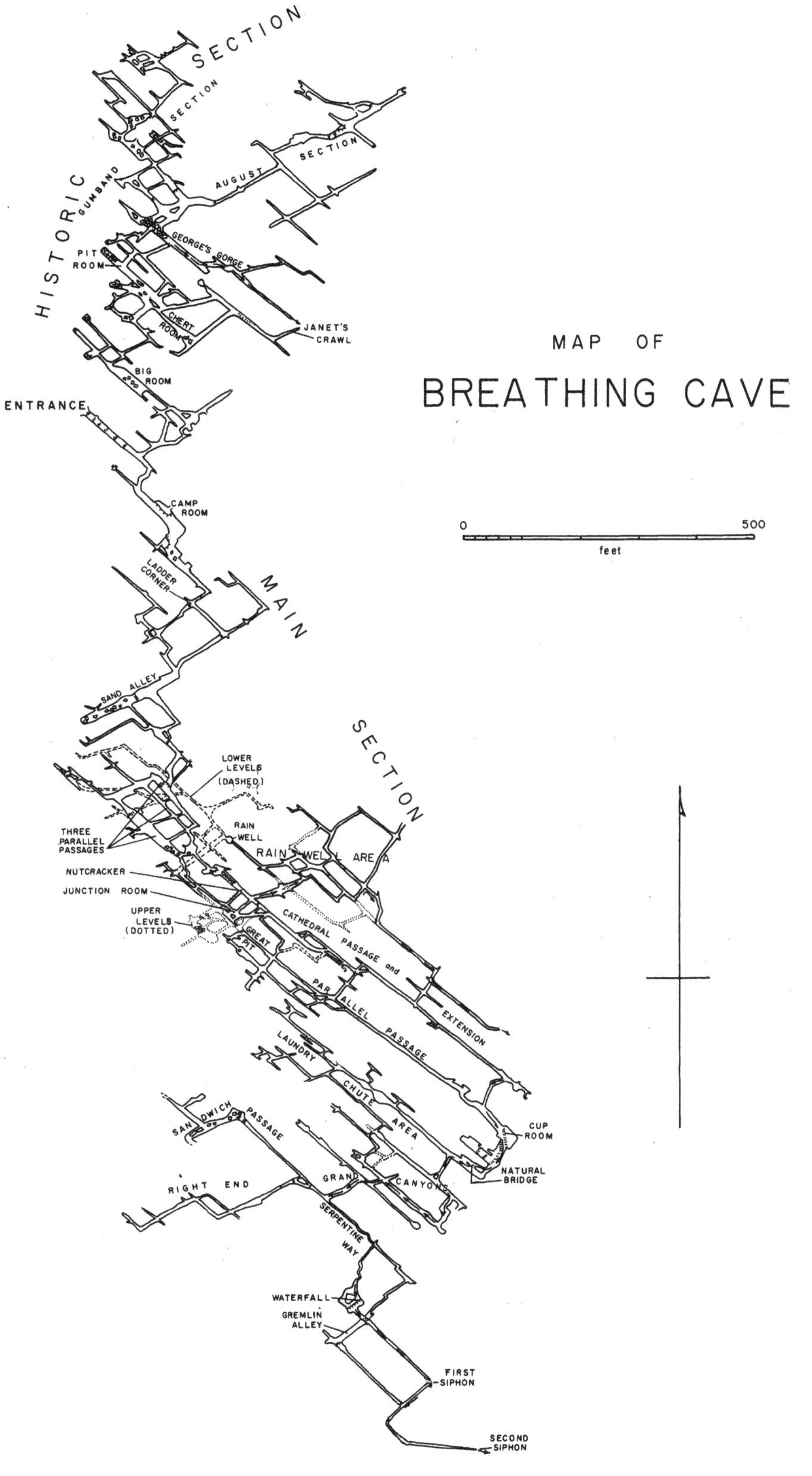

Fig. 18.2 Map of breathing cave. From Deike (1960a)

several routes, and through it a connection made to the Lower Levels.

The Cathedral Passage and its extension lead more than 500 feet southeast. To the south it is paralleled by the Parallel Passage and its ramifications, to which it is connected at several places. There is also a connection back to the Rain Well Area through a passage which parallels the Cathedral Passage on the northeast.

The Cup Room is a direct extension of the Parallel Passage, and is a wide walking passage with a small stream. Around a corner from the Cup Room is the Natural Bridge. Beyond this a small stream traverses a series of rocky crawls connecting high N 50°W joint passages called the Canyons. The Laundry Chute Area is a maze extending northwest from the Canyons and Natural Bridge.

At the end of the Grand Canyon, the passage leads upward to the Right End, a damp, fill-floored passage which ends in a fill plug and to Sandwich Passage, which runs into an unstable breakdown area. In 1959, this breakdown was explored and some new passages were discovered.

The stream in the Grand Canyon follows the very high, narrow Serpentine Way south to the Waterfall and then a N 50°W passage to the First Siphon. A difficult climb to upper level Gremlin Alley bypasses the First Siphon and leads back down over several vertical drops to the stream again. The stream can be followed irregularly south to a keyhole passage which ends at the Second Siphon, about 340 feet below the entrance, and at the time the thesis was written was the end of the cave. Further exploration (see Chap. 2) located a bypass route above and over the Second Siphon and 500 feet of passage that ended at the Third Siphon where the stream disappears in a crack in the Lower Breathing Cave Sandstone.

18.3 Geologic Relations of Breathing Cave

Examination of the cave leaves no doubt that it was developed below the water table. The cave is a network maze developed along joints striking about N 50°W (main passages) and N 60°E (cross passages) and several other less prominent directions. The passages are high and narrow, each developed along one or two joints (Fig. 18.3). The stages of development seen in the cave suggest that initial widening of the joints began along one or more channels which usually coalesced into one high passage. Several reverse faults are exposed on the cave walls, but there has been almost no dissolution along these. The corners at passage intersections are not rounded.

The cave walls show the effects of differential dissolution of the thin shaly limestone beds which range considerably in grain size and sand and clay content. There are no scallops or flutes on the limestone suggesting that the current velocity in the galleries was low. Other subwater-table features include joint-oriented wall cavities, and some spongework. Pockets and spongework are rare because the differences in solubility between the thin beds are much more important to wall form than differences distributed through the mass of the limestone. Continuous rock spans are fairly common. The Natural Bridge (Fig. 18.4) is an interesting example, because it follows a minor fold in the limestone, it's under surface being a bedding plane. It is a remnant of the partition between two joint passages, one of which apparently was blind. The cave also exhibits thin bands of shale

Fig. 18.3 Cathedral passage extension looking southeast. A straight joint passage with a sandstone ceiling (with joint visible) showing differential dissolution of thin bedded limestone. Beds dip away from foreground and to the left

Fig. 18.4 The natural bridge, about 15 feet high, a continuous rock span following a minor fold in the limestone

Fig. 18.5 An exposure of conglomeratic cave fill 250 feet south of the entrance

which fill cracks perpendicular to the bedding and stand in relief because of differential dissolution.

Not all of the joints exposed in the cave have been opened by dissolution, but many have. From northeast to southwest across the cave system there are more than 70 parallel passages in a distance of 1700 feet. Dissolution was accomplished by slowly moving water along most of the joints in this limestone zone.

The original joint passages have been almost completely filled with conglomeratic silt and fill (Fig. 18.5). Cobbles 12 inches. in diameter are common, and the explorable parts of the cave are open as a result of the removal of this fill by small free-surface streams.

The profile of the Cathedral Passage (Fig. 18.6) which is a long main passage in the middle of the main part of the cave south of the entrance, shows the irregular removal of the clastic fill. The roof of the passage is the upper sandstone. Several faults are exposed. The lower sandstone is exposed only in the deep canyon in the fill near the northwest end of the profile where a small stream crosses the Cathedral Passage. The passage is filled to the ceiling at a minor fold, known as the monocline, at the southeast end of the profile. The southern-most passages in the cave have been emptied of fill beyond this fold which lowers the upper sandstone some 80 feet. A small stream cascades down the fold on the lower sandstone to another floor of clastic fill. Other passages, such as Cathedral Passage, also extend farther beyond this big fold but remain filled.

The limits of explorable cave passages are not usually the limits of dissolution. In Fig. 18.7 the passages which end downdip (east) in fill have been extended diagramatically. Those which are extended to the west, or updip, are largely north of the entrance and end in rock falls which may reflect their approach to the surface. The passages which are shown ending in breakdown 500–1000 feet south of the entrance are adjacent to Red Hollow. The map shown in Fig. 18.7 shows structure contour lines drawn on the bottom of the dipping upper sandstone, which is the ceiling of most of the passages. Thus it shows the slope and altitude of the cave roof. The local strike is to the north, at the nose of a subordinate anticline, but at both extremities of the cave the strike is seen to swing toward the regional strike of about N 40°E.

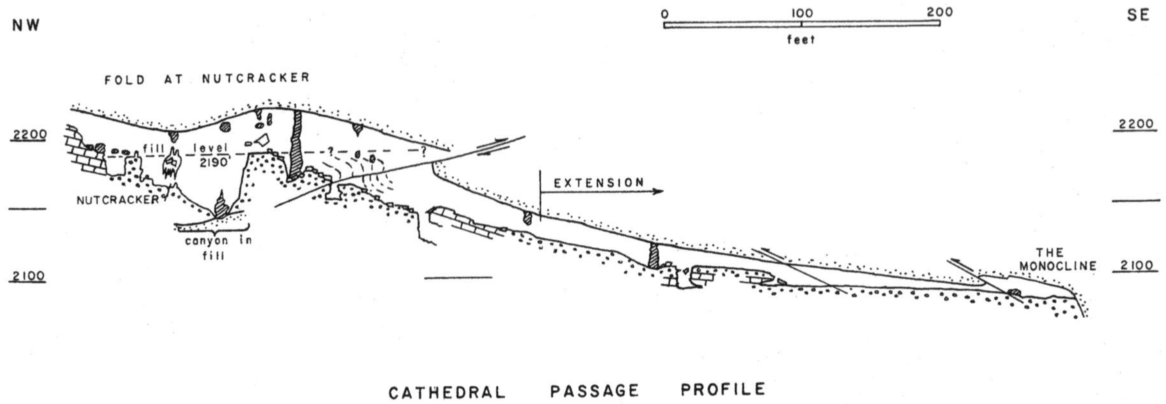

Fig. 18.6 Profile of the cathedral passage and extension. The ceiling is the upper sandstone. Cross passages are shown as *shaded openings*

18.4 Development of Breathing Cave

[**Editor's Note**: Breathing Cave remains something of an enigma. It is a network maze with a north-south orientation. Its passages are blocked by breakdown on the up-dip side and by clastic fills on the down-dip side. The section that follows is the author's interpretation of the cave within the conceptual framework of the late 1950s at a time when little was known of the other caves in the Cove. The subject will be addressed again in Chap. 24 where it will be shown that the speleogenetic history of the caves of Burnsville Cove is both more complex and covers a much larger time span that might have been suspected earlier.]

Many caves are thought to begin as a maze of small openings, later integrated into a few large channels taking the major ground water flow. Although Breathing Cave is still a network, there exists the possibility that water flow in the system would have enlarged some passages preferentially as main conduits. This might be particularly true if major enlargement was accomplished by streams just below the water table. This is the mode of formation postulated by Davies (1960). The main conduits would be expected to be gently sloping, reflecting the slope of the water table. In this case the conduits, confined by sandstone beds, would lie parallel to the strike. Examination of maps and field data indicates that there are no such conduits. The drainage basin uphill from the cave is small, and flow velocities of a few feet per minute could handle the largest floods. As a result, no single large streams were developed in the system. There are some larger passages, mostly trending N 50°W, but these are not connected to other N 50°W passages by large cross passages, and if anything they reflect several preferred paths of flow diagonally down dip along this joint direction.

It appears that most of Breathing Cave was developed simultaneously in this confined horizon by slowly moving water.

Surface erosion has taken place in stages here in the James River drainage much as it has in the Potomac River drainage examined by Davies. This is reflected in well-developed terraces cut on the shale of the Bullpasture and Cowpasture River valleys, and probably by various hill summit levels in the region. These features reflect relatively stable periods in the erosional history of the region. At these times the water table would stabilize also.

Even if ground water flow were too small to dissolve large main conduits in Breathing Cave, the water table must have stood at certain levels in the system during the stable periods, and this should be reflected in the morphology of the cave. Additional upward solution would then be limited by the water surface. This probably has occurred at two levels in Breathing Cave. Less than 250 feet south of the entrance, three joint passages become impassibly narrow and seem to pinch out near an altitude of 2300 feet.

An extended stable water surface seems to explain this sudden narrowing best. Several other passages north of the entrance narrow noticeably at this level, although

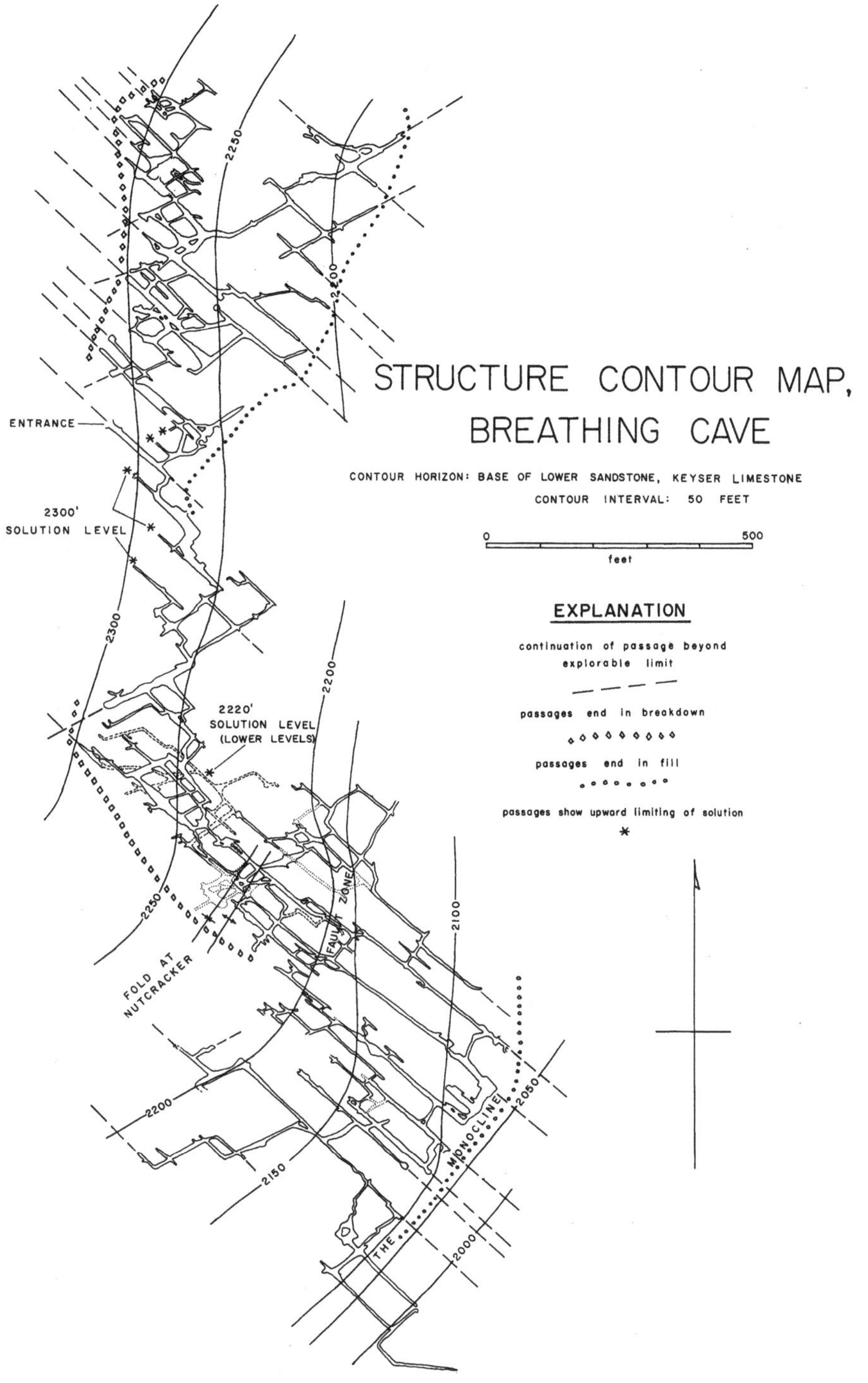

Fig. 18.7 Map of breathing cave superimposed on structure contours

they do not end. These passages were in existence before this stable period. This situation is similar to that observed in Pennsylvania caves by White (1960).

About 650 feet south of the entrance, lower levels are shown on Fig. 18.7 as dashed lines. In these passages, small free-surface streams flow on the lower sandstone, but the passages seem to narrow and pinch out upward at approximates 2220 feet. The main "level" of passages (not truly level) lies above these pinch-outs, and is only connected to the lower passages in a few places. Apparently upward solution in these lower levels was limited by the water table.

It is also possible that concentrated flow of water near the water table would develop a number of passages at this level during stable periods.

Most passages are roofed by the upper sandstone and only those at other horizons in the limestone were examined for concentrations at particular levels. This confined the examination almost entirely to the cross passages, which follow joints other than the "main" N 50°W dip joint.

There are a large number of cross passages near the Cathedral Passage, 900 feet south of the entrance. Some of these are shown in Fig. 18.6. Most of the 18 observed cross passages lie between 2120 and 2170 feet, but few are at any single altitude, and there are commonly several of them one above another along the same joint. It is difficult to imagine that these represent levels. Several passages in the southern end of the cave have ceilings between 2075 and 2095 feet, but again they do not seem to reflect a water surface. If these are levels, then they are numerous and are most likely related to hydrologic conditions within the cave system rather than stable periods of local base level of erosion.

Figure 18.8 presents the relationship of the caves and erosion levels in the Bullpasture Valley. Present base level is the Bullpasture River at an altitude of about 1700 feet where it crosses the limestone outcrop. Above the river the four most prominent terraces are shown diagrammatically. These are often correlated the four glacial stages of the Pleistocene epoch. The two lowest hill summit levels, are reported by Davies (1957) in the Potomac River drainage basin and apparently recognizable in Virginia too, are also shown. The horizontal and vertical extents of several caves are shown diagrammatically with their distance from the river.

Two outlets to the present drainage are shown. One is Blue Spring which rises from a joint passage very like those of Breathing Cave. Diving has shown the passage to extend at least 50 feet below river level. The opening is not confined by a sandstone bed. It is not known whether passage extends down the dip of the beds or becomes horizontal, cutting across the bedding.

The other outlet is Aqua Cave, which is 50 feet above the river. The difference in altitude of these two outlets, which are both in the Keyser limestone on opposite flanks of a syncline, shows that the present active drainage is not confined to a level close to present base level. Aqua Cave, however extends horizontally through dipping beds, crossing the axis of an anticline. It appears to be close below the second river terrace. The cave probably originated as a major outlet to subterranean drainage as this terrace was being formed, in the manner suggested by Davies (1960). Ground water is still using this conduit, presumably for lack of a connection to a lower outlet.

The two definite and two doubtful levels of static water tables in Breathing Cave are shown in Fig. 18.8. The spacing resembles that of the four river terraces, except that these levels are all farther above the present water table than the terraces, and the lowest two levels are questionable. Their exact relationship to erosion levels is therefore uncertain. If the postulated present water table gradient is typical, then the ancient water table would have had to slope toward the summit level at 2150 feet to place the entire Breathing Cave system under water. Its origin would thus date from the time of this erosion level or before.

18.5 Paleomagnetism of Clastic Sediments from Breathing Cave, Virginia, USA by Victor A. Schmidt, Ira D. Sasowsky and Gary D. Storrick

18.5.1 Introduction

This section provides a brief summary and interpretation of the paleomagnetism of several clastic sediment deposits from Breathing Cave, Virginia. On September 23, 1984, Victor Schmidt and Gary Storrick collected 16 reconnaissance paleomagnetic samples from 6 locations in the cave. The results had not been published as of 2009, and in an effort to document them the second author reviewed the available notes and data from the original work, and developed the present report.

18.5.2 Methods

Sample locations were chosen based on the collector's knowledge of the cave. Elevations were determined by

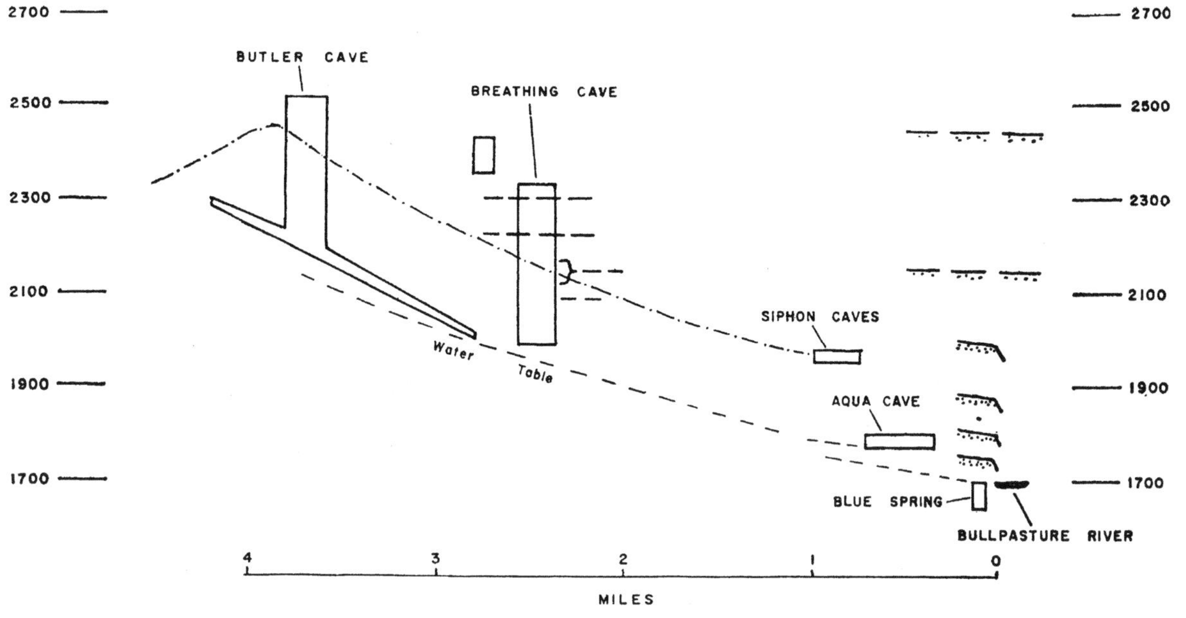

Fig. 18.8 Vertical relations between caves and erosion features. Levels identified within breathing cave are shown by *horizontal dashed lines*

barometric altimetry, employing an American Paulin Surveying Altimeter Model M-1 and an American Paulin Microbarograph Model 23-21. Temperature was compensated, but there may have been a problem with the thermometer used in the cave.

At each sample location the sediments were examined and briefly described. Where needed, they were referenced to an altimeter position. Paleomagnetic samples were collected in pairs (duplicates), generally following the same methods described in Schmidt (1982). A near vertical face was cleared with plastic tools, and a small protruding pedestal was then carved in the face. A small plastic box was slipped over the pedestal, and then the strike, dip, and rotation of the outer face was determined with a Brunton Pocket Transit before the samples was removed. Samples were analyzed at the University of Pittsburgh Paleomagnetic Laboratory. Samples were treated with alternating field demagnetization in steps ranging from 0 (natural remnant magnetization) to 140 mT, and were measured in a vertical cryogenic magnetometer.

18.5.3 Results

The actual magnetometer readings for the analyses were no longer in existence at the time this manuscript was prepared. The results and following discussion were prepared using copies of the original field notes, hand written notes by Schmidt, and computer plots of the demagnetization diagrams. Table 18.1 provides a summary, and Fig. 18.9 shows the sample locations and interpreted polarities of the samples. Six sites were evaluated, and at two of these (site 3 and 6) an upper and lower stratum were sampled; therefore a total of 16 samples were collected, encompassing 8 layers. Except in the cases of sites 3 and 6, the stratigraphic relation between layers is indeterminate. Data for samples XAK030, 033, 034, 036, 037, 038, 040, and 042 were not available. Because site 3 (upper) was sampled by numbers XAK034 and 036, there are no data available on that layer, and only 7 of the 8 sampled layers will be discussed.

Table 18.1 Samples collected for paleomagnetic measurements

Location	Samples	Elev.	Location description	Sediment description	Polarity
1	30, 31	2326	Crawl to left—George's Gorge section	Cobble fill with ~3 inches coarse sand and fine gravel on top	N
2	32, 33	2322	On way down main cave at first sharp bend right after entrance crawl	Very coarse sand	I?
3 lower	34, 36	2245	Beyond dugway to register	Not reported	No data
3 upper	35, 37	2270	Beyond dugway to register	Not reported	N
4	38, 39	2281	At register, in formation section	Medium sand	N?
5	40, 41	2200	Passage down dip from register a bit	Moist clay and silt below old drip zone, on ledge left side	N
6 lower	43, 45	2221	Below climbdown (crawl through below). Jog into next western down-dip passage	In small (10 feet) pit, bottom, gloppy red clay, wet	R
6 upper	42, 44	2231	Below climbdown (crawl through below). Jog into next western down-dip passage	At top of pit, spongy compressible clay	N

Sample numbers are in XAK series
Elevation in feet amsl, determined by altimeter
Polarity key N = normal, R = reversed, I = intermediate, ? = uncertain

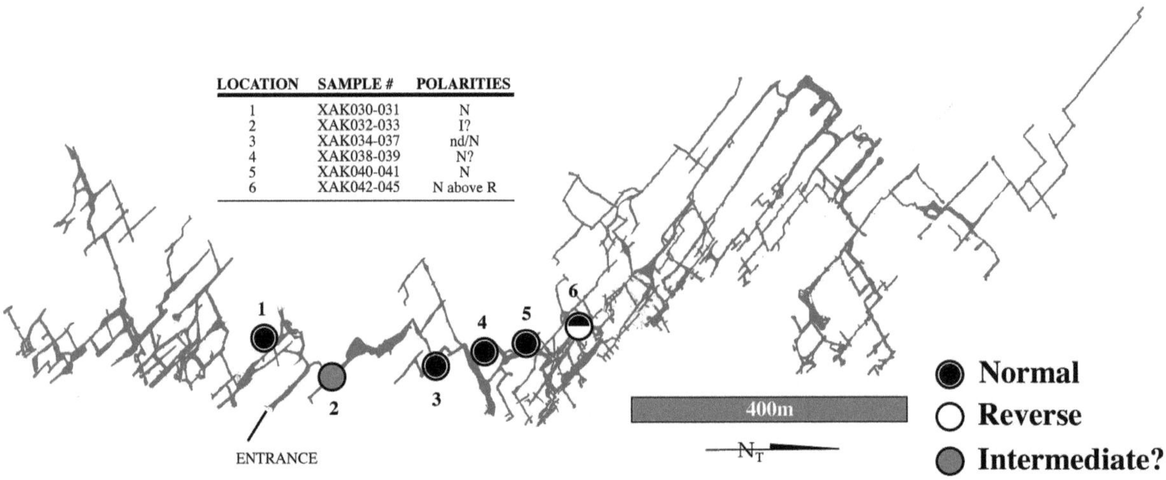

Fig. 18.9 Map of breathing cave, Virginia showing paleomagnetic sampling locations and results. Base map from an original by Fred Wefer (Wefer and Nicholson 1982)

Sediments in the cave contain sand, gravel and cobbles, as well as the clays and silts that are more useful for paleomagnetic analysis. Fine-grained sediments were selected for sampling whenever available. In most cases polarity was apparent from examination of the demagnetograms (Fig. 18.10). Of the 7 layers evaluated, 5 showed normal polarity, 1 was possibly intermediate, and 1 was reversed. The intermediate sample came from a coarse sand deposit and showed paleodirections that were east and downward. The reversed layer occurred at site 6 at the base of a 10-foot deep pit. The material is a wet red clay, and is overlain by a normal polarity springy, compressible clay.

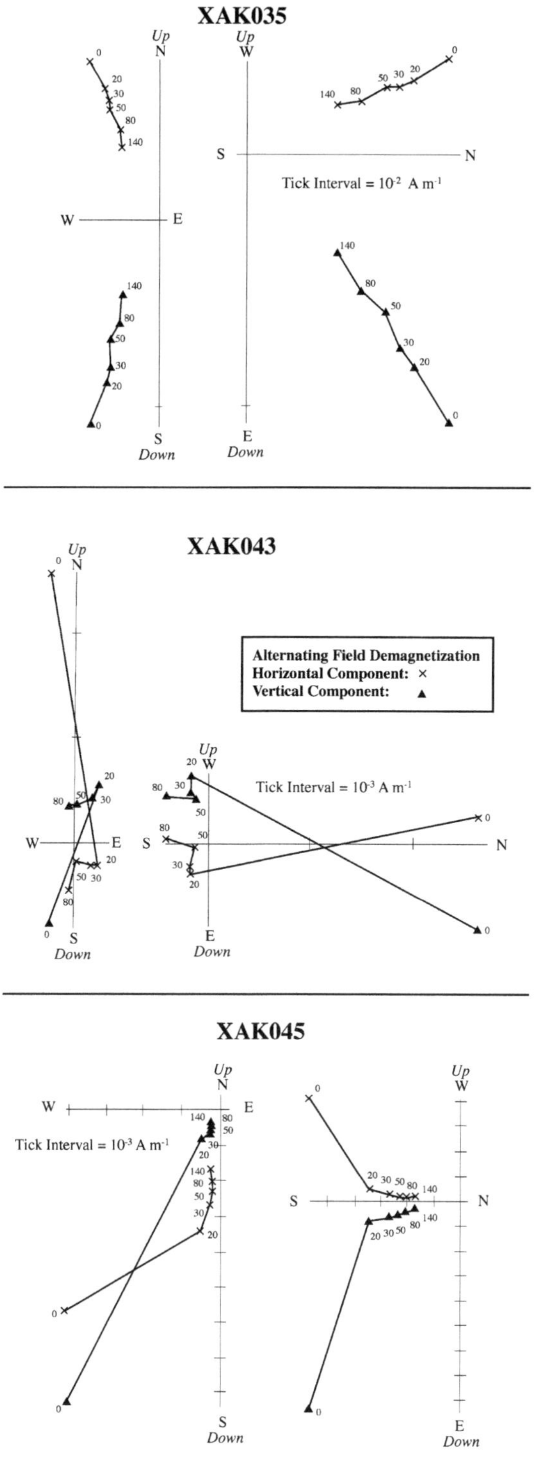

Fig. 18.10 Selected demagnetization diagrams from paleomagnetic study. Sample XAK035 shows classical normal polarity. Samples XAK043 and XAK045 display reverse polarity

18.5.4 Discussion and Conclusions

The clastic fills in this cave are fluvial in origin based upon their grain size, lithology, and mode of occurrence. They originated at up dip sinkholes, and were transported down dip along cave passages, which they now completely occlude in parts. The preponderance of normal polarity sheds little light on the timing of sediment emplacement, as they could have originated during the present normal Brunhes chron (Cande and Kent 1995). The reversed polarity at location 6 provides a minimum age of 780 ka for the cave. However, there are two possible interpretations of the genesis of this "normal over reverse" sequence. One option is purely depositional, and would have the lower layer (samples XAK043, 045) being deposited during a period of reverse polarity, with the upper layer (samples XAK042, 044) deposited later during a period of normal polarity. In this case both layers preserve a detrital remnant magnetization (DRM). The second option would be that both upper and lower layers were deposited under an earlier normal chron, but that then the lower layer underwent diagenetic change during a reverse polarity chron, resulting in acquisition of a reversed chemical remnant magnetization (CRM). Based upon the limited sample descriptions either could be possible. The entire deposit is clay, with the lower part being wet and red. Red color frequently indicates the presence of hematite, which is a common CRM carrier. But the absence of verification such as isothermal remnant magnetization measurements prohibits verification. Therefore, it is not possible to conclude that an actual polarity transition is recorded in this deposit. It is possible to conclude the deposit is older than 780 ka by the presence of the reversed polarity (Cande and Kent 1995). Sample XAK039 at site 4 showed an apparent reverse polarity viscous overprint that was removed during alternating field demagnetization. This is unusual, and could just be a result of the sample storage prior to analysis.

Acknowledgments The authors of the paleomagnetic study thank the owners of Breathing Cave for access to the cave and for permission to remove sediment samples. Aviance Bain prepared figures, for which we are grateful. Phil Lucas and Nevin Davis provided advice on the sample locations.

References

Cande, S.C., and D.V. Kent. 1995. Revised calibrations of the geomagnetic polarity timescale for the Late Cretaceous and Cenozoic. *Journal of Geophysical Research* 100: 6093–6095.

Cournoyer, D.N. 1954. The speleo-barometer. *National Speleological Society Bulletin* 16: 91–93.

Davies, W.E. 1957. Erosion levels in the Potomac drainage system and their relation to cavern development. *D.C. Speleograph* 12(4): 1–5.

Davies, W.E. 1960. Origin of caves in folded limestone. *National Speleological Society Bulletin* 22: 5–18.

Deike, G.H. III. (1960a). Cave development in Burnsville Cove, west-central Virginia, with special reference to Breathing Cave. M.S. thesis, University of Missouri, 155 pp.

Deike, G.H. III. 1960b. Origin and geologic relations of Breathing Cave. *National Speleological Society Bulletin* 22: 30–42.

Douglas, H.H. 1964. *Caves of Virginia*. Falls Church: Virginia Cave Survey. 761 pp.

Schmidt, V.A. 1982. Magnetostratigraphy of sediments in Mammoth Cave, Kentucky. *Science* 217: 827–829.

Wefer, F.L., and I.K. Nicholson. 1982. Exploration and mapping of the sinking creek system. *National Speleological Society Bulletin* 44: 48–63.

White, W.B. 1960. Terminations of passages in Appalachian caves as evidence for a shallow phreatic origin. *National Speleological Society Bulletin* 22: 43–53.

19 Geology of the Butler Cave—Sinking Creek System

William B. White

Abstract
The Butler Cave-Sinking Creek System is composed of a central trunk channel oriented along the axis of the Sinking Creek Syncline with dip-oriented side caves extending mostly up the western flank of the syncline. The overall patterns of the side caves are network mazes with orientations controlled by the local joint pattern. Much of the cave is in the Devonian Tonoloway Limestone with the two interbedded sandstones exerting an important influence. The result is two interconnected tiers of caves with a locally perched drainage system at the downstream end. The cave contains a complex boulder and cobble fill that seems to represent a rapid infilling event of pre-Wisconsinan age. There three streams in the cave all of which ultimately drain to Aqua Spring. The streams are undersaturated with respect to calcite and have low CO_2 concentrations consistent with recharge from mountain runoff and from infiltration through thin organic-poor soils.

19.1 Introduction

Investigations of the geology and mineralogy of Butler Cave began shortly after its discovery in 1958. Some aspects of the geology were reported at meetings but the first formal publication appeared in the special issue of the NSS Bulletin in 1982 (White and Hess 1982). The NSS Bulletin also contained information on the chemistry of the cave waters (Harmon and Hess 1982). There have been continuing studies the cave geology, particularly the clastic sediments (Chess et al. 2010) as well as a variety of other observations. Finally, there came the revisions of the stratigraphic section and the revised placement of the caves in the carbonate rocks, which required some reworking of the geological relations of the cave. All of these have been combined to produce the present chapter.

19.2 Description of the Cave System

In 1985, Lester V. Good compiled all of the survey data on the Butler Cave-Sinking Creek System into a single folio of sectional maps. These maps appear in the electronic files accompanying this volume. The stick map of the entire system provides an index for the master set of maps (Fig. 19.1). Individual areas of the cave discussed in this chapter were all derived by copying appropriate bits of the master map folio. The index map also lists many of the place names that appear in the text.

The backbone of the cave is provided by the Trunk Channel that extends from southwest to northeast

W.B. White (✉)
Department of Geosciences, The Pennsylvania State University, University Park, PA 16802, USA
e-mail: wbw2@psu.edu

W.B. White (ed.), *The Caves of Burnsville Cove, Virginia*,
Cave and Karst Systems of the World, DOI 10.1007/978-3-319-14391-0_19,

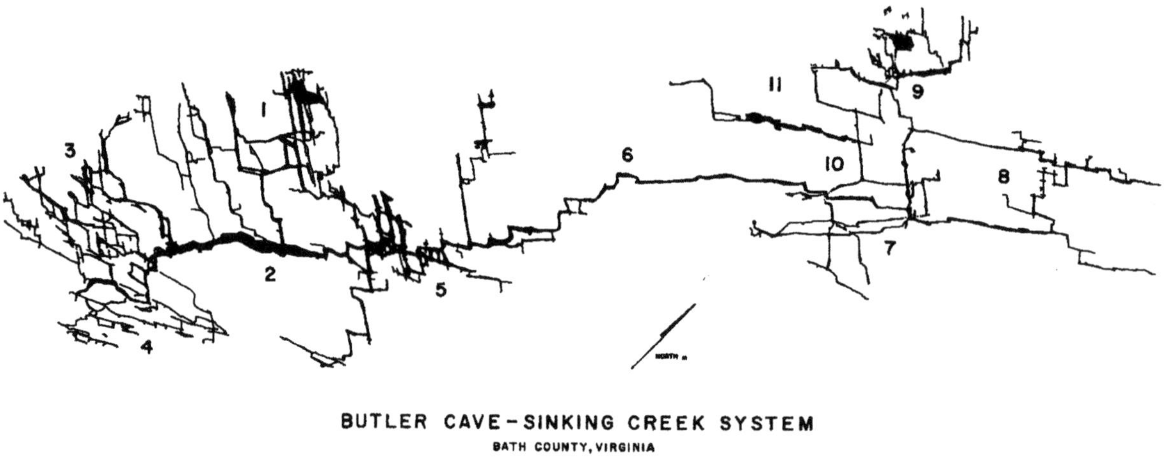

Fig. 19.1 Index map for the Butler Cave-Sinking Creek System. Numbered areas are: *1* Butler Cave Section: Entrance, Difficulty Creek, Complaint Cave, Pennsylvania Cave. *2* Sand Canyon Section: Upstream Trunk, Moon Room Section, Crystal Gallery, Sand Canyon Camp. *3* Huntley's Cave Section: Duke Dump, Birthday Passage, Penn State Lake, Natural Bridge. *4* Beyond the Lake Section: Alphabet Soup, Z-Section, Mbagintao Land, Barking Marshmallow. *5* Sinking Creek Section: Moon Room, Crystal Craters, Pat's Section, Downstream Trunk. *6* Sneaky Creek Section: Sinking Creek Siphon, Silt Crawl, Dry Sumps, Downstream Trunk. *7* Pool Room Section: Crisco Way, French Passage, Evasor Gallery, Pants-Off Crawl. *8* Downstream Loop Section: Dave's Lake, July 6 Room, Last Hope Siphon, Rat's Doom Siphon. *9* Dynamite Section: Frothing Slosh, Christmas Passage, Slippery Creek. *10* Marlboro Country Section-I: Ladder Room, Dude Ranch, Scrog Way, Canoe Passage. *11* Marlboro Country Section-II: Tombstone Territory, Bitter End, Doom Room, Candle Room

closely paralleling the surface valley of Sinking Creek but lying several hundred feet below it. There are a series of tributary caves that slope into the Trunk Channel from both sides of the syncline. The largest of these tributary caves are all on the west side of the synclinal axis and thus slope upward toward Jack Mountain. They have been given individual names, such as "Butler Cave", "Pennsylvania Cave", "Huntley's Cave", "Moon Room Area", "Pat's Section", etc. The tributaries from the eastern side of the syncline, beneath Chestnut Ridge, are generally smaller and do not extend as far up the syncline flank. Breathing Cave is also a side cave to the system except that it lies farther to the northeast and its downslope limits are beyond the terminal sumps of Butler Cave.

Other than the Trunk Channel, most of the cave system consists of isolated sections of network maze, mostly with the individual sections not well interconnected. There is a concentrated area of closely-spaced maze passages south from Natural Bridge. Other maze areas occur at the northern end of the cave system between the Lake Room and the terminal sumps. Both of these areas are made more complicated in map view by the fact that there are two superimposed tiers of caves. At the southern end of the system, the main trunk passage underlies an upper tier of passages called Mbagintao Land. The northern end of the cave system is underlain by a rather complex series of fairly large passages in Marlboro Country. The intermediate connection between these extensive sections of cave passages is by means of a single trunk channel.

The tributary caves on the flank of the syncline are rather elongate network mazes with their largest and best-developed passages extending along the dip of the syncline. These passages are frequently interrupted by minor folds and contortions in the limestone bedding, some of which carry resistant ceiling beds below the level of the passage floor. At such places, the tributary passages are frequently blocked by large infillings of clastic sediments.

The cross-sections of the tributary passages are generally rectangular, much higher than they are wide. A few elliptical tube passages occur, usually as strike-oriented cross passages connecting the main dip passages. The dip passages tend to be canyons 5–20 feet wide and up to 30 or more feet high. The cross passages in the mazes usually have ceiling heights lower than those in the dip passages.

The Trunk Passage from the Natural Bridge to a little below Sand Canyon has a very large cross section. There is an upper, silt-filled level, of which Sand Canyon Camp is a residual terrace, and there is an incised stream channel (Fig. 19.2). This very large cross-section passage 50–100 feet wide, is broken by massive breakdown in the passage segment between Sand Canyon and Natural Bridge. The Natural Bridge itself is a remnant of the upper level portion of the passage. Downstream from Sand Canyon, the Trunk Passage first narrows, breaks into a distributary system (Fig. 19.3), then widens again into another large breakdown complex in the Moon Room area. Northeast of the Moon Room the trunk passage becomes considerably smaller, 10–20 feet wide and 10–30 feet high.

The complex history of cave development can be seen in a traverse of the Butler Cave portion of the system (Fig. 19.4). The Nicholson Entrance to Butler Cave is located directly beneath the Upper Breathing Cave Sandstone. The entrance pit, the Glop Slot, a climb-down, and the God-Is-My-Copilot climb collectively penetrate the entire thickness of the Tonoloway Limestone spanning the Breathing Cave horizon so that the final descent into the top of Breakdown Mountain is through a breach in the lower Breathing Cave sandstone. From Breakdown Mountain the main passage slopes downward following the dip of the bedding for 500 feet until it is blocked by fill. Near the lower end of the passage there is a climb-down to a strike-oriented cross canyon.

Fig. 19.3 Bifurcation of trunk passage at loop downstream from Sand Canyon. WBW photo

Fig. 19.2 Lower canyon upstream from Sand Canyon with breakdown from upper tube. Photo by Joe Kearns

The usual route through the cave leads off from the top of the fill bank below Breakdown Mountain through the Rabbit Hole, a crawlway that connects to the Second Parallel Passage. The Second Parallel Passage is also dip-oriented but instead of blockage by fill at its lower end, there is a cliff that drops about 100 feet into the Bean Room. Just before the drop-off, a short strike passage connects to a breakdown room (the Step-Across shown on Fig. 19.4). Across the breakdown is another dip-oriented passage leading down slope to a cross passage and down a fill bank to the Rimstone Passage. A climb down through the breakdown in the breakdown room connects to the same cross canyon that can be accessed from the main passage. Along the route between these points, the cross canyon crosses the Bean Room on a ledge below the overhang in the Second Parallel Passage above.

A different prospective may be obtained by entering the cave through the SOFA Entrance. The Lower Breathing Cave Sandstone can be seen in the cliff above the entrance. The SOFA Entrance gives direct access to the upper end of Dave's Gallery, also a sloping dip-oriented passage which connects directly with the Rimstone Pool Passage. Although all of the upper level

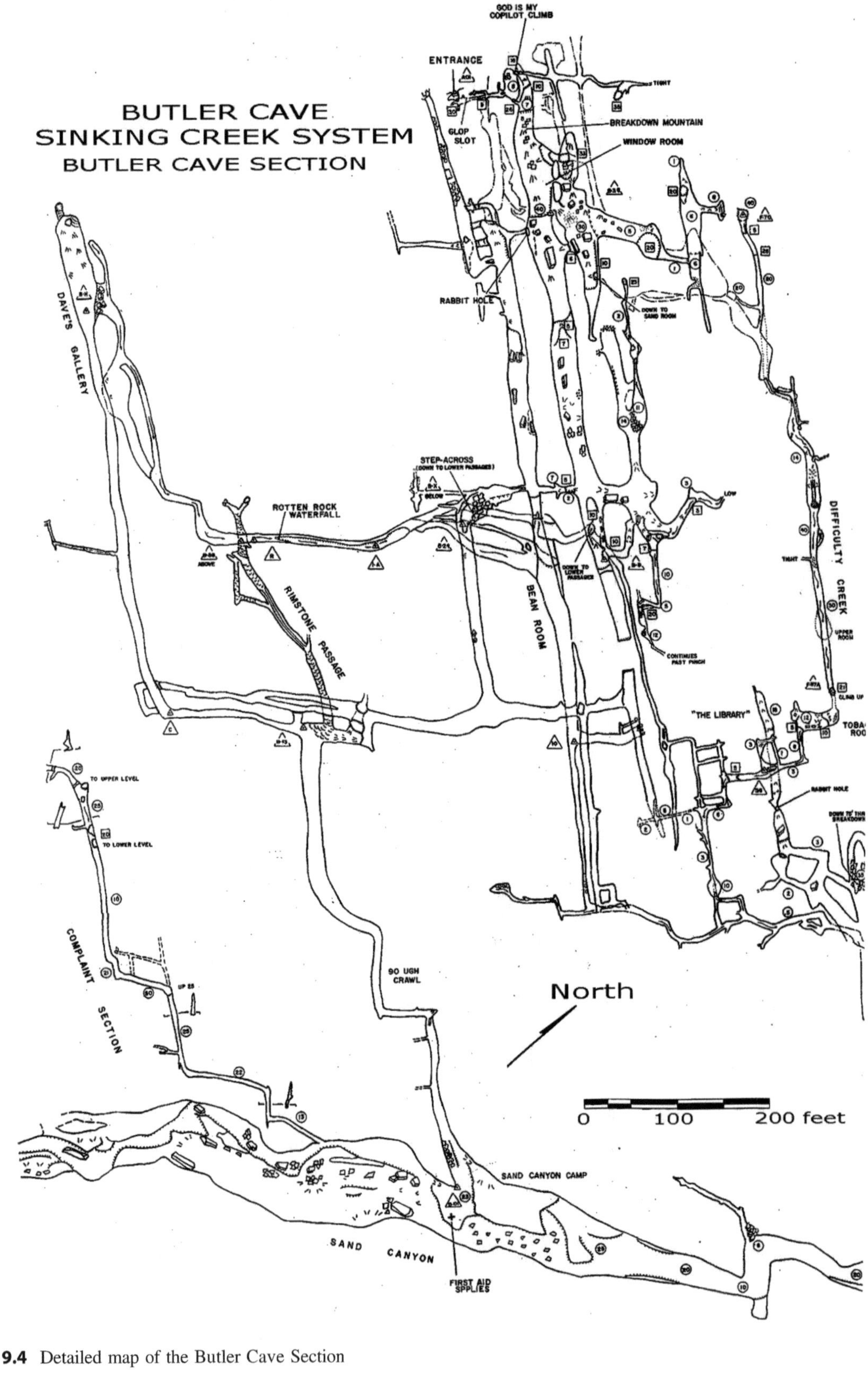

Fig. 19.4 Detailed map of the Butler Cave Section

passages in the Butler Cave Section are dry, the Rimstone Pool Passage is wet with active rimstone pools. It extends up dip several hundred feet to a blockage. The strike passage at the Rimstone Pools connects the lower end of Dave's Gallery with the Fill Bank and the passage from the historic portion of Butler Cave. From the Rimstone Pools, the main passage continues down dip to connect to the Trunk Channel at Sand Canyon. About mid-way along this passage, the cross-section narrows and the ceiling lowers to a crawlway—90 Ugh Crawl. An intriguing—and unanswered—question is: Did bedrock walls of this large cross-section passage really narrow to a crawlway, or was the passage entirely filled with sediment leaving only 90-Ugh Crawl as a small channel in the ceiling?

A short distance inside the SOFA Entrance, it is possible to descend an opening in the side of the passage and reach an active streamway, Rotten Rocks Creek. The stream passage follows a different route and crosses the upper end of the Rimstone Passage suggesting that the water in the Rimstone Passage is leakage from Rotten Rocks Creek, an example of an active streamway crossing an air-filled underlying passage. Downstream from the crossing, Rotten Rocks Creek descends over a waterfall and flows at the bottom of a high canyon, first into the Bean Room and then to join Difficulty Creek. The combined flows enter Sinking Creek near the Moon Room. The passage at the top of Rotten Rocks Waterfall continues as a ledge near the top of the high canyon, then becomes a separate passage beneath the breakdown of the Breakdown Room where it is seen to be the upstream end of the cross canyon first seen from the Main Passage. The cross canyon is revealed as a now-abandoned downstream channel of the ancestral Rotten Rocks Creek.

The sequence of passages from the SOFA Entrance to Difficulty Creek crosses the entire Lower Tonoloway Limestone. The SOFA Entrance is just below the Lower Breathing Cave Sandstone. The passage walls in the downstream section of Rotten Rocks Creek and the lower part of the Bean Room are in the thin-bedded units of the Tonoloway. Observations by Nevin Davis suggest that Difficulty Creek has reached the bottom of the Tonoloway, cut through the underlying thin Wills Creek Formation and is flowing on top of the Williamsport Sandstone.

A rather different pattern appears in the Upstream Maze (Fig. 19.5). The Upstream Maze lies closer to

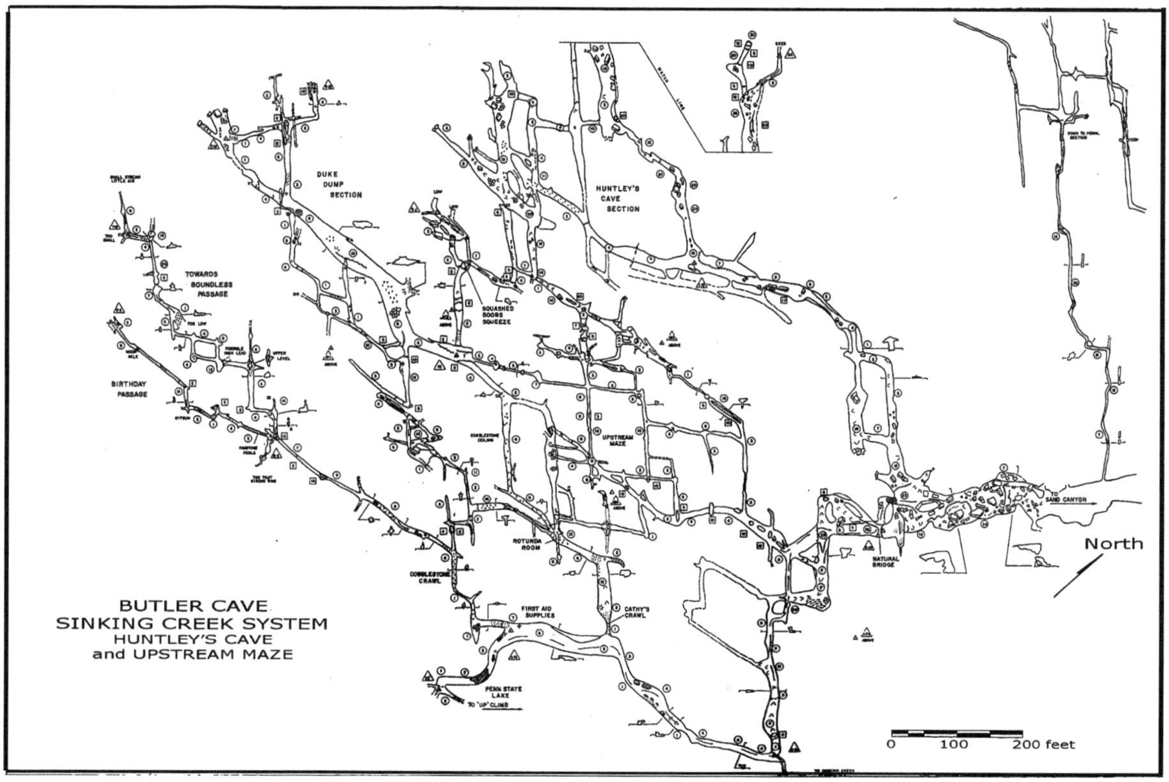

Fig. 19.5 Detailed map of the Upstream Maze Section

the axis of the syncline. Dips are lower and passage gradients less steep. Passages trending along the northwest-southeast joint set are smaller and appear as minor cross passages in the Upstream Maze. Dominant passage trends are on an east-west joint set. The Trunk Channel essentially ends at the Upstream Maze. The single large passage splays into a multiple of passages all cross-linked in the maze. The flood-overflow route of the stream channel skirts the eastern side of the maze to Penn State Lake, a segment of low passage with standing water. At the western side, the Upstream Maze terminates against the side of Burnsville Sink, a large closed depression that is the likely catchment area for the headwaters of Sinking Creek.

On the northern side of the maze, Huntley's Cave is transitional. It partly follows the same northwest-southeast joint set as Butler Cave and the other side caves and partly follows the east-west joint set of the Upstream Maze. Most of the Huntley's Cave passage is dry and contains some small gypsum flowers. A small stream enters at the extreme upper end and is quickly lost to an inaccessible lower level.

The Moon Room Section (Fig. 19.6) is one of considerable complexity. The view, if one were to walk downstream from Sand Canyon to reach the

Fig. 19.7 The internal "spring" where the active flow of Sinking Creek emerges into open cave passage. WBW photo

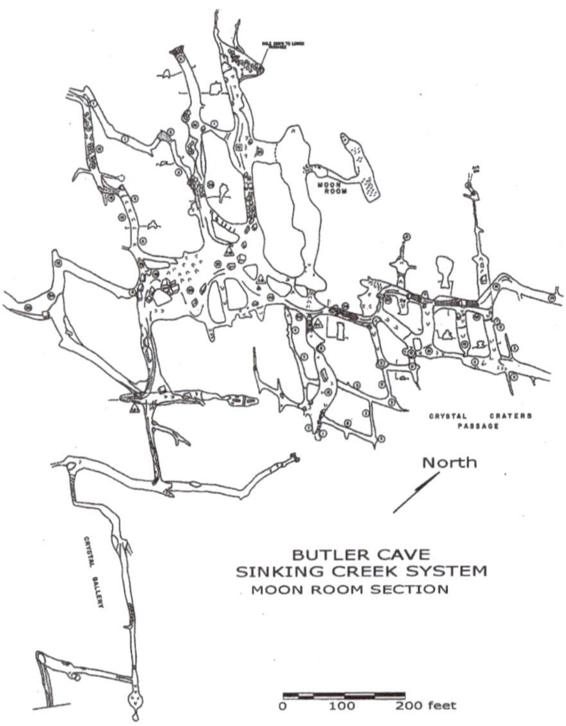

Fig. 19.6 Detailed map of the Moon Room Section

center-left edge of Fig. 19.6, would be the photograph shown in Fig. 19.3. On the lower leg of the Loop, Sinking Creek emerges into the larger cave passage (Fig. 19.7). Near the bottom of Fig. 19.6 is the Crystal Gallery, a well-decorated narrow passage that extends up the southeast side of the syncline to the point in the cave closest to Barberry Cave. The passages to the northwest rise steeply over silt-covered breakdown, some with multiple levels. The Crystal Craters passage lies above the stream channel which is incised deeply below the maze pattern shown in Fig. 19.6, again illustrating the extensive downcutting that has taken place since the primary cave system was formed.

For a considerable distance downstream from the Moon Room Section, the cave consists of only the trunk channel. The gradient of the trunk channel is less that the dip of the plunging syncline so that at the Dry Sumps, the trunk channel climbs up section through the Lower Breathing Cave Sandstone and continues downstream in the Tonoloway beds of the Breathing Cave horizon. The downstream section will be described as part of the hydrologic interpretation.

19.3 Geologic Controls on Passage Development

The most important stratigraphic elements controlling the geometry of the Butler Cave—Sinking Creek system are the interbeds of sandstone which occur within the Tonoloway Limestone sequence. In the earlier report (White and Hess 1982) these were thought to be tongues of the Clifton Forge Sandstone. New mapping (see Chap. 16) shows that Butler Cave is formed in the Tonoloway Limestone which also contains sandstone beds. These beds have no formal name but since they were first identified in Breathing Cave (Deike 1960) they are here informally called the Upper and Lower Breathing Cave sandstones.

The entrance to Butler Cave lies directly below the upper sandstone. The cave descends quickly through the 77 foot interval between the upper and lower sandstones and breaches the lower sandstone at the ceiling of Breakdown Mountain. Breathing Cave lies entirely within the limestone interval between the sandstones. Butler Cave and associated tributaries on the west flank of the syncline are all formed in the Tonoloway Limestone below the lower sandstone. The ceiling of the trunk channel at Sand Canyon is composed of the lower sandstone, so that the cave development essentially follows the bedding plane of the lower sandstone directly beneath it. However, the sandstone is breached in several places.

In the southern end of the cave system, a single narrow passage breaches the lower sandstone to connect to the upper tier of caves known as Mbagintao Land which is formed in the intermediate 77-foot interval of limestone. Downstream to the north, the trunk channel itself breaches the lower sandstone at the dry sumps so that the northern end of the cave including several streams and the Last Hope and Rats Doom sumps are actually perched on top of the lower sandstone. In this area, the lower sandstone is breached again at Kutz Pit and by Crisco Way. By these access routes, one can cross the sandstone and reach the lower tier of cave, Marlboro Country, which is formed in the same stratigraphic interval as the upstream trunk passage and the tributary caves. If one

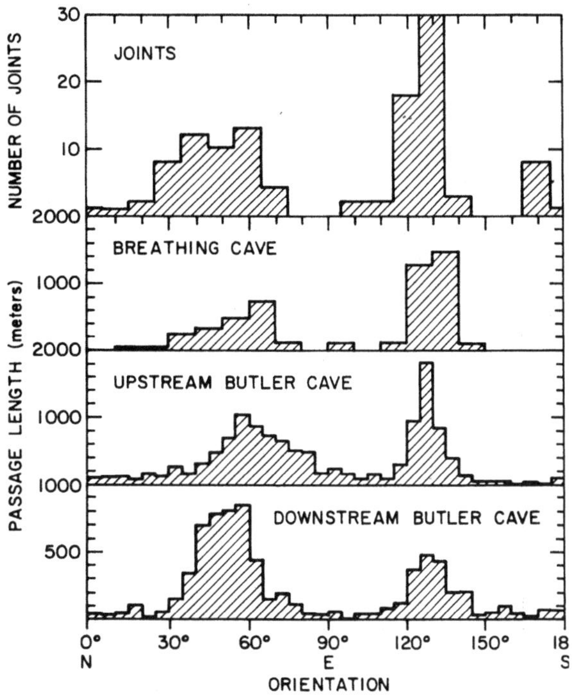

Fig. 19.8 Comparison of passage development with joint patterns. From White and Hess (1982)

views the cave system in long profile, the sandstone is carried down by the plunge of the Sinking Creek Syncline. The cave itself actually slopes at a smaller angle so that the cave system, in effect, crosses the sandstone, developing an additional upper tier at the upstream end and an additional lower tier at the downstream end.

Comparison of cave passage orientation with joint directions (Fig. 19.8) leaves little doubt about the joint control of passages in Butler and Breathing Caves. There are two prominent joint directions, a strike set with a mean orientation of 50° and a dip set with a mean orientation of 130°. The deviation of dip joints about the mean is rather small, whereas the strike joints are broadly distributed from 30° to 70°. There is a similar distribution in the orientation of the cave passages.

Inspection of the Butler Cave map suggests that the passages upstream (south) of the Moon Room have a somewhat different orientation from those downstream. The passage orientation data were therefore plotted in two sets. The dip passage orientations are the same in both sections of the cave and also match those in Breathing Cave and the measured joint pattern. However, the upstream strike passages have a mean orientation of 65° while the downstream passages have a mean orientation of 50° and match the regional strike joints fairly well. The regional joint pattern was mapped by Deike (1960) mostly from outcrops near Breathing Cave. There appears to be a major fracture system that crosses the cave near the Moon Room and this may mark the boundary between two joint blocks, so that the joint pattern south of the Moon Room has a somewhat different orientation from the joint pattern to the north.

The southeasterly dips of the rocks on the western flank of the syncline are broken locally by a large number of minor but highly contorted folds. Often the cave passages on the dip slope cut the folds without any evidence of interaction whereas the cross-passages sometimes are located directly along these minor fold structures. Sometimes the steeply plunging folds bring down the lower sandstone, which then acts as a sediment trap. Dip passages in Breathing Cave are blocked by a fold that Deike referred to as the "monocline". The connection from the Rimstone Pools and Dave's Gallery to Sand Canyon only exists because of the survival of 90-Ugh Crawl which is a short strike-oriented segment connecting two much larger offset dip passages.

19.4 Clastic Sediment Infilling

19.4.1 Description and Classification of the Clastic Sediments

In broad terms, the Butler sediments consist of breakdown, calcite and gypsum speleothems, and fluvial sequences of various kinds. The speleothems are described in Chap. 23. Concern here is with the fluvial deposits—the clastic sediments. These consist mostly of silt, sand, pebbles and cobbles with a wide range of particle sizes and degrees of sorting. The fine grained fraction consists almost entirely of quartz while there is a mix of sandstone and limestone fragments in the large-grained fraction. Discussion of transported detrital sediments in caves in terms of their stratigraphy has not proved to be useful. The facies concept is more helpful. Sediment facies in the Butler Cave-Sinking Creek System are described using the labeling based on grains size and sorting (Fig. 19.9).

Definitions for the facies sketched in cartoon fashion in Fig. 19.9 are:

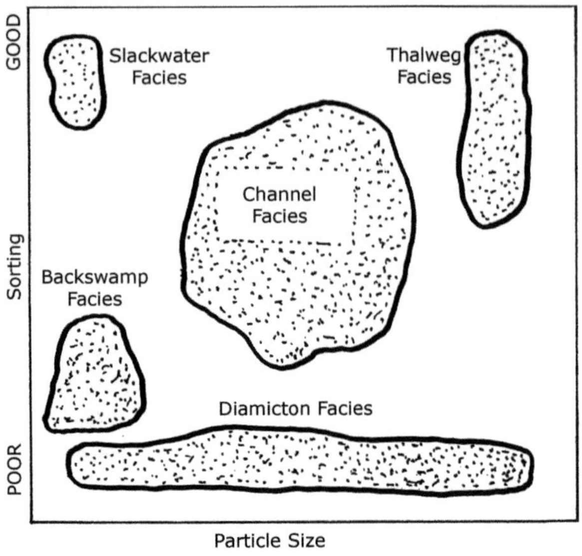

Fig. 19.9 Sketch of clastic sediment facies based on Bosch and White (2004). *Note* The areas between facies types are not blank spaces; they represent a fuzzy transition between the facies. Note also that the diagram has no scale; it is entirely schematic

(i) Slackwater facies: Fine-grained clays and silts that have settled out of muddy flood waters. Often these show a fine layering or varving. Usually they occur at the top of sediment piles.
(ii) Channel or Bank facies: Interbedded sands, silts, possibly with pebbles. Some stratification with substantial sorting between layers. These are most clearly stream deposits that, although stratified, show rapid changes in stratification over short horizontal distances.
(iii) Thalweg facies: Well-winnowed gravel, cobbles and sometimes boulders with most fine-grained material removed. These are stream bed deposits not particularly different from the bed armoring found in surface streams.
(iv) Diamicton facies: chaotic, unsorted mélange of silt, sand, pebbles and cobbles. Little or no statification. Diamicton facies are the remnants of debris flows. They were originally described for high gradient caves in New Guinea (Gillieson 1986).
(v) Backswamp facies: Uncommon in Butler Cave, these are usually fine-grained, poorly stratified muds, silts and sometimes chert fragments that accumulate as the insoluble fraction of the limestone. This facies shows little evidence of transport and the caves in which they are found usually have little evidence of stream action.

Fig. 19.10 Gravel and cobble fills in stream channel near Sand Canyon Camp. WBW photo

Fig. 19.12 Manganese oxide-coated boulders in stream channel near Sand Canyon. WBW photo

Sediments of the thalweg facies are found in the normally dry stream channel from Penn State Lake down to the Sinking Creek Sump. The channel bed is floored with a well-winnowed assortment of cobbles (Fig. 19.10). Some "grains" are in the boulder size range. Both limestone and sandstone boulders occur (Fig. 19.11). The sandstone can be recognized by its coating of black manganese oxides. The Trunk Channel in this reach is a spillover route used only by floods of sufficient magnitude to exceed the carrying capacity of the lower (but unidentified) route of Sinking Creek. The source of the sandstone must be Jack Mountain. Flood flows must have sufficient energy to move these boulders down into the cave and then transport them along the relatively low gradient trunk passage for distances of several thousand feet (Fig. 19.12). The most extreme example of boulder transport in the system was a set of sandstone boulders almost two meters in diameter that had apparently been forced up the lift tube at the drainage outlet in Lockridge Aqua Cave (see Fig. 4.8).

Channel facies occur in many places in the cave but are best displayed 300 feet downstream from Sand Canyon. A deep sediment infilling has been cut by later stream action exposing interbedded sand and gravel (Figs. 19.13 and 19.14). Evidence that the channel was filled with sediment that was later excavated is provided by a column of sediment remaining on top of a large breakdown block (Fig. 19.15). There

Fig. 19.11 Close up of large cobbles in stream channel. Note both sandstone and limestone are present. WBW photo

Fig. 19.13 Channel facies. Bedded sand and cobbles at the upstream end of the Loop. WBW photo

Fig. 19.14 Close-up, bedded sand over cobbles at upstream end of loop. WBW photo

is a large range in particle size with substantial sorting and stratification.

Most remarkable of the Butler sediments are the diamicton facies. These are unsorted and unstratified mixtures of sand, pebbles, and cobbles. These seem to have infilled all of the side caves on the western side of the system. Masses of this sediment occur in pockets along Dave's Gallery (Fig. 19.16). Similar fills have been described in Breathing Cave (Chap. 18). Diamicton facies implies a debris flow. It is not obvious whether the sediment-filled pockets were left behind as the debris swept past or whether they are remnants of a passage infilling that was later excavated. The debris flow sediments are observed mainly in the high gradient dip passages.

Slackwater facies are found in many parts of the cave but occur only as a thin layer of clay and silt overlying much coarser clastic material. In the side caves, the slackwater facies sediments are coated with an extremely thin layer of black material thought to be manganese oxides but might be carbon from decay of a final layer of organic material. They were never analyzed. Most of these delicate coatings no longer exist; they were destroyed by careless cavers who tramped all over the passage floors instead of remaining on the trail. The absence of substantial

Fig. 19.15 Mass of cobble fill remaining on large breakdown block in trunk passage below Sand Canyon. WBW photo

Fig. 19.16 Possible diamicton facies wedged into alcove on wall of Dave's Gallery. WBW photo

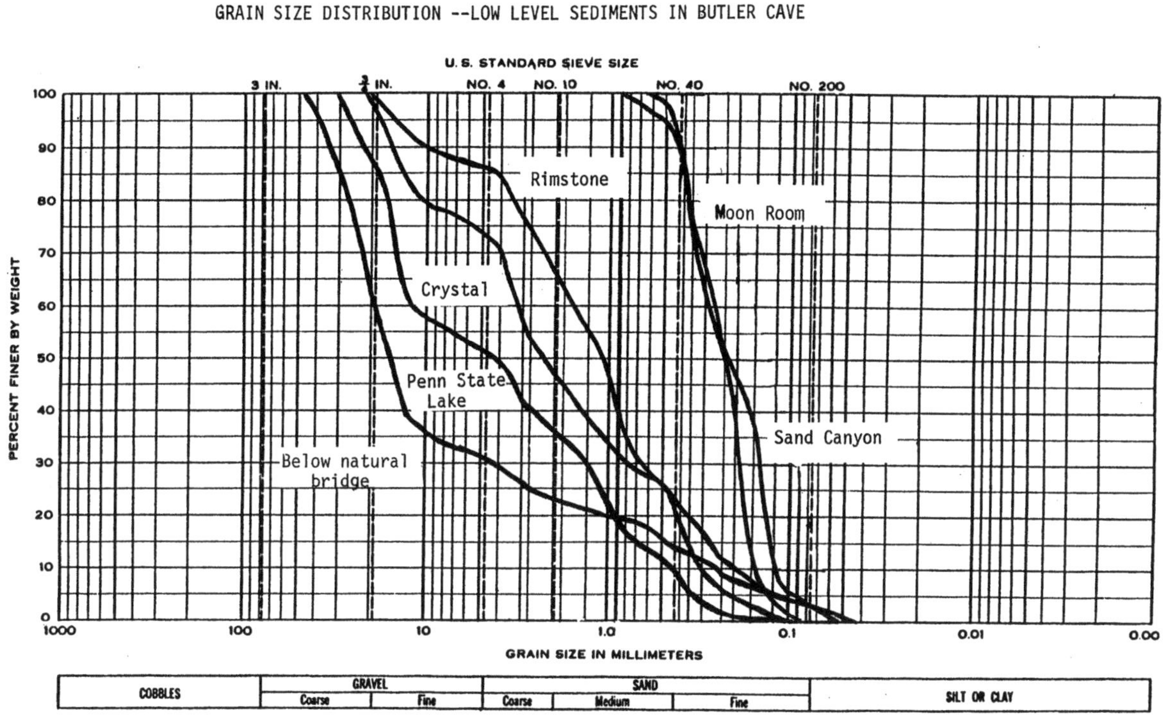

Fig. 19.17 Particle size distribution for low level sediments

slackwater facies development in Butler Cave may be evidence that, in spite of evidence for extensive floodwater action, ponded, muddy floodwater has been relatively uncommon. The caves appear to have drained rapidly during and following flood events with little evidence of ponding. This is in contrast to the Chestnut Ridge System and some of the northern caves where contemporary flooding is common and with it the characteristic fine-grained, sticky muds.

Dan Chess collected thirty two samples of cave sediments from various locations throughout the southern (upstream) and mid-sections of the cave (Chess et al. 2010). These included one eastern-most and one western-most point. Sample locations chosen were from the major passages within the cave and in some cases several sediment samples were taken from the same passage. It was intended that these samples would represent the different sediment facies and perhaps different ages of deposition. The discriminating factor used in sample selection was "low and wet" versus "high and dry" locations. The geologic sources include areas lower in the cave known be the recent depositional environments as expressed by active streams and seeps and higher, drier areas which should be older and removed from present day stream action.

The sediment samples were dried, placed in the top of a sequence of eight sieves with calibrated spacings, and shaken. The material remaining on each of the sieves was then weighed to determine the size distribution (Figs. 19.17 and 19.18).

The low level materials are much more variable than the high level materials. The sediments from the Moon Room and Sand Canyon appear to be well sorted sands, while the others show a wide range of particle sizes. The samples of high level sediments are similar to each other but show a wide range of particle sizes, including a substantial fraction of gravel. The sieve analysis, of course, does not include the cobble to boulder size material found in the thalweg facies.

19.4.2 Paleomagnetic Investigations

Although the morphology of the cave and the sequences of sediment imply a very extensive and complex developmental history, hard dates when these

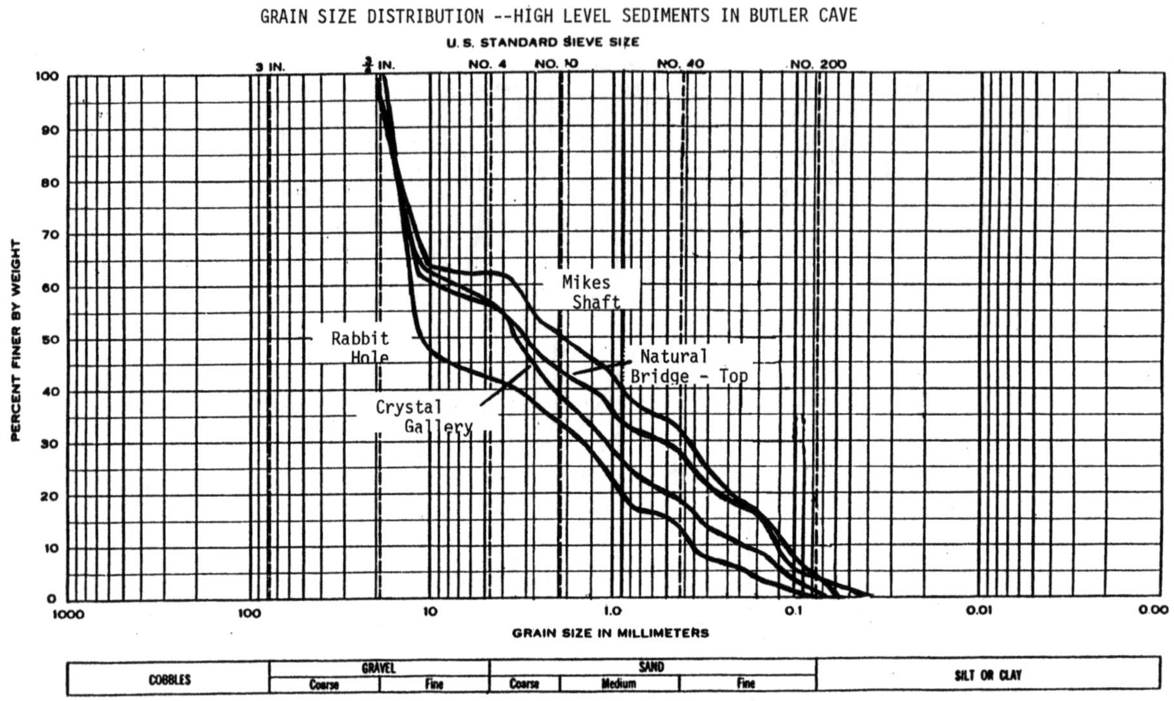

Fig. 19.18 Particle size distribution for high level sediments

events occurred are extremely difficult to obtain. In the mid-1980s Victor Schmidt of the University of Pittsburgh collected sediment samples for paleomagnetic dating in Butler Cave and in Breathing Cave. After Schmidt's untimely death in 1993, Ira Sasowsky reactivated the project and collected further samples in Butler Cave. The sediment project was finally brought to a conclusion (Chess et al. 2010). The following is a summary of the paleomagnetic results.

The principle of paleomagnetic dating is relatively simple although the details are complicated and some very expensive magnetometers are required. The procedure is to carve a one-inch cube of sediment from an exposed bank. A plastic box is slipped over the cube and its orientation carefully marked. The sample boxes with their content of undisturbed sediment were taken to the Paleomagnetism Laboratory at the University of Pittsburgh where influences of current magnetic fields were removed and the average orientation of magnetic particles within the sediment determined in a highly sensitive magnetometer at liquid helium temperature. A great deal of sample preparation and special high-sensitivity equipment is needed because the magnetic signal from cave sediment is usually very weak. The end result of these measurements is a determination of whether the north pole of the sediment sample is aligned with the Earth's present North Pole (called "normal") or aligned with the Earth's present South Pole (called "reversed"). The discovery of reversed sediments means that the sediments were deposited at a time when the Earth's magnetic field was oriented opposite to its present orientation. The chronology of magnetic field reversals is well-established. The problem is to correlate the reversed sediments with a specific period of reversed polarity.

Figure 19.19 shows the locations from which sediment samples were taken and the polarity as measured in the laboratory. The pattern of reversals is not distributed by elevation as would be the case of ordinary sediment layering. Instead, reversed sediments are found high in the dip passages as well as on remnant sediment terraces near present day stream levels. The reversed sediments are located in the massive in-filled sediment. Sediments associated with present day stream activity are normal but normal sediments from abandoned cave passages could easily belong to an

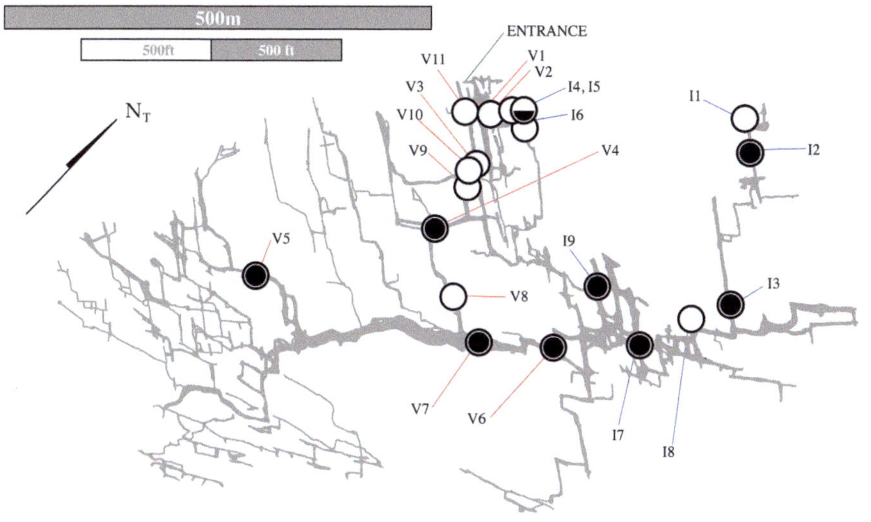

LOCATION	SAMPLE #	POLARITIES
I1	IDS149-164	R
I2	IDS165-166	N
I3	IDS167-168	N
I4	IDS121-126	R
I5	IDS127-132	R above N
I6	IDS133-136	R
I7	IDS137-138	N
I8	IDS139-140	R
I9	IDS141-146	N
V1	XAK001-002	?
V2	XAK003-004	R
V3	XAK005-006	R
V4	XAK007-008	N
V5	XAK009-010	N
V6	XAK011-012	N
V7	XAK013-014	N
V8	XAK015-016	R
V9	XAK017-020	R
V10	XAK021-024	R
V11	XAK025-026	R

● Normal
○ Reverse

Fig. 19.19 Location of sediment samples for paleomagnetic measurement and their magnetic orientation. From Chess et al. (2010)

earlier normal. With only this information, the reversed sediments, at the youngest, would date from sometime in the Jaramillo Reversed Period which ended 780,000 years ago.

At only one location (I-5 in Fig. 19.19) were sediments of different polarity found in continuous stratigraphic sequence. Three layers of clay were sampled of which the upper two layers were reversed while the lower layer was normal. The lower normal layer is, therefore, older than the reversed layers above it which places the minimum age at the Jaramillo/Matayuma boundary, 990,000 years ago.

Because of the limited number of reversals and the lack of tight stratigraphic control, sediment ages range from at least the mid-Pleistocene. They could easily be older.

19.5 Present Day Drainage Patterns

19.5.1 Internal Drainage

The present day drainage through the cave is complex and probably of recent development. Surface streams draining from the flanks of Jack Mountain are the source waters for the infeeder streams such as the Huntley's Cave stream, Rotten Rocks Creek and Difficulty Creek. The headwaters of Sinking Creek are a combination of precipitation collected in and around the Burnsville Depression and a small stream flowing from the flank of Jack Mountain. Sinking Creek flows along the floor of the trunk passage from its rise (Fig. 19.7) to its disappearance into the Sinking Creek sump. The water of Sinking Creek has been dye-traced to a stream in Marlboro Country which can be followed to the Marlboro Sump. Just upstream from the sump, a second stream enters which rises from a sump but may be the same stream that sumps northeast of the Four-Way Stop. Both streams are lost at the Marlboro Sump (Fig. 19.20).

Sneaky Creek enters the system from an unknown source at the Showers where it drains from the ceiling. At this point the trunk passage has breached the lower sandstone so that Sneaky Creek flows in a cave passage above the lower sandstone. Sneaky Creek can be followed through the French Passage, the Lake Room, and the July 6 Room until it is lost in a sump at the Rats Doom Siphon.

A third parallel stream is Slippery Creek which appears near the northwest end of Pants-Off Crawl and flows along the northwestern-most of the downstream passages ultimately to enter Last Hope Siphon and the ultimate sump at the end of the Good News Passage (Fig. 19.21).

Of the various sumps, only Last Hope Siphon has a cross-section large enough to admit a diver. Diving this sump produced some additional passage and then another sump. The other sumps are choked with gravel.

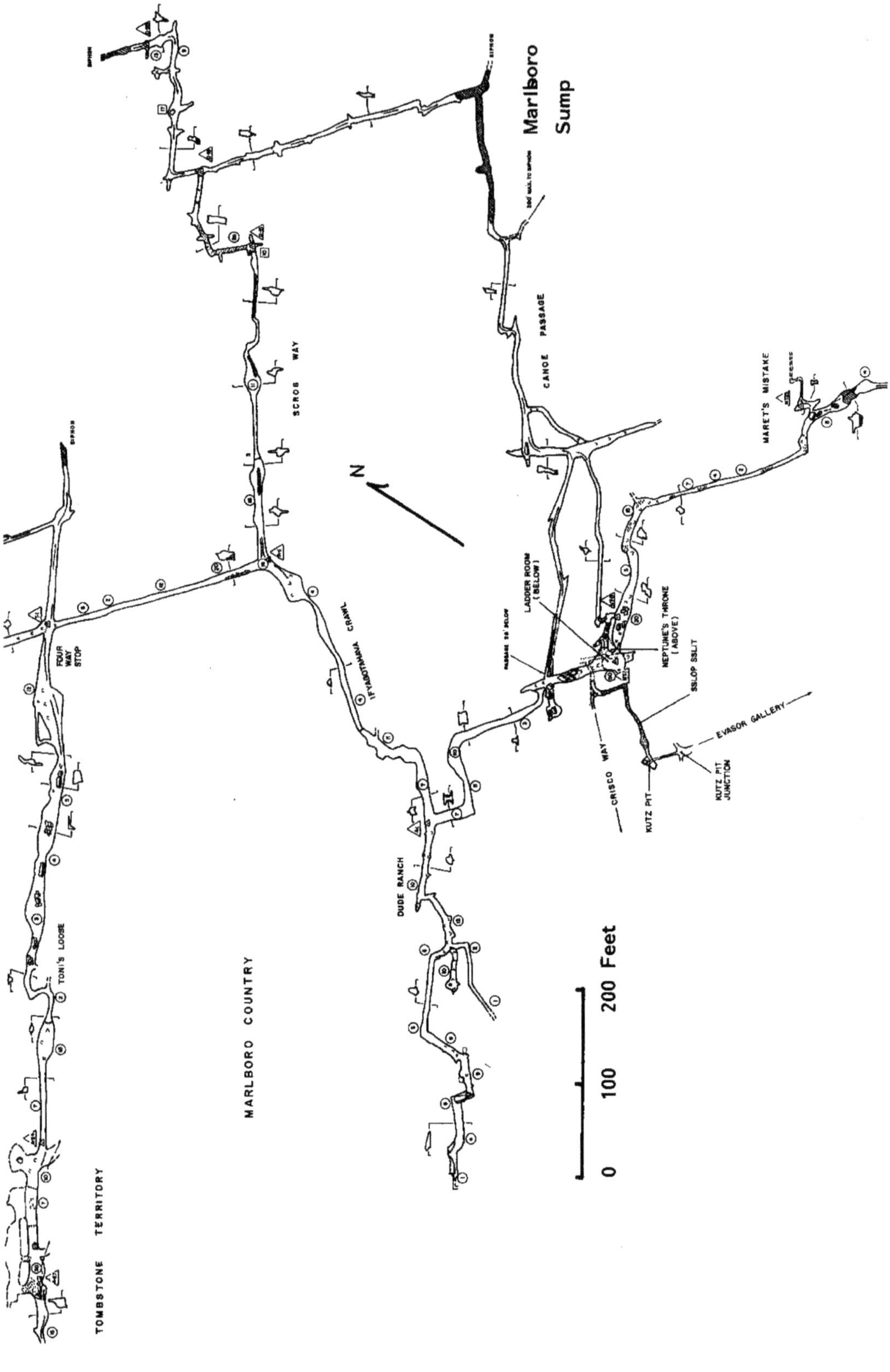

Fig. 19.20 Map of part of Marlboro Country showing the drainage leading to the Marlboro Sump

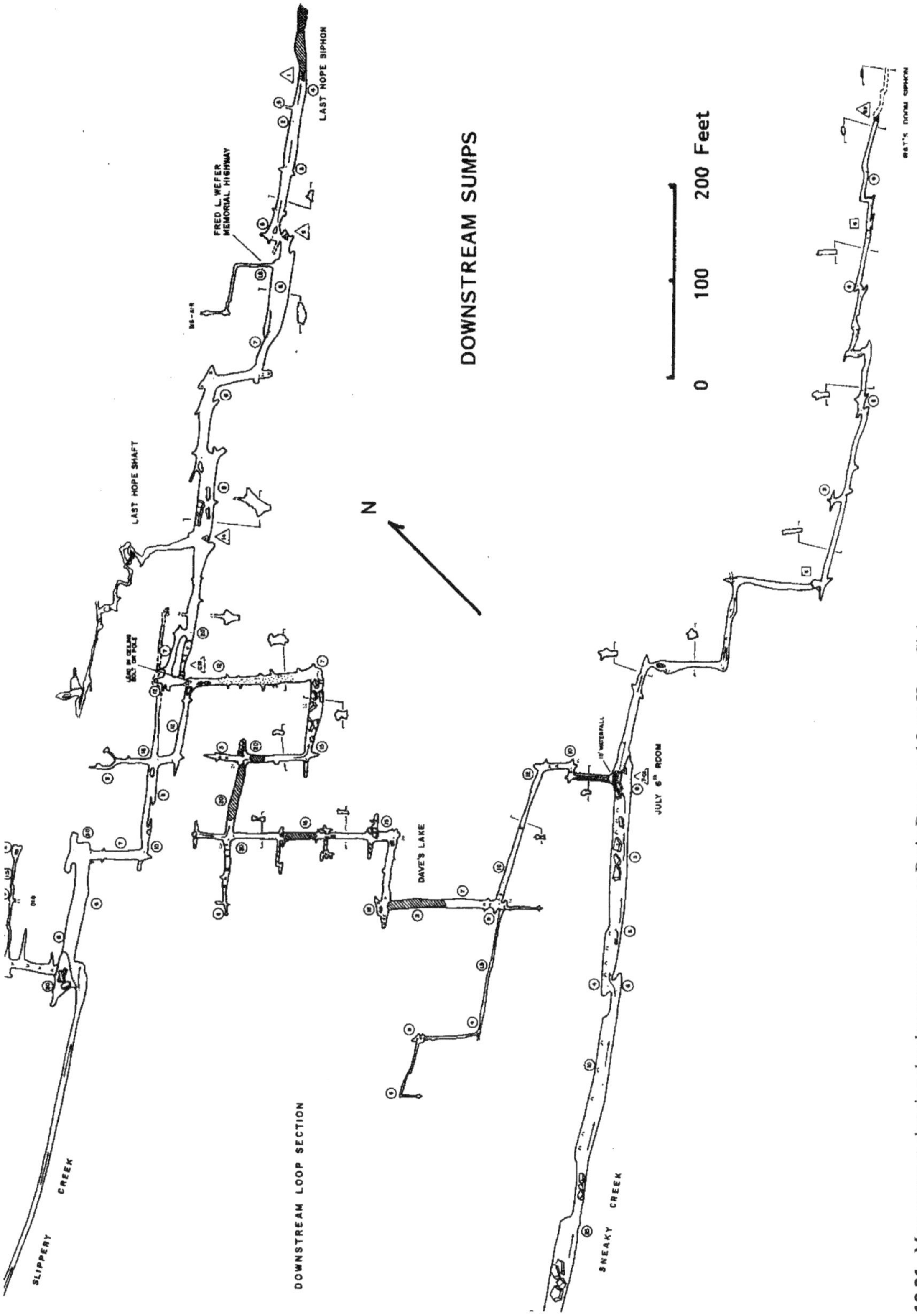

Fig. 19.21 Map segment showing the downstream sumps at Rat's Doom and Last Hope Siphons

Much of the low flow drainage is concealed. Sinking Creek and its confluences with the various in-feeder streams generally occupy small inaccessible passages beneath the trunk channel. During flood flow, the small passages spill over and flood waters flow down the trunk channel. The drainage system raises a question: Are the diversion routes of the trunk channel drainage a recent development or did these low lying passage exist earlier but are nearly choked with sediment so that they only have the capacity to carry low flows?

It is apparent from mud-coatings that the stream passages at the downstream end of the cave fill completely during flood flow. However, little is known of possible diversion and overflow routes during flood flow. Of the hydrologic behavior of Marlboro Country during flood flow, nothing is known.

19.5.2 Elevations and Gradients

There is a great deal of vertical relief both in the Butler Cave-Sinking Creek System and in other caves in the Cove. Interpretation of the development of the cave systems will depend as much on vertical relations of the caves as on the horizontal relations. Cave surveyors usually set their vertical control datum to zero at the cave entrance. There is need for a more useful reference point for inter-cave comparisons. However, elevations above sea level are not as useful as elevations above some local benchmark. The chosen benchmark is the junction of the Bullpasture and Cowpasture Rivers near Williamsville. The wide, flat floodplain of the Cowpasture suggests that this has been a stable base level for a long time. This is also the ultimate low point for Burnsville Cove and vicinity. The benchmark elevation has been set at 1595 feet = zero datum. Elevations of features within the cave are then calculated with respect to this datum.

Table 19.1 lists some locations within the cave giving their elevations with respect to sea level, with respect to the river confluence benchmark, and with respect to the elevation of Aqua Spring. The elevation of Aqua Spring is 1770 feet, 175 feet above the river confluence benchmark. The in-cave elevations are those tabulated on the Butler Cave-Sinking Creek System map folio compiled by Lester Good in 1985. The vertical accuracy of the survey data is not known so that these values should be used with some caution.

The highest point in the cave is the Nicholson Entrance. The Butler Cave section descends steeply to

Table 19.1 Selected elevations within the Butler Cave-Sinking Creek System

Station	Location	Sea level	Benchmark	Spring
	Butler Cave			
A-01	Nicholson Entrance	2536	941	766
B-36	Window Room	2380	785	610
B–H	Dave's Gallery, below SOFA Entrance	2366	771	596
B-24	Cross Canyon	2307	712	537
10	Lower Bean Room	2179	584	409
B-13	Rimstone Pools	2269	674	499
B-01	Sand Canyon	2240	645	470
	Trunk Channel			
CP-1	Rise of Sinking Creek	2200	605	430
J-22	Pat's Section junction	2156	561	386
L-22	Near Sinking Creek Sump	2129	534	359
I-22	Below Dry Sumps	2096	501	326
E-01	Pool Room	2073	478	303
T-01	July 6 Room	2017	422	247
D-19	Feeder to Marlboro Sump at Scrog Way	1981	386	211
69	Rats Doom Siphon	1912	317	142
1	Last Hope Siphon	1976	381	206

Sand Canyon, roughly 300 feet below the entrance. According to the survey data, the lower Bean Room and also the passages downstream from the Tobacco Room lie at lower elevations than the rise of Sinking Creek into the Trunk Channel. This arrangement seems unlikely; the elevation discrepancy is more likely error in the elevation data.

The ultimate sink of Sinking Creek at the Marlboro Sump is at about the same elevation as the Last Hope Siphon. Although Last Hope Siphon is above the lower sandstone and the Marlboro Sump is below the lower sandstone, the Marlboro Sump lies farther upstream along the Trunk Channel and so is higher in the stratigraphic section. If the elevation data are to be believed, the lowest point in the cave is the Rats Doom Siphon which would be more than 60 feet lower than the other two sumps.

The Trunk Channel drops 270 feet over the roughly one mile between Sand Canyon and the downstream sumps, giving a gradient much higher than the 10–50 feet/mile typical of karst aquifers with open conduit permeability. If this gradient were to continue over the roughly two miles between the downstream sumps and Aqua Cave, the water would need to descend more than 500 feet and would have to rise more than 300 feet to reach the spring. Given the deep sumps at the back of Aqua Cave, this rough extrapolation may not be impossible. It would require strong geological controls to force the water into deep phreatic circulation because karst aquifers usually develop a shallow pathway to local base level streams.

19.6 Geochemistry of Cave Waters

19.6.1 Bulk Chemistry

There is a limited amount of information on the chemistry of the waters of the Butler Cave-Sinking Creek System. Table 19.2 give a set of cation analyses (Chess 1987). The karst waters in the Butler Cave-Sinking Creek system are of very high quality. Toxic heavy metals such as Co, Cu, Ni, Pb and Zn are all below limits of detection.

The cation analyses were based on a single set of samples collected in August, 1984. As part of his thesis work, Dan Chess collected samples for anion analysis at different times of the year. Values given in Table 19.3 are based on from three to eight samples. Again, all concentrations are very low. The species of concern is nitrate which might be expected as a contaminant in a rural farming area such as Burnsville Cove. Although nitrate values tend to fluctuate throughout the year, all values are well below the drinking water standard of 45 mg/L.

19.6.2 Carbonate Chemistry

The investigation by Harmon and Hess (1982) provides the only source of information on the geochemistry of carbonate dissolution and precipitation. The tables that follow are extracted from or calculated from the data in the appendix to their paper. Harmon

Table 19.2 Cation analyses of selected streams and springs (mg/L)

Cation	Butler farm Springhouse	Dave's Gallery	Moon Room Passage	Huntley's Cave	Aqua Spring
Al	<0.02	0.03	<0.02	<0.02	0.04
Ca	2.63	17.2	40	21.3	34
Co	<0.02	<0.02	<0.02	<0.02	<0.02
Cu	<0.02	<0.02	<0.02	<0.02	<0.02
Fe	<0.02	0.03	<0.02	<0.02	0.02
K	0.37	0.76	0.65	0.60	0.75
Mg	0.52	0.70	3.41	0.89	3.21
Mn	<0.02	<0.02	<0.02	<0.02	<0.02
Na	0.30	0.39	0.48	0.35	0.36
Ni	<0.02	<0.02	<0.02	<0.02	<0.02
Pb	<0.02	<0.02	0.03	<0.02	<0.02
Sr	<0.02	0.03	0.10	0.06	0.08
Zn	<0.02	<0.02	<0.02	<0.02	<0.02

Table 19.3 Anion analyses of selected surface and cave streams

Location	Cl^-	σ	NO_3^-	σ	SO_4^{2-}	σ
Surface streams						
Butler Farmhouse Spring	0.61	0.08	0.21	0.19	0.96	0.14
White Rock Mountain Stream	0.46	0.08	0.05	0.02	1.83	0/17
Sink of Sinking Creek	1.11	0.32	0.20	0.17	2.43	0.59
Cave streams						
Dave's Gallery	0.73	0.21	1.72	1.14	2.41	0.28
Moon Room	0.95	0.08	6.28	0.04	4.85	0.45
Huntley's Cave	0.65	0.16	1.22	0.81	3.31	0.38
Sinking Creek Resurgence	1.22	0.60	2.25	1.06	3.92	0.26
Sinking Creek Siphon	1.26	0.13	3.08	0.23	4.11	0.02
Sneaky Creek—Pool Room	1.60	0.17	2.76	1.53	3.45	1.76
Springs						
Aqua Spring	0.72	0.20	0.69	0.51	4.74	0.32

Analyses are given in mg/L. σ standard deviation

and Hess measured Ca-ion, Mg-ion, and bicarbonate ion concentrations as well as pH, temperature, and specific conductivity at the springs, and in Butler Cave at a number of sites. Sites were visited from one to five times as indicated by the dates. The temporal data are too sparse for more than a hint at the chemical variability at the sites; certainly no comparisons of chemistry with flow hydrographs can be made. However, the data are adequate to provide a rough idea of the present day chemical behavior of the cave system. Emphasis must be placed on the phase "present day".

The analytical quantities measured by Harmon and Hess permitted the calculation of other parameters that are more helpful in interpreting cave processes. For a more complete derivation and justification of these parameters, see any of several textbooks such as White (1988) and Langmuir (1997).

Hardness: Hardness is a measure of the amount of dissolved carbonate in the water. It is defined in terms of the measured Ca^{2+} and Mg^{2+} concentration and recalculated in units of mg/L as $CaCO_3$. It is a somewhat phony parameter in that both calcium and magnesium are treated as $CaCO_3$ but it does provide a useful measure of total dissolved carbonate.

$$Hd = 100.09\left(\frac{C_{Ca}}{40.08} + \frac{C_{Mg}}{24.31}\right)$$

The C's are concentrations of Ca and Mg in units of mg/L; the numerical values are atomic and molecular weights. Hardness is expressed in units of mg/L as $CaCO_3$.

Saturation Index: A question for any karst water: Is the water at chemical equilibrium with the limestone (or dolomite) bedrock? Is the water undersaturated, meaning that it is capable of dissolving more limestone? Is the water supersaturated, meaning that calcite (speleothems) should be precipitated. Saturation index is calculated from the measured Ca^{2+} and bicarbonate concentrations, the pH, the specific conductance, and the temperature. The saturation index is the logarithm of the ratio of the ions actually in solution to what the solution could hold if it were at equilibrium. Water at equilibrium has a saturation index of zero, positive values indicate supersaturation and negative values indicate undersaturation.

Carbon Dioxide Partial Pressure: Karst waters contain various concentrations of dissolved CO_2 which provides the weak acid that allows the water to dissolve limestone. Thick, organic-rich soils tend to produce high CO_2 concentrations (usually expressed as a partial pressure rather than a concentration) while water infiltrating from bare bedrock usually has a CO_2 pressure equal to or only slightly higher than the atmosphere. The amount of CO_2 in the water can be calculated from the bicarbonate ion concentration, the pH, the specific conductance, and the temperature. CO_2 concentrations in the atmosphere are actually rather low, 0.033 volume percent (330 ppmv) in the 1970s when these data were collected. Rather than giving the results of the calculation as CO_2 pressures,

Table 19.4 Carbonate parameters for streams within the Butler Cave-Sinking Creek System

Location	Date	Hardness	SI_C	P_{CO_2}/Atm CO_2
Rise of Sinking Creek	10/24/70	110	−0.55	8.8
	12/19/70	70	−1.08	11.6
	10/3/70	120	−0.66	15.3
	2/20/71	52	−1.15	3.3
	5/8/71	76	−0.58	3.5
Sinking Creek Sump	10/24/70	141	−0.53	13.0
	12/19/70	81	−0.93	10.8
	10/3/70	132	−0.43	11.4
	2/20/71	55	−1.07	3.2
	5/8/71	86	−0.45	3.3
Slippery Creek	10/24/70	116	−0.78	10.3
	10/3/70	116	−0.77	10.6
Moon Room Stream	12/19/70	117	−0.44	7.3
Rise of Sneaky Creek	12/19/70	124	−0.81	13.0
	12/19/70	133	−0.64	10.8
	10/3/70	186	−0.35	24.9
Huntley's Cave Stream	*2/20/71*	29	−1.66	2.7
Natural Bridge Stream	*2/20/71*	34	−1.39	3.1
Sand Canyon Stream	*2/20/71*	41	−1.28	3.0

Italicized data were collected under high flow conditions

it is convenient to calculate a CO_2 enhancement factor, defined as the ratio of the calculated CO_2 pressure to the CO_2 pressure in the atmosphere and it is this parameter that is listed in the tables.

Calculated carbonate parameters for flowing waters sampled in the cave at various places and times are given in Table 19.4. The data collected on February 20, 1971 were noted as being under "high flow conditions" although no quantitative measure of "high flow" was given.

As representatives of the input waters to the system, the chemical data on the flowing streams are remarkably consistent. All waters are highly undersaturated, meaning that they have been in contact with the limestone for too short a time for the chemical reactions to go to equilibrium. Carbon dioxide pressures are mostly in the range of 3–10 times the atmospheric background which is rather low. This also indicates the cave streams are derived primarily from surface runoff that has not had lengthy contact with organic-rich soils. Much of the water in the cave is mountain runoff. Soils are thin, sandy, and throughput times are rapid. Storm flow further dilutes the already dilute water.

The chemistry of the output waters was investigated by sampling the springs (Table 19.5).

The geochemical parameters for the spring waters are not greatly different from those measured for the streams in the cave. At first glance, this would seem to be entirely reasonable. In reality, these results raise some very difficult questions. All of the cave streams ultimately drain into sumps. The sumps are about two miles from Aqua Spring. The water feeding Aqua Spring rises from deep sumps at the back of Aqua Cave. There is no known source of additional CO_2 at depth in the ground water system so the CO_2 pressure should remain characteristic of surface runoff as it does. However, the deep flow system should be slow moving and provide plenty of time for the water to reach chemical equilibrium. Obviously, that didn't happen. The spring waters are undersaturated at about the same values as the cave streams. The transport of water through the entire system, including the unknown passages between the cave and the spring, must be very rapid and could contain air-filled segments.

A similar situation applies to Emory, Cathedral, and Blue Springs. There is a single measurement at

Table 19.5 Carbonate parameters for the Burnsville Cove springs

Location	Date	Hardness	SI_C	P_{CO_2}/Atm CO_2
Emory Spring	10/24/70	94	−0.39	5.2
	10/4/70	96	−0.85	12.2
	5/8/71	73	−0.50	3.2
	5/2/72	81	−1.06	12.8
Aqua Spring	10/24/70	106	−0.28	4.5
	10/4/70	111	−0.48	19.3
	2/20/71	67	−0.64	2.1
	5/81/71	78	−0.44	3.4
	5/2/72	71	−0.91	7.9
Cathedral Spring	10/24/70	105	−0.48	7.0
	10/4/70	102	−1.16	33.5
	5/8/71	90	−0.35	3.0
Blue Spring	10/4/70	142	−0.40	15.3
	5/8/71	131	−0.12	4.7

Blue Spring where the water approaches equilibrium. From the perspective of the cave explorer, this is good news. The unknown system that feeds Emory Spring must consist mostly of open conduits because the water chemistry is not distinguishably different from that of the other springs.

19.7 Concluding Thoughts

A full interpretation of the geologic history of the Butler Cave-Sinking Creek System must be woven into a broader and more comprehensive picture of the geologic and geomorphic history of entire Burnsville Cove (Chap. 24). There are clearly three, or perhaps four, major epochs in the development of the cave system.

1. The development of the "Old Cave"—the large passages of Butler Cave and certainly the southern portion of the Trunk Channel.
2. A massive sedimentation event that filled many of the passages of the "Old Cave" with sand, gravel and cobbles. Paleomagnetic reversals show that this event took place at least 990,000 years ago, mid-Pleistocene or older.
3. The development of the "New Cave", the removal of much of the earlier fill and the downcutting of much more passage such as the Bean Room and the canyon upstream from Sand Canyon.
4. The invasion of the pre-existing cave by present day drainage and perhaps the development of the present-day drainage pathways beneath the trunk channel. The use of the cave as a pathway carrying runoff from mountain streams to the springs seems an opportunistic use of existing cave passages, possibly with the development of new and immature passages beneath the present system.

References

Bosch, R.F., and W.B. White. 2004. Lithofacies and transport of clastic sediments in karstic aquifers. In *Studies of cave sediments*, ed. I.D. Sasowsky, and J. Mylroie, 1–22. New York: Kluwer Academic/Plenum Press.

Chess, D.L. 1987. Comparison of Microbiology and Selected Anions for Surface and Subsurface Stream Waters for the Aqua Spring Watershed of Burnsville Cove, Virginia. M.S. thesis, The Pennsylvania State University, 116 pp.

Chess, D.L., C.A. Chess, I.D. Sasowsky, V.A. Schmidt, and W. B. White. 2010. Clastic sediments in the Butler Cave-Sinking Creek System, Virginia, USA. *Acta Carsologica* 39: 55–70.

Deike, G.H. 1960. Origin and geologic relations of Breathing Cave, Virginia. *National Speleological Society Bulletin* 22: 30–42.

Gillieson, D. 1986. Cave sedimentation in the New Guinea Highlands. *Earth Surface Processes and Landforms* 11: 533–543.

Harmon, R.S., and J.W. Hess. 1982. Ground water geochemistry of the Burnsville Cove area, Virginia. *National Speleological Society Bulletin* 44: 84–89.

Langmuir, D. 1997. *Aqueous environmental geochemistry*, 600 pp. Upper Saddle River: Prentice-Hall.

White, W.B., and J.W. Hess. 1982. Geomorphology of Burnsville Cove and the geology of the Butler Cave-Sinking Creek System. *National Speleological Society Bulletin* 44: 67–77.

White, W.B. 1988. *Geomorphology and hydrology of karst terrains*, 464 pp. New York: Oxford University Press.

The Geology of the Chestnut Ridge Caves 20

William B. White

Abstract
The Chestnut Ridge caves consist of the Chestnut Ridge System itself with more than 20 miles of surveyed passages and also a line of smaller caves strung out southwestward along the axis of the ridge. The Chestnut Ridge System contains two branches that converge at the northeast end. The eastern branch is a north-south trend that cuts across the structure and contains segments of maze formed on joints oblique to the main trend. At the southwestern end is the Burns section with streams that drain to Cathedral Spring. The western branch is strike-oriented along the trend of Chestnut Ridge. It includes the Burnsville Turnpike, the largest passage in the system, and includes along the trend line the other caves and also the main trunk of Barberry Cave. Steams in this section drain to Aqua Spring and thus the Chestnut Ridge System displays and underground drainage divide. The system has multiple levels and also many shafts and later-stage modifications.

20.1 Introduction

The title to this chapter is given as the geology of the Chestnut Ridge Caves, not the geology of the Chestnut Ridge System. The convention among cavers is that for caves entered by different entrances to be considered part of a system, they must be physically connected underground. The caves accessed through the Bobcat, Blarney Stone and Burns entrances have been interconnected but there are a series of smaller caves strung out along the ridge south of the Chestnut Ridge System that are not (yet!) connected. And there is Barberry Cave which may be a link between the Chestnut Ridge Caves and the Butler Cave-Sinking Creek System. It is apparent from the regional map (Fig. 1.12) that all of the caves so far discovered on Chestnut Ridge are interrelated.

The Chapters on Breathing Cave and Butler Cave draw heavily on a variety of studies that have been undertaken in the caves over the 50 years since their discovery. Further, these two caves are relatively easy to access so that those who wished to undertake scientific investigations could do so with minimum difficulty. The details of the Chestnut Ridge caves, in contrast, were revealed only slowly over many years beginning with the initial explorations through the Bobcat Entrance in 1983 right up to the most recent

Information Compiled by the Editor from Various Sources.
Cave Description by John Rosenfeld.

W.B. White (✉)
Department of Geosciences, The Pennsylvania State University, University Park, PA, USA
e-mail: wbw2@psu.edu

W.B. White (ed.), *The Caves of Burnsville Cove, Virginia*,
Cave and Karst Systems of the World, DOI 10.1007/978-3-319-14391-0_20,

expeditions. The caves are physically difficult. There is no part of the system, with the possible exception of Barberry, where easy geological investigations would be possible. As a result, this chapter is based on observations, maps, and photographs collected by the explorers rather than a lengthy and systematic probing of the geological aspects of this exceedingly complex group of caves.

20.2 Description of the Chestnut Ridge System by John Rosenfeld

[**Editor's Note**: Adapted from the cave description given in the Guidebook for the 1995 NSS Convention, *Underground in the Appalachians*, pp. 14–18. The photos are from the Ron Simmons collection.]

The entrances to the Chestnut Ridge caves are located near the ridge-top and represent the very beginnings of recently formed cave passages connecting to a much larger and older cave system below. Typically, for most caves on Chestnut Ridge, a small entrance opens into a steeply descending fissure heading down dip. Within a short distance of traverse, a tiny stream appears in the floor, lined with small gravel, silt and clay. Pits, from downclimable size to more than 40 feet, occur in small domes formed along the fissure. Most of the passage dimensions are crawl to squeeze size except for the occasional small dome room where one can usually stand up. Then within a few hundred feet, the fissure becomes too tight to follow and this marks the end of passable cave. This description fits most of the known caves on Chestnut Ridge with the important exceptions of the Bobcat, Blarneystone, and Burns entrances, all of which were extended to the underlying paleocave 400 feet below the ridge summit by some extremely strenuous and determined exploration as documented in earlier chapters.

The complete system map reduced to stick map size (Fig. 20.1) shows the overall layout of the Chestnut Ridge Cave System and gives some of the more important place names. The full map on an expandable scale is given in the electronic file that accompanies this volume and segments of the map are used to illustrate specific small areas.

20.2.1 The Bobcat Section

Three main trunk passages have been discovered in Bobcat Cave. The first of these has a north-south orientation for nearly 4000 feet, and is known as the North-South Trunk (or the North-South Lead in some reports). Passage dimensions vary widely and range from 40 to 70 feet wide by 40 feet high to 10-foot high and wide; the floor profile is mostly irregular. The passage is highly decorated. Near the southern end of this large passage is the Camp Room, a 60-foot wide by 250-foot long room where the underground camps were staged for exploration of the far reaches of the cave (Fig. 6.4). From the Camp Room, several passages branch out, including the continuation of the North-South Trunk. Another of the large passages leading from the Camp Room is the passage to Shamrock Dome, a large dome room measuring 40 feet wide by 130 feet long and up to 100 feet high at the dome. From Shamrock Dome, a window lead connects to a series of walking size passages and rooms characterized by successive upclimbs and downclimbs. The last room at the end is Satisfaction Junction. Two passages from this room shortly lead to pits down to sumps. At 722 feet below the entrance, the sump is the deepest explored point in Bobcat Cave.

The Camp Room seems to indicate a major juncture for passages in the Chestnut Ridge System. It has been suggested that the Camp Room is a paleo-sump room where the passages from the north and south brought streams to form a large sump. Below this sump, the remainder of the cave would then have been water-filled. It is characterized by steeply descending and ascending phreatic passages. The characteristics of the passages below the Camp Room to Shamrock Dome and beyond are similar to passages discovered by Ron Simmons in Cathedral Spring. Passages that undulate in the vertical plane are characteristic of sub-water table cave development.

At the north end of the North-South Trunk is SVG Hall (named for the Shenandoah Valley Grotto), a large room measuring 75 feet wide by 160 feet long and 50 feet high. From SVG Hall, a 100-foot high lead climb led to the discovery of the Porpoise Passage. Here the cave passage crosses over the Chestnut Ridge anticline to the western limb and into the Aqua Spring

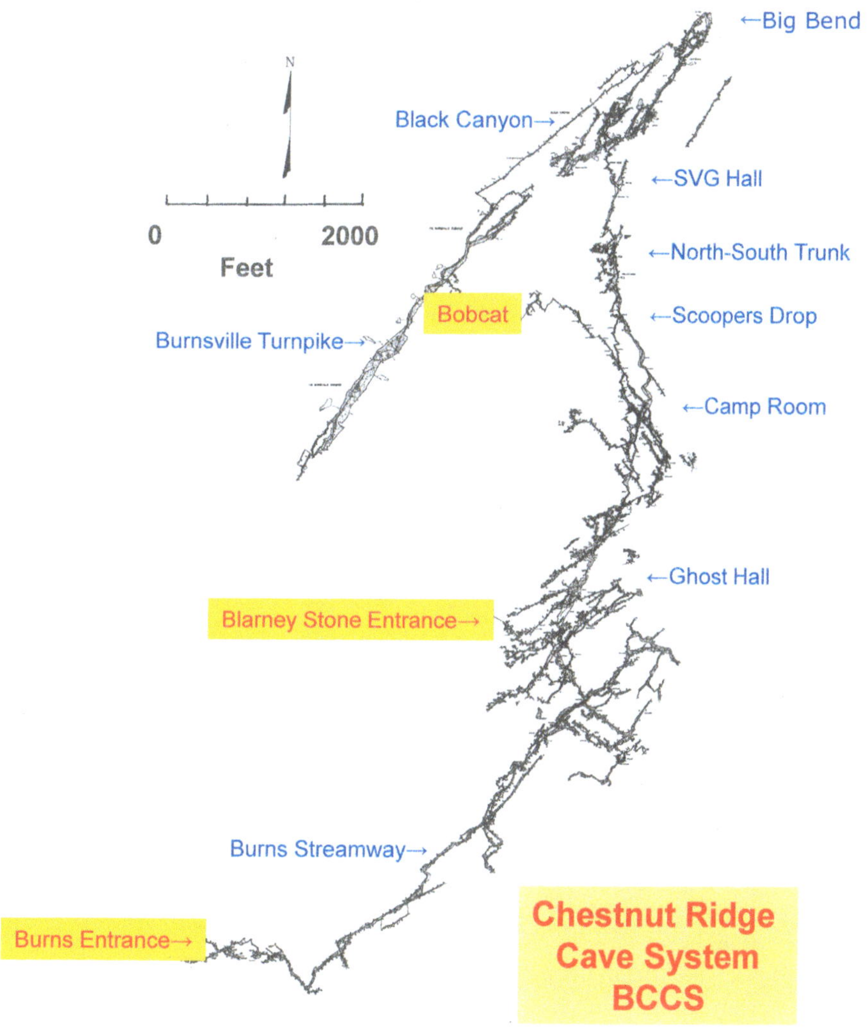

Fig. 20.1 Overview map of chestnut ridge system

drainage. The Porpoise Passage ends at a pit 80 feet deep named Damart Drop. The bottom of the pit is a large canyon 10 feet wide by 60 feet high. This passage shortly terminates at 35-foot deep Polypro Pit. Here begins the second trunk series, made up of the Sixth of July Room and Jewel Cave. These passages are characterized as large, dry, and highly decorated passages 35 feet wide by 20 feet high, with some areas up to 60 feet wide, trending in a northeast-southwest (strike) direction. At the northern end of this trunk is a large loop in the passage almost closing on itself called Big Bend. At the southern end are two significant side leads. One is a short crawl leading to Megalo Junction, a short segment of trunk passage up to 140 feet wide which ends in breakdown, heading back toward Big Bend. The other, Maret's Lead, continues as a series of rooms and walking passages with many upclimbs and downclimbs over slippery clay and silt floor. This continues for approximately 2000 feet to a sump, dye-traced to Aqua Spring. Near the sump, a stream tributary comes in and marks the beginning of Black Canon, a mostly clean washed stream canyon typically 4–6 feet wide by 6 feet high. The limestone appears very dark, hence the name.

Black Canyon can be followed for a half mile to where the canyon climbs up a series of small waterfalls to the third major trunk passage in the cave, the Burnsville Turnpike. This passage is among the largest passages in Virginia. Unfortunately, its remoteness has prevented many cavers from visiting it. The Turnpike heads in a southwesterly direction as a large, low-gradient steam trunk. As one travels upstream, the passage gradually becomes larger until after 2000 feet, the passage dimensions are up to 145 feet wide and over 100 feet high. At this point, the passage floor becomes lined with large breakdown and clay banks, making traverse difficult. After another 2000 feet, the passage floor slopes steeply up into a breakdown choke. A stream passage under the breakdown continues several hundred more feet to a point blocked by flowstone and breakdown. This location is the most remote point in the cave, at 4–6 h travel time from the Camp Room. The major side passage of 40-foot wide by 80-foot dimensions, exits the Turnpike about 20 feet above the floor. This has been explored to a second breakdown choke with good airflow coming from the breakdown. The passage lines up with Megalo Junction, however, it is on a gradient to pass at least 100 feet below.

20.2.2 Blarneystone Cave

The dug entrance begins as a narrow vertical fissure. Like almost all caves on Chestnut Ridge, this passage turned to a narrow stream fissure leading steeply down dip with many challenging downclimbs. After about 1000 and 400 feet in depth, the stream fissure becomes too tight to follow. A short distance from the end is an upclimb leading to an older passage parallel to the active cave. This passage can be characterized as a scramble, having squeezes, crawls, and stoopways. After a couple hundred feet, another climb up to a dry horizontal passage leads to Artz's Attic. Here an aid climb by Mike Artz across a 15-foot wide by 35-foot deep pit was required to continue exploration of the cave. Any trips into the system from the Blarneystone entrance must cross this rebelay traverse.

About a hundred feet beyond Artz's Attic, the main paleotrunk passage of the cave is intersected. The passage is named Ghost Hall for a very large white stalagmite in the middle of the passage (Fig. 7.2). This highly decorated passage leads southwest for a few hundred feet to where it becomes nearly blocked by flowstone. This marks the beginning of the Black Diamond Crawl, a passage characterized by a series of crawls and small rooms through which flows a small stream lined with black gravel. After several hundred feet, at the intersection of several fissures, a pit opens in the floor. This leads down to The Pearly Gates, so named for some beautiful cave pearls. At this location, a canyon passage measuring up to 10 feet wide by 30 feet high continues. Following this leads to a major stream passage called the Moon River. The Moon River goes upstream as a 8-foot by 8-foot walking passage for about 1000 feet where it breaks up into a series of infeeders. All of the infeeders shortly lead into terminal gravel and boulder chokes.

The downstream direction of Moon River continues for about 1000 feet to where the stream becomes too low to follow. A few hundred feet upstream from this point is a large steeply sloping room formed on the left side of the passage with leads heading in both upstream and downstream directions. These passages become too tight after a couple hundred feet. Another lead from the room perpendicular to the other passages leads into a nice size passage that turns back to a downstream direction and terminates after several hundred feet in the Drain Room. About midway along this passage is a crawl on the left that leads to a series of nice walking passages named Slop Holler. Slop Holler can be followed until it reconnects to Moon River. Near the beginning of Slop Holler, a large window lead intersects the passage several feet off the floor. Following it leads upstream back towards the entrance as a series of upclimbs ranging from 4 to 10 feet. This passage, called The Stairway to Heaven, has a vertical extent of 500 feet. It may be the most challenging passage in Chestnut Ridge.

Ghost Hall continues north for several hundred feet to where it ends as a crawl named the Leprechaun Forest. This is one of the most highly decorated and fragile passages in the cave. Clear to white crystalline chandeliers several feet long curve outward from the walls of this passage. After several hundred feet, the passage opens up into a large dry trunk called the Over Forty Passage, having dimensions of up to 80 feet wide by 30 feet high and lined with large breakdown. This leads several hundred more feet to the northwest to where it terminates in breakdown. Continuing in the same direction is a low crawl that soon opens into walking height passage 12 feet wide. After a short distance this leads to Spic and Span Junction, so named for a clean washed pit on one side of the room. From this area, the first attempts for a Bobcat connection were made. In the process, a narrow crevice descending steeply down was discovered. Because of the extremely muddy and wet condition, this passage was named the Poison Passage. All of the fissures at the bottom of several pits leading out of Spic and Span became too tight for traverse, ending any hopes for a connection to Bobcat in this area.

Two significant passages intersect the Over Forty Passage on the left and right. The right leads steeply down to a stream canyon named Rolling Rock Canyon Creek. The left leads almost due north to an area of maze, characterized by high canyons of 6–8 feet wide with climbs up and over breakdown. This can be followed for several hundred feet to an east-west oriented stream canyon named Wobblestone Canyon Creek for the loose unstable breakdown one must walk on for traverse. Wobblestone Creek flows downstream to the east where it becomes very tight. It is the same stream that enters Bobcat Cave near Satisfaction Junction. At the juncture with Wobblestone is a lead continuing to the north that is about ten feet off the floor. This passage requires a short aid climb for access. Once access is gained, one must immediately descend a long sloping canyon and then cross over a pit that reconnects back to Wobblestone Canyon Creek. From here, a 15-foot climb up leads to a small room nearly filled with breakdown. On the left side, a stoopwalking height passage leads to a tight pit in the floor which can be downclimbed to some squeezes and more climbdowns. This leads to the Earth Works, so named for the large dry silt and clay banks lining the floor and walls. At the end of the Earth Works is a small room that would appear to end except for a tight upclimb at the far end. The climb leads up into a small unstable breakdown room which lies directly under the terminal end of the North-South Trunk in Bobcat. During August, 1994, a connection was made between the two caves at this location.

At the southern end of Ghost Hall was found a small opening in flowstone-covered breakdown called the Airblower. Extensive excavation over a period of years opened into another segment of trunk, Anthodite Alley. Pits in Anthodite Alley gave access to a breakdown passage, frequently flooded and thus covered with mud, called the Boulder Dash. Southwest of this passage, an upclimb leads to Opportunity Knocks and the Outer Limits. Half a mile southwest along this twisting passage is the tight muddy crawl connecting the Blarney Stone Section to the underlying main streamway in Burns. This connection extends the drainage system another half mile to the southwest.

20.3 Geological Interpretation

20.3.1 The Old System

Chestnut Ridge survived as a topographic high because of the resistant caprock of Oriskany Sandstone along the crest of the anticline. Even so, dissection of the landscape is well advanced. Much of the caprock has been removed by erosion allowing limestone to be exposed at the surface except for the northeastern end of the ridge. The small gullies tributary to the valley of Sinking Creek on the west and White Oak Draft on the east have cut into the sides of the ridge while the collapse that formed the large closed depression called the Wine Gourd terminated the southern end of the Burnsville Turnpike. What is here called the "Old System" is composed of fragments of large trunk passage that lie deep and presumably predate the present dissection of the landscape.

The most important guiding structure is the Chestnut Ridge Anticline. Its approximate location is shown on the regional cave map (Fig. 1.12) and on the geologic map (Fig. 16.3). However, the anticline contains considerable additional structural detail of tight folds and contorted beds (Figs. 20.2 and 20.3).

The overview map (Fig. 20.1) suggests that the cave can be subdivided into three main parts. The northwest part is an almost linear sequence of passages following the strike along the western flank of the Chestnut Ridge

Fig. 20.2 Contorted bedding in the North-South Trunk, Bobcat Section. Ron Simmons photo

Fig. 20.3 The stone rainbow, Battered Bar Cave. Photo by Philip C. Lucas

Anticline. The main trunk of the Burnsville Turnpike is formed in beds dipping about 30° to the northwest. The Pancake Caves and the main passage in Barberry Cave continue this trend line to the southwest. To the northeast, the Turnpike reaches a blockage, but the trend continues in the tangle of large passages that trend from Megalo Junction through the 6th of July Room to the Big Bend. The passage trend crosses the structure because at the 6th of July Room, the beds are almost horizontal (Fig. 20.4). To the southeast is a nearly linear trend that includes the Burns streamway and parts of the Outer Limits in the Blarney Stone Section. This trend line is roughly along the axis of the White Oak Syncline. The bedding in the Burns streamway is also nearly horizontal (Fig. 8.9).

The middle part of the cave contains the North-South Trunk and many other passages. The North-South Trunk cuts across the structure. From the south end near the connection with the Blarney Stone Section, the passage trends almost due north but with a zig-zag pattern of alternating dip and strike oriented segments. Typical strike-oriented segments are well down on the eastern limb of the anticline with a characteristic triangular cross-section with the sloping ceiling formed by a bedding plane. Segments of passage that cut across the structure have a more rounded, tubular morphology (Figs. 20.5, 20.6, 20.7 and 20.8).

The relationships of the trunk passage can also be displayed as profiles (Fig. 20.9). The Bobcat entrance datum is 2460 feet above sea level. Cave surveys set the entrance datum as zero. The Burnsville Turnpike has a somewhat undulating pattern in the survey which may represent nothing more than the irregularity of the floor due to the massive breakdown in this passage but is otherwise nearly horizontal at an elevation 447 feet below the entrance or 2013 feet above sea level. The North-South Trunk has a more irregular profile but on the average, it too is nearly horizontal from the southern end to SVG Hall lying (on the average) 381 feet below the entrance at an elevation of 2079 feet. These elevations show that the main trunks are at roughly the same elevation or only slightly lower than the valley floors of Sinking Creek and White Oak Draft. Because the passage morphology clearly shows that these portions of the cave formed under sub-water table conditions, the passages must pre-date the present topography by a substantial amount.

20.3.2 The Younger System and More Recent Events

The Chestnut Ridge Cave System is very much a superposition of multiple stages of cave development. The upper level trunk passages in the Bobcat Section in particular, are dry, decorated with aragonite and

Fig. 20.4 Sixth of July Room. Note horizontal bedding. Ron Simmons photo

Fig. 20.5 Passage cross-section with gentle dip at Nevin's Pendant in North-South Trunk. Ron Simmons photo

Fig. 20.6 Cross-structure segment of North South Trunk near Nevin's Pendant. Dave Morrow photo

gypsum crusts. The profile through the system (Fig. 20.9) shows an irregular pattern in the vertical plane but with an average major zone of cave development at about 400 feet below the entrance which would correspond to 2060 feet above sea level. Below and entangled with the level of the old master trunks are passages formed by later evolution of the system. Some of these are vadose canyons cut by free-surface streams after deepening of surface valleys lowered the water table below the level of main passage development. Vertical shafts such as Shamrock Dome (Fig. 6.12) and Damart Drop (Fig. 6.16) are clearly vadose features, dissolved by vertically moving water. As is common in many other cave systems, shafts are a later stage feature with no direct relationship to the primary horizontal passages (Brucker et al. 1972).

The lowest levels contain segments of active streams which usually rise from sumps and end in other

Fig. 20.7 South in North-South Trunk near Camp Room, looking down dip. Ron Simmons photo

Fig. 20.8 Segment of North passage near thundering raccoon trail formed along a steeply-dipping fold. Doug Molyneaux surveying. Photo by Dave Morrow

sumps. One of the longest of these is the Black Canyon in the Bobcat Section which carries an active stream from the end of the Burnsville Turnpike to a sump at the northeast end at 622 feet below the entrance. This streamway has a much steeper gradient than the trunk passages. There are also steep passages which have the rounded forms expected from sub-water table development. Some of these are likely lift tubes but without detailed examination, it is difficult to be sure.

The Bobcat and Burns entrance series, as well as the up-trending passage called Stairway to Heaven, are steep, narrow vadose passages with occasional vertical segments. The size and extreme gradients have been used as evidence to suggest that these infeeder passages are very young and barely developed sufficiently for human exploration. They are certainly juvenile in the sense of passage development but that, in itself, in not conclusive evidence that these passages are young. Although the entrance series passages carry small streams which act as infiltration routes for precipitation on Chestnut Ridge, the catchment areas are very small. An alternative hypothesis is that the infeeder passages are drains from the time when Chestnut Ridge was capped with Oriskany Sandstone. Acidic water draining from the sandstone creates these pathways at the sandstone/limestone contact. With the sandstone long since removed by erosion, the infeeder passages continue to function as drains for their small local catchments.

20.3.3 Stratigraphic Relations

The position of the various components of the Chestnut Ridge System within the stratigraphic column are poorly known. With several depths of more than 700 feet from entrance to lowest point, overall the system must span much of the thickness of the carbonate section. Photographs taken in the entrance series of both the Bobcat and Burns entrances do not show the prominent chert units characteristic of the Licking Creek Limestone. This unit seems to have been removed from the crest of Chestnut Ridge along with the Oriskany Sandstone caprock. More and more of the limestone is removed to the south. At the southern end of the sequence of caves, the stream on the floor of the main passage in Barberry Cave appears to be flowing on the Williamsport Sandstone which marks the base of the carbonate sequence. The complex fold structure takes the Williamsport Sandstone to at least the 2440-foot contour above Buckwheat and Blind Faith Caves. Because the structure plunges to the northeast, the main trunk passages of the system must be formed well up into the Tonoloway or Keyser limestones. The interbedded sandstones within the Tonoloway and also the Clifton Forge Sandstone are likely to have been of importance in controlling passage development but no information is available.

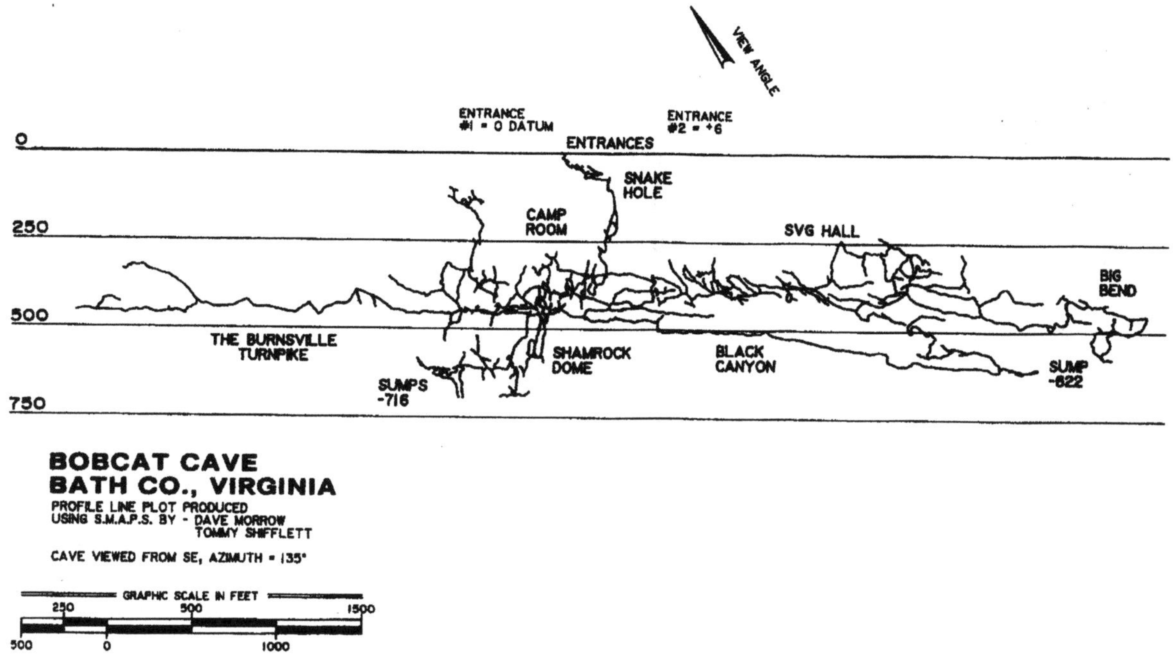

Fig. 20.9 Profile of Bobcat Cave. The elevation lines are given in feet below the entrance. This profile predates the discovery and connection of the Blarney Stone Section

20.4 Features of the Chestnut Ridge Caves

20.4.1 Breakdown

More evidence that the Chestnut Ridge System is drifting into old age is the widely distributed and extensive breakdown. Large passages such as the Camp Room, the 6th of July Room, the Burnsville Turnpike and many others are littered with large breakdown blocks. Large trunk passages are terminated by massive breakdowns. The shapes of these passages suggest that they were formed below the local water table. It is possible that they simply enlarged to a size where they became mechanically unstable. When the passages were drained by lowered base levels, the removal of buoyant support caused them to collapse. It is also possible that these passages formed when the land surface was much higher than at present and when much of the ridge top was protected by the Oriskany Sandstone caprock. When surface erosion removed the caprock, infiltration of surface water enlarged joint and bedding plane surfaces, destabilizing ceilings and initiating collapse. The mechanisms are not mutually exclusive.

20.4.2 Clastic Sediments

Little has been written about clastic sediments in the Chestnut Ridge Cave System. Much of the material appears to be sandy and silty material derived from the Oriskany Sandstone caprock and as residual insoluble material remaining from the dissolution of the limestone. No measurements have been made of particle size distribution, mineralogy, or possible paleomagnetic records. Passages such as the Boulder Dash are mud-coated from periodic flooding.

Fig. 20.10 Stalactites with aragonite crystals growing from the side. Paul Winter photo

Fig. 20.11 Anthodites near Camp Room. Paul Winter photo

20.4.3 Speleothems

The Chestnut Ridge Caves contain a rich, diverse, and somewhat surprising display of speleothems. There are the usual stalactites, stalagmites, and flowstone located mostly in the southern sections of the system where the caprock has been longest removed and where thick limestone soils have developed on the land surface above. The most extensive dripstone occurs in Barberry Cave (Figs. 10.15, 10.16, 10.17, and 10.18) but good displays of these speleothems are also found in Battered Bar, Fuhl's Paradise, and other Pancake Caves. In the main Chestnut Ridge System, dripstone deposits occur mostly in the Blarney Stone Section.

Gypsum occurs in the Sahara Pipeline and other high and dry passages in the northern part of the system. It takes the form of cave cotton, thin gypsum crusts and occasional gypsum flowers. Gypsum crusts likely occur farther south along the North-South Trunk but no detailed mineralogical analyses have been made.

The most extensive, but least expected, mineralization in the system are the aragonite crystals which are found all along the North-South Trunk and south in the Blarney Stone Section. Examples, discovered only in 2013 in passages near the Camp Room, are shown in Figs. 20.10, 20.11 and 20.12.

Aragonite occurs in other Appalachian caves and in other Helderberg Limestone caves but seldom in such profusion as the deposits in the Chestnut Ridge System. Aragonite spelethems are also found in the Butler Cave-Sinking Creek System but the deposits

Fig. 20.12 Aragonite crystals on breakdown block. Nathan Farrar photo

Table 20.1 Elevation data for the Chestnut Ridge caves

	Depth	Elevation	Above benchmark
Springs			
Blue spring		1704	109
Cathedral spring		1713	118
Cathedral dive	150	1563	−32
Aqua Cave entrance		1770	175
Siphon Lake		1795	200
French Lake		1799	204
French Lake dive		1543	−52
Chestnut ridge caves			
Bobcat entrance		2460	865
North-South Trunk	380	2080	485
Burnsville turnpike	447	2013	418
Shamrock dome sump	722	1738	143
Black canyon sump	622	1838	243
Blarney Stone entrance		2421	826
Burns entrance		2520	925
Burns Streamway sump	755	1765	170
Burns deep sump	782	1738	143
Burns upper sump	738	1782	187
Barberry Cave entrance		2441	846
Barberry main trunk	261	2180	585
Barberry sump	328	2113	518
Battered bar entrance		2373	
Upper Trunk	230	2143	548
Bodaceous boulevard	320	2053	458
Sump	410	1963	368

are much less extensive than those found in Chestnut Ridge. Speleothems are discussed more extensively in Chap. 23.

20.5 The Chestnut Ridge Caves in Regional Context

There are two active drainage systems in the Chestnut Ridge Cave System, one draining along the west flank of the anticline to Aqua Spring and the other draining along the eastern flank of the anticline to Cathedral Spring. One would have expected two independent cave systems, each with its own drainage basin. But higher in the Chestnut Ridge System is the North-South Trunk which cuts across the Chestnut Ridge Anticline. Later and deeper cave development below the North-South Trunk allow human passage between the two drainage basins underground.

Of interest in attempting to disentangle the relations between the present-day cave system and other caves and surface features are the vertical relationships between the features. As with the caves in the Sinking Creek drainage, the chosen benchmark was the wide floodplain at the confluence of the Bullpasture and Cowpasture Rivers at an elevation of 1595 feet above sea level. Surveyed depths below cave entrances were converted to mean sea level elevations and these in turn to elevations with respect to the 1595 foot benchmark (Table 20.1).

Reference

Brucker, R.W., J.W. Hess, and W.B. White. 1972. Role of vertical shafts in movement of ground water in carbonate aquifers. *Ground Water* 10(6): 5–13.

Geology of the Caves in the Northern Cove 21

William B. White

Abstract

The structure plunges to the northeast, carrying the limestone below the Oriskany Sandstone and the Millboro Shale. There are three large caves in this region, Water Sinks Cave that marks the downstream termination of the Burnsville Cove Drainage, Helictite Cave just north of the contact, and Wishing Well Cave, completely beneath the caprock. Much of the drainage from the Cove converges to the Emerald Pool in the Water Sinks Subway and from there follows a major lineament eastward to Aqua Cave and discharges at Aqua Spring. Helictite Cave and Wishing Well Cave are an abandoned part of the Emory Spring drainage but developed deeper below the non-karstic sandstones and shales than might have been expected. Present day recharge along the flanks of Jack Mountain must flow at considerable depth and almost right angles to the structure to reach Emory Spring.

21.1 Introduction

At the northern end of Burnsville Cove, where it opens into the valley of the Bullpasture River, the carbonate rocks plunge beneath the Devonian sandstones and shales. In the early years of exploration, the water sinks depression, at the edge of the sandstone, marked the downstream end of the Sinking Creek Valley and the northern limit of cave development. The water sinks depression contained two small caves, known then as Siphon #1 and Siphon #2, and there was Aqua Cave, accessible through the Aqua Spring. Then Phil and Charlotte Lucas bought the Water Sinks property and spearheaded intensive exploration. First came the discovery of Helictite Cave, a complex maze cave apparently unrelated to the other drainage systems in the Cove. This was followed by the discovery of Water Sinks Subway and the caves of the water sinks depression suddenly became more significant. Finally, and most recently, there was the excavation into the Wishing Well. The discovery and exploration of these caves are described extensively in Chaps. 11, 12, and 13 and also in Phil Lucas' book (Lucas 2012). A description of Aqua Cave, known since the early exploration, is found in Chap. 4.

The northern caves add a whole new layer of complexity to the interpretation of cave development in Burnsville Cove. The objective of the present chapter is to describe the geologic setting of these caves and fit them into the overall pattern of cave development. Much of the geologic detail has already appeared in the description chapters. The present chapter draws heavily on the cave descriptions and on

Compiled by the Editor.

W.B. White (✉)
Department of Geosciences, The Pennsylvania State University, University Park, PA, USA
e-mail: wbw2@psu.edu

W.B. White (ed.), *The Caves of Burnsville Cove, Virginia*,
Cave and Karst Systems of the World, DOI 10.1007/978-3-319-14391-0_21,

the revised Cove geology as given in Chap. 16. Mostly, it is an attempt to reconfigure the available information to highlight the geological story.

21.2 Geology of the Northern Cove

The northeastward plunging Sinking Creek Syncline carries the carbonate rocks deeper and exposes younger rocks at the land surface. The hilly upland north of the water sinks depression is underlain by the Oriskany Sandstone and the incised valley that descends into the main valley of the Bullpasture exposes the Needmore and Millboro Shales (Fig. 21.1). The entrance locations of the three caves are shown by the red dots; the location of Emory (along the river to the north) and Aqua (on Mill Run to the south) springs by green dots.

Important structural features are not well defined near the Bullpasture Gorge although the best available interpretation is shown in Fig. 21.1. The Sinking Creek Syncline continues to the northeast paralleling Jack Mountain. The plunge of the syncline takes the limestone beneath the sandstone and shale just north of the Highland County line. To the west, the only surface exposure of limestone is in a narrow band along the foot of Jack Mountain. Limestone is exposed along the Chestnut Ridge Anticline as far north as Water Sinks where the limestone is carried below the sandstone by the plunging structure. The Chestnut Ridge Anticline is shown with a jog near Water Sinks and then it appears to be lost near Wishing Well Cave. An extension of the trend of the Chestnut Ridge Anticline into the shale ridges to the northeast is shown as a syncline in Fig. 21.1. The White Oak Syncline veers more to the northeast leaving a wide region of poorly defined structure along the Bullpasture Gorge.

There is clearly a structural feature of different character and different orientation also present. A southeast-northwest line can be drawn across the northeastern end of Tower Hill Mountain, through a segment of the Bullpasture Gorge, up the valley of Mill Run past Aqua Spring and on to the Water Sinks. The nature of this structure is unclear except that it is almost at right angles to the anticlinal and synclinal fold axes of the basic Appalachian pattern. It may be a master lineament of the sort known elsewhere in the Appalachians,

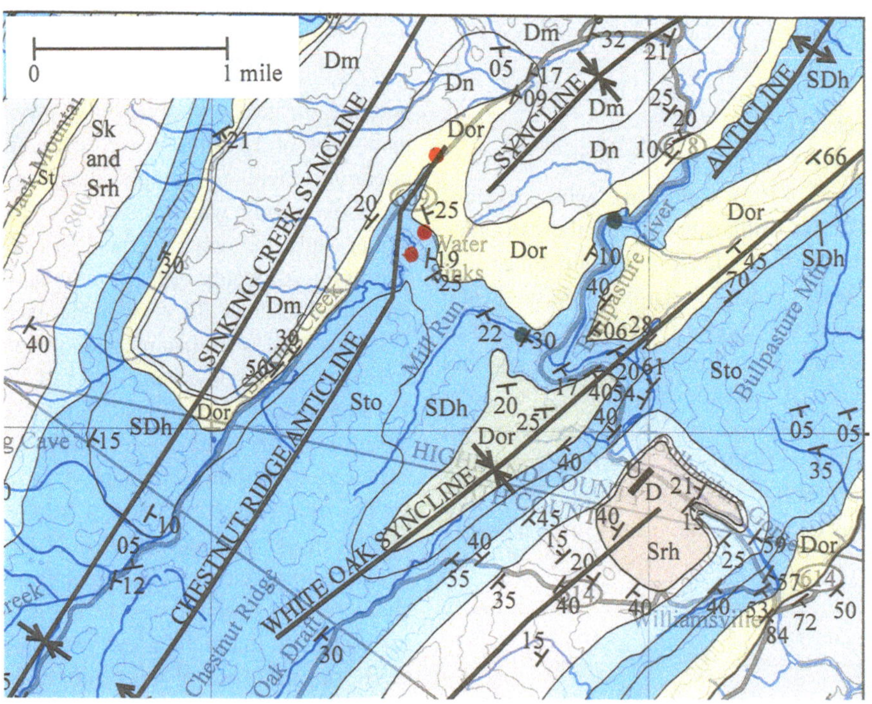

Fig. 21.1 Section of the geologic map of Burnsville Cove taken from Fig. 16.3 showing the location of the Water Sinks Caves, Helictite Cave, and Wishing Well Cave (*red dots*) and Emory and Aqua Springs (*green dots*)

for example the one through the Greenbrier Karst of Greenbrier and Monroe Counties West Virginia (Lessing 1979) and the one that guided the development of Simmons-Mingo Cave in Randolph County, West Virginia (Medville and Storage 1986). This structure is intimately connected to the drainage pathways that connect the Sinking Creek and Chestnut Ridge active drainage to Aqua Spring. Aqua Cave is developed along this feature with its main passage trends perpendicular to the regional strike. This feature will be referred to as the Water Sinks Lineament.

Many of John Haynes' stratigraphic studies were concentrated in and around the water sinks depression and as a result we have much better information on the stratigraphic setting of the northern caves than on many others in the Cove. Because of the plunge of the regional structure to the northeast, most of the caves are in the upper part of the section. The exposed section at the water sinks depression is shown in Fig. 21.2.

21.3 Geologic Setting of the Caves

21.3.1 The Water Sinks Caves

According to caver's conventions, Owl Cave is separate from upper Water Sinks Cave because the two have no known underground connection. Upper Water Sinks Cave and the Water Sinks Subway are part of one cave because one can traverse from one of the upper Water Sinks Cave's entrances to the entrance pipe of the Subway without getting outside the dripline. According to the hydrology and geomorphology of the system, it's the other way around.

Upper Water Sinks Cave and Owl Cave are fragments of network maze formed mostly in the Jersey Shore Member of the Keyser Limestone. The definitive stromatoporoid reef bed occurs immediately above the entrances to the two caves. See Figs. 16.19 and 16.20. Further extension of the caves is limited on three sides

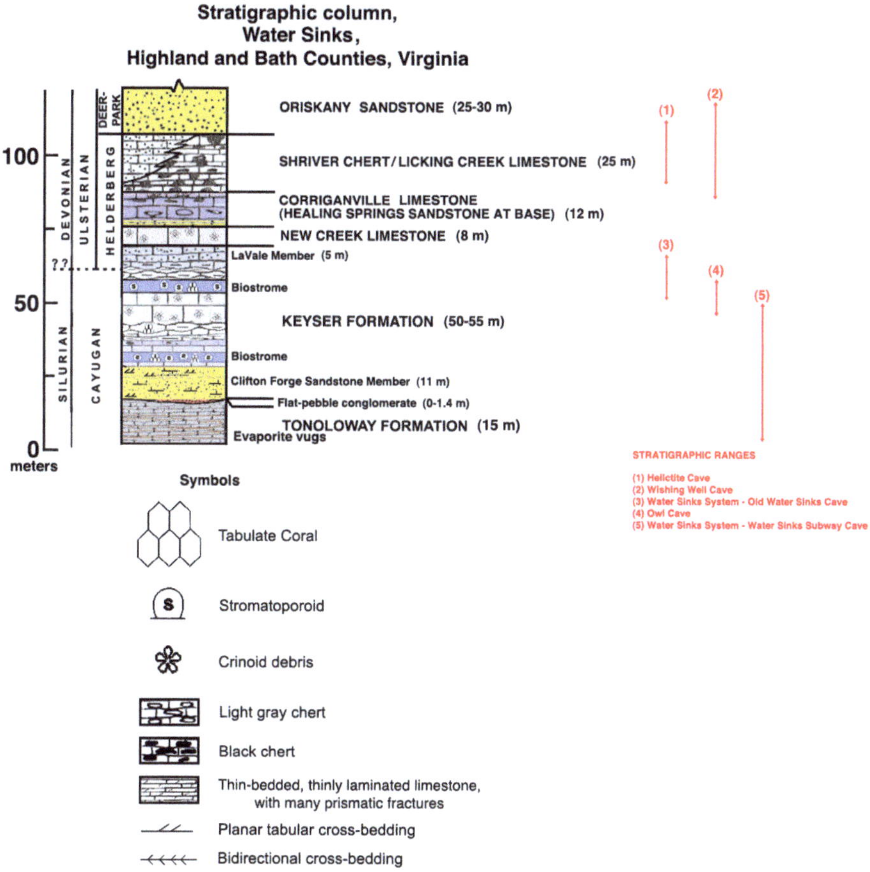

Fig. 21.2 Stratigraphic section for the water sinks depression as measured by John Haynes and his students

by the edges of Cave Hill, an erosional remnant formed by the dissection of Chestnut Ridge. Whether there is a further extension of Owl Cave into Chestnut Ridge at the talus cone dig remains to be seen.

The Water Sinks Subway, presently located below the level of the clastic sediment that forms the bottom of Water Sinks, is also a maze cave but has a somewhat different morphology than the upper caves. The main subway passage begins at the collapse at the edge of Water Sinks and continues southwest as a very large cross-section trunk (see Figs. 11.12, 11.13 and 11.14) which splits into multiple smaller passages. The other passages that make up the maze are even smaller. The subway section has all of the characteristics of a floodwater maze as described by Palmer (1975). As Phil Lucas points out in his description of the cave, prior to the sediment infilling of the water sinks depression, the main subway passage may have been an open insurgence cave, taking the flow of Water Sinks Creek. When the creek was in flood, the entire system would back up, creating a high hydrostatic head that would force water into every available joint and bedding plane parting. The result is a master passage with many auxiliary maze passages. The Brush Creek Caves in the Uinta Mountains of Utah show a similar development (White 1979).

The subway section breaks through the Clifton Forge Sandstone so that many of the passages are in the thin-bedded upper Tonoloway Limestone (Fig. 11.19). Dips are very low throughout the Water Sinks Caves so that the terminal sump, 155 feet below the entrance should be in the middle Tonoloway beds.

That the Water Sinks Caves are formed in tiers is apparent from the profile shown in Fig. 21.3. The upper maze caves–Upper Water Sinks Cave and Owl Cave–form what appears to be a double tier with passages 50–60 feet above the zero datum and at 10–20 feet above the zero datum. The subway section has two tiers, one sharply defined at 20 feet below datum and the other 44 feet below datum. The upper tier has a pronounced dip to the northeast, presumably following the structural plunge. A third possible tier is represented by Roaring River which rises from Emerald Pool at 140 feet below datum and disappears at the terminal sump at 155 feet below datum.

21.3.2 Aqua Cave

Aqua Spring at the head of Mill Run is about 50 feet above the elevation of the Bullpasture River at this location. The stream passage that feeds Aqua Spring, described in some detail in Chap. 4, can be followed upstream from the spring to the Siphon Lake. The average passage trend is N 50°W, close to the trend of the Water Sinks lineament. The stream passage is developed in the Keyser Limestone. The Clifton Forge Sandstone has been identified in the upstream stream passage by John Haynes and his students, suggesting that the cave is perched on the Clifton Forge Sandstone and this may account for the location of the spring, perched above the river. The deep submerged passages below the sumps must cross the Clifton Forge Sandstone so that the lowest points reached by

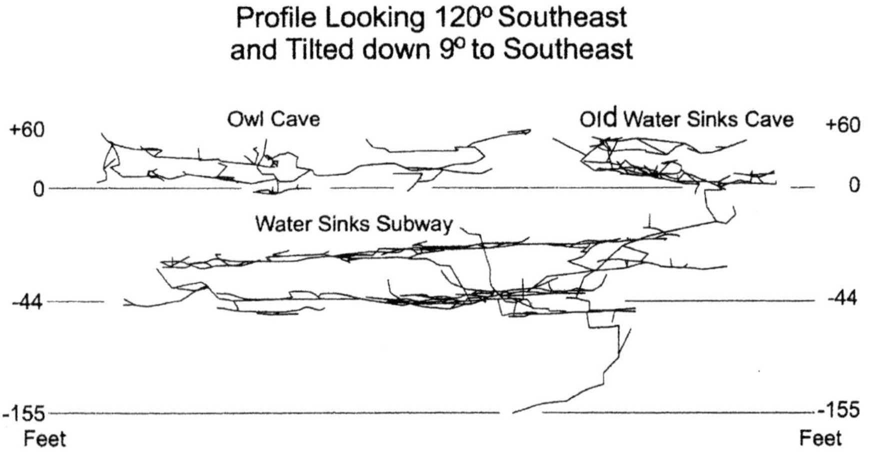

Fig. 21.3 Profiles of the Water Sinks Caves. Extracted from the Water Sinks map by Philip C. Lucas

divers in Siphon Lake and French Lake must be well down into the Tonoloway Limestone.

21.3.3 Helictite Cave

Helictite Cave has three components (Fig. 21.4). The Entrance Series is a tangle of tightly-spaced passages on multiple levels that descend to the Streamway, a well developed but short length of elliptical conduit (Fig. 12.16). The Streamway trends almost due north-south. The third, and most extensive component, is the Canyon Maze. The cave is developed in the cherty Licking Creek Limestone and the chert beds have a strong controlling influence on passage morphology. Except for the area immediately around the entrance, the cave lies beneath the Oriskany Sandstone that forms the caprock on Helictite Hill.

The orientations of the joints that provided the primary ground water pathways for the development of Helictite Cave were determined by measuring passage trends rather than by constructing a detailed passage rosette. The main passage trends are N 37°E and N 46°W (or S 134°E). The northeast joint set is more or less along the orientation of the Appalachian folding and may be called strike joints. The northwest joint set is roughly perpendicular to the regional folding and may be called dip joints. The joint set that controlled the Canyon Maze may be compared with the joint set that guided the development of the Butler Cave-Sinking Creek System (Fig. 19.8). The dip joints in Breathing Cave, and upstream and downstream Butler Cave have a mean orientation of 129°, close to the 134° of Helictite Cave. The strike joints in upstream Butler Cave are oriented at 63° but with a much wider distribution of values than the dip joints. The downstream passages in Butler have strike orientations of 52°. At Helictite Cave, the strike orientation has shifted to 36°. This distinct shift in the strike-joint orientation may represent a different structural pattern northeast of the water sinks lineament or may represent the different orientations of the Sinking Creek Syncline and the Chestnut Ridge Anticline.

The horizontal extension of Helictite Cave is to some extent limited by the limits of Helictite Hill. However, it is curious that no passage has been discovered east of the Streamway. The southwestern side of the maze is cut off abruptly. A line drawn along the ends of these southwest passage terminations has an orientation of N 42°W, roughly parallel to the orientation of the dip joints. This cutoff may represent the influence of the Water Sinks Lineament, a hypothesis supported by the extensive faulting observed in the Entrance Series part of the cave.

21.3.4 Wishing Well Cave

The discovery of Wishing Well Cave is a triumph of caver's intuition and persistence over logical geological deduction. It had been understood for a long time that the water sinks depression marked the northern limit of cave development. Underground drainage flowing northeast down Burnsville Cove reached the Water Sinks Lineament, made a right turn, and flowed to Aqua Spring. Searching for new caves farther to the northeast, well out onto the Oriskany Sandstone caprock, would seem a complete waste of time. But there it was, a persistent airflow from a sandstone rubble-filled sinkhole. At the end of an incredible exercise in excavation, documented in Chap. 13, the master cave was discovered, completely beneath the sandstone caprock.

The cave appears to be formed mainly in the Licking Creek Limestone. Chert beds are prominent in the passage walls. The excavated entrance shaft is entirely in a rubble zone of Oriskany Sandstone as is the excavated crawlway connecting the base of the shaft to the Doodlebug Room. The contact between the Oriskany and the Licking Creek Limestone is in the crawlway leading from the base of the Doodlebug Hole to the top of Companion Canyon. The contact is roughly 100 feet below datum. There are ledges of Oriskany Sandstone along Route 609 showing that some sandstone was eroded to form the shallow valley where the entrance is located. Nonetheless, the combined pit excavation and the descent through the Doodlebug Hole must span much of the thickness of the Oriskany Sandstone. It seems likely that a large cave chamber existed in the limestone below the entrance. Collapse and upward stoping through the sandstone produced the rubble zone that allowed escape of the air that called attention to the site and also permitted excavation without mining through solid rock.

Wishing Well Cave is mainly a set of conduit fragments—fragments of a drainage system entirely

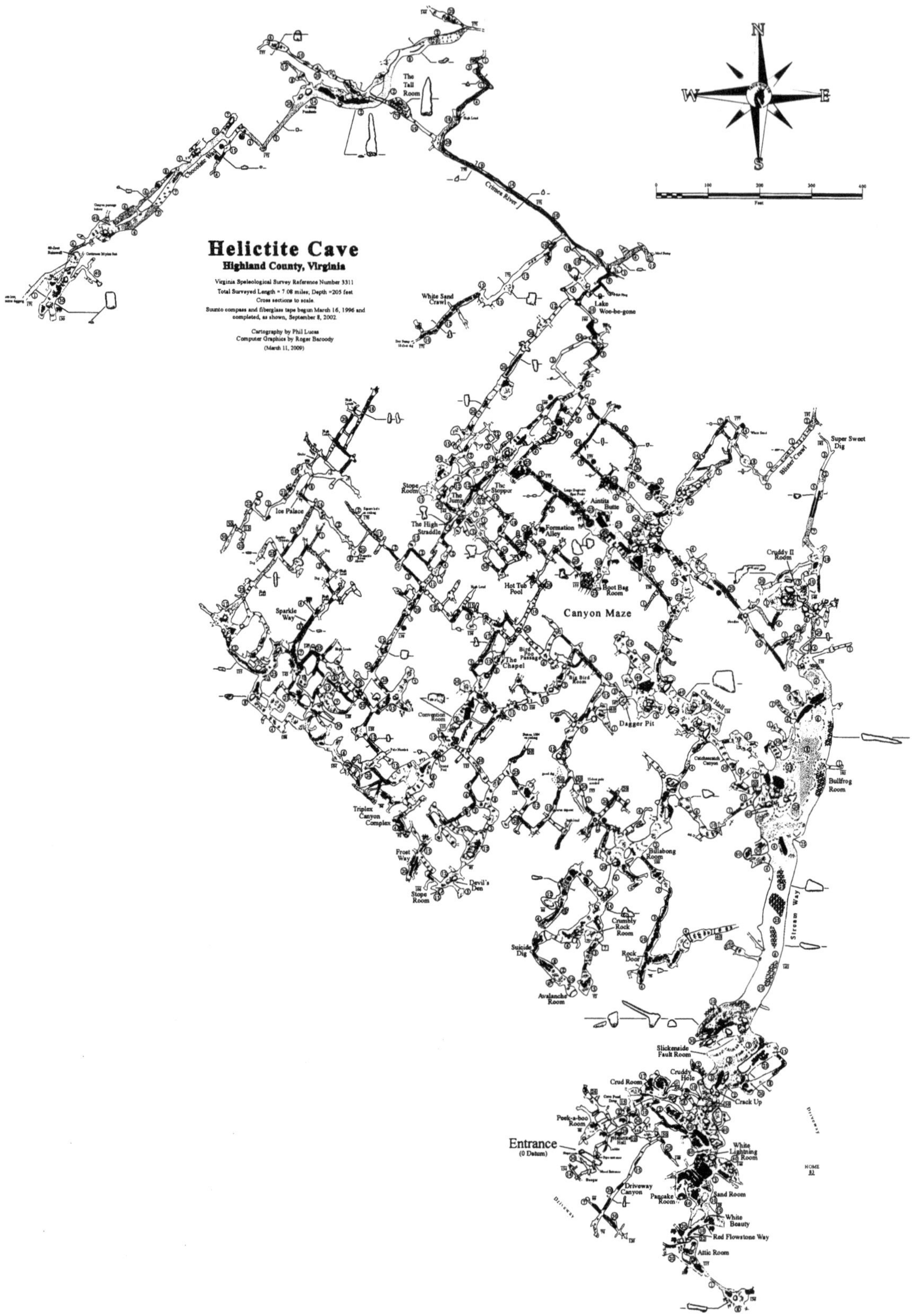

Fig. 21.4 The Lucas map for Helictite Cave in outline. An expandable copy of this map is in the electronic file

separate from the systems farther south in Burnsville Cove. It has several oddities. For one, the general orientation of the cave is north-south which cuts across the structural grain much like the North-South Trunk in the Chestnut Ridge System. A second oddity is the profile of Sugar Run, the master trunk passage of the system. The cave system, reduced to passage outline, is shown with some key elevations in Fig. 21.5. According to the interpretation of the explorers (Chap. 13) Sugar Run originates from the complex of passages at Echo Junction at the extreme southern end of the system at an elevation of about 180 feet below the entrance. At the Declivity the passage rises 40 feet as a lift tube to an elevation of about 135 feet at which it remains to the Angel Wing Room and the intersection with the Northwest Passage. Sugar Run then drops back to about 180 feet and continues downward to what was interpreted as the downstream end in the sump at 228 feet below the entrance. The Northwest Passage can be regarded as a south-flowing tributary, nearly horizontal, at an elevation of about 170 feet except at the downstream junction where the Hourglass lift tube segment is necessary to bring the passage up to the level of Sugar Run. These undulations in the vertical plane provide strong evidence that the cave system developed deep below local base levels as they existed at the time of passage development.

The main conduit is oriented more or less northeast-southwest following the crest of what should be the extension of the Chestnut Ridge Anticline. The Northwest Passage cuts almost at right angles to the regional structure and, because the passage is very nearly horizontal must also cut across the bedding. The presence of stream cobbles in Sugar Run demonstrates that the passage functioned as a master drain carrying water at velocities high enough to move the cobbles. It is less obvious where the water was coming from and where it was going. The geologic map (Fig. 21.1) illustrates the problem. The source for the Northwest Passage has been suspected to be the band of limestone exposed along the base of Jack Mountain. Recharge from Jack Mountain would have to pass down and under the Sinking Creek Syncline in order to reach the Northwest Passage. If the Sugar Run sump does indeed represent the downstream end of the system, the passage is trending to the northeast, following the anticline but there is no outlet in that direction. The plunging structure is taking the limestone deeper beneath the sandstone and shale. To reach the obvious outlet, Emory Spring, the drainage must turn eastward and pass under the unnamed syncline shown on the geologic map. Discovering more segments of the cave would definitely be helpful.

21.4 Karst Drainage in the Northern Cove

For the most part, the discoveries of major new caves in the northern Cove created more problems than they solved. To quote former Defense Secretary Donald Rumsfeld, there are known knowns, there are known unknowns, and—most importantly—there are unknown unknowns. The known knowns make up the substance of this book in the form of maps, photographs, and written text. With regard to the northern Cove, the known unknowns are discussed in the following sections. But it is more than likely that there are important unknown unknowns that will force major revisions of everything that has been said.

Elevations for important components of the northern Cove caves were calculated with respect to the established base level at 1595 feet, the confluence of the Bullpasture and Cowpasture Rivers (Table 21.1).

21.4.1 The Mysterious Feeder System for Aqua Spring

The underground drainage in Burnsville Cove is mostly southwest to northeast following the trend of the geologic structure. Because of the fold axes and more importantly because of the interbedded sandstones in the Tonoloway Limestone, the drainage takes the form of more or less parallel drainage lines, the various streams in the Butler-Sinking Creek System being good examples. What is observed are segments of streams, rising from sumps and ending in sumps, so that it is difficult to string together any complete drainage line.

The larger sources draining to Aqua Spring include the Butler-Sinking Creek streams, last seen at their sumps, the steam in Better Forgotten Cave, the Black Canyon stream in the Chestnut Ridge System, last seen at the 622 Sump, the surface overflow route of Sinking Creek at the Sinking Creek Swallet, and Water Sinks Creek which sinks at several points near Twisting Sister and Castle Rock with its final swallet in the water sinks

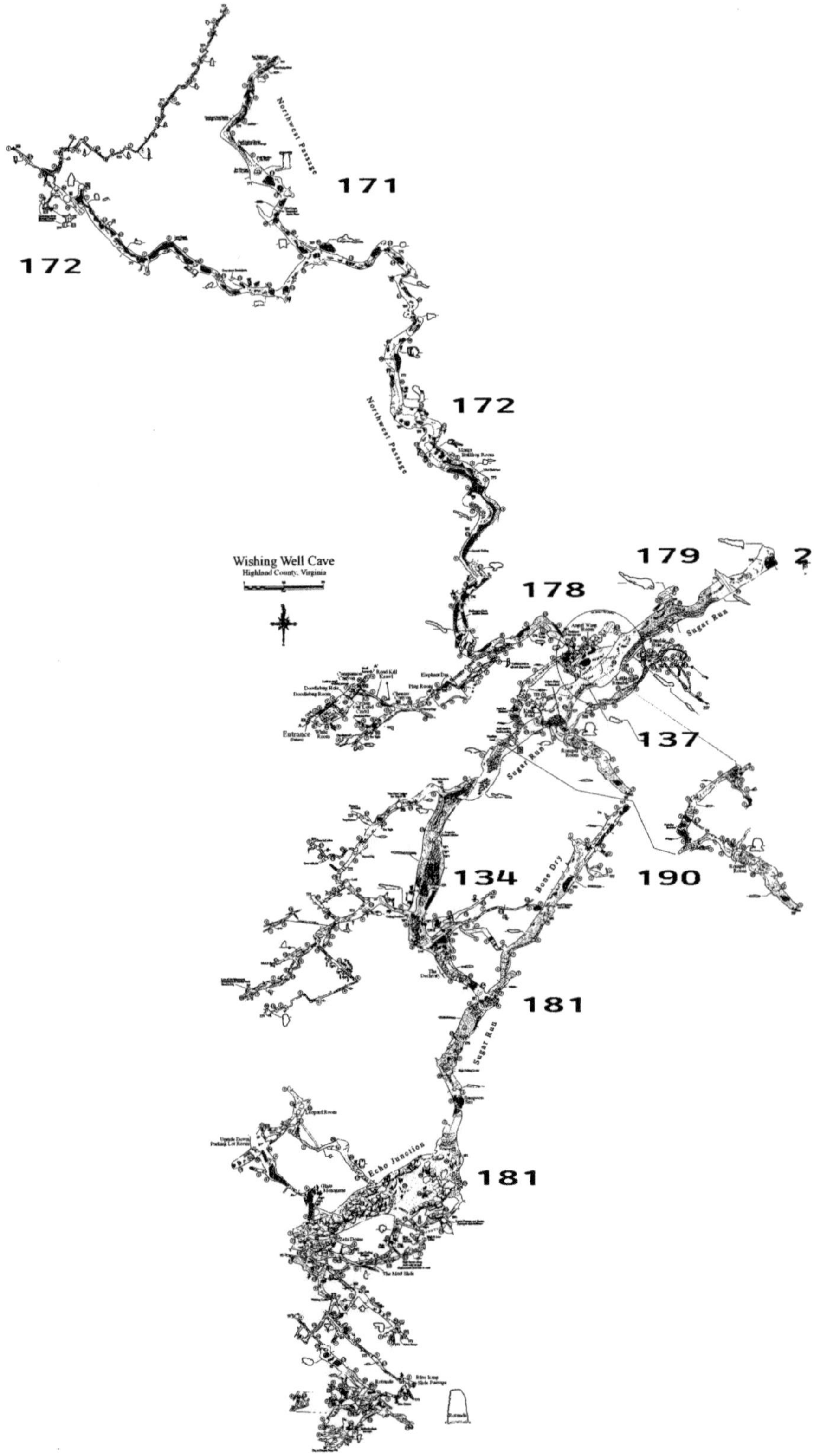

Fig. 21.5 Outline of the Lucas map of Wishing Well Cave. An expandable map is given in the electronic file. Passage elevations are given as depth below the zero datum at the cave entrance

Table 21.1 Elevation data for major features in the Northern Cove Caves

Location	Below datum	Above sea level	Above base level
Aqua Spring		1770	175
Siphon Lake	−25	1795	200
French Lake	−29	1799	204
French Lake Dive	256	1543	−52
Emory Spring		1754	159
Pancake Fields		2030	435
Water Sinks		1946	351
Subway passage	20	1926	331
Emerald Pool	155	1799	204
Owl Cave		1986	391
Helictite Cave		2068	473
Streamway	165	1903	308
Wishing Well		2005	410
Northwest Passage	170	1835	240
Angel Wing Junction	135	1870	275
Sugar Run	180	1825	230
Sugar Run Sump	228	1777	182

depression. Where and how these tributaries merge to form the aqua stream is not known. Water Sinks Creek, the Sinking Creek swallet, and the Butler streams appear at Emerald Pool but the upstream insurgences do not. In the 1600 feet separating the Roaring River Sump from the Siphon Lake and French Lake rise pools in Aqua Cave, the stream must pick up the other tributaries, descend to at least the depth reached in the Aqua Cave dives, rise stratigraphically from the middle Tonoloway Limestone to breach the Clifton Forge Sandstone at the base of the Keyser Limestone, and finally become the Aqua stream.

Down-loops in conduits usually form in response to structural constraints. What these structures might be along the water sinks lineament are unknown. It is possible that Roaring River is perched on the Upper Breathing Cave Sandstone and that the Emerald Pool rise is a break through the sandstone.

One way to probe the system would be to make careful flow rate measurements on the input streams during a dry period when water levels are low and stable. A water balance calculation would then show if the discharge at Aqua Spring does indeed account for all of the infeeders. The perching of Aqua Spring at the head of Mill Run is curious. If the water feeding the spring rises through lift tubes beneath the lakes, the lakes would provide a substantial hydrostatic head tending to force a new pathway closer to river level. Springs near river level or under the river might already exist, fed by leakage from the perched underground lakes in Aqua Cave.

21.4.2 The Equally Mysterious Drainage Basin of Emory Spring

According to the dye traces described in Chap. 17, Emory Spring has a large catchment area that extends several miles north up the Bullpasture Valley. The storm response of the spring (see Figs. 17.2 and 17.3) clearly indicates that the spring is fed by a conduit system in the Licking Creek Limestone. The few geochemical data available for the spring (Table 21.2) (Harmon and Hess 1982) support the hypothesis of a conduit system extending from the only available catchment along Jack Mountain to the spring on the Bullpasture River in spite of stratigraphic and structural barriers.

The concentration of dissolved limestone (hardness) is low compared with typical carbonate waters. The saturation index is strongly negative, showing that the water is far below equilibrium and has moved along the flow path from recharge area to the spring without coming into equilibrium with the limestone

Table 21.2 Geochemical data for Emory Spring

Date	Temperature	Hardness	SI-Calcite	P_{CO_2} enhancement
October 4, 1970	10.2	96	−0.85	12.2
October 24, 1970	11.0	94	−0.39	5.2
May 8, 1971	10.5	73	−0.50	3.2
May 2, 1972	11.2	81	−1.06	12.8

Temperature is in °C and hardness is in mg/L as $CaCO_3$. P_{CO_2} enhancement is the ratio of the CO_2 concentration calculated for the spring water to the atmospheric CO_2 concentration which, in 1972, was 326 parts per million walls of the conduit. The concentration of carbon dioxide is enhanced by at most a factor of ten above the atmospheric background. The CO_2 concentrations are supportive of the hypothesis that the recharge is mountain runoff from the sandstone slopes of Jack Mountain sinking at the limestone contact. The geochemical data imply that there is indeed a conduit system and moreover that it is an open and highly efficient conduit system. This is consistent with the observation of stream cobbles in the Sugar Run Passage of Wishing Well Cave.

For streams sinking along Jack Mountain, the flow path to Emory Spring is at right angles to the structure. It must cross under the Sinking Creek Syncline, over the Chestnut Ridge Anticline, and under the unnamed Syncline shown on the geologic map just west of Emory Spring. The dye traces (Chap. 17) show that this flow path is indeed possible. A dye trace in the Clover Creek Basin, the next drainage basin to the north, reached Clover Creek Spring by a path which required flow beneath the Bullpasture River.

It seems reasonable that the Streamway in Helictite Cave and the Northwest Passage and Sugar Run in Wishing Well Cave are fragments of paleodrainage trunks that were part of the Emory Spring drainage basin. The active drain—the long-sought Emory River—must be down there somewhere. Exploration from Wishing Well Cave, on the crest of the anticline, has the best possibility of finding an air-filled segment of what is otherwise likely to be completely submerged passages.

References

Harmon, R.S., and J.W. Hess. 1982. Ground water geochemistry of the Burnsville Cove area, Virginia. *National Speleological Society Bulletin* 44: 84–89.

Lessing, P. 1979. Karst subsidence and linear features, Greenbrier and Monroe Counties, West Virginia. *West Virginia Geological Survey Environmental Geology Bulletin* EGB-17, map sheet.

Lucas, P.C. 2012. *Caves and Karst of the Water Sinks Area.* Privately published, 355 pp.

Medville, D.M., and W.K. Storage. 1986. Structural and stratigraphic influences on the development of solution conduits in the upper Elk River Valley, West Virginia. *National Speleological Society Bulletin* 48: 8–25.

Palmer, A.N. 1975. The origin of maze caves. *National Speleological Society Bulletin* 37: 56–76.

White, W.B. 1979. Karst landforms in the Wasatch and Uinta Mountains, Utah. *National Speleological Society Bulletin* 41: 80–88.

Meteorology of Butler Cave

22

Fred L. Wefer and Philip C. Lucas

Abstract

Precise measurements of temperature and relative humidity were made along a traverse from the Butler Cave entrance to Sand Canyon and along the Butler trunk channel. Summer temperatures decrease with distance into the cave but stabilize at the cave ambient only 800–1000 feet inside. Winter temperatures rise more rapidly and reach cave ambient within 100 feet. Temperature rises by a small amount along the Trunk Channel with the highest temperature at the downstream end. The temperature rise with depth along the Trunk Channel is consistent with the geothermal gradient in the area. In a separate investigation, a fan was used to introduce an oscillating air current to one cave entrance which a sensitive anemometer and data-logger used to record air flow from a difference cave entrance. The recorded square-wave pattern confirmed connections between caves over considerable distances.

[**Editor's Note**: Until his untimely death, Fred Wefer maintained a series of observation stations in Butler Cave. He visited these stations periodically and made very careful meteorological measurements. The chapter that follows is constructed from the very detailed results that Fred published in The Proceedings of the Appalachian Karst Symposium (Radford, VA, 1991) and in the BCCS Newsletter.]

This chapter is more blatantly ghost-written than the others in this book. Fred wrote extensively and in detail about his meteorological measurements, totaling at least 128 pages. Rather than reproducing this mass of material, the editor has attempted to extract the key results and present them in a single coherent narrative.

In a curious twist, Fred chose temperature, water vapor pressure, and relative humidity as the meteorological properties to measure. These, as expected, turn out to be very nearly constant. What was not measured, and what is definitely not constant, is the wind. It was the wind that attracted interest to Breathing Cave many years ago. It is the wind, often an elusive breeze whispering from a pile of breakdown, that keeps explorers going. And, in a new study, the wind can be used as a much more direct exploration tool. That study is presented at the end of this chapter.

With an addendum by Philip C. Lucas, Frank Marks, Jr. and Nevin W. Davis. Fred L. Wefer is deceased.

P.C. Lucas (✉)
587 Limestone Lane, Williamsville, VA 24487, USA
e-mail: Lucas@VirginiaCaves.org

W.B. White (ed.), *The Caves of Burnsville Cove, Virginia*,
Cave and Karst Systems of the World, DOI 10.1007/978-3-319-14391-0_22,

22.1 Introduction

22.1.1 What Do We Mean by "Cave Weather?"

Cave weather really means much the same thing as "surface weather". Any point in the atmosphere, at any given moment, has a temperature, a barometric pressure, a rate of air movement (wind) and a certain chemical composition. These variables are continuously changing. It is this assemblage of parameters observed over time that constitute the "weather." There are also extreme excursions of the atmospheric variables, resulting in rainstorms, thunderstorms, tornados, ice storms, and blizzards. Chemically, the atmosphere is three quarters nitrogen which is not usually of much interest, and one quarter oxygen and a long list of minor components. The chemical variables of interest are the concentrations of water vapor, oxygen, and carbon dioxide. Of these, the air in Burnsville Cove caves is usually fresh, meaning that oxygen partial pressures are close to atmospheric normal. The CO_2 partial pressure is of importance in discussing the dissolution of limestone and the precipitation of speleothems but is otherwise not considered a "weather" variable. That leaves water vapor as the chemical variable to be measured. The concentration of water vapor can be expressed either as a partial pressure or as relative humidity.

The temperatures of cave atmospheres are often said to be constant, reflecting the mean annual temperature of the region. As will be demonstrated, this statement is not exactly true, although the temperature variations are not large. Relative humidity in caves tends to be high, often near saturation with respect to liquid water (100 % relative humidity). The barometric pressure in caves varies because the coupling between the cave atmosphere and the surface atmosphere results in cave winds. Cave weather is much less extreme than surface weather; there are no thunderstorms, blizzards, or tornadoes. However, in some locations and under some circumstances, cave temperatures fall below the freezing point of water and either seasonal or permanent ice may be formed. There has been much interest in caves containing ice—known as glacières or freezing caverns—such as Fossil Mountain Ice Cave in the Rockies and the Eisenriesenwelt and Rieseneishöhle in Austria. Permanent ice such as that in the Scărişoara Glacier Cave in Romania is now being investigated as a climate record. In temperate climate caves such as those of Burnsville Cove, ice is seasonal and limited to entrance areas.

There have been relatively few cave meteorological projects in the United States although a systematic literature search turned up 219 papers dealing with some aspect of the subject (Wefer 1991a). Data have been published for Lehman Caves, Nevada (Bamberg 1973) and Wind Cave, South Dakota (Nepstad and Pisarowicz 1989). Closest to Burnsville Cove are Cropley's (1965) investigations of Organ Cave and Ludington's Cave in Greenbrier County, West Virginia. The meteorology of caves was considered an important subject by early 20th Century European speleologists and early text books such as Kyrle in 1923 and Trombe in 1952 devote entire chapters to the subject. The background principles are discussed in some detail by Wigley and Brown (1976).

22.1.2 The Butler Cave Meteorology Project

The project was initiated in April, 1984 and by the time of the last publication in March, 1993 had accumulated 680 measurements. The study was concerned with temperature and relative humidity variations both at different points within the cave and also with variations with time. One sequence of measurements was made from the Nicholson entrance to Sand Canyon and another along the Trunk Channel from Penn State Lake to the 6th of July Room, a traverse of about 8000 feet. Additional measurements at Sand Canyon provided more time-series data.

Most of the results were published as a series of progress reports some of which also contain the raw data (Wefer 1984, 1985, 1988, 1989a, 1990, 1992). There was a report on the measurement technique (Wefer 1989b) and one formal paper concerning the measurements along the Trunk Channel (Wefer 1991b).

22.2 Measurement Techniques

Water as a vapor phase in air has a limited partial pressure above which liquid water condenses. This limiting pressure is known as the saturation pressure and is a function of temperature. The water vapor partial pressure over liquid water increases with temperature until it reaches the atmospheric pressure, at which temperature water boils. If the water vapor partial pressure in the atmosphere increases to the saturation pressure, liquid water condenses as rain (large droplets) or fog (small droplets). Alternatively, if the temperature and thus the saturation pressure decrease to where the saturation pressure equals the existing water vapor partial pressure, condensation will also occur. The temperature at which moisture begins to condense out of the atmosphere is known as the dew point. If liquid water is in contact with an atmosphere with a water vapor pressure less than the saturation value, the liquid water will evaporate at a rate that depends both on the temperature and on the difference between the water vapor partial pressure in the atmosphere and the saturation pressure. The evaporation of water requires heat so that evaporating water has a cooling effect. This combination of phenomena provides the basis for the practical measurement of water vapor in the atmosphere.

Atmospheric water vapor can be measured as either a partial pressure or as relative humidity. Relative humidity is defined as the ratio the water vapor partial pressure in the atmosphere to the saturation pressure expressed as percent. Partial pressure is the physically more meaningful parameter but relative humidity is the number that tells you if you're going to feel hot and sticky on a warm summer day. Water vapor partial pressures are given in inches of mercury (Hg) in the plots.

[Editor's Note: For reasons that seem mysterious for a person with a PhD in astronomy, Fred chose the totally obsolete unit of inches of mercury. The metric version of the same unit is millimeters of mercury, called the Torr (short for Torricelli who invented the mercury barometer in the early 1600s). The correct SI unit is the Pascal (Pa). One inch of mercury = 25.4 Torr = 3386 Pa.]

Relative humidity was measured with a psychrometer. The device consists of two thermometers side-by-side. One, the dry bulb thermometer, is simply exposed to the air. The other, the wet bulb thermometer, has a cloth sleeve over the thermometer bulb that is kept wet with water. Water will evaporate from the wet bulb causing a lowering of the wet bulb temperature. Psychrometric tables allow the calculation of relative humidity from the dry bulb temperature and the temperature difference. If the relative humidity is 100 %, there will be no evaporation and both thermometers will read the same temperature. The device used in these measurements had a small battery-driven fan that blew a steady stream of air over the wet bulb. The dry bulb thermometer, of course, gives the cave air temperature directly. Both thermometers were calibrated in the Fahrenheit (°F) scale and all temperatures are plotted on this scale.

Because relative humidity is high in caves and because temperature variations are expected to be small, great care was taken to obtain accurate measurements. The procedure was to unpack the psychrometer, place it at the pre-designated observation station, turn on the fan, and then stand away and wait for the instrument to come to equilibrium. The observer was to approach the instrument quickly, using a flashlight (not a carbide lamp), read the thermometers to the nearest 0.1 °F, and then retreat. Return and do the readings again until successive readings give the same values.

22.3 Cave Weather in the Butler Cave Entrance Series

Temperature and humidity profiles were measured between the entrance and Sand Canyon at a series of specific observation stations (Table 22.1).

Representative data sets for summer and for winter are compared in Fig. 22.1a, b. On the day of the summer data, the cave was drawing in warm outside air. The air temperature (the dry bulb temperature) fell over the distance along the cave passages, but a distance of 1000 feet was required before the temperature fell to cave ambient. In contrast the relative humidity rose rapidly because the temperature fell below the dew point only 100 feet inside the entrance. Relative humidity remained nearly constant along the remainder of the traverse. Although the temperature continued to fall, the excess moisture condensed out locking in the relative humidity.

During the winter measurements air was blowing out of the cave and the outside temperature was only about 10 degrees cooler than the cave. As a result, the air temperature rose to cave ambient at the bottom of the entrance pit. Relative humidity also rose rapidly

Table 22.1 Observation stations for the Butler Cave entrance series

Station	Location	Elevation	Distance
1	Outside of entrance	2535	0
2	Bottom of entrance pit	2501	40
3	Top of 6 foot drop below Glop Slot	2481	80
4	Top of God-Is-My-Copilot climb	2475	130
5	Top of Breakdown Mountain	2438	175
6	Top of clay bank beyond window	2380	335
7	Top of bank beyond Rabbit Hole	2357	400
8	Intersection near Bean Room Overlook	2313	690
9	T-intersection at crawl to Rimstone Pools	2291	1000
10	Rimstone Pools	2275	1200
11	Middle of 90-Ugh Crawl	2245	1545
12	Sand Canyon	2240	1815

Elevations are in feet above mean sea level; distances are in feet beyond the entrance

and reached near saturation values at the bottom of the entrance pit. Dew point for the outside air was low and was never a factor in the behavior of the cave atmosphere. The key parameter in the contrasting behavior between summer and winter conditions seems to be the direction of air flow into or out of the cave.

The complete set of meteorological traverses between the entrance and Sand Canyon are shown in a single plot (Fig. 22.2). The first point represents the outside air temperature. The curves for all traverses converge with distance within the cave but only at the Bean Room overlook, almost 700 feet into the cave, do all of the curves converge to essentially a common value. When the outside air temperature is near or below cave temperature, the curves converge immediately inside the cave.

An alternative display of the same information is by means of the Cropley plot (Cropley 1965) (Fig. 22.3). In this display, the air temperature appears as a curved surface with time and distance from the entrance as the x- and y-coordinates. The seasonal effect on damping of air temperature with distance into the cave shows up clearly as the double-humped surface for two years of data.

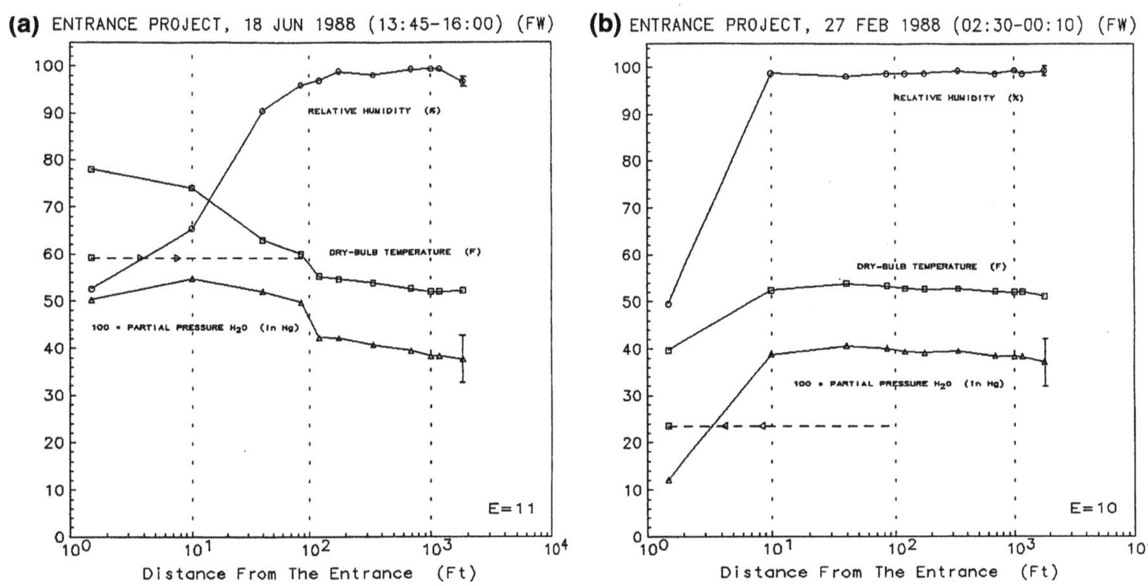

Fig. 22.1 Plots of relative humidity, air temperature, and water vapor partial pressure as a function of distance into Butler Cave from the entrance. Note that distance is plotted on a log scale. **a** Summer conditions—June 18, 1988. **b** Winter conditions—February 27, 1988. The *horizontal dashed line* indicates the dew point for outside air. The *arrows* on the dew point line show the direction of air movement at the cave entrance

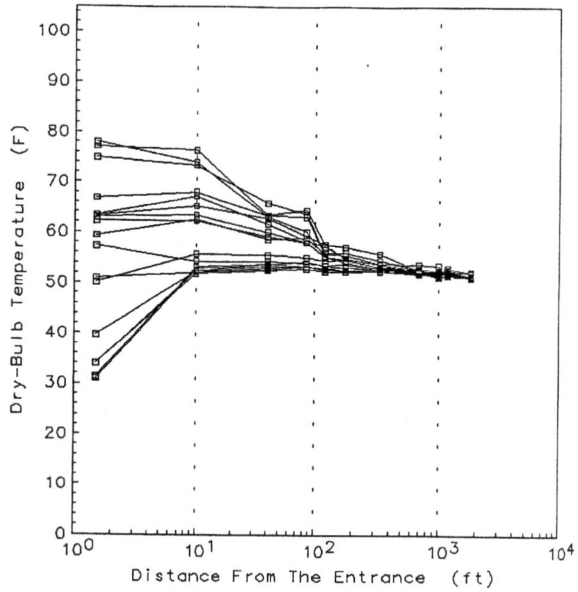

Fig. 22.2 Temperature plot for 12 traverses between the Butler Cave entrance and Sand Canyon

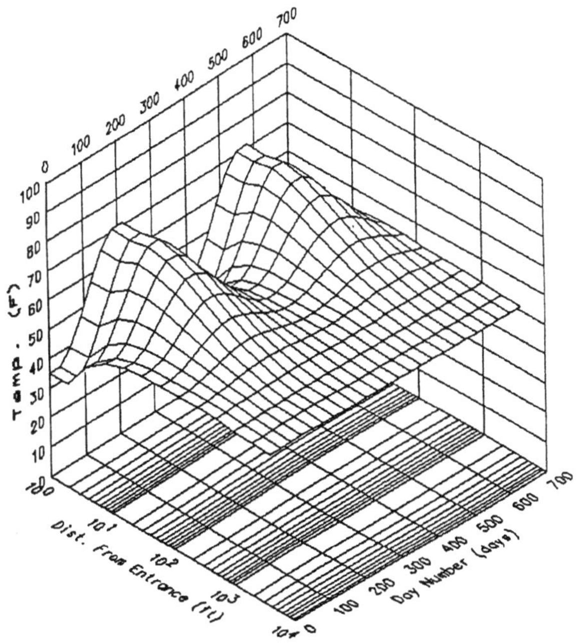

Fig. 22.3 Cropley plot for the entrance to Sand Canyon traverses. Two years of data were used in construction of the plot with day zero being January 1 of the first year of data. Distance from the entrance is again plotted on a log scale. Temperature is plotted on the vertical axis in °F

22.4 Cave Weather Along the Sinking Creek Trunk Passage

The 16 stations used for measurements along an 8000 foot traverse in the Trunk Passage are shown in Fig. 22.4. All stations are more than 1000 feet from any known air entry points. The depth below the surface varies from 150 to 300 feet. From Penn State Lake at the upstream end to the Moon Room area, the trunk passage is dry except during extreme storm events. Except for a segment at the Dry Sumps, the remainder of the Trunk Passage carries a stream, Sinking Creek in the upstream part and Sneaky Creek in the lower part. Stream flow rates and water temperatures were not measured although these data would be important for more quantitative modeling of the meteorological data.

Figure 22.5 shows one set of traverse data taken on January 26, 1991. Relative humidity is close to 100 % except for a few observations taken, curiously, along Sneaky Creek. The temperature rises slightly in the downstream direction. There is some variability but the overall trend is distinct. Note that the wind direction from both downstream and upstream of Sand

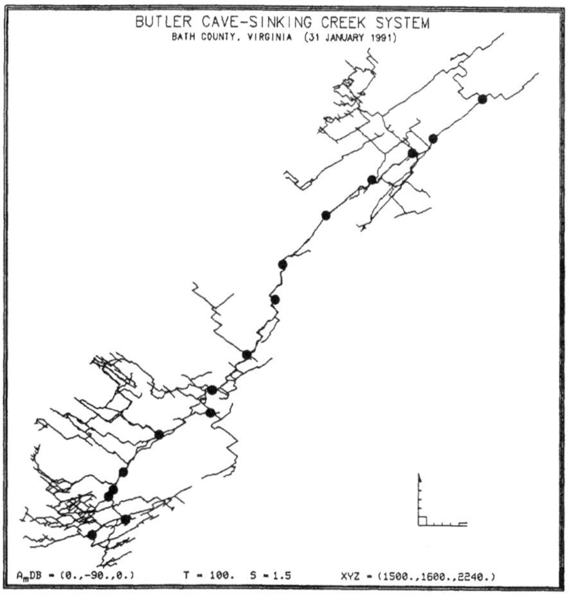

Fig. 22.4 Line map of the Butler Cave-Sinking Creek System showing location of measurement stations for the Trunk Passage

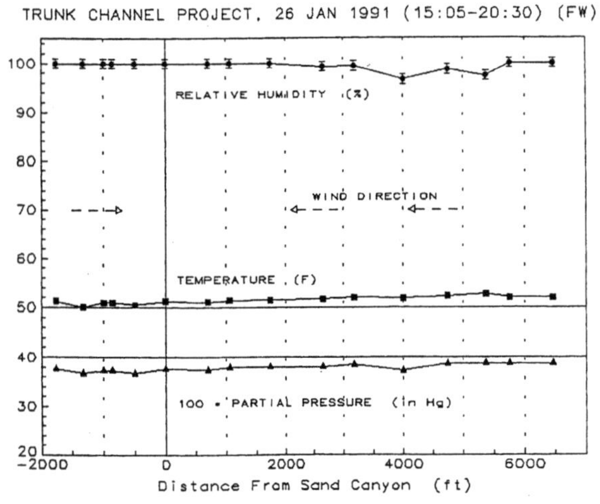

Fig. 22.5 Temperature, relative humidity, and water vapor pressure measurements taken along the Trunk Passage. *Note arrows* showing air flow direction

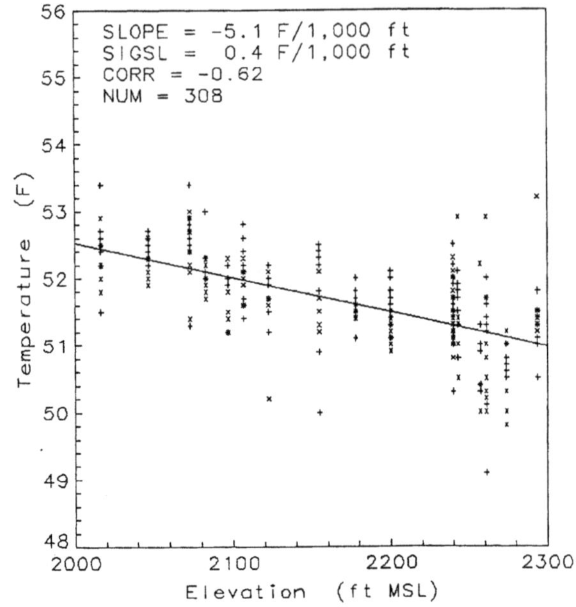

Fig. 22.6 Temperature data along the trunk passages as a function of elevation. Note that downstream is to the left. Data collected during the warmer months are plotted as +. Data collected during the cooler months are plotted as x

Canyon is toward Sand Canyon. This seems good evidence that the Nicholson Entrance was indeed the vent for the entire cave system with the air flow out of the entrance in these January measurements.

The complete Trunk Passage data set is shown plotted as temperature vs elevation in Fig. 22.6. For a total of 308 observations, there is considerable variation at each observation station but the trend line is significant. Because these data are plotted as a function of elevation, the 2300 foot elevation points are at Penn State Lake and the 2000 foot points are at the 6th of July Room. The warmest part of the traverse is at the downstream end. The Trunk Passage follows the surface valley of Sinking Creek, both loosing elevation in the downstream direction so that the depth below the surface does not change much.

The regression line fitted through the temperature data gives a thermal gradient of 5.1 °F/1000 feet. Converting the units, the gradient becomes 9.2 °C/km, a number which is in the right range, although a little low, to be the geothermal gradient in the tectonically stable Burnsville Cove.

22.5 Long Term Observations at Sand Canyon

Sand Canyon is an easily accessed measurement station that lies on both the entrance series traverses and the Trunk Passage traverses. As the entrance series data show, it is well beyond the zone of influence of in-flowing surface air. As such, Sand Canyon provides a useful data set for examining the seasonal variability of cave meteorological parameters (Fig. 22.7).

Sand Canyon is regarded as a dry area in the cave. It was the chosen camp site for the early explorers before shorter and more time efficient routes were discovered. "Dry" is a relative term. For many of the observations, the relative humidity at Sand Canyon was near 100 %. "Dry" in a cave means relative humidity in the range of 96–98 %.

[**Editor's Note**: Turner Avenue in the Flint Ridge section of Mammoth Cave is extremely dry and dusty with speleothems made of water-soluble sodium salts. The lowest value obtained in a relative humidity traverse was 85 %.]

Although the variability is low, it is not appropriate to say that the cave temperature is constant. Ninety temperature measurements at Sand Canyon are shown on an expanded scale in Fig. 22.8. The precision of the temperature measurements, limited by the scale interval of the dry bulb thermometer, is about 0.1 °F so the

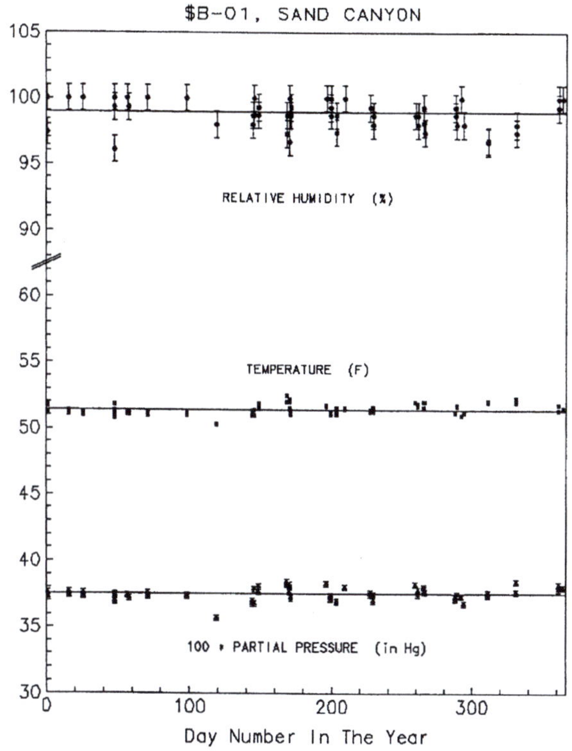

Fig. 22.7 Relative humidity, temperature, and water vapor partial pressure data taken at Sand Canyon as a function of day of the year. Day 1 is January 1. Dates arranged in this fashion are generally known as Julian days

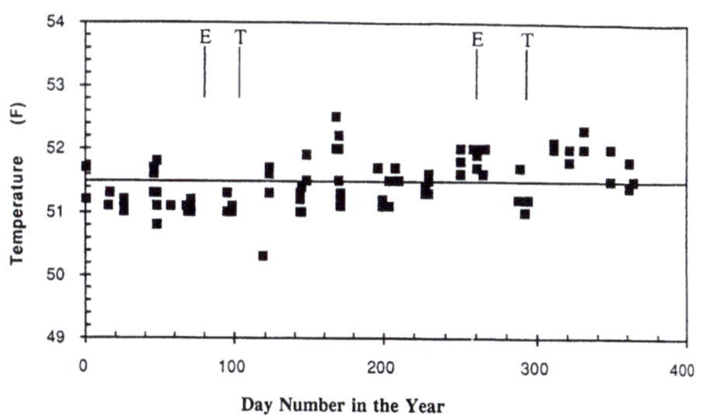

Fig. 22.8 Ninety dry bulb temperatures at Sand Canyon plotted on an expanded scale

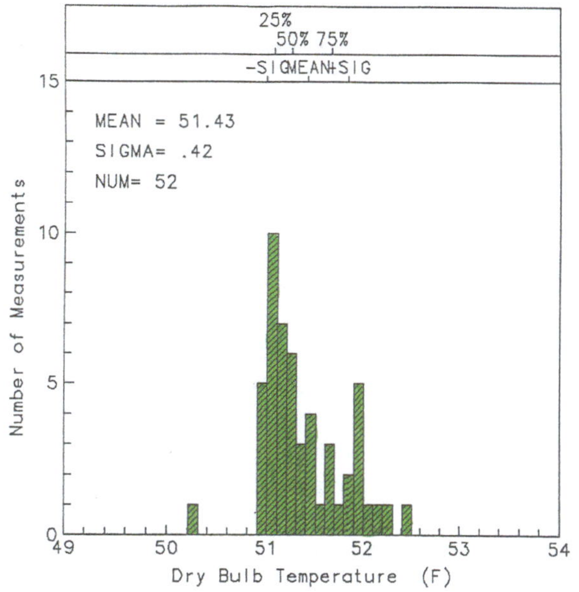

Fig. 22.9 Temperature at Sand Canyon

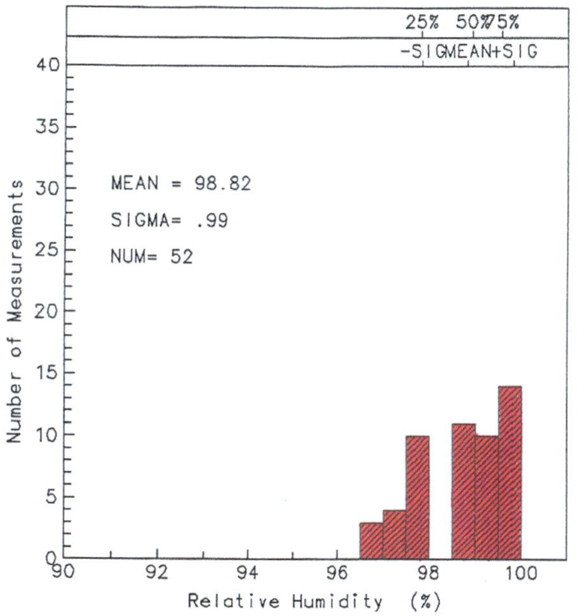

Fig. 22.10 Relative humidity at Sand Canyon

variations shown in Fig. 22.9 represent real temperature variations, not measurement uncertainty.

The 52 data points for temperature and relative humidity collected by 1989 are shown in histogram form in Figs. 22.9 and 22.10. The mean temperature at Sand Canyon is 51.4 °F and the relative humidity is 98.8. The distributions do not have a smooth Gaussian form suggesting that individual values depend on some combination of factors which vary independently.

22.6 Conclusions and Implications Written by the Editor

Fred's medical condition interrupted his measurement plan so that he never had a chance to draw conclusions from his very extensive data base. What was established is that even in a relatively dry, stable cave such as Butler, there are temperature and water vapor variations that extend deep into the cave. Surface influence on temperature extends at least 800 feet from the entrance. Along the Trunk Passage deep in the cave, temperatures can vary by more than one degree Fahrenheit. Relative humidity fluctuates with temperature as well as with water vapor partial pressure. Air current velocity and air flow direction are the dominant controlling variables but the amount and temperature of water flowing through the cave likely plays some role particularly in the deep cave measurements.

So why should we care? Although there are measureable variations in cave environmental parameters, these variations are small and do not deviate much from the mean values. Why we should care is that caves are becoming extremely important paleoclimate archives. Speleothems that contain trace amounts of uranium can be assigned absolute ages by measuring the isotopes ^{234}U, ^{238}U and the daughter isotope ^{230}Th. A great amount of effort is being expended on the measurement of oxygen isotope ratios in speleothems and using these data along with the ages to calculate the cave temperature at the time when the speleothem was deposited (see Fairchild and Baker 2012 for a recent summary of this type of research). The basic assumption is that the cave temperature represents annual mean of the surface temperatures. By establishing the cave temperature from speleothem isotope measurements, the mean surface temperature, a key climatic variable, can be established. Studies such as Fred's are useful in establishing the uncertainty in the basic assumption that cave temperature equals mean surface temperature.

22.7 Addendum: A Method for Detecting Cave Connections by Induced Air Flow by Philip C. Lucas, Frank Marks Jr., Nevin W. Davis

[**Editors Note**: The following short paper was extracted from the Proceedings of the 15th International Congress of Speleology, Kerrville, TX, 2009.]

22.8 Introduction to the Addendum

"Follow the Air" has certainly been the creed of cavers, rewarding them in many instances. The feel of a cave breeze blowing from some unknown source is intriguing and exciting. Can the source be from some vast cave system not yet discovered? Is it possible that a "far distant blowhole" is somehow connected to a known cave? The possibility of finding such a connection becomes an exciting prospect especially if it might lead to the discovery of virgin cave passages. But air flow can be elusive and trying to follow a tantalizing, barely discernable draft, can be exasperating. Many times the question is raised "Are we on the right track or is this just a local circulation?" "Should we dig here?" or "Is that dome worth a bolt climb?" To have the knowledge of the source or ultimate destination of air flow can be very useful. To determine that a cave is connected, at least by air currents, to another cave 6 km down the valley is an exciting prospect.

In the past various methods have been used to attempt to trace air currents. These range from injecting some type of smoke into the air stream such as white phosphorus, burning wood, leaves or even tires. Ethanethiol, an air scent tracer producing an "essence of skunk" odor has been used with limited success. But none of these methods has proven to be effective in detecting a cave's air flow over long distances.

The method described in this paper is simple. It is based on the premise that under certain conditions if two entrances are connected, then mechanically blowing air into (and out of) one entrance will rapidly change the air flow velocity at the other entrance.

22.9 Instrumentation and Equipment Used

1. One 10.16 cm wind run meter modified to act as an optically detected bidirectional anemometer with associated electronics. Startup wind speed 0.27 m/s and a maximum wind speed in or out of 4 m/s.
2. Four 10.16 cm optically detected bidirectional anemometers with associated electronics. Startup wind speed 0.36 m/s and a maximum wind speed in or out of 4 m/s. This instrument was designed and constructed by the authors.
3. One 480 watt 5 blade fan (76 cm diameter) modified to reverse direction every 65, 100 or 200 s. which includes a 9 s. off period while reversing.
4. Five HOBO U12 4-Channel Data Loggers.
5. One ultrasonic bidirectional anemometer with associated electronics. The anemometer is able to measure wind speed as low as 0.045 m/s with no practical limit on the high end. The output is recorded on two channels of a data logger. The math necessary to calculate a wind velocity is performed in an Excel Spreadsheet. This instrument was designed and constructed by an author.

22.10 Summary of Results

22.10.1 Butler Cave—Nicholson and Sofa Entrances

Butler Cave has 27 km of passages many of which are of large volume. It has two known entrances. The Nicholson entrance is located 144 m to the north and 41 m higher from the SOFA entrance. The two entrances are separated by about 478 m of passages generally large in volume (6 m by 15 m) but with many turns and several restrictions (crawlways). The upper Nicholson entrance has an air current consistent with the direction of a convection flow. However the SOFA entrance generally has a neutral or an oscillating air current indicating it is an intermediate (in elevation) entrance and that a lower, presently unknown, entrance, exists elsewhere. Figure 22.11 shows a 480 W reversing fan (76 cm diameter) placed in the SOFA entrance (a rectangular opening reduced to the fan's dimensions). Power was supplied by a generator. A signal is sent from the fan to a recording data logger when the fan changes direction. For redundancy an anemometer is also placed at the fan to record the air flow and direction. Part of the Nicholson entrance is a small opening that allows bats access. The anemometers were placed in this opening during the testing (Fig. 22.12). All anemometers were connected to data loggers and the air velocities recorded at one second intervals.

Fig. 22.11 The *rectangular* doorway Butler Cave's SOFA entrance has been sealed around a fan. The fan blows alternately in and out of the cave in even intervals. A signal is sent from the fan to a recording data logger when the fan changes direction (*photo* by P. C. Lucas)

Figure 22.13 shows the air velocity at the Nicholson Entrance. It includes a 500 s period before the fan operation (base-line) and the next 500 s period during the fan's operation of alternately blowing in and out of the SOFA Entrance. The measurements are summarized as follows:

- The "base-line" air flow is blowing out from the cave (the direction of a convection air current) but with small variations. These variations are assumed to be caused by small changes in barometric

Fig. 22.12 An ultrasonic anemometer (*left*) and a mechanical anemometer (*right*) recording data are located in the bat access hole to measure the rate of air flow (*photo* by P. C. Lucas)

pressure (atmospheric waves). They are referred to as "noise" in this paper. Much of the noise is caused by surface winds and it increases with increasing surface wind speeds.

- With each fan reversal there is a corresponding change in the air velocity at the Nicholson entrance.
- The change in air flow takes place 3 s after the fan is reversed.

Note that during the period of 600–650 s, the anemometer was being re-aligned. Being out of the main air stream the blades anemometer slowed their rotation and the effect is seen as lower velocities during that 50 s.

22.10.2 Big Bucks Entrance, Buckwheat Cave, Backyard Cave, and Basswood Cave

Barberry Cave, about 5.5 km in length, has a large passage extending roughly along a small valley trending northeast. Big Bucks Pit, its northern-most entrance, is unusual in that it is located in a small building used as an apple shed. The entrance is a 43 m pit that extends down from the floor of the apple shed. The building is sealed to the top of the pit. The fan was placed in the apple shed's door which provided the same sealed fan attachment as at the Butler site (Fig. 22.14). There are three other cave entrances that lie to the northeast of the Big Bucks Pit entrance, Basswood, Backyard, and Buckwheat.

The Basswood Cave entrance, a 76 cm metal culvert, is located just 524 m northeast of Big Bucks Pit. There is only moderate air flow at the entrance. The cave is formed in upper limestone units above

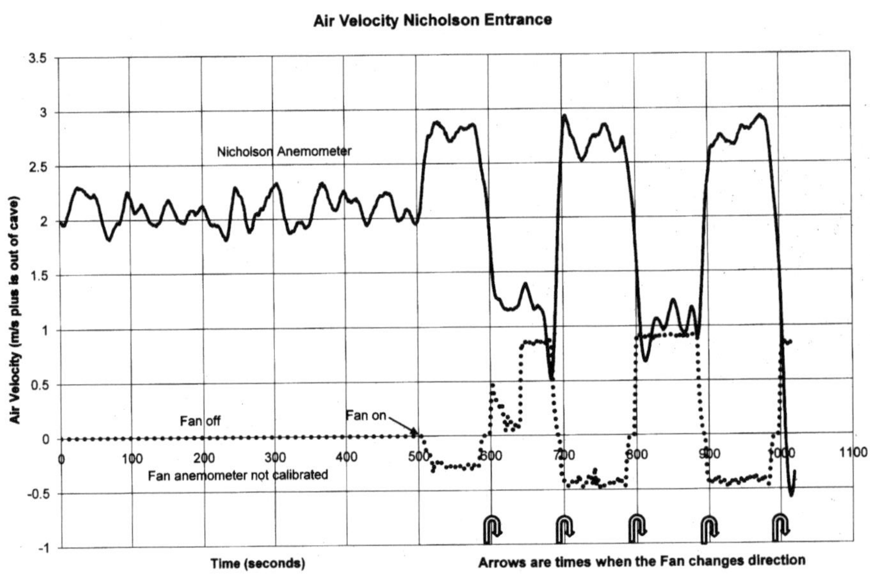

Fig. 22.13 Butler Cave entrances air flow test. With the fan placed in the SOFA Entrance and an anemometer placed at the Nicholson Entrance, this chart shows the result of changing air velocities before and during the fan's operation. When the fan begins running for each 100 s interval there is a distinct signal (change in air velocity) measured by the anemometer. At the end of each interval, the fan stops for 9 s and begins rotating in the opposite direction (*chart* prepared by N. W. Davis)

Fig. 22.14 Inside this shed used for apple storage is Big Bucks Pit, a 43 m drop into Barberry Cave. The pit is sealed to the floor of the building. The door to the building has been reduced to the size of the fan. Accordingly, as the fan blows in and out of the building the air is exchanged in the cave (*photo* by P. C. Lucas)

Fig. 22.15 Cardboard has been placed over the entrance (a *vertical* culvert pipe) to Basswood Cave. The anemometer placed over a hole in the cardboard records the rate of air flow (*photo* by P. C. Lucas)

Barberry Cave so a connection to Barberry Cave seemed unlikely. Backyard cave is a 61 cm metal culvert located 772 m northeast of Big Bucks Pit. Although it has strong air currents and is in upper limestone units it was not suspected to be connected. The Buckwheat entrance is a 61 cm metal culvert and is located 773 m to the northeast of Big Bucks Pit. Prior to this investigation it was suspected that Buckwheat Cave might have some connection to Barberry Cave through an area of breakdown although no human sized opening has been found.

The test consisted of anemometers being placed in each of these three cave entrances to record changing air velocities when the fan at Big Bucks Pit was in operation. At Backyard and Buckwheat the anemometers were placed inside the metal culvert entrances that had strong air currents. However at Basswood where air flow was weaker, cardboard was placed on top of the entrance pipe. A small hole in the cardboard accommodating the anemometer concentrated the air flow. Figure 22.15 shows this arrangement.

Figure 22.16 shows the result of an air flow test between Big Bucks and Basswood where the signal is present but is weaker than the signals found in the Butler entrances test. Shown are nine cycles of the reversing fan operation and the resulting air velocity as measured at the cardboard restriction in the Basswood Cave entrance. Identifying the weaker signal against the background "noise" is still possible with this chart but it demonstrates that weak signals can become buried in the noise. There is a simple mathematical method of suppressing the random noise in a signal when the period of the repetitive signal, the fan, is known. If we add the periods together, first second to first second, to next second, over all nine periods (in this case), the signal will be enhanced and the noise, being random, will be suppressed. We can make an average out of this by dividing by the number of periods. Figure 22.17 shows the effect of this method. It shows the nine cycles of the air flow velocity combined to show the maximum effect from any changes from the fan operation. A dark line representing the overall average air velocity for the entire test separates the periods for the fan blowing in each direction. This simple method clearly shows that the velocity is significantly altered from the overall average by the fan's influence and demonstrates the two entrances being connected.

Results of all the air flow tests at these three entrances show that all are connected to Barberry Cave (Big Bucks Pit). Buckwheat's anemometer/logger recorded the strongest signal from the fan's operation at Big Bucks Pit. The delay time was 13 s. Backyard Cave had a weak but definite response and had a 13 s delay time. Basswood Cave had a definite response but one that was delayed by 12 s. This curious delay

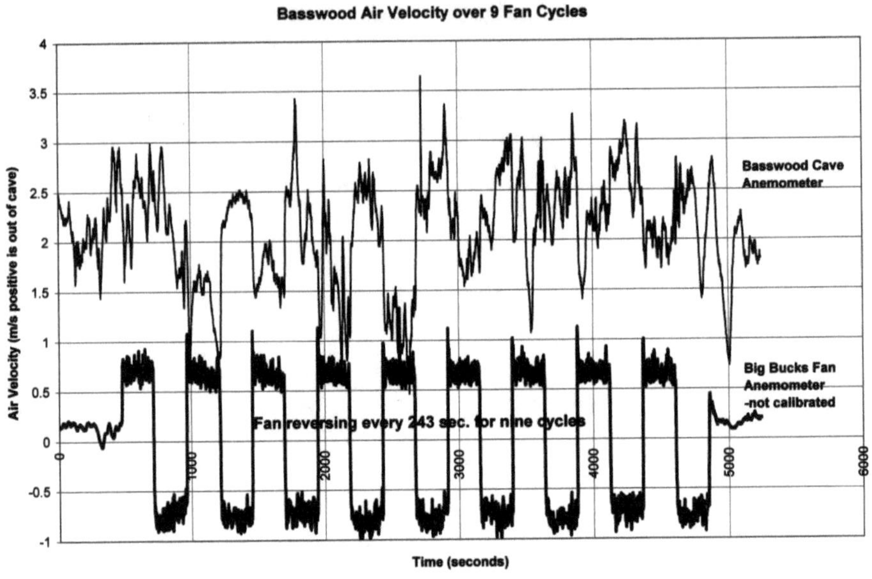

Fig. 22.16 Big bucks to Basswood Cave air flow test. This shows nine cycles with the fan placed at Big Bucks Pit and anemometer at Basswood Cave. In this instance the signal was much weaker than at Butler Cave entrances (*chart* prepared by N. W. Davis)

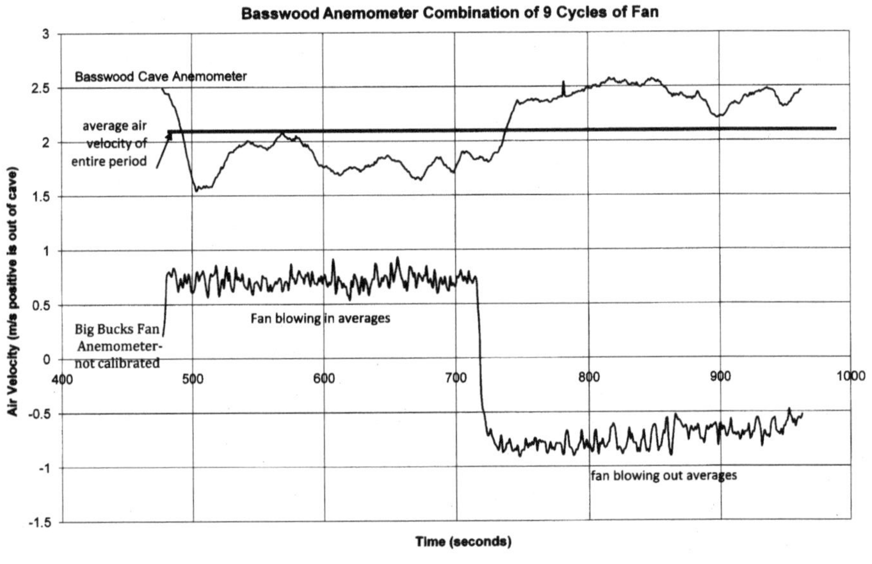

Fig. 22.17 Big bucks to Basswood average. This shows the combined average air velocity for the nine cycles of fan operation compared to the overall average. This calculation produces a clearer signal by cancelling much of the background noise (*chart* prepared by N. W. Davis)

might mean that the connection is a long route despite the two entrances being only separated by 524 m.

22.10.3 Helictite Cave to Subway

Helictite is a compact maze cave with over 11 km of canyon passages averaging 6 m high by 2 m wide. The cave has only one known entrance. At this site the fan was placed directly into the 76 cm steel culvert entrance to Helictite Cave. The anemometer was placed at an entrance to the Subway (a major section of the Water Sinks Cave). The two entrances are only 159 m apart but are on opposite sides of a large sinkhole. The air flow testing shows no detectable signal between entrances.

22.10.4 Helictite to a Blowhole

With the fan placed in the Helictite entrance, an anemometer was placed at a surface blowhole 880 m to the north. Previously some digging has taken place at the blowhole in an attempt to find a possible cave. As a result of the digging the air flow was coming up through an area of broken loose rock measuring about 4 m^2 on one side of the excavation. To concentrate the air flow, a tarp was placed over the loose rock with the anemometer placed in a small hole cut in the tarp (Fig. 22.18). The result of this test shows a connection exists between Helictite Cave and the blowhole. The effect at the blowhole was slight and delayed by 40 s.

Fig. 22.18 A tarp was fastened to one side of an excavation to cover an area where air was passing through a zone of rocks. Cardboard was taped to the tarp to provide a rigid surface and a hole was cut in the cardboard and tarp. An anemometer was placed in this hole where the air flow was then concentrated (*photo* by P. C. Lucas)

The blowhole is now the entrance to Wishing Well Cave.

An interesting observation was made during the Helictite tests. A strong outflow of air would follow after the fan had been blowing into the cave for a ten minute period. Even though the entrance displays a convection air flow, the fan's operation was enough to create increased pressure within the cave that it far exceeded the outflow for the convection air flow.

22.11 Conclusions

The results from these tests vary from strong changes in air flow to weak changes or no detectable changes. The time it took for the effect to occur changed from just a few seconds to nearly 30 s. As this method of testing was being developed and preliminary testing was performed, it was apparent that testing done on windy days was not satisfactory. Gusty winds over hills and ridges create many pulses that tend to be amplified at entrances and this "noise" overwhelmed the signal the fan introduces. Also the clearest signals were obtained when the outside temperatures and cave temperatures are close to each other thereby creating only weak convection currents. All of the caves tested with possibly one exception are thought to have convection as the primary energy driving the air flow. Helictite Cave's airflow might have a strong barometric component as determined by another test (not described in this paper). Accordingly additional testing using this method between barometric caves entrances will be interesting. Although, these tests reveals a connection between some cave entrances it does not provide information about the nature of the connection or its location. It does not say whether this connection is humanly traversable. The different time delays suggest that some connections are more distant, longer in length or more restricting than others but a more examination of this is needed. This testing requires some preparation of the placement of the equipment, fan, and anemometer/loggers along with a power source. But the outcome can provide a valuable insight about the relationship of caves in a karst area. This method of air tracing is certainly less polluting to the cave than past methods.

References

Bamberg, S.A. 1973. Environments in Lehman Caves, Nevada. *National Speleological Society Bulletin* 35: 35–47.

Cropley, J.B. 1965. Influence of surface conditions on temperatures in large cave systems. *National Speleological Society Bulletin* 27: 1–10.

Fairchild, I.J., and A. Baker. 2012. *Speleothem science*. Chichester, UK: Wiley-Blackwell. 432 pp.

Nepstad, J., and J. Pisarowicz. 1989. Wind Cave, South Dakota: temperature and humidity variations. *National Speleological Society Bulletin* 51: 125–128.

Wefer, F.L. 1984. Cave meteorology. *BCCS Newsletter* 10: 33–49.

Wefer, F.L. 1985. More on cave meteorology. *BCCS Newsletter* 11: 25–56.

Wefer, F.L. 1988. Still more on cave meteorology. *BCCS Newsletter* 14: 76–84.

Wefer, F.L. 1989a. Yet still more on cave meteorology. *BCCS Newsletter* 15: 18–38.

Wefer, F.L. 1989b. On the measurement of relative humidity in cave meteorology projects. *Nittany Grotto News* 36(1): 6–14.

Wefer, F.L. 1990. Even yet still more on cave meteorology. *BCCS Newsletter* 16: 20–41.

Wefer, F.L. 1991a. An annotated bibliography of cave meteorology. *Cave Geology* 2: 83–119.

Wefer, F.L. 1991b. Meteorology of the butler cave-sinking creek system. In *Appalachian Karst*, ed. E.H. Kastning, and K.M. Kastning, 65–74, Huntsville: National Speleological Society.

Wefer, F.L. 1992. Meteorological conditions far inside a large cave system. *BCCS Newsletter* 18: 36–41.

Wigley, T.M.L., and M.C. Brown. 1976. The physics of caves. Chap. 9. In *The science of speleology*, ed. T.D. Ford, and C.H.D. Cullingford, 329–358. London: Academic Press.

Minerals and Speleothems in Burnsville Cove Caves

23

William B. White

Abstract

The caves of Burnsville Cove contain widely distributed, often sparse but sometimes spectacular, secondary mineral deposits. Massive flowstone and dripstone occur mainly in localized areas, perhaps reflecting the protective influence of the various interbedded sandstones and the overlying Oriskany Sandstone. Helictites, anthodites, nodular speleothems, cave pearls, crusts, and other mineral deposits occur widely in many parts of the caves. An unusual feature is the wide occurrence of aragonite speleothems, especially in the Chestnut Ridge System, the percentage of aragonite being considerably higher than in typical Appalachian caves. Other minerals include gypsum, which seems to be associated with shaley layers in the Tonoloway Limestone, moonmilk, and a suite of phosphate minerals.

23.1 Introduction

The caves of Burnsville Cove contain widely distributed, often sparse but sometimes spectacular, secondary mineral deposits. Massive flowstone and dripstone occur mainly in localized areas, perhaps reflecting the protective influence of the various interbedded sandstones and the overlying Oriskany Sandstone. Helictites, anthodites, nodular speleothems, cave pearls, crusts, and other mineral deposits occur widely in many parts of the caves.

The calcite speleothems found in the caves of Burnsville Cove are typical of those found in other Appalachian caves. An unusual feature is the wide occurrence of aragonite speleothems, the percentage of aragonite being considerably higher than in typical Appalachian caves. Other minerals include gypsum, which seems to be associated with shaley layers in the Tonoloway Limestone, moonmilk, and a suite of phosphate minerals.

The purpose of this chapter is to describe the speleothems found in Burnsville Cove caves and provide some information on their compositions, mineralogy, and mode of deposition. The chapter contains most of the quantitative analytical work reported in the earlier account of minerals in the Butler Cave-Sinking Creek System (White 1982) and also new information that has accumulated since the earlier report. The results summarized in this chapter are based on analyses of 66 specimens described in the earlier report plus additional samples collected more recently, particularly from the Chestnut Ridge caves. It should be noted that most of the specimens collected were loose fragments, generally less than 10 g. Mineral investigations can be conducted with little or no damage to the caves. For background information and a comprehensive review of cave mineralogy, see Hill and Forti (1997).

W.B. White (✉)
Department of Geosciences, The Pennsylvania State University, University Park, PA 16802, USA
e-mail: wbw2@psu.edu

W.B. White (ed.), *The Caves of Burnsville Cove, Virginia*,
Cave and Karst Systems of the World, DOI 10.1007/978-3-319-14391-0_23,

23.2 Carbonate Speleothems: Flowstone and Dripstone

The dominant speleothems are stalactites, stalagmites and flowstone composed mainly of calcite. Their external forms are determined in large part by the patterns of dripping or flowing water that carries dissolved ions to the site of deposition. There are, throughout the cave systems, well decorated areas separated by sections of cave with only sparse speleothems. The sections that follow describe some of the more interesting areas.

Fig. 23.1 Thin section under polarized light of a calcite stalagmite from Huntley's Cave. The colors are due to interference, not the true color of the grains. WBW photo

23.2.1 The Butler Cave-Sinking Creek System

The Butler Cave Section of the Sinking Creek System and much of the southern portion of the cave lie below the Lower Breathing Cave Sandstone. The sandstone is generally impermeable so that the cave passages have little drip water and thus few speleothems. These occur where the sandstone has been fractured. Concentrations of dripstone and flowstone speleothems occur in Huntley's Cave, in the Moon Room area, and in the French Passage.

The stalactites and stalagmites in Huntley's Cave tend to be dried with a powdery weathered layer on the outside. The core is dense, almost transparent, white calcite. X-ray powder diffraction shows that the weathered residue remains calcite. One small stalagmite, found loose in Huntley's Cave, was sectioned for microscopic examination. The grains are small, but very densely packed (Fig. 23.1).

There is some evidence that a master fracture crosses the cave in the vicinity of the Moon Room. The fracture allows access to vadose water which has produced a variety of speleothems. The largest stalactite in the cave (Fig. 23.2) is in one of these passages.

The Moon Room itself is a high level cross-passage with an active flowstone deposition area. Active flowstone is formed on the walls (Fig. 23.3). New stalagmites are being formed on the clay floor of the passage (Fig. 23.4). The Moon Room has a flowstone floor and a considerable amount of flowstone and other decoration on the walls (Fig. 23.5). The passage has been flagged to discourage visitors from tracking up the flowstone.

Fig. 23.2 Large stalactite near the Moon Room. Photo by Philip C. Lucas

The Crystal Passage is a narrow canyon on the Chestnut Ridge side of the Butler Trunk Passage that branches off just downstream from the rise of Sinking

Fig. 23.3 White flowstone in the Moon Room. WBW photo

Fig. 23.4 New stalagmites forming in splash pit. WBW photo

Creek. There is some dripstone (Fig. 23.6), mostly coated with brown material and apparently inactive. Other active dripstone occurs in the French Passage and other downstream locations.

23.2.2 The Chestnut Ridge System

Many of the upper level passages in the Bobcat Section are dry with gypsum and other crystal coatings, but are generally lacking in dripstone and flowstone. The best speleothem development is found in the southern, Blarney Stone Section. Here, there is clear evidence for multiple episodes of speleothem deposition. In Ghost Hall, there is a massive column (Fig. 23.7) that has a deep brown coating suggesting that the speleothem has been flooded at sometime in the past. This may be compared with the nearby "ghost" in Ghost Hall (see Fig. 7.2), a large, completely white stalagmite.

The brown outer coatings on many of the speleothems do appear to be exactly that—a coating. In the few instances where the interior can be observed on a broken surface, the interior consists of lightly colored, massive crystals. This poses an interesting question concerning the sequence of events that produced the speleothems seen today. There was at least one episode of undisturbed speleothem deposition followed by a period when the passages occasionally flooded produces the brown coatings, followed by a second episode of speleothem deposition to produce the clean white deposits such as the ghost. A further question is whether these episodes correlate with glacial and interglacial periods. These questions could be answered by obtaining a set of uranium-thorium dates but this would be a difficult and expensive program not likely to be undertaken in the near future.

In Bull-in-a-China-Closet, fresh, active dripstone occurs in a low passage, requiring exceptional care in exploration (Fig. 23.8). There is some interesting timing here. Note that white flowstone is forming over the mud cracks on the cave floor. There is a long interval between the last flooding event that deposited the mud, the drying out of the mud to produce the cracks, and the deposition of the flowstone. Whether this mud is related to the brown coatings on other speleothems in this section of the cave is also an open question.

23.2.3 Barberry Cave

The caprock has been removed for a long time over Barberry Cave, at the southern end of Chestnut Ridge. As a result, vadose water containing the chemical signature of the overlying limestone soils can reach the

Fig. 23.5 Glistening flowstone floor in the Moon Room. Photo by Philip C. Lucas

cave and deposit what is volumetrically the most extensive flowstone in any of the Cove caves. Chapter 10 contains photographs of a selection of these speleothems. The Great White Wow (Fig. 10.18) is probably the largest speleothem in the Cove. Because Barberry Cave is not protected by caprock, the passages are wet and dripping with speleothem deposition still in progress.

23.2.4 Helictite Cave

Helictite Cave as the name implies has the finest display of helictites in Burnsville Cove. In addition, there are areas of flowstone and dripstone deposits (Figs. 23.9 and 23.10). These two images also offer a contrast. The stalactite and stalagmite in Fig. 23.10 is white, lacking in either included clay particles or other pigment. The stalactite cluster in Fig. 23.9 is colored the oranges and tans usually associated with humic substances derived from overlying soils.

23.2.5 Wishing Well Cave

Both Wishing Well Cave and nearby Helictite Cave are formed in the Licking Creek Limestone which has extensive interbedded chert. Both were entranceless caves so that growing speleothems were not subject to drying by low-humidity air currents from the surface or had their growth surfaces contaminated by air-born dust from the surface. Lack of open surface connection likely was an important factor in producing the exceptional mineralization in these two caves (Figs. 23.11 and 23.12).

23.3 Carbonate Speleothems: Erratic Forms

The external forms of all speleothems represent the outcome of a contest between the natural growth habit of the component minerals and the fluid flow patterns of the water carrying the dissolved ions to the site of

Fig. 23.6 Stalactites in the Crystal Passage. WBW photo

deposition. When flow dynamics dominates, the result is a deposit with a smooth streamlined pattern such as seen in dripstone and flowstone. When crystal growth is in control, especially when growth habits are modified by impurities, very complex external shapes can result. Those forms that are found in the Burnsville Cove caves are described below arranged by their commonly accepted names.

23.3.1 Helictites

Helictites in Butler Cave take on several forms. Some are massive clear calcite with a sugary surface texture (Fig. 23.13). Others are more filiform with many twisting filaments of small diameter (Fig. 23.14). These also appear to be composed of calcite; however, no loose fragments were found that could be used for analysis.

The filiform helictites are similar to those described by Gèze (1957) from the Moulis Cave in France.

Fig. 23.7 Column in Ghost Hall. Photo by Nevin W. Davis

Fig. 23.8 Dripstone in Bull-in-a-China-Closet. Photo by Nevin W. Davis

These forms are of interest because their diameters are less than the 5 mm calculated by Curl (1972) to be the minimum diameters of stalactites formed by deposition from free-hanging drops. The small sizes of the filiform helictites are evidence that growth takes place

Fig. 23.9 Corner of the Attic, Helictite Cave. Photo by Philip C. Lucas

Fig. 23.10 White dripstone, Helictite Cave. Photo by Philip C. Lucas

Fig. 23.11 Soda straw stalactites near Echo Junction, Wishing Well. Photo by Yvonne Droms

Fig. 23.12 Dripstone in Wishing Well Cave. Photo by Philip C. Lucas

Fig. 23.13 Massive helictite with coarsely crystalline (sugary) texture. Near entrance to Crystal Passage. WBW Photo

Fig. 23.14 Filiform helictites in Helictite Passage, Butler Cave Section. *Note* Width of soda stream in comparison with diameter of helictite. The ruler is extended 9 inches. WBW Photo

by slow seepage along the central canal without the formation of free drops. The contrast between the helictites and straw stalactites growing from the same feeder system is apparent in Fig. 23.14. However, the flow must be continuous for the helictite to be bathed in fluid so that the continuity of the calcite crystal is maintained.

The helictites in Helictite Cave (Fig. 23.15) and in Wishing Well Cave (Fig. 23.16) represent a third type in which the transition from straw stalactite to helictite is more continuous. The theoretical 5 mm diameter of the soda straw is more in evidence.

The three styles of helictite described above span most of the helictite morphologies that have been described. Explanations have not been forthcoming. The fast growth direction for calcite is along the c-axis which, in many soda-straw stalactites corresponds to the long axis of the straw. The growth interface is easily poisoned by a variety of impurities causing new

Fig. 23.15 Helictites in Helictite Cave. Photo by Philip C. Lucas

Fig. 23.16 Helictites in Wishing Well Cave. Photo by Philip C. Lucas

nucleation and growth in some different direction. However, the nature of the impurities and how they function has not been determined.

23.3.2 Anthodites

The term "anthodite" was first applied to the radiating sprays of crystals that occur in Skyline Caverns, Virginia (Henderson 1949). It was assumed from the external shape of the Skyline anthodites that the constituent mineral was aragonite and the same name has since been applied to similar speleothems in many other caves. Skyline Caverns has turned out to be a poor type locality. A scientific assessment of the Skyline anthodites (White 1994) found that although the speleothems may have originally been made of aragonite, much of the aragonite had inverted to calcite. Aragonite is metastable in the cave environment and eventually inverts back to calcite. "Eventually" may be up to several million years depending on the impurities in the aragonite. There was a discussion concerning the definition of the term by Donald Davis who argued that the term "frostwork" has precedence. What will be done here is retain "anthodite" for larger radiating clusters of aragonite crystals and "frostwork" for smaller, fuzzier, aragonite needles.

Anthodites are uncommon in the Butler Cave-Sinking Creek System except for a few examples in the Crystal Passage. In the Chestnut Ridge System, the North-South Trunk, the Jewel Passage, and other locations in the Bobcat Section contain a rich display of anthodites and related speleothems (Figs. 23.17 and 23.18). No systematic study has been undertaken of the mineralogy of the Chestnut Ridge Cave system but 8 samples collected by Greg Clemmer in 1992 were analyzed by X-ray diffraction. The acicular crystals are indeed aragonite as expected but two of the samples produced X-ray patterns that do not match either calcite or aragonite.

By far, the most spectacular examples of anthodites occur in the Blarney Stone Section of the Chestnut Ridge System. Representative aragonite crystals from Pride of Skyline (Fig. 23.19) show some of the forms.

Not all aragonite crystals occur as visible clusters of crystals. Some thin white coatings on cave walls turn out to be masses of fuzzy aragonite needles (Fig. 23.20). The aragonite needles are themselves composed of much smaller, fibrous crystals (Fig. 23.21).

Although stalactites, stalagmites, and flowstone composed of aragonite do occur, they are not common. The dominant habit for the crystal growth of aragonite in caves is the form of long acicular, branching crystals. In contrast, calcite tends to grow as blocky crystals forming the solid bulk of calcite speleothems. Calcite incorporates humic and fulvic acid molecules within its structure which, along with other impurities and inclusions, give calcite speleothems their characteristic colors of yellows, oranges, tans, and browns. Aragonite crystals, which do not include these substances, are typically white or water clear.

In the Pride of Skyline is what appears to be a small column (Fig. 23.22). Although the bulk of the column appears to be calcite, close examination (Fig. 23.23) reveals a frostwork of aragonite crystals.

Fig. 23.17 Clusters of crystals in Cotton Ball Hall. Photo by Ron Simmons

Fig. 23.18 An aragonite tree near the bottom of Scoopers Drop. Photo by Ron Simmons

Further examples from Pride of Skyline (Fig. 23.24) and from Wishing Well Cave (Figs. 23.25 and 23.26) illustrate the problems of interpretation. We do not, usually, know the chemical composition of the water that deposited the speleothem. We do not know the internal structure—mineral composition, grain size, and grain orientation—of the speleothem or the trace element content of the minerals. It is a gigantic step from examination of the speleothem from the distance of a camera lens to a complete determination of its structure and chemistry. This step has rarely been taken because of effort required, because of cost, and—perhaps most importantly—because obtaining the information would destroy the speleothem. We prefer to stand back, admire, and speculate.

Fig. 23.19 A selection of anthodites from Pride of Skyline, Blarney Stone Section, Chestnut Ridge Cave System. Photos by Nevin W. Davis

Fig. 23.20 Scanning electron microscope image of a mat of aragonite needles coating the limestone wall of the Crystal Passage in Butler Cave. Magnification 135X

Fig. 23.21 Scanning electron microscope image of the tip of an aragonite needle. Magnification 2675X

Fig. 23.22 Small calcite column with overgrowth of aragonite. Photo by Nevin W. Davis

Fig. 23.23 Close-up of aragonite crystals on column surface. Photo by Nevin W. Davis

Chemical factors controlling which mineral precipitates are very subtle and not at all understood. Aragonite is not thermodynamically stable in the cave environment, yet there it is. Very small changes in the drip-water chemistry can cause the growing speleothem to switch from calcite to aragonite and back again. Mg^{2+} and Sr^{2+} are thought to be the most important impurity ions. Mg^{2+} poisons the precipitation of calcite which permits supersaturation to build up until it exceeds the critical supersaturation for the nucleation of aragonite. Sr^{2+} can form strontium-rich nuclei which have the aragonite structure and which serve as templates for the further growth of aragonite. Details of these mechanisms have been worked out using the atomic force microscope. For details see a review article (White 2012).

There have been very few chemical analyses of the speleothems from the Burnsville Cove caves. These are given in Table 23.1 (taken from White 1982). The aragonite speleothems have higher strontium concentrations than the calcite speleothems but otherwise the data are too sparse for conclusions.

23.3.3 Nodular Speleothems

There is no universally accepted name for the small nodular speleothems found in many caves. Such names as “globulites”, “cave popcorn”, “cave coral”, and “cave grape” have been applied, none of which are universally accepted. Some nodular speleothems consist entirely of calcite (Fig. 23.27). These are usually layered in the same manner as stalactites are layered but the origin of the layer is the point of attachment of the nodule to the wall. Others are mineralogically complex and contain more than one carbonate mineral.

The nodular speleothems that line the walls of the Crystal Passage in Butler Cave are examples of the complex forms. Some are smooth nodules—referred to as “little schmoos” by the early explorers. Others have clusters of aragonite needles growing from them (Fig. 23.28). Still others have the aragonite needles covered with clumps of moonmilk (Fig. 23.29).

23.3.4 Moonmilk

Moonmilk occurs as a dry powder dusted over the surfaces of stalactites and some bedrock surfaces in Butler Cave in such locations as Huntley’s Cave and 90-Ugh Crawl. It also occurs as sticky white blobs associated with acicular aragonite (Fig. 23.29). All specimens of moonmilk examined yielded the X-ray diffraction patterns of hydromagnesite, $4MgCO_3 \cdot Mg(OH)_2 \cdot 4H_2O$. Samples were transported to the

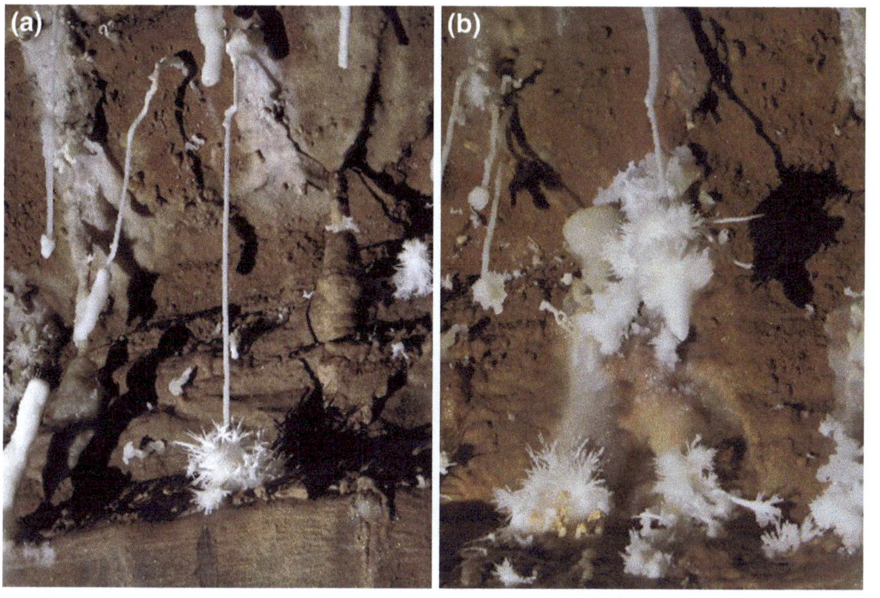

Fig. 23.24 Mixed calcite/aragonite speleothems from Pride of Skyline. Photos by Nevin W. Davis

Fig. 23.25 Cluster of fibrous aragonite crystals from Wishing Well Cave. Photo by Philip C. Lucas

Fig. 23.26 Small aragonite bush in a cluster of nodular speleothems from Wishing Well Cave. Photo by Philip C. Lucas

laboratory in plastic vials at ambient temperature. They had the pattern of hydromagnesite when first examined and the same pattern a year later. Hydromagnesite is stable under surface conditions of temperature and relative humidity but it is not possible to say for certain that no changes took place between the cave environment and the laboratory environment. The crystals of hydromagnesite are in the range of 1–10 μm. Some are well-developed rhombs and some are irregular grains (Fig. 23.30).

The observed association of hydromagnesite with aragonite is also common in other caves. There is a mineral sequence in some of these erratic speleothems: a calcite core, an outer sheaf of aragonite, and tufts of hydromagnesite as the final product. The chemistry of drip water measured in Butler Cave averages 9 mg/L Mg^{2+} and 47 mg/L Ca^{2+} or 16 % Mg^{2+} (Harmon and

Table 23.1 Compositions of selected speleothems from the Butler Cave-Sinking Creek System

Specimen	Location	Mineral	SiO_2	Al_2O_3	MgO	SrO
Stalagmite	Natural bridge	Calcite	<0.01	0.002	1.42	1.00
Soda straw	Sneaky creek	Calcite	0.10	0.005	0.85	0.10
Massive crystals	Christmas passage	Calcite	0.02	0.005	0.08	<0.01
Nodular speleothem	90-Ugh crawl	Aragonite	2.90	0.01	2.15	4.42
Nodular crust	Crystal passage	Aragonite	9.85	0.002	5.15	2.75

All concentrations are given as weight percent as the specified oxide

Fig. 23.27 Nodular speleothem from Butler Cave. Photo by Philip C. Lucas

Fig. 23.28 Nodular speleothems with aragonite needles. Crystal Passage, Butler Cave—Sinking Creek system. WBW photo

Fig. 23.29 Nodular speleothems with clumps of moonmilk (hydromagnesite) coating the outer surface. Crystal Passage, Butler Cave—Sinking Creek System. WBW photo

Hess 1982). Cave calcites tend to have low magnesium contents, typically 1 % or less. When the calcite core is being deposited, the excess Mg^{2+} builds up in the residual solution, reaching a concentration where it inhibits the growth of calcite and allows the necessary supersaturation to build up for the growth of aragonite. Mg^{2+} is even less soluble in the aragonite structure so the final step is the deposition of the residual magnesium as hydromagnesite. Although hydromagnesite is thermodynamically stable in the cave environment (White 1997), the small grain size and pasty, "cream cheese" texture suggests that microbial processes may also be involved (Northup and Lavoie 2001).

Fig. 23.30 Scanning electron microscope image of hydromagnesite moonmilk from Crystal Passage. Magnification: 1000X

23.4 Carbonate Speleothems: Pool Deposits

23.4.1 Crystal Linings

When calcite and other mineral crystals grow completely submerged in water, the crystal faces are free to develop without constraint. Some crystals grow on the surface of the pools; other form as linings around the bowl of the pool (Fig. 23.31).

One of the first pool deposits discovered during the early exploration of Butler Cave was the Crystal Craters. The Crystal Craters are in an upper level passage near the Moon Room. These were at one time shallow pools, now dry, which have accumulated thick deposits of white calcite crystals (Fig. 23.32). The crystals are scalenohedral with lengths up to half an inch. The deposition of these pool deposits required water at high CO_2 pressure that had the opportunity to de-gas over long periods of time. The water would have had to be continuously replenished to account for the size of the crystals, given the low overall solubility of calcium carbonate.

Fig. 23.31 Calcite crystals growing in a shallow pool, Moon Room area, Butler Cave. WBW photo

Fig. 23.32 The Crystal Craters. Photo from the Oscar Estes Collection

Although other crystal-lined pools have been found in Butler and in the Chestnut Ridge System, it was the exploration of Helictite Cave and Wishing Well Cave that revealed some of the most interesting deposits. In Helictite Cave is a spectacular pool lining of scalenohedral calcite (Fig. 23.33). The scalenohedron is the most common form for free-growing calcite in the cave environment. These crystals are exceptionally clean and well formed. They are also more tapered than typical scalenohedral crystals. An even more unusual pool deposit was found in Wishing Well Cave (Fig. 23.34). In addition to being exceptionally clear, the larger crystals in this cluster have a prismatic rather than a scalenohedral habit.

In Helictite Cave was found a pool lining of hollow calcite crystals (Fig. 23.35). These are known in the crystal growth literature as hopper crystals. The triangular outline reflects the 3-fold symmetry of the calcite structure which can also be seen in the 3-fold symmetry of scalenohedral crystals. Under most circumstances, the 3-fold axis (the crystallographic c-axis) of calcite is the fast growth direction thus forming the elongate scalenohedra. In the hopper crystals, growth along the c-axis was completely suppressed so that the edges continued to grow but

Fig. 23.33 Scalenohedral calcite from Helictite Cave. Photo by Philip C. Lucas

Fig. 23.34 Cluster of calcite crystals from Wishing Well Cave in which the scalenohedral habit has been modified into a prismatic habit. Photo by Philip C. Lucas

Fig. 23.35 Calcite hopper crystals, Helictite Cave. Photo by Philip C. Lucas

Fig. 23.36 Cave pearls in shallow pool. WBW Photo

what should have been the fast axis of the crystal did not. Although hopper crystals occur with some frequency in the experimental growth of crystals in the laboratory, they are uncommon in nature and even more uncommon for the mineral calcite. One other cave occurrence, an unnamed cave in northern Cuba, is recorded by Huizing et al. (2003).

23.4.2 Cave Pearls

Cave pearls are a well known cave pool deposit. In their crude form, cave pearls are calcite coatings over some nucleus—a sand grain or a bit of gravel—and may range in shape from irregular to almost perfectly spherical. There are irregular cave pearls in the Moon

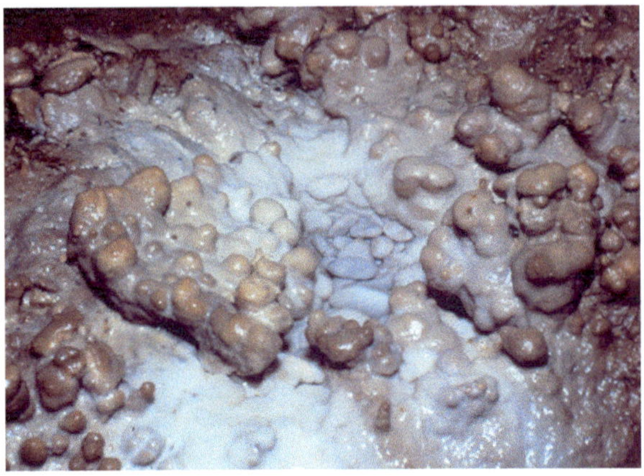

Fig. 23.37 Cave pearls mostly cemented. WBW Photo

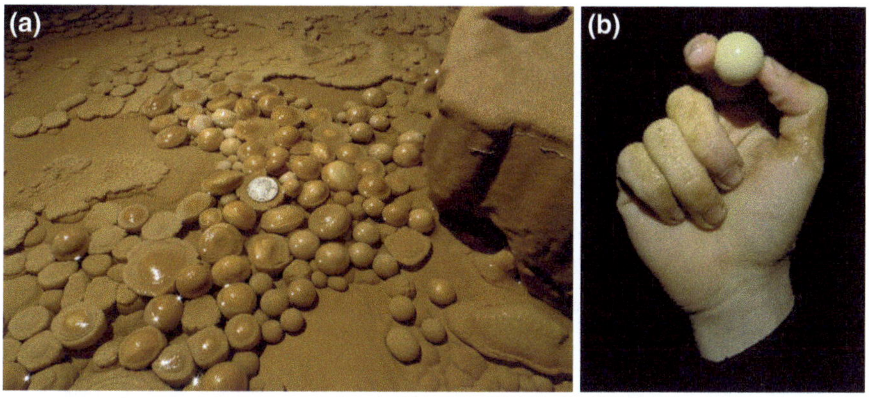

Fig. 23.38 Cave pearls in Barberry Cave. Photos by Nevin W. Davis

Room Section of the Butler Cave-Sinking Creek System. Some are loose pearls and some are cemented to the flowstone pool lining (Figs. 23.36 and 23.37).

A particularly fine nest of cave pearls is found in Barberry Cave (Fig. 23.38). Some of these are nearly perfect spheres; other have what has been called a ravioli shape.

23.4.3 Transitional Forms: Rimstone Dams

Ponded water that is highly supersaturated with calcium carbonate has a tendency to construct dams that may separate a stream of flowing water into a sequence of pools. Rimstone is usually calcite although aragonite may be intermixed. The dams tend to be somewhat porous compared to dripstone and flowstone. Their mechanism of formation is not definitely established. Rimstone occurs sparsely throughout the cave systems. The example given in Fig. 23.39 is the Rimstone Pool Passage in Butler Cave.

23.5 Speleothems Composed of Non-carbonate Minerals

23.5.1 Gypsum

Gypsum, $CaSO_4 \cdot 2H_2O$, is common in the dry passages of the Burnsville Cove caves. Gypsum is about ten times more soluble than calcite and is quickly dissolved and removed when liquid water is present.

In the Butler Cave-Sinking Creek System, gypsum crusts occur on the cave walls, and there are thin gypsum crusts mixed in with other coatings on the surface of the soils in many places. Small gypsum flowers occur in Huntley's Cave. Many of the wall crusts are thickest on certain beds and follow the irregularities in the bedding (Fig. 23.40). The limestone wall rock is thin-bedded and strongly folded. The gypsum crusts follow all details in the folds.

The gypsum deposits shown in Fig. 23.40 were noticed in the early explorations of Butler Cave when the suggestion was made by John Haas that the primary source of gypsum is from the oxidation of minor amounts of sulfide minerals contained in the shale partings within the limestone. The Tonoloway Limestone has many of these. The strongly folded section in Butler Cave just above the 90-Ugh Crawl, has many red beds with dark shales interbedded in the limestone. As the bedrock is removed by dissolution, the fine-grained sulfides are exposed to oxygen-carrying water. The reaction between the sulfides (likely pyrite, although the actual mineral is unknown) and water may be enhanced by the presence of sulfur-oxidizing bacteria.

White coatings occur widely in the dry upper levels of the Chestnut Ridge Cave System. Very few of these have been analyzed so the distribution between gypsum coatings and aragonite coatings is unknown. Deposits in the Sahara Pipeline seem quite definitely to be gypsum because both selenite needles and cave cotton have been found. Cave cotton is a fibrous form of gypsum consisting of long but extremely thin gypsum crystals (Fig. 23.41). As in Butler Cave, the gypsum appears to be localized along certain bedding planes.

The mechanism for the precipitation of gypsum in the Burnsville Cove caves is not intrinsically different from the mechanism that produced the much more

Fig. 23.39 The Rimstone Pools in Butler Cave. Photo by Philip C. Lucas

Fig. 23.40 Gypsum stringers marking bedding planes in highly contorted bedding. Passage just above 90-Ugh Crawl in the Butler Cave Section. WBW photo

extensive gypsum speleothems in the Mammoth Cave System (White and White 2003). In contrast to Mammoth Cave where the sulfide source minerals are either present at some depth within the limestone or are transported from overlying beds, the gypsum deposition in Butler Cave and, by inference, the other

Fig. 23.41 Cave cotton and gypsum along bedding planes in the Sahara Pipeline passage in the Chestnut Ridge Cave System. Photo by Ron Simmons

Cove caves is a process that occurs in the immediate cave walls.

Further support for the sulfide mineral origin for the gypsum is provided by sulfur isotope ratios measured by Christopher Swezey of the U.S. Geological Survey (Swezey and Piatak 2003). The handbook value for the atomic weight of sulfur is 32.064 which says that naturally occurring sulfur consists mainly of a sulfur isotope with atomic weight 32 (written ^{32}S) but also contains minor amounts of a heavier isotope, ^{34}S. The exact mix of sulfur isotopes varies somewhat from source to source and this information can be used to track the source of specific sulfur-bearing minerals. The convention is to measure the deviation from a standard. In the case of sulfur the standard is an iron sulfide mineral (troilite) from the Canyon Diablo

Table 23.2 Sulfur isotope analysis of gypsum

Sample number	$\delta^{34}S$	Speleothem
BUT-1-CS	−5.2	Crust
BUT-2-CS	−1.0	Crust
BUT-3-CS	−4.1	Crust
BUT-4-CS	−4.1	Needles

meteorite. The deviation is expressed in parts per thousand by the equation

$$\delta^{34}S = 1000\left[\frac{\left(\frac{^{34}S}{^{32}S}\right)_{Sample} - \left(\frac{^{34}S}{^{32}S}\right)_{Standard}}{\left(\frac{^{34}S}{^{32}S}\right)_{Standard}}\right]$$

Swezey and Piatek collected four gypsum samples from the Butler Cave Section and from the Sand Canyon Section in 2001. The results of the isotope analysis are given in Table 23.2. These values are in the range expected if the sulfur in the gypsum is derived from the oxidation of pyrite. If the source had been sedimentary gypsum beds within the limestone, the values would have been in the range of +10 to +25.

23.5.2 Iron, Manganese and Other Oxide Minerals

Yellow to red-brown platy masses were found in the Huntley's Cave Section. A specimen from Evasor Gallery has a complex, layered structure, an alteration of deep brown, homogeneous layers that had a vitreous luster and broke with a conchoidal fracture and yellow layers with some internal banding and a satiny texture. The Huntley's Cave sample had a few weak X-ray diffraction peaks that matched the pattern of goethite. The Evasor Gallery sample was amorphous to X-rays. Near-infrared diffuse reflectance spectra of both specimens contained a broad absorption band near 900 nm which is characteristic of the hydrated iron oxides. It would probably more accurate to refer to these samples as ferrihydrite rather than goethite because both samples were poorly crystallized.

A mass of deep brown iron-containing mineral about 2-inches long was collected from the Blarney Stone section near Walnut Tree Junction. It has a shiny, flaky appearance and is clearly an iron oxide hydrate. Ferric iron usually precipitates as a highly disordered but highly insoluble $Fe(OH)_3$ which would be one of several varieties of ferrihydrite. Partial loss of water produces goethite, FeO(OH), and a mixture of phases is quite likely in the cave environment.

Black coatings occur on stream cobbles throughout the Burnsville Cove caves. At the time of its discovery, there was a very thin black coating that formed the uppermost layer on the crusts that cap the clastic sediments in the Butler Cave Section. Most of these have now been trampled by explorers not keeping on the designated paths. Neither the sediment coating nor the coatings on the stream cobbles have been identified or analyzed. The black stream cobble coatings are very similar to coatings found on stream cobbles in other caves, many of which have been identified as the manganese oxide mineral birnessite, (Na, Ca, Mn^{2+}) $Mn_7O_{14} \cdot 2.8H_2O$ (White et al. 2009).

One of the more enigmatic discoveries has been quartz crystals in the cave sediments. Early in the exploration of the Butler Cave—Sinking Creek System, Dick Kutz found a large (2–3 inch) euhedral quartz crystal in the Trunk Channel stream sediments. By "euhedral" is meant that the crystal had sharp edges and flat crystal faces. Other crystals have since been found. It is doubtful if they were formed under present cave conditions but, likewise, they couldn't have traveled far. Tumbling along with the other stream sediments during flood flow would have quickly rounded off the sharp edges between the crystal faces. These crystals are an important bit of evidence for hydrothermal processes in the early development stages of the Burnsville Cove caves. See Chap. 24 for more discussion of this hypothesis.

23.5.3 Phosphate Minerals

Phosphate minerals occur in many cave localities and are usually associated with guano deposits. Leachates from the guano react with limestone wall rock to produce a suite of calcium phosphates such as brushite, whitlockite, crandallite, and hydroxyapatite along with magnesium phosphates, ammonium compounds, and some organic minerals such as urea and guanine. Detailed descriptions of these assemblages have been published for caves in Puerto Rico (Kaye 1959), Western Australia (Bridge 1971, 1973a, b, 1974) and South Africa (Martini 1978).

The phosphate minerals that have been found in Burnsville Cove are unusual in that they are primarily hydrated aluminum phosphates, and they are not associated with identifiable guano deposits. The Butler Cave-Sinking Creek System had no known natural entrances and very few bats have been found in the cave even after the opening of the Nicholson Entrance. No fossil guano deposits have been discovered. The minerals identified are taranakite, sasaite, crandallite and hydroxyapatite. All of these were found in the Butler Cave Section of the system.

Taranakite occurs as a wall coating in a small side passage just off the Second Parallel Passage, across from the crawlway that connects the Second Parallel Passage with the Mountain Room. It has the form of a pasty white coating on the wall of the passage. The coating extends about three feet up from the floor and is about 0.2 inches thick. The material resembles moonmilk. As it occurs in the cave, it is wet and pasty, and large water droplets accumulated inside the sample bottle when the material was removed from the cave. Samples of the coating gave sharp, well-defined X-ray diffraction patterns that agreed well with the patterns published by Murray and Dietrich (1956) for taranakite from Pig Hole Cave, Virginia and by Balenzano et al. (1976) for taranakite from Castellana Cave in Italy. More recent data for several Italian caves are given by Fiore and Lavanio (1991).

The *Handbook of Mineralogy* (Anthony et al. 2000) lists more than a dozen localities for taranakite distributed all over the world, almost entirely cave sites. The composition was originally given as $H_6K_3Al_5(PO_4)_8 \cdot 18H_2O$. The *Handbook of Mineralogy* gives the composition as $(K, Na)_3(Al, Fe^{3+})_5(PO_4)_2(PO_3OH)_6 \cdot 18H_2O$. The two formulae are identical except that the Handbook formula recognizes the possibility of Na and Fe substitution and also that the phosphate anion consists of two phosphate groups and six monohydrogen phosphate groups.

The drops of water which collected in the sample bottle suggest dehydration of the sample. However, the X-ray pattern of material air-dried for one year was the same as that of the original except for some small changes in peak intensity. Both fresh and dried material had the characteristic 15.8 Å basal spacing of taranakite. There was no evidence for dehydration to francoannellite with a 13.7 Å basal spacing, as was observed in the Castellana Cave minerals. Because the Butler Cave taranakite seemed to be exceptionally well-crystallized, refined X-ray powder diffraction data were collected. Comparisons of the X-ray diffraction data are given in the previous report (White 1982).

Phosphate minerals occur in isolated pockets in the relatively dry sandy soils of the Second Parallel Passage. One of these was a pocket about an inch in diameter filled with loose deep brown to black material. The only identifiable reflections in the X-ray diffraction pattern could be assigned to poorly crystallized hydroxyapatite, $Ca_5(PO_4)_3(OH)$. Almost certainly, other material is present that is amorphous to X-rays.

Crandallite, $CaAl_3(PO_4)_2(OH)_5 \cdot H_2O$, occurs as white earthy nodules mixed with a black material in loose masses in the clastic sediment at the top of the fill bank at the head of the Second Parallel Passage. The X-ray diffraction pattern is sharp and well-resolved, indicating a well-crystallized material. SEM images of a sample show a nodular texture (Fig. 23.42). Energy dispersive X-ray spectroscopy of the nodules indicates a calcium-aluminum-phosphate of quite uniform composition. Scattered through the nodules are small acicular crystals which have essentially the same chemical composition as the bulk crandallite (Fig. 23.43). The X-ray diffraction pattern is also a fairly good match to the pattern for woodhouseite, $CaAl_3(PO_4)(SO_4)(OH)_6$, which is both chemically and structurally very similar to crandallite. However, the energy dispersive X-ray spectrum reveals no trace of sulfur, thus limiting the choice to crandallite.

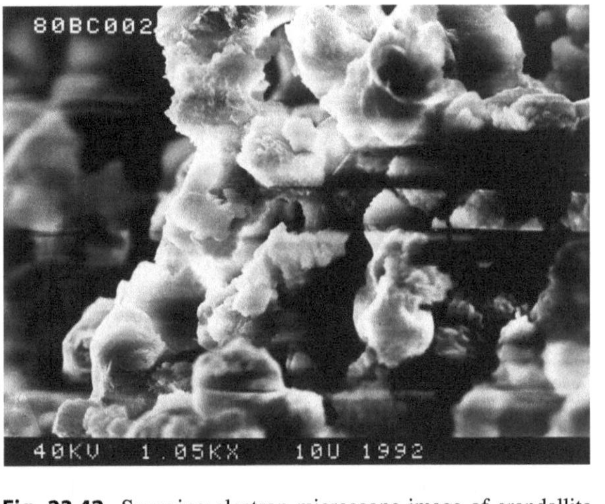

Fig. 23.42 Scanning electron microscope image of crandallite nodules. Magnification 1050X. Individual nodules are 5–10 μm across

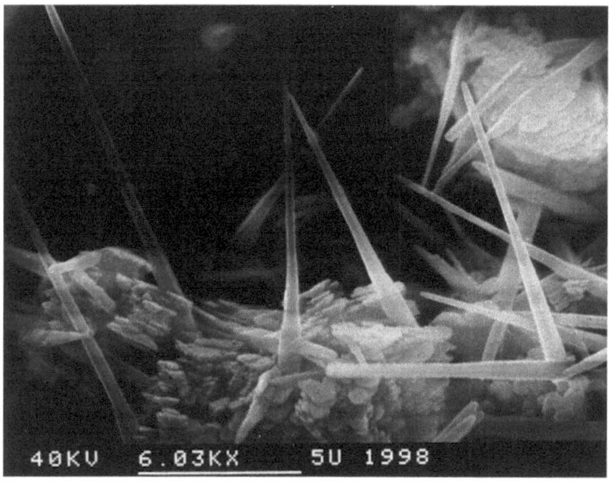

Fig. 23.43 Scanning electron microscope image of acicular crystals embedded in crandallite nodules. Magnification 6030X. Needles are about 5 μm in length

Fig. 23.44 Exfoliated crystals of sasaite. Magnification 800X. Each "crystal" is about 50 μm in diameter

An unusual phosphate mineral occurs in a small pocket in partially indurated sandy soil. This material was a homogeneous white powder either mixed with sand grains or filling small cracks and voids in the bedded sediments. It appears to be the mineral sasaite first described as a new mineral from the West Dreifontein Cave, South Africa (Martini 1978). However, the identification of this mineral is less certain than those described above. If the Butler Cave material really is sasaite, it was only the second observation of this mineral at the time the original report was published in 1982. The name is from SASA—the South African Speleological Association—and may be the only example of a mineral named for a caving group.

The X-ray diffraction pattern is not definitive. The major diffraction peaks match those reported by Martini for the West Dreifontein Cave material but not precisely. The diffraction pattern is superimposed on a broad maximum indicative of a considerable fraction of material that is amorphous to X-rays. Sasaite is a highly hydrated mineral and some of the water is easily lost. Varying water content would also vary the positions of the X-ray diffraction peaks. More definitive evidence for the correct identification of sasaite is the morphology of the crystals. Sasaite apparently has an expandable layer structure so that, as water is lost, the layers unfold like a wet book that has been allowed to dry (Figs. 23.44 and 23.45). The SEM images of the Butler samples are very similar to SEM images published by Martini.

Fig. 23.45 Closeup of an exfoliated sasaite crystal. Magnification 2000X

There is also an unresolved issue concerning the composition of the Butler sasaite. Martini's original published composition was $Al_{14}(PO_4)_{11}(SO_4)(OH)_7 \cdot 83H_2O$. One out of twelve of the phosphate ions is replaced by a sulfate ion. Martini claimed that the sulfate was uniformly present in all of his samples. The *Handbook of Mineralogy* now gives the composition as $(Al, Fe^{3+})_6(PO_4, SO_4)_5(OH)_3 \cdot 35\text{-}36H_2O$. The ions are present in slightly different ratios, there is the possibility of some iron substitution for aluminum, there is less water although the phase is still highly hydrated, and the sulfate content is variable. No sulfur was detected in the Butler sample by energy dispersive

Table 23.3 Minerals identified in Burnsville Cove Caves

Mineral	Composition	Occurrence
Aragonite	$CaCO_3$	Coatings, anthodites
Birnessite (?)	$(Na, Ca, Mn^{2+})Mn_7O_{14} \cdot 2.8H_2O$	Coatings on stream pebbles
Calcite	$CaCO_3$	Dripstone, flowstone
Crandallite	$CaAl_3(PO_4)(PO_3OH)(OH)_6$	Nodules in clastic sediment
Goethite	FeOOH	Wall crusts and nodules
Gypsum	$CaSO_4 \cdot 2H_2O$	Crusts and rare gypsum flowers
Hydromagnesite	$Mg_5(CO_3)_4(OH)_2 \cdot 4H_2O$	Moonmilk
Hydroxyapatite	$Ca_5(PO_4)_3(OH)$	Nodules in clastic sediments
Sasaite	$(Al, Fe^{3+})_6(PO_4, SO_4)_5(OH)_3 \cdot 36H_2O$	Veins in sediments
Taranakite	$(K, Na)_3(Al, Fe^{3+})_5(PO_4)_2(PO_3OH)_6 \cdot 18H_2O$	Wall crust

X-ray spectroscopy. The current (2011) mineral data website refers back to the Martini description. Sasaite has only been reported from cave localities; clearly questions remain concerning both its structure and its composition.

Although many phosphate minerals have been found in caves associated with guano deposits, aluminum phosphates are less common. Murray and Dietrich (1956) originally discovered taranakite in Pig Hole Cave, Virginia where they ascribed its origin to the reaction of leachate from the guano with clay minerals in the cave soils. The guano provided the phosphorus and potassium while the aluminum was supplied by the clays. Other investigators have essentially agreed with Murray and Dietrich's hypothesis. The difficulty with the Butler Cave occurrences of the aluminum phosphate minerals is that they occur as nodules within the clastic sediments and as well coatings. There is no associated organic material and certainly no guano deposits in the cave at present.

The phosphate minerals in Burnsville Cove caves were all found within a single small area in and near the Second Parallel Passage in the Butler Cave Section of the Butler Cave-Sinking Creek System. It is now established that the upper level passages on the west flank of the Sinking Creek Syncline are very old, certainly early Pleistocene, possibly older. These are steeply sloping passages that have received considerable runoff from Jack Mountain at times past. It is possible that at one time there was an open cave entrance, perhaps much like present day Breathing Cave, at the upper end of the Second Parallel Passage. The cave could have harbored a large colony of bats. With climate change and with the collapse of the cave entrance, the Second Parallel Passage was closed off. The bats disappeared and over a very long period of time so did their guano. All that remains are the aluminum phosphate interaction products which are very stable in the cave environment.

23.6 Conclusions

The minerals that have been identified in Burnsville Cove caves are summarized in Table 23.3. The list is very unlikely to be complete. Although the Butler Cave-Sinking Creek System was examined in some detail, there are crusts and coatings that were not analyzed. The prominent black coatings that are common on both present day stream cobbles and in the old clastic sediments have not been analyzed. The quite spectacular mineralization of the Chestnut Ridge System has been barely touched. A very profitable thesis could be undertaken on the calcite/aragonite speleothems and the various coatings by a graduate student who was able to deal with the rigors of actually entering the cave system.

Acknowledgments The mineralogical study made use of the facilities Materials Research Institute at the Pennsylvania State University. Judy Marks is thanked for preparing many of the X-ray patterns. Samples were collected by Gregg Clemmer and Nevin Davis. Photographs were provided by Philip C. Lucas, Nevin W. Davis and from the Ron Simmons collection.

References

Anthony, J.W., R.A. Bideaux, K.W. Bladh, and M.C. Nichols. 2000. *Handbook of mineralogy, arsenates, phosphates, vanadates*, vol. 4, 680. Tucson: Mineral Data Publishing.

Balenzano, F., L. dell'Anna and M. di Perro. 1976. *Francoanellite, $H_6K_3Al_5(PO_4)_8.13H_2O$, a new mineral from the caves of Castellana*, 49–57. Puglia, Southern Italy: Jahrbuch für Mineralogie, Monatshefte.

Bridge, P.J. 1971. Analyses of altered struvite from Skipton, Victoria. *Mineralogical Magazine* 38: 381–382.

Bridge, P.J. 1973a. Urea, a new mineral, and neotype phosphammite from Western Australia. *Mineralogical Magazine* 39: 346–348.

Bridge, P.J. 1973b. Guano minerals from Murra-el-elevyn Cave, Western Australia. *Mineralogical Magazine* 39: 467–469.

Bridge, P.J. 1974. Guanine and uricite, two new organic minerals from Peru and Western Australia. *Mineralogical Magazine* 39: 889–890.

Curl, R.L. 1972. Minimum diameter stalactites. *National Speleological Society Bulletin* 34: 129–141.

Fiore, S., and R. Laviano. 1991. Brushite, hydroxylapatite, and taranakite from Apulian caves (southern Italy): New mineralogical data. *American Mineralogist* 76: 1722–1727.

Gèze, B. 1957. *Les cristallisations excentriques de la Grotte de Moulis*, 90. Paris: Ateliers D'Impressions D'Art Jean Brunissen.

Harmon, R.H., and J.W. Hess. 1982. Ground water geochemistry of the Burnsville Cove area, Virginia. *National Speleological Society Bulletin* 44: 84–89.

Henderson, E.P. 1949. Some unusual formations in skyline caverns, Va. *National Speleological Society Bulletin* 11: 31–34.

Hill, C.A., and P. Forti. 1997. *Cave minerals of the world*, 2nd ed, 463. Huntsville: National Speleological Society.

Huizing, T., M. Jarnot, G. Neumeier, R.P. Richards, and G. Staebler. 2003. *Calcite—The mineral with the most forms*, 114. East Hampton: Lapis International.

Kaye, C.A. 1959. Geology of Isla de Mona, Puerto Rico, and notes on the age of mona passage. *U.S. Geological Survey Professional Paper* 317-C: 141–178.

Martini, J. 1978. Sasaite, a new phosphate mineral from West Driefontein Cave, Transvaal, South Africa. *Mineralogical Magazine* 42: 401–404.

Murray, J.W., and R.V. Dietrich. 1956. Brushite and taranakite from Pig hole Cave, Giles County, Virginia. *American Mineralogist* 41: 616–626.

Northup, D.E., and K.H. Lavoie. 2001. Geomicrobiology of caves: A review. *Geomicrobiology Journal* 18: 199–222.

Swezey, C., and N. Piatak. 2003. Gypsum in the Butler Cave—Sinking Creek Cave System, Bath County, Virginia. *BCCS Newsletter* 28: 15–22. GSA Abstract (2002).

White, W.B. 1982. Mineralogy of the Butler Cave—Sinking Creek System. *National Speleological Society Bulletin* 44: 90–97.

White, W.B. 1994. The anthodites from Skyline Caverns, Virginia: The type locality. *National Speleological Society Bulletin* 56: 23–26. (Discussion by D.G. Davis and reply, 57: 52–55; more discussion and reply 58: 54–55).

White, W.B. 1997. Thermodynamic equilibrium, kinetics, activation barriers, and reaction mechanisms for chemical reactions in karst terrains. *Environmental Geology* 30: 46–58.

White, W.B., and E.L. White. 2003. Gypsum wedging and cavern breakdown: Studies in the mammoth cave system, Kentucky. *Journal of Cave and Karst Studies* 65: 43–52.

White, W.B., C. Vito, and B.E. Scheetz. 2009. The mineralogy and trace element chemistry of black manganese oxide deposits from caves. *Journal of Cave and Karst Studies* 71: 136–143.

White, W.B. 2012. Speleothem microstructure/speleothem ontogeny: A review of Western contributions. *International Journal of Speleology* 41: 331–360.

Geomorphic Evolution of the Burnsville Cove Caves

24

William B. White

Abstract

The cave descriptions and maps were used to produce a highly tentative chronology for the development of the caves of Burnsville Cove. It begins with the establishment in the late Cretaceous to mid-Tertiary of the Schooley erosion surface. There may have been early hydrothermal development of deep solution pathways by the Eocene volcanism. Cave development accelerated in the Miocene by south-flowing water that created the upper level trunk passages in Chestnut Ridge. The Bullpasture River was captured along the Water Sinks lineament producing a reversal of gradients and a reversal of drainage to its present discharge points in the Bullpasture River springs. There was massive sediment in-filling in the mid-Pleistocene followed by removal of the sediment and further modification of the cave systems from the mid-Pleistocene to the present.

[**Editors Note**: This chapter must be considered a very tentative interpretation of the geomorphic events in Burnsville Cove over a very long period of time. The early time events are those proposed in the paper that Benjamin Schwartz and Dan Doctor presented at the International Congress of Speleology in 2009 (Schwartz and Doctor 2009). The Schwartz-Doctor hypothesis is a highly innovative interpretation of the earliest phases in the development of the Burnsville Cove caves. To bring the story to more recent times, I have attempted to paste in some interpretation of the karst geomorphology based on the 1982 Butler paper and on some recent work on karst geomorphology elsewhere in the Appalachians. There are no reliable time markers so the described sequence of events contains a great deal of guesswork.]

24.1 Introduction

With all of the information compiled in previous chapters, the objective is to attempt the construction of a coherent model for the development of the caves and karst of Burnsville Cove. This entails fitting together all of the major cave systems, their associated drainage and the present day topography. It also entails fitting the specific case of Burnsville Cove into the broader perspective of the karst landscape evolution in the Appalachian Mountains.

The cave geologist is faced with much the same problems as a scholar attempting a translation and compilation of—let us say—the Dead Sea Scrolls or the Nag Hamadi Library. The record is torn and fragmented. Many pieces are missing. Some bits of

With a contribution by Benjamin F. Schwartz and Daniel H. Doctor.

W.B. White (✉)
Department of Geosciences, The Pennsylvania State University, University Park, PA 16802, USA
e-mail: wbw2@psu.edu

W.B. White (ed.), *The Caves of Burnsville Cove, Virginia*,
Cave and Karst Systems of the World, DOI 10.1007/978-3-319-14391-0_24,

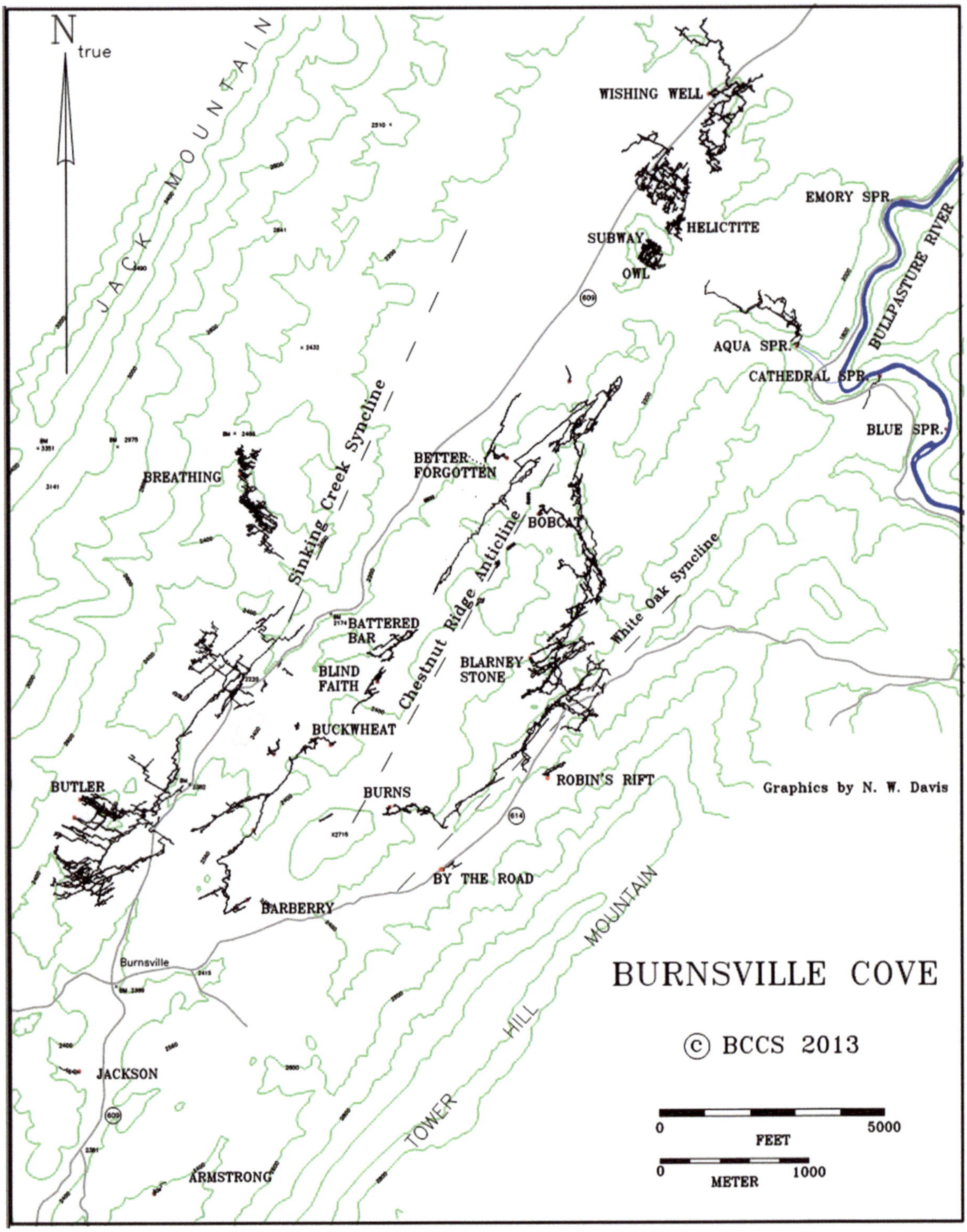

Fig. 24.1 The caves of Burnsville Cave. Map with more detailed topography is in the electronic file. Map by Nevin W. Davis

parchment or papyrus have been overwritten by later writers. And, of course, the scholar must understand the language in which the records were written. A cave is much the same. We see only fragments—segments of passage that happen to be accessible for exploration. We do not see the passages as they were originally created because much of the original conduit system is blocked by collapse, or by clastic fills, or by speleothems, or has been eroded away. Further, the passages often have been overprinted by later processes and younger passage development. Thus, the cave geologist has the double task, first of reconstructing the passages that are accessible, and then reconstructing the entire pattern of cave development by fitting together the passage fragments that are accessible and making reasonable guesses about the passage fragments that are inaccessible or eroded away.

24.2 Puzzle Pieces: The Caves of Burnsville Cove

In Chap. 1 readers were introduced to Burnsville Cove with a map that showed the caves in their proper spatial relationship. The closing chapter of the book returns to this map to guide the interpretation of how all of the cave fragments are related to present day topography (Fig. 24.1). The topography itself, of course, is a snapshot—the land surface as it exists at the present moment of time. It has evolved from previous topographies in the past and will èvolve into different topographies in the future.

The following sections summarize the puzzle pieces presently known in Burnsville Cove. The larger cave systems have been described in detail in previous chapters. To these are added two additional fragments, Better Forgotten Cave and By-the-Road Cave.

24.2.1 The Butler Cave—Sinking Creek System

The Butler Cave-Sinking Creek System drains northeastward, closely following the axis of the Sinking Creek Syncline. Although the route from the entrance to the main trunk passage cuts down through the strata from the Upper Breathing Cave Sandstone to below the Lower Breathing Cave Sandstone, the side caves, including Breathing Cave are mainly tributaries to the trunk drainage. The present day trunk passage is a single master passage with a set of active streams that closely follows the axis of the Sinking Creek Syncline. There are multiple sumps at somewhat different elevations because of the influence of structure and the Lower Breathing Cave Sandstone. If there are abandoned upper level trunks, they have yet to be discovered. The Sinking Creek trunk is in the upstream, southern, portion of the Cove with a large gap in both horizontal distance and in elevation between the downstream sumps and the reappearance of the water in the Aqua Cave outlet system.

24.2.2 The Chestnut Ridge Cave System

There is a clear trend in the Barberry—Buckwheat—Blind Faith—Battered Bar—Burnsville Turnpike as a single sequence forming the western arm of the Chestnut Ridge System. This trend extends for 2.6 miles along the western flank of the Chestnut Ridge anticline but it converges toward the anticline axis rather than being parallel to it. The dips are modest in Barberry Cave, closer to the axis of the Sinking Creek Syncline, become steeper in the Burnsville Turnpike, and flatten again at the northern end on the axis of the Chestnut Ridge Anticline.

The steamways in Burns and the southern portion of the Blarney Stone section of the Chestnut Ridge System trend northeastward closely following the White Oak syncline until the streams are lost in sumps. The higher levels veer away from the syncline axis to trend almost due north along strike as the North-South Trunk which eventually merges with the western arm at the Big Bend.

24.2.3 Better Forgotten Cave

Better Forgotten Cave (see description of exploration in Chap. 2) is a one third-mile segment of streamway on the northwestern side of Chestnut Ridge (Fig. 24.2). It drains to Aqua Spring but has no known connection with either the Butler drainage or the Chestnut Ridge drainage.

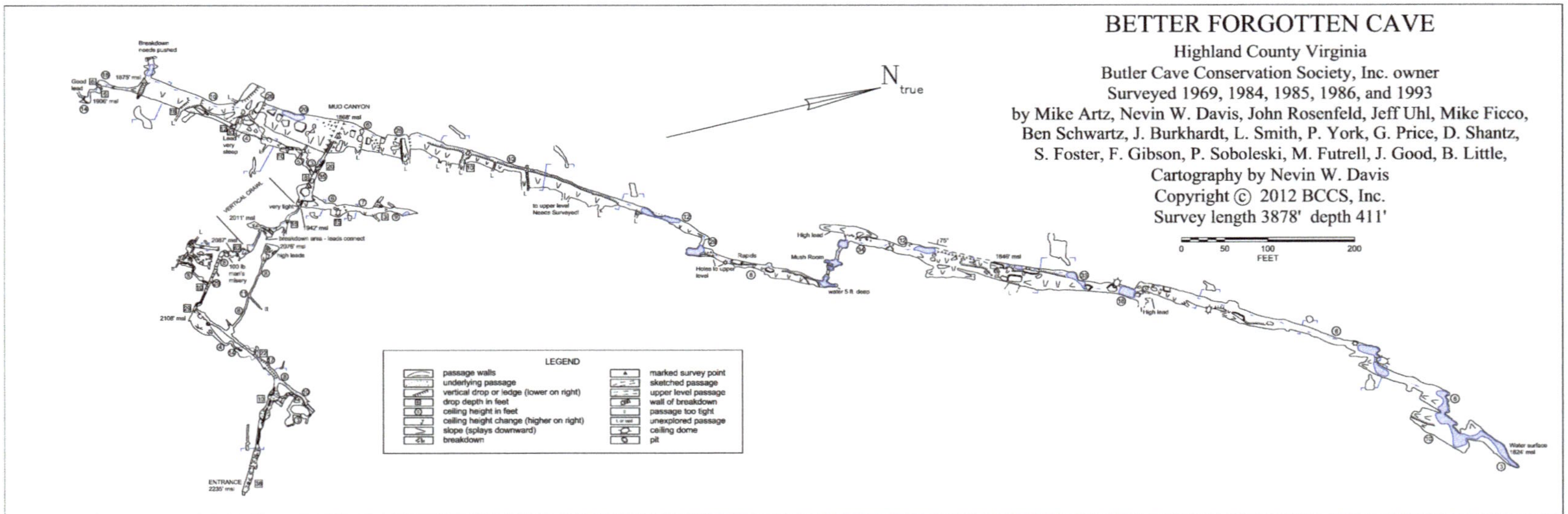

Fig. 24.2 Better Forgotten Cave

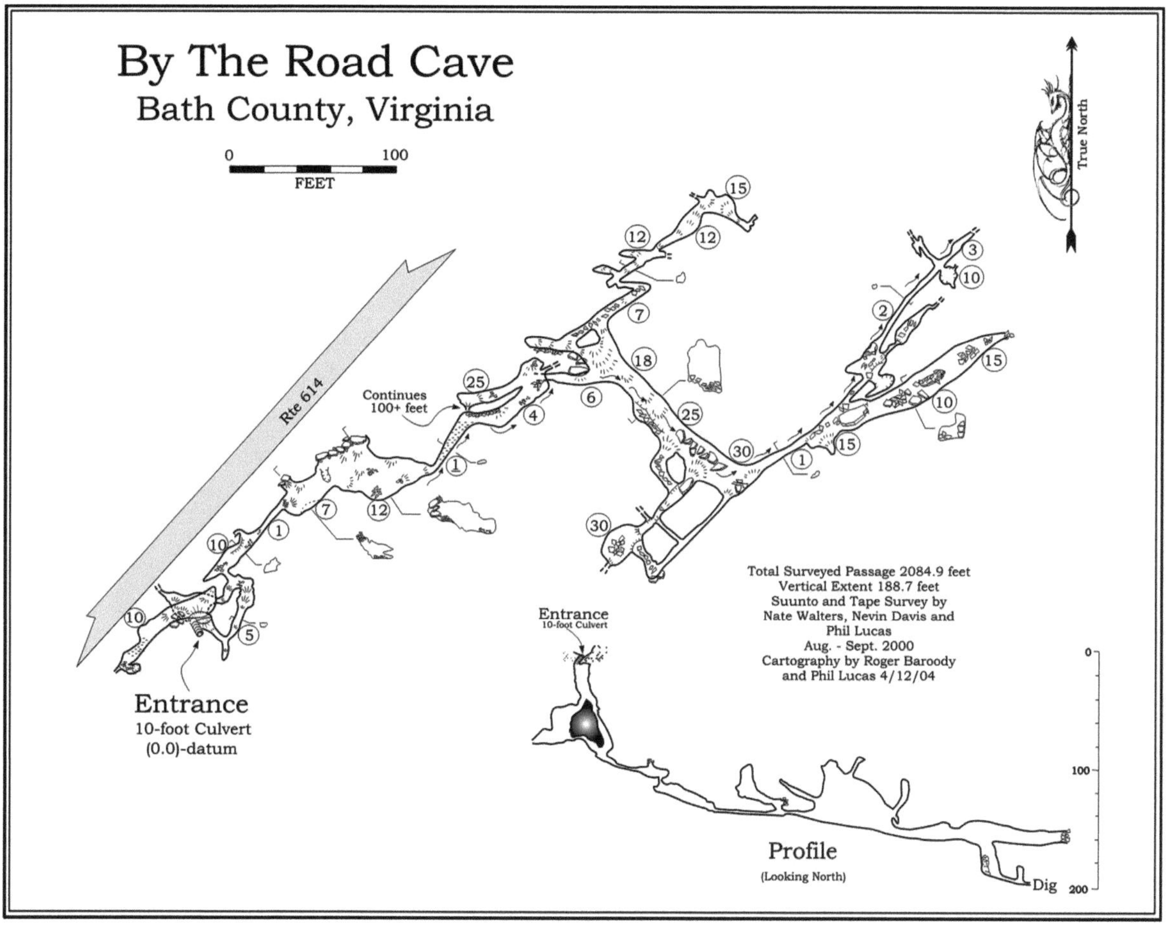

Fig. 24.3 By-The-Road Cave

24.2.4 By-The-Road Cave

By The Road Cave is a fragment of cave (Fig. 24.3) near the base of Tower Hill Mountain. It is aligned with Robins Rift and with the drainage line to Blue Spring assuming that the drainage follows the strike. Curiously, although the cave should be well up on the eastern flank of the White Oak Syncline, the beds in the cave have a low dip.

24.3 The Pre-pleistocene Evolution of the Burnsville Karst Drainage by Benjamin F. Schwartz and Daniel H. Doctor

[**Editor's Note**: This section reproduces the paper by Benjamin Schwartz and Daniel Doctor, Geomorphic and Hydrogeologic Evolution of Karst in the Burnsville Cove, Virginia, USA: New Evidence and Perspectives, originally published in the *Proceedings of the 2009 International Congress of Speleology.* Omitted are Sects. 1 and 2 which give an introduction and background information of the Cove. The references have been merged with the overall chapter references.]

24.3.1 Earlier Interpretations and Difficulties Therewith: Paleo-Flow Direction Prior to Most Recent Incision of the Bullpasture River

Some of the most prominent karst features in the Cove are the four major springs which collectively drain the karst and discharge at or near river level in the

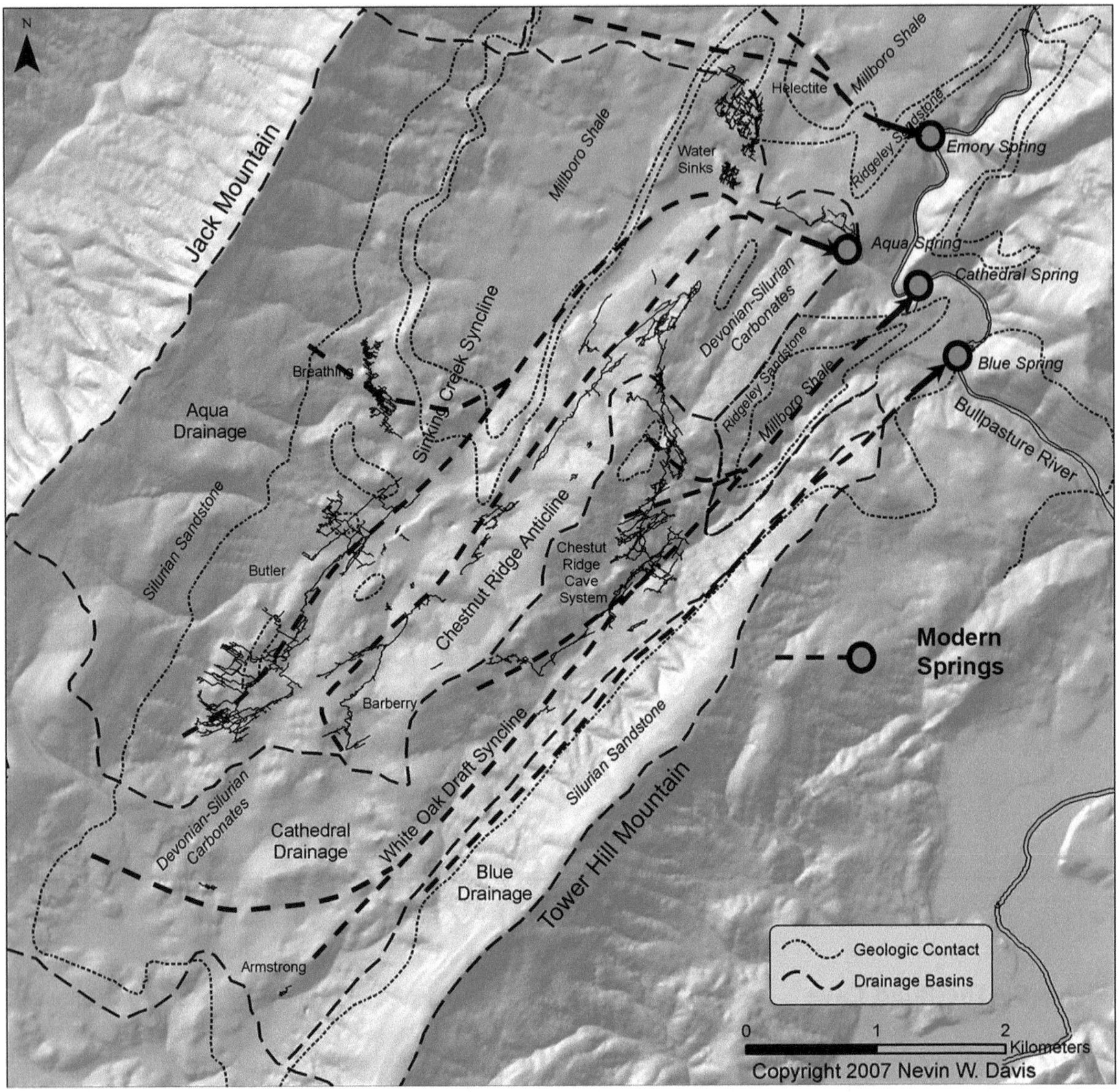

Fig. 24.4 *Shaded* relief detail map of the Burnsville Cove showing major caves, modern drainage basins, geologic contacts, modern springs and flowpaths

Bullpasture River Gorge (Fig. 24.4). Surveyed cave passages containing active streams all drain towards these springs and carry water generally from the southwest to the northeast. However, there are several pieces of evidence which indicate that a similar drainage arrangement in the past is unlikely.

First, the elevation of a regionally extensive abandoned phreatic trunk passage in the Chestnut Ridge System is inconsistent with its current position within the hydrogeologic setting of the Cove. For such large-scale phreatic development to have occurred in Chestnut Ridge, the water table must have been at least as high as the uppermost portions of this passage: currently $\sim$400 feet higher than both the sumps in the cave system and the corresponding spring elevations. Therefore, the regional base level was at least 400 feet higher than at present, which means that the Bullpasture River's channel was also $\sim$400 feet higher. If this were the case, what is now the Bullpasture River Gorge would have been largely or entirely covered by substantial amounts of Oriskany Sandstone (~80 feet) and impermeable Millboro Shale (~1100 feet).

In a similar geologic setting, a breccia body associated with volcanic rocks near Monterey, Virginia, in Highland County, contains chips of the now-eroded overlying Millboro shale. This has been interpreted to represent a history of at least 1100 feet of erosion since Eocene volcanism (Tso and Surber 2006). Although erosion rates remain unconstrained in the Burnsville Cove, such a cap of siliciclastic rocks would have prevented the formation of large springs in locations similar to the modern Emory, Aqua and Cathedral Springs. It would not, however, have prevented deep circulation and dissolution from occurring. With no outlet available near the location of modern springs, water would have been forced to find the most hydraulically efficient discharge point. There is no concrete evidence indicating exactly where the paleo spring may have been, but there is evidence in the Chestnut Ridge System which indicates that the most probable location is in the vicinity of Blue Spring. Blue Spring currently drains a long narrow strip of steeply dipping limestone which is bounded on the southeast by the Silurian siliciclastics which form Tower Hill Mountain. This strip of nearly vertical carbonates may have been the only outcrop exposed in the early Bullpasture River channel and may also have been the most efficient discharge point. In Chestnut Ridge System, a large horizontal trunk passage has been explored to a sediment fill just east of the axis of the White Oak Draft Syncline (Fig. 24.5). This trunk passage extends for approximately 2 miles along strike from the nose of Chestnut Ridge to White Oak Draft where further exploration has been stopped by the sediment plug. It seems probable that this passage curves to the north beyond this obstacle and continues along the strike toward a paleo-resurgence near the modern Blue Spring.

Unfortunately, scallops indicating the paleo-flow direction have not been found in the large phreatic horizontal trunk passage in the Chestnut Ridge System. This is probably due to the combined effect of two factors: extremely low flow velocities and highly deformed and thin-bedded limestone interbedded with sub-mm to mm-scale shale beds. The first definitive evidence of a flow direction was recently discovered in an area of the Chestnut Ridge System known as Leprechaun Forest. Here, the large trunk passage diminishes in size to a crawlway for ~100 feet, likely owing to sediment fill in a descending loop in the passage. On the southeastern side of this restriction, a large mound of sediment was identified as a deposit caused by flow and transport from generally north to south, which is away from the modern springs. As water moves through a restriction it is able to transport sediment. When the passage volume increases and flow velocities decrease, this sediment load drops out (Fig. 24.6). The morphology of the sediment pile and a small excavation in the side of it confirmed the direction of flow.

24.3.2 Questions About the Earliest Phreatic Development

One of the most interesting characteristics of Breathing Cave and much of the upstream portion of Butler Cave is that many of the passages clearly formed under phreatic conditions, perhaps even with ascending water. Such a phreatic hypothesis was previously put forth for Breathing Cave (Deike 1960). Upper wall and ceiling morphologies in both caves preserve records of phreatic conditions with very low velocities. From the evidence in Butler Cave, we know that these passages are at least 760 feet higher than current spring elevations. Upper-level passages in Butler Cave show a strong dependence on geologic structure for their development, both along strike-oriented bedding planes and folds, and along dip-oriented joints. The lack of horizontal levels of passages that cut across the structural planes of weakness provides no indication that these passages were formed near the water table. In fact, passages formed along joints that parallel the dip on the western flank of the syncline exhibit morphologies that indicate water once flowed upward against the dip (Fig. 24.7).

This is a distinctly different setting from the one that currently exists, which is an extensive network of abandoned dry passages with a few active vadose tributary passages carrying water down toward the axis of the Sinking Creek Syncline. In Butler Cave, a large trunk passage roughly follows the axis of the syncline. Water flows to the northeast in this trunk,

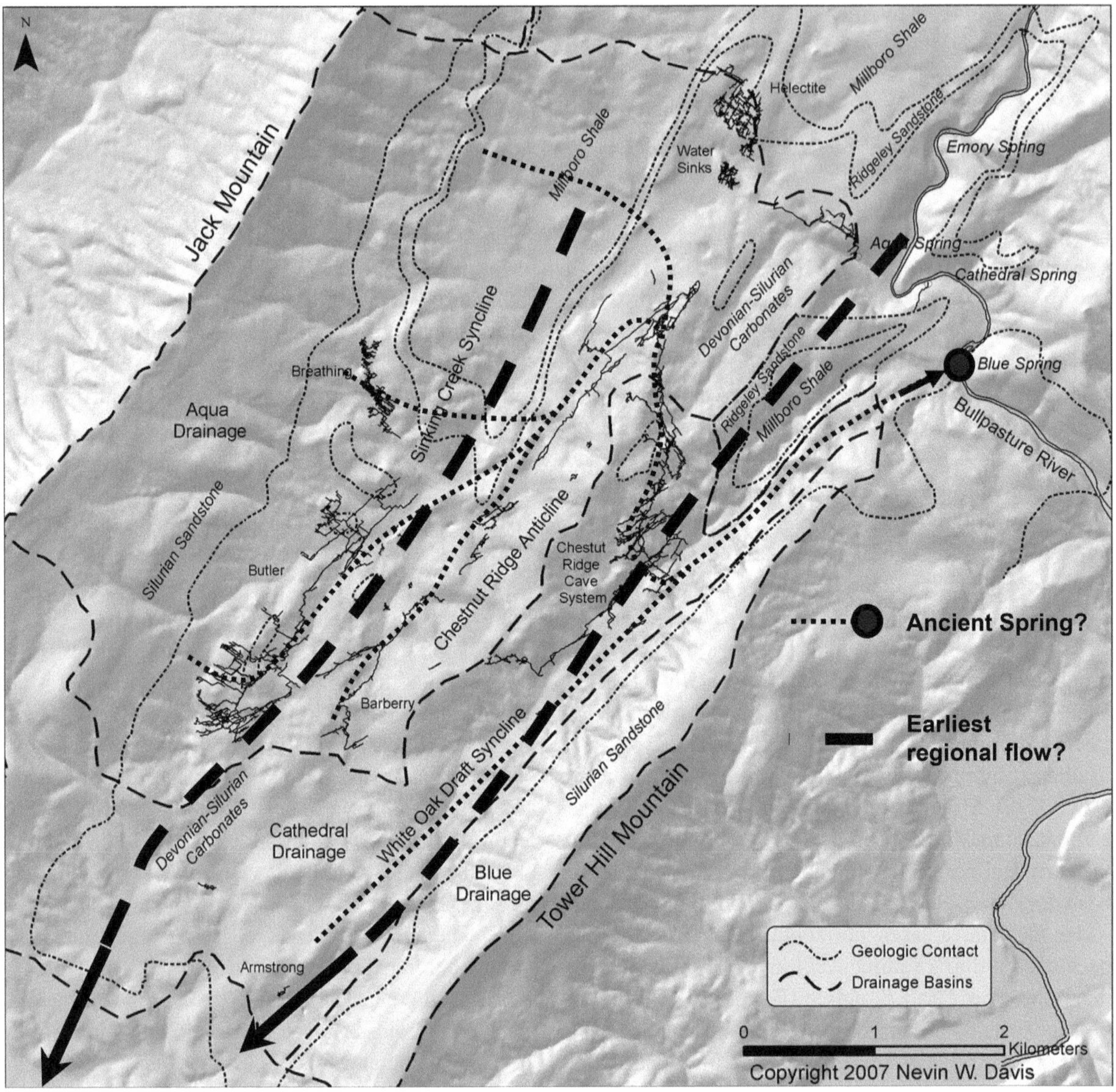

Fig. 24.5 Hypothesized ancient spring and earliest regional flowpath

alternately following the trunk passage and disappearing into inaccessible parallel passages below and to the southeast of the main trunk. Ultimately, all streams sump at the northeastern terminus of the explored cave. A small stream in Breathing Cave also sumps as it drains towards the axis of the syncline. Both Breathing and Butler caves contain abundant evidence of more recent modification by vadose action, including directional scallops and erosional features at floor level. Many passages that were once completely filled with clastic material derived from Jack Mountain have been and continue to be re-excavated by vadose stream action.

In addition to evidence for early deep-phreatic flow and passage development, other geologic evidence in the region indicates that hydrothermal fluid migration affected the rocks that host caves in Burnsville Cove. Numerous Eocene igneous intrusions occur in northern Highland County, about 12 miles north of the Burnsville Cove (Tso et al. 2004). Breccia bodies associated with some of these volcanic intrusions are interpreted as having formed via a mechanism of hydrovolcanic

Fig. 24.6 The sediment pile in the chestnut ridge system which provided the first evidence of paleo flow to the southeast in the abandoned phreatic trunk passage. Flow was from left to right in the picture. Photo by Benjamin Schwartz

Fig. 24.7 A set of rising cupolas in Butler Cave. These features formed along a dip-oriented joint in the Dave's Gallery passage and indicate phreatic flow upwards along the dip. Boulder chockstone is approximately 0.5 m in width. Photo by Dan Doctor

diatreme emplacement in which dikes propagated upward along joints, encounter groundwater, and explosively form diatremes (Tso and Surber 2006).

Recently, an igneous intrusion was found in a cave in southern Bullpasture Mountain (Schwartz 2003; Rossi et al. 2013; Helsley et al. 2013). This is a considerable southern extension to the known range of igneous intrusions in Highland County, and places confirmed igneous activity within 8 miles of Butler Cave. Additionally, a lithium-bearing manganese oxide mineral [Lithiophorite $(Al, Li)Mn^{4+}O_2(OH)_2)$] has recently been identified in Butler Cave (Schwartz et al. 2008). This mineral is most commonly associated with hydrothermal deposits. Herkimer Diamonds and milky quartz veins are also associated with hydrothermal fluid migration and are abundant in the highly deformed Tonoloway Limestone. The link between karstification and the hydrothermal activity that affected the carbonate rocks is as yet unclear.

If extensive networks of passages that are now >760 feet above base level were clearly developed under phreatic conditions, then this is evidence of a karst system which developed under conditions which were much different than those we observe today. Since that time, several pieces of new evidence support this hypothesis of deep-phreatic flow. The first is that explored conduits in Aqua and Cathedral Springs rise from depths of at least 256 and 150 feet, respectively (see Chap. 4). In both cases, the conduits continue and exploration was halted due to diving logistics. The second piece of evidence is from a karst system which lies adjacent to and north of the Emory Spring drainage. Here, the Chestnut Ridge anticline plunges to the north and probably flattens out into the much larger Bullpasture Valley syncline. This syncline exposes the Helderberg Group carbonates on the flanks of Jack Mountain to the west and Bullpasture Mountain to the east. The Bullpasture Valley is floored with the stratigraphically higher Millboro Shale. Dye tracing from high on the flank of Jack Mountain revealed that water sinking there flows under the Bullpasture Valley, rises at a large karst spring on the eastern flank of the valley before flowing back to the west and into the Bullpasture River (see Chap. 13). A third piece of evidence is discharge observations at Emory Spring, which has no known associated cave passage although drainage from both Helictite and Wishing Well Caves have been traced to it. Emory Spring responds differently than Aqua and Cathedral Springs to flood events. Although discharge increases significantly after a rain event, large volumes of clear water discharge before muddy water arrives at the spring. In contrast, Aqua and Cathedral Spring begin discharging muddy water much more rapidly. Combined, these pieces of evidence tell us that not only it is probable that deep-phreatic development occurred in the region, but that these conditions still exist.

Based on geomorphic patterns (Fig. 24.5) and evidence found in the caves, we believe that the system may have originally drained towards an outlet farther to the south near Dry Run. Deike (1960) noted that, "Breathing Cave, as explored, does tend to parallel the local strike southward, and this may reflect a tendency for the water to go around the structures, parallel to the strike and hence to the structure contours, rather than directly beneath the anticlines and synclines." Currently, the southern end of the Burnsville Cove karst system is bounded by the erosive exposure of underlying sandstones that prevent karst development. However, this exposure creates a relatively narrow divide between two karst systems: the Burnsville Cove and the Dry Run karst system to the south. Dry Run, as its name implies, is dry for most of the year downstream from where the streambed encounters carbonates. Below this point, a deeply incised gorge carries floodwaters to the Cowpasture River while the karst system discharges from one or more springs to the west of the Cowpasture River. While we are still in the initial stages of investigation, a basin-size vs. gorge depth analysis suggests that the Dry Run Gorge is oversized and over-incised for its current size. The Dry Run basin drains ~ 30 mi^2 through its gorge while the Bullpasture River basin drains ~ 130 mi^2 through its gorge. In essence, the Dry Run basin contains <25 % of the drainage area in the Bullpasture River basin, yet it has a developed a gorge of similar dimensions and in a similar geologic setting. Both topographic features and an elevation profile along the Cowpasture River valley show that the Cowpasture River has a steeper channel gradient and is incising a gorge near its confluence with Dry Run Gorge. While this is still speculative, it may indicate that this portion of the Cowpasture River has experienced a relatively recent increase in erosion—perhaps because of an increase in erosive power after capturing the Bullpasture River.

24.3.3 The Evolutionary Model

We hypothesize that ancient drainage of both the early karst system and the Bullpasture River was toward the southwest where the nearly abandoned Dry Run Gorge intersects the Cowpasture River ~ 10 miles SW of the Burnsville Cove. During the time of maximum cave enlargement, widespread deep-phreatic development occurred along structural features, and possibly along preexisting hydrothermally-formed proto-conduits, under low-velocity conditions. Mineralogical and geological evidence of hydrothermal fluid migration and volcanism in the area has recently been recognized and deep (>300 feet) phreatic pathways still exist in the northern portion of the Cove's drainage, as well as farther to the north in the Bullpasture River Valley where water sinks high on the western flank of the valley and follows deep flow paths to discharge at a spring on the east side of the valley. Incision of the Bullpasture River Gorge, which continues today, significantly rearranged regional surface and subsurface drainage. As the gorge incised, it created new and more efficient discharge points along the ancestral Bullpasture Gorge, which drained phreatic passage networks and induced free-surface stream flow and sediment transport in many of the caves. There may have been an extended period of relative stability in the Cove during which time a large phreatic trunk passage formed along strike around the Chestnut Ridge anticline. Drainage rearranged again as the gorge probably experienced a second episode of rapid incision. This second period of incision may have been related to either climate variability or the erosive removal of geologic controls (i.e., resistant sandstone beds in the river channel). Much of the karst system as it exists today is in a state of disequilibrium as streams in the cave modify and/or abandon earlier passages in favor of newer vadose flow paths.

24.4 The Geomorphic Evolution of the Burnsville Karst

The Schwartz-Doctor hypothesis argues that initial cave development completely predates present day topography and requires a time when base levels were 400–700 feet higher than at present, comparable or higher than the present day ridge lines of Chestnut Ridge and Bullpasture Mountain. This old topography has been eroded until it eventually reached the topography seen today. The next question addressed is whether this scheme can be fitted into the time line for the geomorphic evolution of the Appalachian Mountains. The construction that follows is very rough and tentative, but, perhaps, is not entirely geofantasy.

24.4.1 Erosion Surfaces, Cave Levels, and the Time Sequence of Events

The present-day relationships can best be illustrated with a vertical profile (Fig. 24.8) oriented southwest-northeast along the Cove. The zero-point for the profile is a line drawn perpendicular to the structure at the divide between the northeast drainage of Sinking Creek and White Oak Draft and the south-flowing drainage of Dry Run. The surface divide is about half a mile north of Burnsville but the underground divide is about a mile south of Burnsville. Valley profiles for Sinking Creek and for the ridge line of Chestnut Ridge were scaled from the Burnsville and Williamsville topographic maps. Cave and spring elevations were obtained from Tables 19.1 and 20.1. The profiles are plotted as elevation with respect to the Bullpasture-Cowpasture River confluence at 1595 feet above sea level to give a measure of elevation with respect to present day base levels.

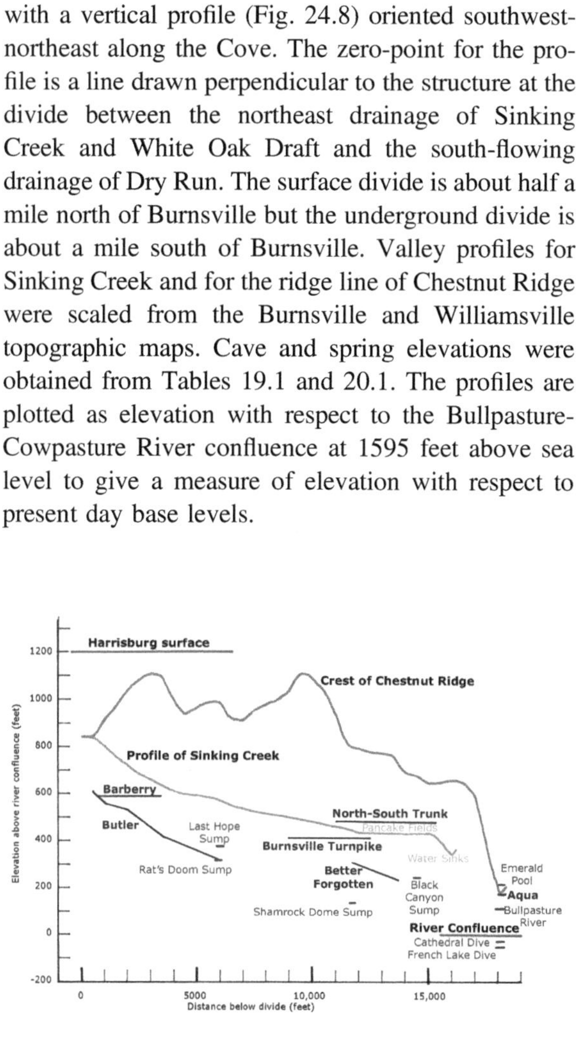

Fig. 24.8 Profile along Burnsville Cove from southwest to northeast. The Chestnut Ridge profile follows the highest ridgeline from the divide to the incised valley of Mill Run which it intersects near the entrance of Aqua Cave. The profile of the Sinking Creek Valley is drawn from the divide to Water Sinks. Also shown are profiles of major cave trunk passages and some of the most important sumps

The oldest recognized (at least by some) geomorphic feature in the Appalachians is the Schooley erosion surface which corresponds to the roughly accordant summits of the Appalachian ridges. The age of the Schooley surface is poorly determined with estimated ages ranging from the Cretaceous to the mid Tertiary (Thornbury 1965). Because the Appalachian ridge tops are supported by resistant sandstones and quartzites, erosion rates are low—typically 2–5 m/Ma—so that the lowering of the ridge lines since the mid-Tertiary would be no more than a few hundred feet. Various estimates from other karst areas in the Appalachians suggest that the dissection of the Schooley surface began 10–14 million years ago, mid-Miocene time (White 2009).

The geologic cross section (Fig. 16.32) shows that Burnsville Cove is essentially a synclinal trough with the Chestnut Ridge anticline as a secondary feature in the center. If the younger age for the Schooley surface is accepted, the Eocene volcanism with its associated hypogenetic karstic activity would have occurred when the carbonate rocks were buried under at least 1000 feet of other sediments.

There is better evidence for the later but less extensive Harrisburg erosion surface in many parts of the central Appalachians. In the Greenbrier limestone karst of West Virginia to the west of Burnsville Cove, the Little Levels in Pocahontas County and the Great Savannah in Greenbrier and Monroe Counties occur near the 2500 level. There are remnants of a doline karst in the Swago Creek Basin in Pocahontas County also at this elevation. It was argued (White and White 1991) that these karstic surfaces are manifestations of the Harrisburg Erosion Surface. The Burnsville divide with its large closed depressions occurs at about the 2400 foot elevation. The crests of Chestnut Ridge and of Bullpasture Mountain where numerous large closed depressions occur are only slightly higher. Likewise, there are remnants of the 2400–2500-foot level preserved by the foothills to Jack Mountain particularly near Breathing Cave. These upland features in Burnsville Cove may well also represent remnants of the Harrisburg surface.

The Harrisburg surface of the folded Appalachians is usually assumed to be equivalent to the Highland Rim surface of central Tennessee. There is good evidence from cosmogenic isotope dating of cave sediments that the dissection of the Highland Rim began about 3.5 million years ago (Anthony and Granger 2004, 2007). Denudation rates of limestone terrain vary considerably but 30 m/million years is typical for humid temperate terrains (White 2009). This value is consistent with Tso and Surber's (2006) estimate of 1100 feet of erosion since the emplacement of the 35 Ma volcanic diatreme at Trimble Knob near Monterey. If a denudation rate of 30 m/million years is accepted for Burnsville Cove, the limestones exposed on the Burnsville divide would have lowered by roughly 300 feet since beginning of dissection. This would place the original Harrisburg Surface at 2700–2800 feet, the elevation of the high points on Chestnut Ridge. The crest of Bullpasture Mountain may also be a remnant of the Harrisburg Surface.

According to the geologic map (Fig. 16.3) the base of the limestone is exposed about a mile south of Burnsville. However, Nevin Davis reports that the Williamsport Sandstone crops out between the 2440 and 2480 foot contours above Blind Faith Cave and at 2440 feet above Buckwheat Cave. These outcrops, several miles northeast of what is nominally the limit of the limestone must represent structural highs that would also represent the locations where the limestone would first be exposed by the dissection of the Harrisburg surface. The total thickness of carbonate rocks varies from 700 to 900 feet with the lower number more likely for Burnsville Cove. Before erosion, the top of the limestone near the Burnsville Divide would have been near 3100, 1500 feet about present day base level and well above the estimated position of the Harrisburg surface.

If the high level master trunks of the Chestnut Ridge System developed below the Harrisburg surface, there would have been limestone exposed at the surface near and south of Burnsville while the ancestral Bullpasture River, meandering on the Harrisburg surface, would be flowing over hundreds of feet of clastic rocks. The only recharge areas to the north would be the limestone exposures along the flanks of Jack Mountain and the near-vertical beds of limestone on Bullpasture Mountain. Bullpasture Mountain, of course, would not have been a mountain. The present day ridge line of Bullpasture Mountain would have been part of the floor of a wide valley possibly extending from Jack Mountain to Shenandoah Mountain. Tower Hill Mountain would have been an isolated ridge rising above the valley floor.

At the time of maximum development of the Harrisburg Surface, mid- to late Miocene, it is possible that the ancestral Bullpasture river flowed south through Burnsville Cove, around the eastern side of Warm Springs Mountain, and joined the ancestral Cowpasture River near what would later be the mouth of the Dry Run canyon. Inspection of the Google Earth image (Fig. 1.1) shows the possible pathway. With the onset of dissection of the Harrisburg surface in the late Pliocene, the structural weakness of the Water Sinks lineament could have allowed the diversion of the Bullpasture around the northern end of Tower Hill Mountain leading to the incision of the present day Bullpasture Gorge. The deepening gorge provided the hydraulic gradient for the reversal of drainage in the karst system.

The master trunk in the Butler Cave—Sinking Creek System has a higher gradient to the north than the main trunk passages in Chestnut Ridge. However, it should be noted that the large cross section passage that extends from Sand Canyon to the Natural Bridge may be a fragment of the old south-flowing drainage line which would also include much of the higher portions of the side caves.

It remains unclear whether the development of the high level trunk passages are related to the Harrisburg surface and thus date from the late Miocene to early Pliocene or whether they relate to the dissection of the Schooley surface and are thus much older. What is clear is that cave development in Burnsville Cove has been operating for a very long time and that erosion on the surface and breakdown in the caves has made the early record very hard to read.

24.4.2 Contemporary Valley Development

[**Editor's Note**: The section that follows is an edited and revised segment of text taken from White and Hess (1982).]

The most recent geomorphic features are the contemporary valleys of Sinking Creek and White Oak Draft. North of the Burnsville divide various tributary streams on the flanks of Jack Mountain form the headwaters of Sinking Creek. Without exception these streams sink during dry weather along the limestone contact, and these and many smaller tributaries without surface expression form the various streams seen in the cave. The surface channel of Sinking Creek, however, is maintained for a distance of 3 miles along the axis of the valley, more or less the axis of the syncline, to its ultimate swallet at the southern edge of Pancake Fields. The surface channel of Sinking Creek carries water only during periods of high runoff—spring snow melt or exceptional rains. Most seasons of the year the main stream bed is dry throughout its length.

Some insight into the development of the Sinking Creek Valley can be obtained by fitting the valley profile to an exponential function of the form:

$$\left(E - E_{ref}\right) = E_0 e^{-KL}$$

In this equation, E is elevation in meters, scaled from topographic maps. E_{ref} is a reference datum taken as the previously defined intersection of the Bullpasture and Cowpasture rivers at an elevation of 486 m. E_0 is a fitting parameter, K is a characteristic slope function, and L is the distance in meters along the valley from a defined origin. The origin chosen for convenience was the junction of two tributaries of Sinking Creek flowing from Jack Mountain with the main valley thalweg at an elevation of 680 m (2230 feet).

When the valley profile of Sinking Creek is plotted (Fig. 24.9), the individual segments of valley thalweg fit the simple exponential model rather well. The two mountain tributaries have similar slope factors and appear as straight line segments in spite of the different

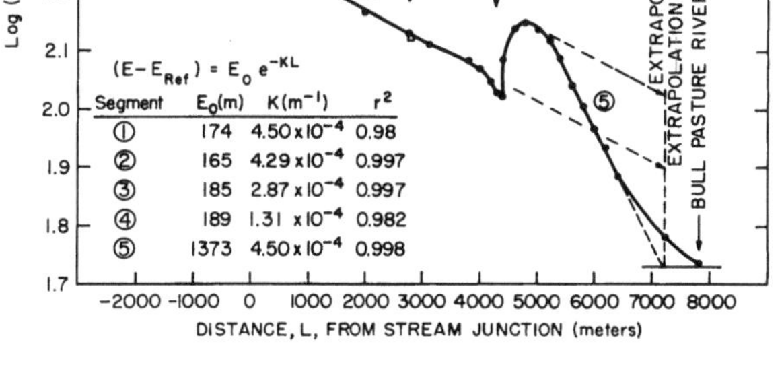

Fig. 24.9 Profile of sinking creek along Burnsville cove. Numbered segments have been least-squares fitted to exponential functions with the parameters shown in the lower left corner of the diagram. Note that the various fittings were calculated in the metric system in the original publication

sequences of rock lithology over which the streams flow. The main thalweg of Sinking Creek is also of simple exponential form from the stream junction to the Water Sinks Depression. The present sink of Sinking Creek must be a relatively recent development because there is no measureable break in the valley profile at this point. As expected, there is a large discontinuity in the valley profile at the Water Sinks Depression. The valley downstream from the sink has a distinctly steeper slope than the valley on the upstream side. Likewise, the upstream end of the Cove, from the Burnsville divide to the stream junction is steepened with respect to the main reach of the valley.

The confluence of the Cove with the Bullpasture River is difficult to project because the river at this point has moved to the eastern side of the valley where it begins its descent into the Bullpasture Gorge. If profile (5), the lower end of the valley is simply extended down to the elevation of the river, it defines a length, 7200 m, as the distance from the stream junction to the river. If the main valley profile (4) were extended out to this distance, it would intersect the ancestral Bullpasture River at an elevation of 566 m (1850 feet). At this elevation, the topographic maps show a well-defined terrace level into which the present day floodplain of the Bullpasture has been cut. It appears that the main Sinking Creek Valley floor has been left perched when the surface drainage was diverted underground. The present day valley has a fossil profile, graded toward the position that the river had when the diversion took place.

If the crest of the Water Sinks saddle is extrapolated downstream to the Bullpasture, it intersects the river position at 594 m (1950 feet), an elevation that corresponds to a series of accordant hilltops along the Bullpasture Valley to the north.

The saddle formed between the Water Sinks Depression and the downstream continuation of the valley is located at an elevation of 628 m (2060 feet) only 120 feet above the main valley profile extrapolated under the saddle. However, this is in the downstream reach where the exponential profiles are flattening out. If the crest of the saddle is extrapolated upstream to the Burnsville divide parallel to the main valley profile (Segment 4), it reaches the divide at 777 m (2550 feet) elevation, well above the present elevation of the divide at 750 m (2460 feet) but still below the calculated elevation of the Harrisburg surface. However, the uncertainties in these numbers do not allow much significance to be attached to the discrepancies.

24.4.3 The Pancake Field Paleolake by Benjamin F. Schwartz

The Water Sinks preserves what may be one of the most significant Appalachian and Eastern US deposits from the Last Glacial Maximum (LGM) period in the Late Pleistocene. The reason it does is that for at least some extended period of time prior to the LGM, the Water Sinks drain was apparently plugged, and what are now known as Water Sinks and Owl Caves were not hydrologically active in draining surface waters in the Cove. This resulted in the formation of a large lake that likely extended more than 1.5 miles to the SW of the current drain in the Water Sinks, and existed for an unknown period of time (Schwartz et al. 2009).

The area known as the Pancake Field is actually the top of a >70-feet thick deposit of paleo-lake-bed sediments. From two separate auger cores, and a ~ 40-foot tall outcrop in the modern incised stream channel, we know that the upper ~35 feet are generally sands, silts, and clays that appear to have been deposited by a low-gradient (and possibly braided) stream system alternating with periods of inundation as the lake slowly filled with sediment eroded from the upstream reaches of the watershed. Below these is a thick sequence of laminated grey clays that preserve biological remains in excellent condition (woody and herbaceous plant materials, terrestrial and aquatic invertebrate remains, and pollen). These clays represent a time when the lake was continuously filled with relatively deep water. Laminated clays probably indicate annual 'varves', and several drop-stones suggest that seasonal melting carried pebbles and cobbles in ice-rafts out into the lake before thawing. Beneath these clays are more sands and gravels on top of coarse gravel and cobble thalweg deposits. These may represent the original streambed prior to the blockage in The Water Sinks, but this is unknown; the drill rig was unable to penetrate this deposit and reach a bedrock surface.

Two separate pieces of wood recovered from the laminated clays produced ^{14}C dates of 22,000 BP ±100. Ongoing work continues to reveal details about what types of vegetation existed in the watershed, and tries to determine the period(s) during which the Pancake Field Lake existed. Preliminary analysis of preserved pollen and plant evidence indicates that the watershed was dominated by a boreal forest at that time; similar to what can be found in southern Canada today.

Although the extraordinary sedimentary record preserved in the Water Sinks lake bed is still in the early stages of study and interpretation, we can say that these sediments preserve a relatively recent but important piece of the story of landscape evolution in the Burnsville Cove.

24.5 Final Summary: The Chronology of Burnsville Cove

The long and complex history of karst development in Burnsville Cove can be summarized by the following events. The dates are extremely rough but it seems likely that the sequence is correct.

Late Cretaceous to mid-Tertiary: Establishment of the Schooley erosion surface (assuming that there really was such a thing).

Mid-Tertiary—late Eocene—35 Ma: Volcanism in northern Highland County with hydrothermal fluids producing pathways in the limestone deep below the Schooley surface.

Mid-Miocene—14–10 Ma: Dissection of the Schooley surface with breaching of the limestone at the southern end of the Cove. Establishment of the Harrisburg surface.

Late Miocene or early Pliocene: Development of high level trunk passages by south-flowing water according to the Schwartz-Doctor hypothesis. Beginning of dissection of the Harrisburg Surface at 3.5 Ma. Possible development of ancestral Butler and Breathing Caves.

Late Pliocene: Capture of the Bullpasture River along the Water Sinks lineament. Incision of the Bullpasture Gorge and drainage reversal within the cave systems.

Mid Pleistocene: Massive infilling of cave systems with clastic sediment. May be related to the sedimentation event at 0.85 Ma identified by cosmogenic isotope dating in Mammoth Cave and on the Cumberland Plateau. Certainly pre-Jaramillo (0.78 Ma) in Burnsville Cove.

Mid-Pleistocene to Present: Removal of clastic sediment by cave streams. Downcutting, development of new passages, and modification of caves by formation of shafts and canyons. Establishment of present day drainage to spring locations guided by structure, by sandstone interbeds, and by the Water Sinks master lineament.

Late Pleistocene. Formation of Pancake Field Lake which persisted at least 22,000 years ago.

Late Holocene: Arrival of settlers with conversion of land from forest to agriculture. Very late event—arrival of cave explorers.

References

Anthony, D.M., and D.E. Granger. 2004. A late tertiary origin for multilevel caves along the western escarpment of the Cumberland Plateau, Tennessee and Kentucky, established by cosmogenic ^{26}Al and ^{10}Be. *Journal of Cave and Karst Studies* 66: 46–55.

Anthony, D.M., and D.E. Granger. 2007. A new chronology for the age of appalachian erosional surfaces determined by cosmogenic nuclides in cave sediments. *Earth Surface Processes and Landforms* 32: 874–877.

Deike, G.H. 1960. Origin and geologic relations of Breathing Cave, Virginia. *National Speleological Society Bulletin* 22: 30–42.

Helsley, J.T., E.A. Johnson, R.J. Bodnar, C. Farley, and B.F. Schwartz. 2013. *Ultrahigh temperatures recorded in lower crustal xenoliths from the Shenandoah Valley, VA*. Denver: Geological Society of America Annual Meeting.

Rossi, N.P., E.A. Johnson, and B.F. Schwartz. 2013. *Geochronology of Grenvillian and Late Jurassic lower crustal xenoliths in the Shenandoah Valley*. Denver: Geological Society of America Meeting.

Schwartz, B.F. 2003. Preliminary analysis of an igneous intrusion and metamorphic xenoliths found in Highland County, Virginia. Unpublished Report, Geology Department Radford University, Radford, Va, p. 17.

Schwartz, B.F., D.H. Doctor, C.S. Swezey, N.M. Piatak. 2008. Geologic and geomorphic controls on karst conduit development and drainage rearrangement in Burnsville Cove, Bath County, Virginia. *Proceedings of Geological Society of America Annual Meeting*, Houston, Texas, p. 194.

Schwartz, B.F., D.H. Doctor. 2009. Geomorphic and hydrogeologic evolution of karst in the Burnsville Cove, Virginia, USA: New evidence and perspectives. *Proceedings of the 15th International Congress of Speleology*, Kerrville, TX, pp. 984–990.

Schwartz, B.F., G.R. Upchurch, D.A. Willard, C.E. Bernhardt, P.A. Valella. 2009. A new late glacial site in the central Appalachians of Virginia. Preliminary findings from paleobotany, palynology, and geomorphology. *Eos Transactions American Geophysical Union*. Fall Meeting Supplement, Abstract PP31C-1374.

Thornbury, W.D. 1965. *Regional geomorphology of the United States*. New York: Wiley. 609 pp.

Tso, J.L., R.R. McDowell, K.L. Avary, D.L. Matchen, and G. P. Wilkes. 2004. Middle Eocene igneous rocks in the Valley and Ridge of Virginia and West Virginia. *U.S. Geological Survey Circular, Trip 4* 1264: 137–161.

Tso, J.L., and J.D. Surber. 2006. Eocene igneous rocks near Monterey, Virginia: a field study. *Virginia Minerals* 49(3–4): 9–24.

White, W.B., and J.W. Hess. 1982. Geomorphology of burnsville cove and the geology of the butler cave-sinking creek system. *National Speleological Society Bulletin* 44: 67–77.

White, W.B., E.L. White. 1991. Karst erosion surfaces in the appalachian highlands. In *Proceedings of the Appalachian Karst Symposium*, ed. Kastning, E.H., Kastning, K.M., 1–10. Huntsville: National Speleological Society.

White, W.B. 2009. The evolution of Appalachian fluviokarst: competition between stream erosion, cave development, surface denudation, and tectonic uplift. *Journal of Cave and Karst Studies* 71: 159–167.

Appendix A
The Stratigrapic Nomenclature of Burnsville Cove, Bath and Highland Counties, Virginia

Christopher S. Swezey and John T. Haynes

The Paleozoic rocks of the Appalachian region have been the target of geologists' attention since the earliest days of geology in the United States. Not surprisingly, two centuries of attention have resulted in a long and complicated discourse concerning descriptions of rock units, what names to give to them, and how to define the boundaries between them. Chapter 16 presents geologic maps, sections, and descriptions of the rocks in Burnsville Cove. The objective of this appendix is to provide additional information following the rules of stratigraphic nomenclature as described in North American Commission on Stratigraphic Nomenclature (1983) and in Salvador (1994). Specifically, this appendix describes the history of stratigraphic nomenclature use in Bath and Highland Counties, the type sections of the stratigraphic units, and the contacts between units.

For easy reference, the lithostratigraphic and chronostratigraphic columns from Chap. 16 are duplicated (Fig. A.1). In this figure, the Middle Ordovician-Upper Ordovician boundary is placed at the base of the Nealmont Formation, although this location is somewhat controversial. Le Van and Rader (1983) placed the Middle Ordovician-Upper Ordovician boundary at the contact between the Lincolnshire Limestone and the overlying Ward Cove Limestone (both of these limestone units are located well below the Nealmont Limestone of this paper), whereas Harris et al. (1994) and Ryder et al. (1996) placed the Middle Ordovician-Upper Ordovician boundary within the basal part of the Reedsville Shale. A more recent publication by Young et al. (2005), however, reported conodont and isotope data that assign the Dolly Ridge Formation to the Chatfieldian Stage and the underlying Nealmont Formation to the Chatfieldian-Turinian Stages, and Taylor et al. (2013) indicate that these two stages are within the Upper Ordovician Series. The Ordovician-Silurian boundary is customarily placed at the contact between the Juniata Formation and the overlying Tuscarora Formation (Diecchio 1985), although in some publications this boundary is placed at a prominent unconformity within the Tuscarora Formation (Dorsch et al. 1994; Dorsch and Driese 1995).

The Silurian Period is divided into the following four series (from oldest to youngest): Llandovery, Wenlock, Ludlow, and Pridoli. The older two series are designated informally as lower Silurian, whereas the younger two series are designated informally as upper Silurian (Gradstein et al. 2004). Le Van and Rader (1983) and Diecchio and Dennison (1996) placed the lower Silurian-upper Silurian boundary at the contact between the Rose Hill Formation and the overlying Keefer Formation, whereas Harris et al. (1994) placed the boundary at the contact between the Clinton Group (e.g., Keefer Formation) and the overlying McKenzie Formation. The Silurian-Devonian boundary is located within the upper part of the Keyser Limestone of the Helderberg Group (Denkler and Harris 1988a, b; Harris et al. 1994; Rodríguez 2005), and the Lower Devonian-Middle Devonian boundary is located within the Needmore Shale (Rossbach and Dennison 1994; Harris et al. 1994). Using uranium-lead dating techniques, Roden et al. (1990) obtained an absolute age of 390 ± 0.5 Ma from the Tioga Ash Bed. Finally, the Middle Devonian-Upper Devonian boundary is placed at the top of the Millboro Shale, but the uppermost few feet of the Millboro Shale may be Upper Devonian (Rossbach and Dennison 1994; Harris et al. 1994).

W.B. White (ed.), *The Caves of Burnsville Cove, Virginia*,
Cave and Karst Systems of the World, DOI 10.1007/978-3-319-14391-0,

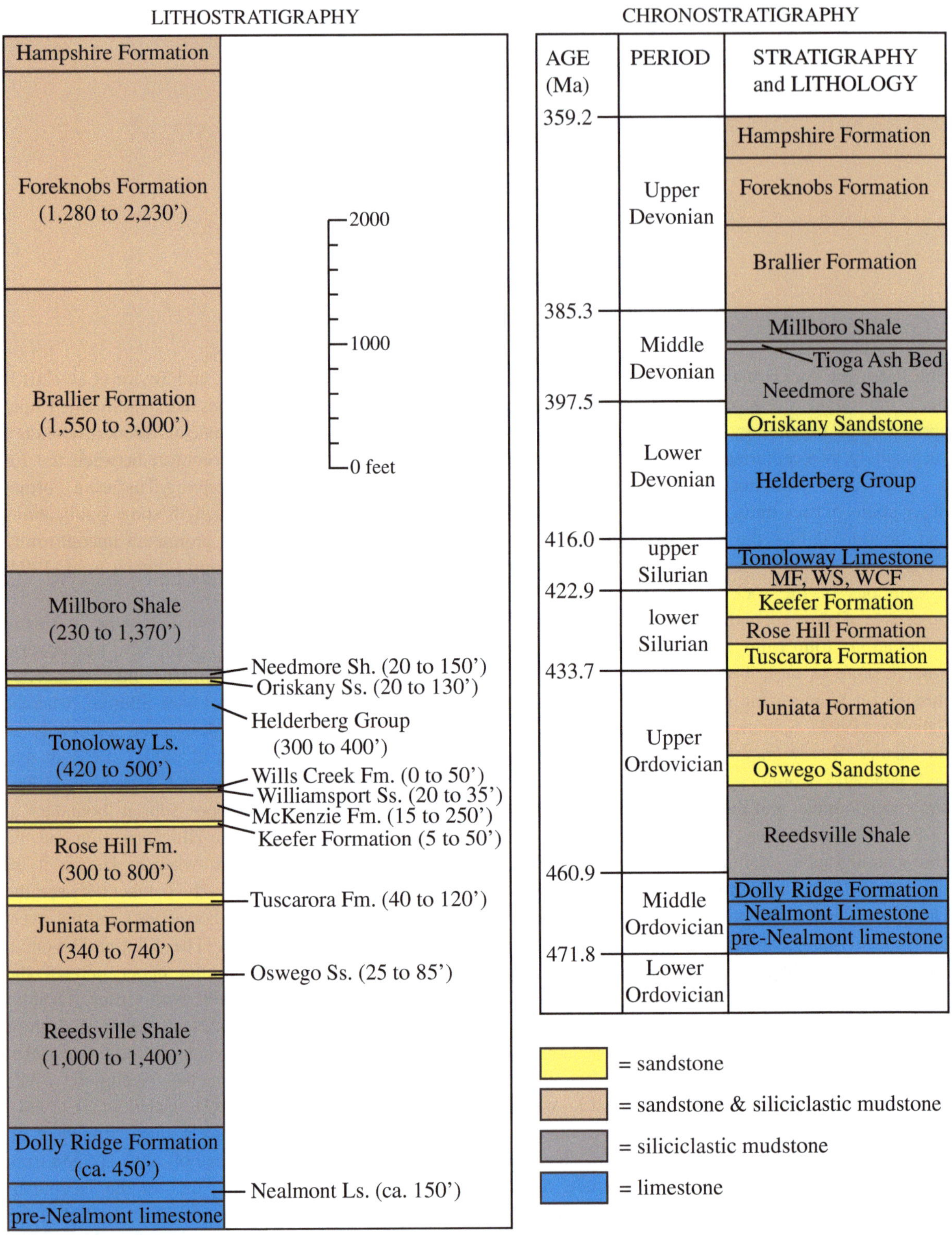

Fig. A.1 Lithostratigraphic and chronostratigraphic charts for strata in the vicinity of Burnsville Cove, Bath and Highland Counties, Virginia. Time scale is from Gradstein et al. (2004). *Fm.* Formation; *Ls.* Limestone; *Ss.* Sandstone; *MF* McKenzie Formation; *WCF* Wills Creek Formation; *WS* Williamsport Sandstone

Ordovician Nealmont Limestone

In the vicinity of Burnsville Cove, the Nealmont Limestone is a 150 ft thick unit of gray limestone. In Bath and Highland Counties, these strata were mapped by Darton (1899) as part of the Shenandoah Limestone. Butts (1933) later mapped these strata in Bath and Highland Counties as part of the Lowville Limestone of the Black River Group, whereas Butts (1940) mapped them as the "Lowville-Moccasin Limestone" of the Black River Group. Furthermore, Butts (1940) stated that in Highland County and Bath County these strata are a true limestone named the Lowville Limestone, but further south these strata grade into a red argillaceous limestone named the Moccasin Formation. Woodward (1951) mapped these strata in Highland County as the Nealmont Formation, and he stated that the strata can be traced south into the Eggleston-Moccasin Formation. Kay (1956) mapped these strata as the Nealmont Formation in Highland County, although he used the names "Moccasin-Nealmont Formation" and "Nealmont-Moccasin Formation" in Bath County. At an outcrop near Bolar (near western edge of Burnsville Cove geologic map; Fig. 16.3), he described these strata as limestone with a few red argillaceous beds in the upper part of the unit. Bick (1962) later mapped these strata in Bath and Highland Counties as the Moccasin Formation, but subsequent publications have used the name Nealmont Limestone or Nealmont Formation (Read 1980; Le Van and Rader 1983; Diecchio 1991). At present, these strata in the vicinity of Burnsville Cove are mapped as the Nealmont Limestone. The Nealmont Limestone is named for the village of Nealmont in Blair County, Pennsylvania, but the type section is at Union Furnace in Huntingdon County, Pennsylvania (Kay 1941, 1944a, b).

Ordovician Dolly Ridge Formation

In the vicinity of Burnsville Cove, the Dolly Ridge Formation is a 450 ft thick unit of black to gray limestone. In Bath and Highland Counties, these strata were mapped by Darton (1899) as part of the Shenandoah Limestone. Butts (1933) later mapped these strata in Bath and Highland Counties as part of the Lowville Limestone of the Black River Group. In a subsequent publication, although he did not specifically mention Bath or Highland County, Butts (1940) suggested that these strata in Virginia may be younger than the strata of Lowville age, and that they might be mapped as the Eggleston Limestone or the Chambersburg Limestone. Woodward (1951) later mapped these strata in Highland County as the Salona Formation, which he stated is the northeastern extension of the Eggleston facies of the Trenton Limestone in southwestern Virginia, and is comparable with the Oranda Formation and the basal Martinsburg Shale in the Shenandoah Valley. Kay (1956) mapped these strata in Highland County and northern Bath County as the Onego ("Oranda") Formation (or Onego Member of the Salona Formation), and he indicated that these strata grade into a bed of quartz sandstone overlain by limestone (mapped collectively as the Eggleston Formation) in southern Bath County. Bick (1962) later mapped these strata in Bath and Highland Counties as the Edinburg Formation. Perry (1972) noted that the equivalent strata in Pendleton County (West Virginia) do not resemble the Onego Member, Salona Formation, or Edinburg Formation at their respective type sections, and therefore he proposed the name Dolly Ridge Formation for these strata in Pendleton County. Subsequent publications have followed Perry (1972) and used the name Dolly Ridge Formation for these strata in Bath and Highland Counties (Le Van and Rader 1983; Diecchio 1991; Rader and Wilkes 2001). At present, these strata in the vicinity of Burnsville Cove are mapped as the Dolly Ridge Formation. The type section of the Dolly Ridge Formation is on the southeastern side of Dolly Ridge, 1.3 km south of Riverton in Pendleton County, West Virginia (Perry 1972).

Ordovician Reedsville Shale

In the vicinity of Burnsville Cove, the Reedsville Shale is a 1,000–1,400 ft thick unit of gray shale. In Bath and Highland Counties, these strata were mapped by Darton (1899) as part of the Martinsburg Shale. Butts (1933, 1940), Woodward (1951), Kay (1956), and Bick (1962) mapped these strata in Bath and Highland Counties as the Martinsburg Shale or the Martinsburg Formation. Later, Diecchio (1991) stated that the name Reedsville Shale should be given to these strata on the west side of the North Mountain Front (a major fault on the west side of the Shenandoah Valley) and that the name Martinsburg Formation should be given to equivalent strata on the east side of the North Mountain Front. Subsequent publications have followed Diecchio (1991) and used the name Reedsville Shale for these strata in Bath and Highland Counties (Diecchio 1985, 1991, 1993; Rader and Wilkes 2001). At present, these strata in the vicinity of Burnsville Cove are mapped as the Reedsville Shale. The type section of the Reedsville Shale is at Reedsville in Mifflin County, Pennsylvania (Ulrich 1911; Keroher 1966).

Ordovician Oswego Sandstone

In the vicinity of Burnsville Cove, the Oswego Sandstone is a 25–80 ft thick unit of gray to green sandstone. In Bath and Highland Counties, these strata were mapped by Darton (1899) as part of the Juniata Formation. Butts (1933) did not identify these strata as a separate unit in Bath or Highland County, but later Butts (1940) identified a thin outcrop of the Oswego Sandstone in northwestern Highland County, and he stated that these strata may be absent south of Highland County. Likewise, Woodward (1951) stated that the Oswego Sandstone is not found south of Monterey in Highland County. Bick (1962) did not recognize these strata in Bath or Highland County. Several subsequent publications, however, have described these strata in Bath and Highland Counties, and mapped them as the Oswego Sandstone (Crowder 1980; Rader and Gathright 1984; Diecchio 1985; Avary et al. 1999). At present, these strata in the vicinity of Burnsville Cove are mapped as the Oswego Sandstone. The Oswego Sandstone is named for outcrops in Oswego County, New York, although a specific type section was not specified (Prosser 1888).

Ordovician Juniata Formation

In the vicinity of Burnsville Cove, the Juniata Formation is a 340–740 ft thick unit of red sandstone and mudstone (shale). In Bath and Highland Counties, these strata were mapped by Darton (1899) as part of the Juniata Formation. Subsequent publications have continued to map these strata in Bath and Highland Counties as the Juniata Formation (Butts 1933, 1940; Bick 1962; Woodward 1951; Diecchio 1985; Rader and Wilkes 2001). At present, these strata in the vicinity of Burnsville Cove are mapped as the Juniata Formation. The Juniata Formation was named by Darton and Taff (1896) in West Virginia and Maryland. It was named for the Juniata River in central Pennsylvania.

Silurian Tuscarora Formation

In the vicinity of Burnsville Cove, the Tuscarora Formation is a 40–120 ft thick unit of white to yellow-gray sandstone. In Bath and Highland Counties, these strata were mapped by Darton (1899) as the Tuscarora Quartzite. Butts (1933) later mapped these strata in Highland County and in northern Bath County as the Tuscarora Quartzite, and he mapped these strata in southern Bath County as the Clinton Formation. In a subsequent publication, Butts (1940) indicated that the name Tuscarora Sandstone should be applied to these strata from Pennsylvania in the north to 38° North latitude in Virginia (encompassing the area of Burnsville Cove), and that the name Clinch Sandstone should be applied to these strata south of 38° North latitude. Woodward (1941) mapped these strata in Bath and Highland Counties as the Tuscarora Sandstone, but Bick (1962) mapped them as the Clinch Sandstone. Subsequent publications have mapped these strata in Bath and Highland Counties as the Tuscarora Formation (Kozak 1965; Sweet 1981; Rader and Wilkes 2001). At present, these strata in the vicinity of Burnsville Cove are mapped as the Tuscarora Formation. The type section of Tuscarora Formation is at Tuscarora Mountain in Pennsylvania (Darton and Taff 1896; Clark 1897).

Silurian Rose Hill Formation

In the vicinity of Burnsville Cove, the Rose Hill Formation is a 300–800 ft thick unit of red sandstone and yellow to olive siliciclastic mudstone (shale). In Bath and Highland Counties, these strata were mapped by Darton (1899) as the Cacapon Sandstone. Butts (1933) later mapped these strata in Bath and Highland Counties as part of the Clinton Formation. In a subsequent publication, however, Butts (1940) mapped these strata as the Cacapon division of the Clinton Formation (or "Iron Gate facies" of the Clinton Formation), and he suggested that these strata may be equivalent to the Rose Hill Formation of Maryland. Woodward (1941) later stated that the strata previously mapped as the Cacapon Sandstone in western Virginia should be mapped as the Rose Hill Formation, although Bick (1962) and Kozak (1965) mapped these strata in Bath and Highland Counties as the Cacapon Member of the Clinton Formation. Subsequent publications have mapped these strata in Bath and Highland Counties as the Rose Hill Formation (Helfrich 1975, 1980; Diecchio and Dennison 1996; Rader and Wilkes 2001). At present, these strata in the vicinity of Burnsville Cove are mapped as the Rose Hill Formation. The type section of the Rose Hill Formation is at Rose Hill in the city of Cumberland, Allegany County, Maryland (Swartz 1923).

Silurian Keefer Formation

In the vicinity of Burnsville Cove, the Keefer Formation is a 5–50 ft thick unit of white to yellow-gray sandstone. In Bath and Highland Counties, these strata were mapped by Darton (1899) as the Rockwood Formation. Butts (1933) later mapped these strata in Bath and Highland Counties as part of the Clinton Formation. In a subsequent publication, Butts (1940) mapped these strata in Bath and Highland Counties as the Keefer Sandstone Member of the Clinton Formation. Woodward (1941) mapped these strata in Bath and Highland Counties as the Keefer Sandstone, although Bick (1962) and Kozak (1965) mapped these strata as the Keefer Member of the Clinton Formation. Helfrich (1975, 1980) mapped these strata near Monterey as the informal lower hematitic member and the overlying Cosner Gap Member of the Mifflintown Formation, and he stated that the Cosner Gap Member is a limey equivalent of the Keefer Sandstone. More recently, Rader and Wilkes (2001) mapped these strata in Bath and Highland Counties as the Keefer Formation. At present, these strata in the vicinity of Burnsville Cove are mapped as the Keefer Formation. The type section of the Keefer Formation is at Keefer Mountain, a few miles northeast of Hancock in Washington County, Maryland (Ulrich 1911; Keroher 1966).

Silurian McKenzie Formation

In the vicinity of Burnsville Cove, the McKenzie Formation is a 15–230 ft thick unit of gray shale and white to yellow-gray sandstone, with minor beds of yellow-orange to black sandy and oolitic limestone and gray limestone. In Bath and Highland Counties, these strata were mapped by Darton (1899) as part of the Lewistown Limestone. Butts (1933) later mapped these strata in Bath and Highland Counties as the McKenzie Formation, whereas Butts (1940) mapped this unit as the McKenzie Limestone of the Cayuga Group. Woodward (1941) mapped these strata in Bath and Highland Counties as the McKenzie Formation. Bick (1962) mapped these strata in Bath and Highland Counties as the McKensie [*sic*] Limestone of the Cayuga Group, whereas Kozak (1965) simply mapped these strata in Bath County as part of the Cayuga Group. Helfrich (1975, 1980) mapped these strata in Highland County as the McKenzie Member of the Mifflintown Formation. The name "Cayuga Group" has been abandoned as a lithostratigraphic term (http://ngmdb.usgs.gov/Geolex/NewRefsmry/sumry_937.html), and these strata have since been mapped in Highland County as the McKenzie Formation (Diecchio and Dennison 1996). At present, these strata in the vicinity of Burnsville Cove are mapped as the McKenzie Formation. The type section of the McKenzie Formation is at McKenzie Station on the Baltimore and Ohio Railroad in Alleghany County, Maryland (Ulrich 1911; Keroher 1966).

Silurian Williamsport Sandstone

In the vicinity of Burnsville Cove, the Williamsport Sandstone is a 20–35 ft thick unit of sandstone. In Bath and Highland Counties, these strata were mapped by Darton (1899) as part of the Lewistown Limestone. Butts (1933, 1940) later mapped these strata in Bath and Highland Counties as part of the Wills Creek Formation, which he defined as all of the strata between the McKenzie Limestone below and the Tonoloway Limestone above. Woodward (1941) first identified these strata in Highland County as a separate unit, which he mapped as the Williamsport Sandstone. Bick (1962), however, mapped these strata in Bath and Highland County as part of the Wills Creek Formation of the Cayuga Group, whereas Kozak (1965) simply mapped these strata in Bath County as part of the Cayuga Group. The name "Cayuga Group" has since been abandoned as a lithostratigraphic term (http://ngmdb.usgs.gov/Geolex/NewRefsmry/sumry_937.html), and subsequent publications have mapped these strata in Highland County as the Williamsport Sandstone (Helfrich 1975; Diecchio and Dennison 1996; Rader and Wilkes 2001). At present, these strata in the vicinity of Burnsville Cove are mapped as the Williamsport Sandstone. The type section of the Williamsport Sandstone is on a branch of Patterson Creek, 0.6 miles east of Williamsport in Grant County, West Virginia (Reger 1924).

Silurian Wills Creek Formation

In the vicinity of Burnsville Cove, the Wills Creek Formation is a 0–212 ft thick unit of brown to green sandstone, sandy limestone, and limestone. In Bath and Highland Counties, these strata were mapped by Darton (1899) as part of the Lewistown Limestone. Butts (1933) later mapped these strata in Bath and Highland Counties as the Wills Creek Sandstone. In a subsequent publication, however, Butts (1940) mapped these strata in Bath and Highland Counties as the Wills Creek Formation, which he defined as all of the strata between the McKenzie Limestone below and the Tonoloway Limestone above (this definition would include the Williamsport Sandstone). Woodward (1941) mapped these strata in Highland County as the Wills Creek Limestone. Bick (1962) later mapped these strata in Bath and Highland County as the Wills Creek Formation, whereas Kozak (1965) simply mapped these strata in Bath County as part of the Cayuga Group. The name "Cayuga Group" has since been abandoned as a lithostratigraphic term (http://ngmdb.usgs.gov/Geolex/NewRefsmry/sumry_937.html), and subsequent publications have mapped these strata in Highland County as the Wills Creek Formation (Helfrich 1975; Diecchio and Dennison 1996). At present, these strata in the vicinity of Burnsville Cove are mapped as the Wills Creek Formation. The type section of the Wills Creek Formation is at Wills Creek in Cumberland, Allegany County, Maryland (Uhler 1905).

Silurian Tonoloway Limestone

In the vicinity of Burnsville Cove, the Tonoloway Limestone is a 420–500 ft thick unit of gray to blue limestone and dolomitic limestone, with some beds of fine-grained sandstone and siliciclastic mudstone (shale). In Bath and Highland Counties, these strata were mapped by Darton (1899) as part of the Lewistown Limestone. Swartz (1930) later mapped these strata in Bath and Highland Counties as the Tonoloway Limestone of the Cayuga Group, whereas Butts (1933, 1940) mapped these strata in Bath and Highland Counties as the Tonoloway Limestone. Woodward (1941) mapped these strata in Highland County as the Tonoloway Limestone. Bick (1962) later mapped these strata in Bath and Highland Counties as the Tonoloway Limestone of the Cayuga Group, whereas Kozak (1965) simply mapped these strata in Bath County as part of the Cayuga Group. The name "Cayuga Group" has since been abandoned as a lithostratigraphic term (http://ngmdb.usgs.gov/Geolex/NewRefsmry/sumry_937.html). Helfrich (1975) mapped these strata in Highland County as the Tonoloway Formation, and Diecchio and Dennison (1996) mapped these strata in Highland County as the Tonoloway Limestone. At present, these strata in the vicinity of Burnsville Cove are mapped as the Tonoloway Limestone. The type section of the Tonoloway Limestone is at Tonoloway Ridge in Washington County, Maryland (Ulrich 1911; Keroher 1966).

Silurian-Devonian Helderberg Group

In the vicinity of Burnsville Cove, the Helderberg Group is a 300–400 ft thick unit of limestone with some beds of sandstone, siliciclastic mudstone (shale), and cherty limestone. In Bath and Highland Counties, these strata were mapped by Darton (1899) as part of the Lewistown Limestone. Swartz (1930) later mapped these strata in Bath and Highland Counties as the Helderberg Group (which included the Keyser Limestone). Butts (1933, 1940) mapped these strata in Bath and Highland Counties as the Helderberg Limestone (which included the Keyser Limestone Member), and Woodward (1943) mapped these strata in Bath and Highland Counties as the Helderberg Group (which included the Keyser Limestone). Bick (1962) and Kozak (1965) mapped the upper part of these strata in Bath and Highland Counties as the Helderberg Group, but they considered the Keyser Limestone to be a separate formation below the Helderberg Group. Most subsequent publications have mapped these strata in Bath and Highland Counties as the Helderberg Group, which includes the Keyser Limestone (Smosna and Warshauer 1979; Smosna 1984; Dorobek and Read 1986; Linn et al. 1990). Rader and Wilkes (2001), however, mapped these strata in Bath and Highland Counties as "Devonian and Silurian rocks, undivided," but they mentioned the Helderberg Group, and they indicated that the Keyser Limestone is not part of the Helderberg Group. At present, these strata in the vicinity of Burnsville Cove are mapped as the Helderberg Group, which includes the Keyser Limestone. The type section of the Helderberg Group is located at the Helderberg Mountains in Albany County, New York (Conrad 1839).

Silurian-Devonian Keyser Limestone of the Helderberg Group

In the vicinity of Burnsville Cove, the Keyser Limestone is a 100–200 ft thick unit of limestone, sandstone, and siliciclastic mudstone (shale). In Bath and Highland Counties, these strata were mapped by Darton (1899) as part of the Lewistown Limestone. Swartz (1930) later mapped these strata in Bath and Highland Counties as the Keyser Limestone of the Helderberg Group. Butts (1933, 1940) mapped this unit in Bath and Highland Counties as the Keyser Limestone Member of the Helderberg Limestone, whereas Woodward (1943) mapped this unit in Bath and Highland Counties as the Keyser Limestone of the Helderberg Group. In contrast, Bick (1962) and Kozak (1965) mapped this unit in Bath and Highland Counties as the Keyser Limestone, which they considered to be a separate formation below the Helderberg Group. Most subsequent publications have mapped these strata in Bath and Highland Counties as the Keyser Limestone of the Helderberg Group (Smosna and Warshauer 1979; Smosna 1984; Dorobek and Read 1986; Linn et al. 1990). Rader and Wilkes (2001), however, mapped these strata in Bath and Highland Counties as "Devonian and Silurian rocks, undivided," but they mentioned the Keyser Limestone, and they indicated that it is not part of the Helderberg Group. At present, these strata in the vicinity of Burnsville Cove are mapped as Keyser Limestone of the Helderberg Group. The type section of the Keyser Limestone is at a quarry near the town of Keyser in Mineral County, West Virginia (Ulrich 1911; Schuchert et al. 1913; Keroher 1966).

Byers Island Limestone Member of the Keyser Limestone of the Helderberg Group

In the vicinity of Burnsville Cove, the Byers Island Limestone Member of the Keyser Limestone is a 0–75 ft thick unit of gray to blue-gray to pink limestone and argillaceous limestone, with some beds of sandstone, sandy limestone, and nodules of black chert. Many publications have indicated that the Keyser Limestone in Bath and Highland Counties has a lower unit of limestone, a middle unit of sandstone and siliciclastic mudstone (shale), and an upper unit of limestone (Swartz 1930; Butts 1933, 1940; Woodward 1943; Deike 1960a; Bick 1962; Kozak 1965; Dorobek and Read 1986), but Head (1972) first named this lower limestone unit as the Byers Island Limestone Member of the Keyser Limestone of the Helderberg Group. Diecchio and Dennison (1996) later applied the name Byers Island Limestone Member of the Keyser Formation to an outcrop of these strata in Highland County. At present, these strata in the vicinity of Burnsville Cove are mapped as the Byers Island Limestone Member of the Keyser Formation of the Helderberg Group. The type section of the Byers Island Limestone Member of the Keyser Limestone is a series of outcrops along the Susquehanna River northeast of Selinsgrove in Snyder County, Pennsylvania (Head 1972).

Clifton Forge Sandstone Member of the Keyser Limestone of the Helderberg Group

In the vicinity of Burnsville Cove, the Clifton Forge Sandstone Member of the Keyser Limestone is a 10–100 ft thick unit of white to gray sandstone. Swartz (1930) first identified these strata in Bath County as a separate unit, which he named the Clifton Forge Sandstone Member of the Keyser Limestone of the Helderberg Group. North of the type section in Alleghany County (Virginia), the Clifton Forge Sandstone Member is reported to split into two or three sandy beds (Swartz 1930), with the lower of these sandy beds merging towards the north with the Big Mountain Shale Member of the Keyser Limestone, and the upper of these sandy beds extending north into Highland County where it presumably pinches out. Butts (1933) later mapped these strata in southern Bath County as the Clifton Forge Sandstone Member of the Keyser Limestone, and he stated that the sandstone is replaced by shale in northern Highland County. In a subsequent publication, Butts (1940) mapped these strata in Bath County as the Clifton Forge Sandstone of the Keyser Limestone Member of the Helderberg Limestone. Woodward (1943) mapped these strata in Bath and Highland Counties as the Clifton Forge Sandstone of the Keyser Limestone of the Helderberg Group. Bick (1962) and Kozak (1965) mapped these strata in Bath and Highland Counties as the Clifton Forge Sandstone Member of the Keyser Limestone (which they did not consider to be part of the

Helderberg Group). Dorobek and Read (1986) later mapped this unit in Bath and Highland Counties as the Clifton Forge Sandstone of the Keyser Limestone of the Helderberg Group. At present, these strata in the vicinity of Burnsville Cove are mapped as the Clifton Forge Sandstone Member of the Keyser Limestone of the Helderberg Group. The type section of the Clifton Forge Sandstone Member of the Keyser Limestone is at Clifton Forge in Alleghany County, Virginia (Swartz 1930).

Deike (1960a, b) considered the 12 ft thick sandstone beds that form the floor and ceiling of Breathing Cave to be the same two sandstone beds that Swartz (1930) and Woodward (1943) identified as tongues of the Clifton Forge Sandstone Member, and he informally named these beds the "lower Breathing sandstone" and the "upper Breathing sandstone." White and Hess (1982) later applied the name lower Clifton Forge Sandstone to the lower bed and the name upper Clifton Forge Sandstone to the upper bed in Breathing Cave. Recent work by John Haynes, Rick Lambert, and Phil Lucas, however, suggests that the "upper Breathing sandstone" and "lower Breathing sandstone" are both located within the Tonoloway Limestone, and that they are not tongues of the Clifton Forge Sandstone Member of the Keyser Limestone.

Big Mountain Shale Member of the Keyser Limestone of the Helderberg Group

The Big Mountain Shale Member of the Keyser Limestone is absent in Burnsville Cove, but further north it is a 10–30 ft thick unit of olive-gray shale, with some thin beds of sandstone and limestone. Swartz (1930) first identified these strata as a separate, mappable unit within the Keyser Limestone in Highland County, and he indicated that the shale of these strata is replaced towards the south by sandstone that is mapped as the Clifton Forge Sandstone Member of the Keyser Limestone of the Helderberg Group. Butts (1933, 1940) also identified these shale-rich strata within the Keyser Limestone in Highland County, but did not give a separate name to the strata. Woodward (1943) later mapped these strata in Highland County as the Big Mountain Shale Member of the Keyser Limestone, and he stated that the unit passes south into the lower portion of the Clifton Forge Sandstone Member of the Keyser Limestone. Deike (1960a) stated specifically that the Big Mountain Shale is not present in Burnsville Cove. Likewise, Bick (1962) and Kozak (1965) did not identify these strata in their areas of study. Dorobek and Read (1986) mapped this unit in Highland County as the Big Mountain Shale of the Keyser Limestone of the Helderberg Group, whereas Diecchio and Dennison (1996) mapped this unit in northern Highland County as the Big Mountain Shale Member of the Keyser Formation. At present, these strata in northern Highland County (north of Burnsville Cove) are mapped as the Big Mountain Shale Member of the Keyser Limestone of the Helderberg Group. The type section of the Big Mountain Shale Member of the Keyser Limestone is at Big Mountain, approximately 1.5 miles west of the community of Upper Tract in Pendleton County, West Virginia (Swartz 1930).

Jersey Shore Limestone Member of the Keyser Limestone of the Helderberg Group

In the vicinity of Burnsville Cove, the Jersey Shore Limestone Member of the Keyser Limestone is a 25–150 ft thick unit of gray to blue-gray to pink limestone and argillaceous limestone, with some beds of sandy limestone. Many publications have indicated that the Keyser Limestone in Bath and Highland Counties has a lower unit of limestone, a middle unit of sandstone and siliciclastic mudstone (shale), and an upper unit of limestone (Swartz 1930; Butts 1933, 1940; Woodward 1943; Deike 1960a; Bick 1962; Kozak 1965; Dorobek and Read 1986), but Head (1972) first gave the name Jersey Shore Limestone Member of the Keyser Limestone of the Helderberg Group to the limestone unit immediately above the Clifton Forge Sandstone Member and (or) the Big Mountain Shale Member. Diecchio and Dennison (1996) and Dennison et al. (1997) later mapped these strata at an outcrop near McDowell (Highland County) as the Jersey Shore Limestone Member of the Keyser Formation, and Dennison et al. (1997) applied the same nomenclature to these strata at an outcrop at Mustoe (Highland County). At present, these strata in the vicinity of Burnsville Cove are mapped as the Jersey Shore Limestone Member of the Keyser Limestone of the Helderberg Group. The type section of the Jersey Shore Limestone Member of the Keyser Limestone is near Jersey Shore in Lycoming County, Pennsylvania (Head 1972).

LaVale Limestone Member of the Keyser Limestone of the Helderberg Group

In the vicinity of Burnsville Cove, the LaVale Limestone Member of the Keyser Limestone is a 3–30 ft thick unit of laminated gray sandy limestone to calcite-cemented sandstone. Many publications have indicated that the Keyser Limestone in Bath and Highland Counties has a lower unit of limestone, a middle unit of sandstone and siliciclastic mudstone (shale), and an upper unit of limestone (Swartz 1930; Butts 1933, 1940; Woodward 1943; Deike 1960a; Bick 1962; Kozak 1965; Dorobek and Read 1986), but Head (1972) first gave the name LaVale Limestone Member of the Keyser Limestone of the Helderberg Group to the uppermost strata of the Keyser Limestone. Stock (1996) and Dennison et al. (1997) later mapped these strata at an outcrop at Mustoe (Highland County) as the LaVale Limestone Member of the Keyser Limestone. At present, these strata in the vicinity of Burnsville Cove are mapped as the LaVale Limestone Member of the Keyser Limestone of the Helderberg Group. The type section of the LaVale Limestone Member of the Keyser Limestone is at the Corriganville Quarry near the town of LaVale in Alleghany County, Maryland (Head 1972).

Devonian New Creek Limestone of the Helderberg Group

In the vicinity of Burnsville Cove, the New Creek Limestone is an 8–60 ft thick unit of gray to pink limestone. In Bath and Highland Counties, these strata were mapped by Darton (1899) as part of the Lewistown Limestone. Swartz (1930) later mapped these strata in Bath and Highland Counties as the Coeymans Limestone of the Helderberg Group. Butts (1933) stated that this unit is difficult to recognize in Virginia, but in a subsequent publication Butts (1940) mapped these strata in Bath and Highland Counties as the Coeymans Limestone Member of the Helderberg Limestone. Woodward (1943) mapped these strata in Bath and Highland Counties as the Coeymans Formation of the Helderberg Group, whereas Bick (1962) and Kozak (1965) mapped these strata in Bath and Highland Counties as the Coeymans Limestone of the Helderberg Group. Bowen (1967) later demonstrated that these strata in Virginia cannot be traced continuously through the intervening areas to the type section of the Coeymans Limestone in New York. As a result, the name Coeymans Limestone has been discontinued in areas south of central Pennsylvania, and the name New Creek Limestone is used instead (Bowen 1967). Subsequently, these strata in Bath and Highland Counties have been mapped as the New Creek Limestone of the Helderberg Group (Dorobek and Read 1986). At present, these strata in the vicinity of Burnsville Cove are mapped as the New Creek Limestone of the Helderberg Group. The type section of the New Creek Limestone is a quarry on the north side of U.S. Route 50, approximately 0.5 miles south of the town of New Creek in Mineral County, West Virginia. This quarry is 100 yards east of a stream called New Creek where U.S. Route 50 crosses New Creek Mountain (Bowen 1967).

Devonian Corriganville Limestone of the Helderberg Group

In the vicinity of Burnsville Cove, the Corriganville Limestone is a 5–30 ft thick unit of gray limestone and cherty limestone. In places, the unit contains a basal bed of gray sandstone (0–20 ft thick) that is mapped as the Healing Springs Sandstone Member of the Corriganville Limestone. In Bath and Highland Counties, these strata were mapped by Darton (1899) as part of the Lewistown Limestone. Swartz (1930) later mapped these strata in Bath and Highland Counties as the New Scotland Limestone of the Helderberg Group. Butts (1933, 1940), however, mapped these strata in Bath and Highland Counties as the New Scotland Limestone Member of the Helderberg Limestone. Woodward (1943), Bick (1962), and Kozak (1965) mapped these strata in Bath and Highland Counties as the New Scotland Limestone of the Helderberg Group. Head (1972) later demonstrated that these strata in Virginia cannot be traced continuously through the intervening areas to the type section of the New Scotland Limestone in New York. As a result, the name New Scotland Limestone has been discontinued in areas south of central Pennsylvania, and the name Corriganville Limestone is used instead. Subsequently, these strata in Bath and Highland Counties have been mapped as the Corriganville Limestone of the Helderberg Group (Dorobek and Read 1986). At present, these strata in the vicinity of Burnsville Cove are mapped as the Corriganville Limestone of the Helderberg Group. The type section of the Corriganville Limestone is at a railroad cut 0.3 miles southeast of the town of Corriganville in Alleghany County, Maryland (Head 1972).

Devonian Healing Springs Sandstone Member of the Corriganville Limestone of the Helderberg Group

In the vicinity of Burnsville Cove, the Healing Springs Sandstone Member of the Corriganville Limestone is a 0–20 ft thick unit of gray sandstone. Swartz (1930) first identified these strata in Bath County as a separate unit, which he named the Healing Springs Sandstone Member of the New Scotland Limestone of the Helderberg Group. Butts (1933, 1940) later mapped these strata in Bath and Highland Counties as the Healing Springs Sandstone of the New Scotland Limestone Member of the Helderberg Limestone. Woodward (1943) and Bick (1962) mapped these strata as the Healing Springs Sandstone Member of the New Scotland Limestone of the Helderberg Group. Since Head (1972) indicated that the name Corriganville Limestone should be used instead of New Scotland Limestone in Virginia, these strata have been mapped in Bath County as the Healing Springs Sandstone Member of the Corriganville Limestone of the Helderberg Group (Dorobek and Read 1986). At present, these strata in the vicinity of Burnsville Cove are mapped as the Healing Springs Sandstone Member of the Corriganville Limestone of the Helderberg Group. The type section of the Healing Springs Sandstone Member of the Corriganville Limestone is at Healing Springs in Bath County, Virginia (Swartz 1930).

Devonian Licking Creek Limestone of the Helderberg Group

In the vicinity of Burnsville Cove, the Licking Creek Limestone is an 85–200 ft thick unit of gray cherty limestone and argillaceous limestone, overlain by gray limestone and sandy limestone. In Bath and Highland Counties, these strata were mapped by Darton (1899) as part of the Lewistown Limestone. Swartz (1930) later mapped these strata in Bath and Highland Counties as the Becraft Limestone of the Helderberg Group. Butts (1933, 1940) mapped these strata in Bath County as the Becraft Limestone Member of the Helderberg Limestone. Woodward (1943) mapped these strata in Bath and Highland Counties as the Port Jervis Limestone and Chert, which he stated was equivalent in part to the Becraft Limestone and Shriver Chert of previous reports. Woodward (1943) also stated that the Port Jervis Limestone at Monterey (Highland County) was previously mapped by F.M. Swartz as the Licking Creek Limestone. Bick (1962) and Kozak (1965) mapped these strata in Bath and Highland Counties as the Licking Creek Limestone of the Helderberg Group. Later, Head (1974) redefined the Licking Creek Limestone so as to comprise all limestone and cherty limestone between the Corriganville Limestone [New Creek Limestone] and the Oriskany Sandstone, and he stated that the Licking Creek Limestone passes laterally into cherty strata that are mapped as the Shriver Chert of the Helderberg Group. Subsequently, these strata in Bath and Highland Counties have been mapped as the Licking Creek Limestone of the Helderberg Group (Dorobek and Read 1986). At present, these strata in the vicinity of Burnsville Cove are mapped as the Licking Creek Limestone of the Helderberg Group. The type section of the Licking Creek Limestone is at a bluff on the south side of Licking Creek approximately 1 mile east of Warren Point in Franklin County, Pennsylvania (Swartz 1939).

Devonian Shriver Chert of the Helderberg Group

In the vicinity of Burnsville Cove, the Shriver Chert is a 5–40 ft thick unit of gray to black chert, cherty limestone, and siliciclastic mudstone (shale). In Bath and Highland Counties, these strata were mapped by Darton (1899) as part of the Monterey Sandstone, which he stated was approximately equivalent to the Oriskany Sandstone. Swartz (1930) later mapped these strata in Bath and Highland Counties in some places as the Shriver Chert of the Oriskany Group (e.g., near McDowell) and in other places as the Shriver Chert of the Helderberg Group (e.g., Monterey and Back Creek Mountain west of Warm Springs). Butts (1933, 1940) mapped these strata in Bath and Highland Counties as the Becraft Limestone Member of the Helderberg Limestone. Woodward (1943) mapped some of these strata in Bath and Highland Counties as the Port Jervis Limestone and Chert, which he stated was equivalent in part to the Becraft Limestone and Shriver Chert of previous reports. Woodward (1943) also mapped some of these strata in Bath and Highland Counties as the Port Ewen Shale and Chert, which he stated was equivalent in part to the Shriver Chert of previous reports. Later, Head (1974) redefined the Shriver Chert so as to comprise the cherty silty mudstone, siltstone, and calcite-cemented siltstone above the Mandata Shale [which lies above the Corriganville Limestone in West Virginia, but is not present in Virginia], and below the Oriskany Sandstone. Head (1974) also indicated that all limestone and cherty limestone strata between the Corriganville Limestone [New Creek Limestone] and the Oriskany Sandstone should be given the name Licking Creek Limestone of the Helderberg Group, and he noted that the Licking Creek Limestone passes laterally into cherty strata that are mapped as the Shriver Chert of the Helderberg Group. Subsequently, these strata in Bath and Highland Counties have been mapped as the Shriver Chert of the Helderberg Group (Dorobek and Read 1986). At present, these strata in the vicinity of Burnsville Cove are mapped as the Shriver Chert of the Helderberg Group. The type section of the Shriver Chert is at Shriver Ridge, near the town of Cumberland, Maryland (Schuchert et al. 1913).

Devonian Oriskany Sandstone

In the vicinity of Burnsville Cove, the Oriskany Sandstone is a 20–130 ft thick unit of white to yellow to red-brown sandstone. In Bath and Highland Counties, these strata were mapped by Darton (1899) as the Monterey Sandstone, which he stated was approximately equivalent to the Oriskany Sandstone. Kindle (1911) later mapped these strata in Bath County as the Oriskany Sandstone. Schuchert et al. (1913) mapped equivalent strata in West Virginia and Maryland as the Ridgeley Sandstone Member of the Oriskany Formation, and they stated that the Oriskany Formation contained a lower member named the Shriver Chert Member of the Oriskany Formation. They designated the type section of the Ridgeley Sandstone Member at the town of Ridgely [spelling later changed to "Ridgeley"] in Mineral County, West Virginia. Swartz (1930) later mapped these strata in Bath and Highland Counties as the Ridgeley Sandstone of the Oriskany Group. However, in comprehensive publications on the geology of western Virginia, Butts (1933, 1940) mapped these strata as the Oriskany Sandstone. Furthermore, Butts (1940) stated that the Oriskany Sandstone "corresponds exactly with the Ridgely Sandstone" (p. 291) and he also indicated that the Oriskany Sandstone is the same as the Monterey Sandstone. Bick (1962) mapped these strata in Bath and Highland Counties as the Ridgeley Sandstone, whereas Kozak (1965) mapped these strata in Bath County as the Oriskany Sandstone. Avary and Dennison (1980) mapped these strata in Highland County as the Oriskany Sandstone. In many subsequent publications, however, the name Ridgeley Sandstone has persisted for these strata in Bath and Highland Counties (Sweet 1981; Sweet and Wilkes 1986; Rader and Wilkes 2001). However, the rules of stratigraphic nomenclature (North American Commission on Stratigraphic Nomenclature 1983; Salvador 1994) dictate that that name Oriskany Sandstone should be given to these strata in Bath and Highland Counties. Thus, at present, these strata in the vicinity of Burnsville Cove are mapped as the Oriskany Sandstone. The type section of the Oriskany Sandstone is at Oriskany Falls in New York (Vanuxem 1839).

Devonian Needmore Shale

In the vicinity of Burnsville Cove, the Needmore Shale is a 20–150 ft thick unit of gray to black siliciclastic mudstone (shale), with some beds of gray limestone. In Bath and Highland Counties, these strata were mapped by Darton (1899) as part of the Romney Shale. Kindle (1911) later mapped these strata in Bath County as the Onondaga Limestone, but stated that these strata consist primarily of shale rather than limestone. Butts (1933) followed Darton (1899), and he mapped these strata in Bath and Highland Counties as part of the Romney Shale. However, in a subsequent publication, Butts (1940) stated that the name Romney Shale should be abandoned, and he mapped these strata at Bullpasture Mountain in Highland County as the Onondaga Formation. Woodward (1943) later mapped these strata in Bath and Highland Counties as the Needmore Shale of the Onondaga Group. Bick (1962) described these strata in Bath and Highland Counties as predominantly shale, which he mapped as the Onondaga Formation, whereas Kozak (1965) mapped these strata in Bath County as the Needmore Shale. Subsequent publications have mapped these strata in Bath and Highland Counties as the Needmore Shale (Dennison and Hasson 1977; Avary and Dennison 1980). At present, these strata in the vicinity of Burnsville Cove are mapped as the Needmore Shale. The type section of the Needmore Shale is located between the towns of Needmore and Warfordsburg in southern Fulton County, Pennsylvania (Willard 1939).

Where the lower part of the Needmore Shale is notably darker than the rest of the unit, Dennison (1961) mapped these strata in West Virginia and Virginia (including Bath and Highland Counties) as the Beaver Dam black shale subfacies of the Needmore Shale. Hasson and Dennison (1988) later mapped these strata in Bath and Highland Counties as the Beaverdam Shale Member of the Needmore Shale. At present, however, the U.S. Geological Survey does not use this nomenclature in Virginia (http://ngmdb.usgs.gov/Geolex/).

The Beaverdam Shale Member was originally named the Beaver Dam Black Shale Member of the Needmore Shale, after a brook (Beaverdam Run) near outcrops on the Pennsylvania Railroad [now the Norfolk Southern Railway] approximately 1.5 miles northeast of Newton Hamilton in Mifflin County, Pennsylvania (Willard 1939). Willard's original definition of the unit was discarded by Swain (1958), who redefined an 18–20 ft thick unit of gray to black shale above the Ridgeley (Oriskany) Sandstone in central Pennsylvania as the Beaverdam Run Member of the Newton Hamilton Formation of the Onondaga Group. Klemic et al. (1963) renamed these strata in Pennsylvania as the Beaverdam Run Member of the Catskill Formation, and he stated that these strata are very similar to the Trimmers Rock Sandstone. In a subsequent publication, Wood (1973) renamed these strata in Pennsylvania as the Beaverdam Run Tongue of the Trimmers Rock Sandstone of the Susquehanna Group.

Devonian Tioga Ash Bed

In the vicinity of Burnsville Cove, the Tioga Ash Bed is a 3–20 ft thick unit of volcanic ash, silty tuff, and tuffaceous shale. Dennison and Textoris (1971) first identified these strata in Bath County at a roadside shale quarry 2.0 miles south of Williamsville. They described these strata as a separate and mappable unit, which they referred to as the Tioga Bentonite. Dennison and Hasson (1977) and Avary and Dennison (1980) later identified other outcrops of this unit in Bath and Highland Counties, and they also referred to it as the Tioga Bentonite. The type section of this unit (which was initially named the Tioga Bentonite Bed) is located at the Tioga gas field in Tioga County, Pennsylvania (Ebright et al. 1949). The name of this unit, however, has since been changed to the Tioga Ash Bed (Roen and Hosterman 1982).

Devonian Millboro Shale

In the vicinity of Burnsville Cove, the Millboro Shale is a 230–1,370 ft thick unit of black siliciclastic mudstone (shale). In Bath and Highland Counties, these strata were mapped by Darton (1899) as part of the Romney Shale. Butts (1933) also mapped these strata in Bath and Highland Counties as part of the Romney Shale. However, in a subsequent publication, Butts (1940) stated that the name Romney Shale should be abandoned, and he mapped the upper portion of the Romney Shale in Bath and Highland Counties as a separate formation named the Millboro Shale. Elsewhere in the Appalachian region, these strata are mapped as the Marcellus Shale, Mahantango Formation, Harrell Shale, and (or) Hamilton Formation, but Butts (1940) had difficulty differentiating these units in Virginia, and thus he proposed the name Millboro Shale for these strata in Virginia (Butts 1940; Hasson and Dennison 1968; Dennison 1970). Woodward (1943) used the terms Millboro Shale and Harrell Shale for these strata in Bath and Highland Counties, but he also stated that Butts proposed the name Millboro Shale instead. Subsequent publications have mapped these strata in Bath and Highland Counties as the Millboro Shale (Bick 1962; Kozak 1965; Dennison and Hasson 1977; Avary and Dennison 1980; Rader and Wilkes 2001). At present, these strata in the vicinity of Burnsville Cove are mapped as the Millboro Shale. The type section of the Millboro Shale is at Millboro Springs in Bath County, Virginia (Cooper 1939; Butts 1940; Hasson and Dennison 1968; Rossbach and Dennison 1994).

Devonian Tully Limestone

In the vicinity of Burnsville Cove, the Tully Limestone is a 1–8 foot thick interval of discontinuous limestone nodules (that are not present at every outcrop). Although the Tully Limestone is a clearly identifiable bed of limestone to the north of Highland County, most publications do not identify the Tully Limestone in Bath or Highland Counties. Avary and Dennison (1980), however, described a discontinuous interval of limestone nodules (which they mapped as the Tully Limestone) in Bath and Highland Counties. Hasson and Dennison (1988) later described limestone nodules near the top of the Millboro Shale in Bath and Highland Counties, and they indicated that these nodules are equivalent to the Tully Limestone. At present, these strata in the vicinity of Burnsville Cove are mapped as the Tully Limestone. The type section of the Tully Limestone is at the town of Tully in Onondaga County, New York (Vanuxem 1839).

Devonian Brallier Formation

In the vicinity of Burnsville Cove, the Brallier Formation is a 1,550–3,000 ft thick unit of gray to green to yellow-brown siliciclastic mudstone and sandstone. In Bath and Highland Counties, these strata were mapped by Darton (1899) as part of the Jennings Formation. Butts (1933, 1940) later designated the lower half of the Jennings Formation as a distinct formation, which he named the Brallier Shale. Woodward (1943) used the term Brallier Shale for these strata in Bath and Highland Counties. Bick (1962) mapped these strata in Bath and Highland Counties as the Brallier Shale, whereas Kozak (1965) mapped these strata in Bath County as the Brallier Formation. In subsequent publications, Dennison (1970), Dennison and Hasson (1977), Avary and Dennison (1980), and Rader and Wilkes (2001) mapped these strata in Bath and Highland Counties as the Brallier Formation. At present, these strata in the vicinity of Burnsville Cove are mapped as the Brallier Formation. The type section of the Brallier Formation is at the Brallier railroad station, located approximately 6 miles northeast of Everett in Bedford County, Pennsylvania (Butts 1918).

Devonian Foreknobs Formation

In the vicinity of Burnsville Cove, the Foreknobs Formation is a 1,280–2,230 ft thick unit of white to brown sandstone and green to brown siliciclastic mudstone (shale). In Bath and Highland Counties, these strata were mapped by Darton (1899) as part of the Jennings Formation, which he stated was approximately equivalent to the Chemung Formation. Butts (1933, 1940) later mapped these strata in Bath and Highland Counties as the Chemung Formation, and Woodward (1943) used the term Chemung Formation for these strata in Bath and Highland Counties. Bick (1962) and Kozak (1965) also mapped these strata in Bath and Highland Counties as the Chemung Formation. Dennison (1970), however, stated that the name "Chemung" should be abandoned, and he proposed the name Foreknobs Formation of the Greenland Gap Group. Subsequent publications have mapped these strata in Bath and Highland Counties as the Foreknobs Formation (Avary and Dennison 1980; Rader and Wilkes 2001). At present, these strata in the vicinity of Burnsville Cove are mapped as the Foreknobs Formation. The Foreknobs Formation is named for a topographic feature called the Fore Knobs of Allegheny Front in Grant County, West Virginia, and the type section is near the community of Scherr (Dennison 1970). The Foreknobs Formation is overlain by the Devonian Hampshire Formation, which is not present in the immediate vicinity of Burnsville Cove.

References

Avary, K.L., and J.M. Dennison. 1980. Back Creek siltstone member of devonian brallier formation in Virginia and West Virginia. *Southeastern Geology* 21: 121–153.

Avary, K.L., R.R. McDowell, D.L. Matchen, and R.J. Diecchio. 1999. Previously unmapped upper ordovician reedsville formation and Oswego Sandstone, Jack Mountain, Valley and Ridge, Highland County, Virginia [Abstract]. *Geological Society of America, Abstracts with Programs* 31(3): 2.

Bick, K.F. 1962. Geology of the williamsville quadrangle. *Virginia Division of Mineral Resources Report of Investigations* 2: 40.

Bowen, Z.P. 1967. Brachiopoda of the keyser limestone (Silurian—Devonian) of maryland and adjacent areas. *Geological Society of America Memoir* 102: 103.

Butts, C. 1918. Geologic section of Blair and Huntingdon counties, central Pennsylvania. *American Journal of Science* 46: 523–537.

Butts, C. 1933. Geologic map of the Appalachian Valley in Virginia with explanatory text. *Virginia Geological Survey Bulletin* 42: 56.

Butts, C. 1940. Geology of the Appalachian Valley in Virginia. *Virginia Geological Survey Bulletin* 52(pt. I): 568.

Clark, W.B. 1897. Outline of present knowledge of the physical features of Maryland. *Maryland Geological Survey Volume Series* 1(pt. 3): 172–188.

Conrad, T.A. 1839. Second annual report of T.A. Conrad, on the Palaeontological Department of the Survey. New York Geological Survey, Annual Report 3, pp. 57–66.

Cooper, B.N. 1939. Geology of the draper mountain area [Wythe and Pulaski Counties], Virginia. *Virginia Geological Survey Bulletin* 55: 98.

Crowder, R.K. 1980. Deposition of upper ordovician red beds in western Virginia. *The Compass of Sigma Gamma Epsilon* 57 (4): 101–111.

Darton, N.H. 1899. Monterey folio, Virginia-West Virginia. *U. S. Geological Survey Geologic Atlas of the United States, Folio* 61: 9.

Darton, N.H. and J.A. Taff. 1896. Piedmont Folio, West Virginia-Maryland. *U.S. Geological Survey Geologic Atlas of the United States, Folio* 28: 6.

Deike, G.H., III. 1960a. Cave development in Burnsville Cove, west-central Virginia, with special reference to Breathing Cave, 155 pp. M.A: Thesis, University of Missouri (Columbia).

Deike, G.H., III. 1960b. Origin and geologic relations of Breathing Cave, Virginia. *National Speleological Society (NSS) Bulletin* 22(1): 30-42.

Denkler, K.E., and A.G. Harris. 1988a. Conodont-based determination of the Silurian-Devonian boundary in the Valley and Ridge Province, northern and central Appalachians. *U.S. Geological Survey Bulletin* 1837-B: B1–B13.

Denkler, K.E., and A.G. Harris. 1988b. *Homeognathodus peniculus* (Conodonta), a new earliest Pridolian index species, and the Ludlovian-Pridolian boundary in the central Appalachian Basin. *U.S. Geological Survey Bulletin* 1837-C: C1–C8.

Dennison, J.M. 1961. Stratigraphy of the Onesquethaw stage of Devonian in West Virginia and bordering states. *West Virginia Geological Survey Bulletin* 22: 87.

Dennison, J.M. 1970. Stratigraphic divisions of upper Devonian Greenland Gap Group ("Chemung Formation") along Allegheny front in West Virginia, Maryland, and Highland County. *Virginia. Southeastern Geology* 12(1): 53–82.

Dennison, J.M., and K.O. Hasson. 1977. Stratigraphic cross section of Devonian shales along the Allegheny Front from Maryland to Highland County, Virginia. *Proceedings of the West Virginia Academy of Science* 49(2–4): 103–110.

Dennison, J.M., and D.A. Textoris. 1971. Appendix I—Stop 8; Tioga Bentonite, 2.0 miles southwest of Williamsville, Virginia. In: Field Trip to the Igneous Rocks of Augusta, Rockingham, Highland, and Bath Counties, Virginia, ed. Johnson, R.W. Jr., Milton, C., and J.M. Dennison (*Virginia Division of Mineral Resources Information Circular* 16: 47–50).

Dennison, J.M., F.M. Linn, and D.A. Textoris. 1997. Late Silurian buildups related to syntectonic sedimentation across flexure in Virginia and West Virginia [Abstract]. *Geological Society of America, Abstracts with Programs* 29(3): 13.

Diecchio, R.J. 1985. Post-Martinsburg Ordovician stratigraphy of Virginia and West Virginia. *Virginia Division of Mineral Resources Publication* 57: 77.

Diecchio, R.J. 1991. Taconian sedimentary basins of the Appalachians. In *Advances in Ordovician Geology*, ed. Barnes, C.R., and S.H. Williams, Geological Survey of Canada Paper vol. 90–9, pp. 225–234.

Diecchio, R.J. 1993. Stratigraphic interpretation of the Ordovician of the Appalachian Basin and implications for Taconian flexural modeling. *Tectonics* 12(6): 1410–1419.

Diecchio, R.J., and J.M. Dennison. 1996. Silurian stratigraphy of central and northern Virginia and adjacent West Virginia. *Studies in Geology (University of Tennessee at Knoxville)* 26: 107–127.

Dorobek, S.L., and J.F. Read. 1986. Sedimentology and basin evolution of the Siluro-Devonian Helderberg Group, central Appalachians. *Journal of Sedimentary Petrology* 56: 601–613.

Dorsch, J., and S.G. Driese. 1995. The Taconic foredeep as sediment sink and sediment exporter: Implications for the origin of the white quartzarenite blanket (Upper Ordovician-Lower Silurian) of the central and southern Appalachians. *American Journal of Science* 295(2): 201–243.

Dorsch, J., R.K. Bambach, and S.G. Driese. 1994. Basin-rebound origin for the "Tuscarora Unconformity" in southwestern Virginia and its bearing on the nature of the Taconic Orogeny. *American Journal of Science* 294: 237–255.

Ebright, J.R., C.R. Fettke, and A.I. Ingham. 1949. East Fork-Wharton gas field, Potter County, Pennsylvania. *Pennsylvania Geological Survey Mineral Resource Report, 4th Series* 30: 43.

Gradstein, F., J. Ogg, and A. Smith. 2004. *A Geologic Time Scale 2004*. Cambridge (UK): Cambridge University Press. 589 pp.

Harris, A.G., N.R. Stamm, D.J. Weary, J.E. Repetski, R.G. Stamm, and R.A. Parker (1994) Conodont color alteration index (CAI) map and conodont-based age determinations for the Winchester 30' × 60' quadrangle and adjacent area, Virginia, West Virginia, and Maryland. *U.S. Geological Survey Miscellaneous Field Studies Map* MF-2239, 1 sheet (1:100,000 scale map) and explanatory pamphlet (40 p).

Hasson, K.O., and J.M. Dennison. 1968. Relation of type Millboro Shale in Virginia to Romney Group (Devonian) of West Virginia and Maryland [Abstract]. *Geological Society of America Special Publication* 115: 476–477.

Hasson, K.O., and J.M. Dennison. 1988. Devonian shale lithostratigraphy, central Appalachians, U.S.A. In: *Devonian of the World*, ed. McMillan, N.J., Embry, A.F., and D. J. Glass. Canadian Society of Petroleum Geologists Memoir 14, vol. 2, 157–177.

Head, J.W. 1972. Upper Silurian—Lower Devonian stratigraphy and nomenclature in the central Appalachians. In *Stratigraphy, Sedimentology, and Structure of Silurian and Devonian Rocks along the Allegheny Front in Bedford County, Pennsylvania, Allegany County, Maryland, and Mineral and Grant Counties, West Virginia*, ed. Dennison, J.M., de Witt, W. Jr., Hasson, K.O., Hoskins, D.M. and J. W. Head, III, field trip leaders. Guidebook for the 37th Annual Field Conference of Pennsylvania Geologists, 96–103.

Head, J.W. 1974. Correlation and paleogeography of upper part of Helderberg Group (Lower Devonian) of central Appalachians. *American Association of Petroleum Geologists (AAPG) Bulletin* 58(2): 247–259.

Helfrich, C.T. 1975. Silurian conodonts from Wills Mountain anticline, Virginia, West Virginia, and Maryland. *Geological Society of America Special Paper* 161: 82.

Helfrich, C.T. 1980. Late Llandovery-early Wenlock conodonts from the upper part of the Rose Hill and basal part of the Mifflintown Formations, Virginia, West Virginia, and Maryland. *Journal of Paleontology* 54(3): 557–569.

Kay, G.M. 1941. Middle Ordovician formations in central Pennsylvania. *Geological Society of America Bulletin* 52(12) (pt. 2): 1969.

Kay, G.M. 1944a. Middle Ordovician of central Pennsylvania. *Journal of Geology* 52(1): 1–23.

Kay, G.M. 1944b. Middle Ordovician of central Pennsylvania. Part II. Later Mohawkian (Trenton) formations. *Journal of Geology* 52(2): 97–116.

Kay, G.M. 1956. Ordovician limestones in the western anticlines of the Appalachians in West Virginia and Virginia northeast of the New River. *Geological Society of America Bulletin* 67: 55–106.

Keroher, G.C. 1966. Lexicon of geologic names of the United States for 1936–1960, Parts 1–3, 4341 pp.

Klemic, H., Warman, J.C., and A.R. Taylor 1963. Geology and uranium occurrences of the northern half of the Lehighton, Pennsylvania Quadrangle and adjoining areas. *U.S. Geological Survey Bulletin* 1138: 97.

Kindle, E.M. 1911. The recurrence of *Tropidoleptus carinatus* in the Chemung fauna of Virginia. *The Journal of Geology* 19: 346–357.

Kozak, S.J. 1965. Geology of the Millboro Quadrangle, Virginia. *Virginia Division of Mineral Resources Report of Investigations* 8: 19.

Le Van, D.C., and E.K. Rader 1983. Relationship of stratigraphy to occurrences of oil and gas in western Virginia. *Virginia Division of Mineral Resources Publication* 43 (1 sheet).

Linn, F.M., D.A. Textoris, and J.A. Dennison. 1990. Syndepositional tectonic influence on lithofacies of the Helderberg Group (Silurian-Devonian) of the central Appalachians. *Appalachian Basin Industrial Associates Fall Program* 17: 190–213.

North American Commission on Stratigraphic Nomenclature. 1983. North American Stratigraphic Code. *American Association of Petroleum Geologists (AAPG) Bulletin* 67(5): 841–875.

Perry, W.J., Jr. 1972. The Trenton Group of Nittany Anticlinorium, eastern West Virginia. *West Virginia Geological and Economic Survey Circular* 13: 30.

Prosser, C.S. 1888. Exploration for gas in central New York. *American Institute of Mining and Metallurgical Engineers, Transactions* 16: 940–951.

Rader, E.K., and T.M. Gathright II. 1984. Road log. In: *Stratigraphy and structure in the thermal springs area of the western anticlines*, ed. Rader, E.K. and T.M. Gathright II. Virginia Geologic Field Conference Guidebook no. 16, 23–55.

Rader, E.K., and G.P. Wilkes. 2001. Geologic map of the Virginia portion of the Staunton 30 × 60 minute quadrangle. *Virginia Division of Mineral Resources Publication* 163: (1:100,000 scale), 1 sheet.

Read, J.F. 1980. Carbonate ramp-to-basin transitions and foreland basin evolution, Middle Ordovician, Virginia Appalachians. *American Association of Petroleum Geologists (AAPG) Bulletin* 64(10): 1575–1612.

Reger, D.B. 1924. Mineral and Grant Counties [West Virginia]. West Virginia Geological Survey, 866.

Roden, M.K., R.R. Parrish, and D.S. Miller. 1990. The absolute age of the Eifelian Tioga Ash Bed, Pennsylvania. *Journal of Geology* 98: 282–285.

Rodríguez, N.B. 2005. Stratigraphy of the Silurian-Devonian Upper Helderberg Group in Northeastern West Virginia (USA), 133 pp. Mayagüez: M.S. thesis, The University of Puerto Rico-Mayagüez.

Roen, J.B., and J.W. Hosterman. 1982. Misuse of the term "bentonite" for ash beds of Devonian age in the Appalachian Basin. *Geological Society of America Bulletin* 93(9): 921–925.

Rossbach, T.J., and J.M. Dennison. 1994. Devonian strata of Catawba syncline, near Salem, Virginia. In: *Fieldguides to southern Appalachian structure, stratigraphy, and engineering geology*, ed. Schultz, A. and B. Henika, Virginia Polytechnic Institute and State University, Department of Geological Sciences Guidebook, no. 10, 95–126.

Ryder, R.T., J.E. Repetski, and A.G. Harris. 1996. Stratigraphic framework of Cambrian and Ordovician rocks in the central Appalachian Basin from Fayette County, Ohio, to Botetourt County, Virginia. *U.S. Geological Survey Miscellaneous Investigations Series Map* I-2495: (1 sheet).

Salvador, A., Ed. 1994. *International Stratigraphic Guide* (2nd edition). The International Union of Geological Sciences (Trondheim, Norway) and The Geological Society of America (Boulder, Colorado, USA), 214.

Schuchert, C., C.K. Swartz, T.P. Maynard, and R.B. Rowe. 1913. The Lower Devonian deposits of Maryland. In *Lower Devonian—Text*, ed. C.K. Swartz, C. Schuchert, and C.S. Prosser, 67–132. Maryland Geological Survey: The Johns Hopkins Press, Baltimore.

Smosna, R.A. 1984. Diagenesis of a stromatoporoid patch reef. *Journal of Sedimentary Petrology* 54(3): 1000–1011.

Smosna, R.A., and S.M. Warshauer. 1979. A very early Devonian patch reef and its ecological setting. *Journal of Paleontology* 53(1): 142–152.

Stock, C.W. 1996. Stop 5. Mustoe. In: *Silurian stratigraphy of central and northern Virginia and adjacent West Virginia*, ed. Diecchio, R.J., and J.M. Dennison, Studies in Geology (University of Tennessee at Knoxville) vol. 26, 119–121.

Swain, F.M. 1958. Organic materials of early Middle Devonian, Mt. Union area, Pennsylvania. *American Association of Petroleum Geologists (AAPG) Bulletin* 42(12): 2858–2891.

Swartz, C.K. 1923. Stratigraphic and paleontologic relations of the Silurian strata of Maryland. In: *Silurian*, ed. C.K. Swartz, Maryland Geological Survey, Baltimore, Maryland, 25–52.

Swartz, F.M. 1930. The Helderberg Group of parts of West Virginia and Virginia. *U.S. Geological Survey Professional Paper* 158-C, 27–75.

Swartz, F.M. 1939. The Keyser Limestone and Helderberg Group. In *The Devonian of Pennsylvania*, ed. B. Willard, Pennsylvania Geological Survey, 4th Series, Bulletin G-19, 29–91.

Sweet, P.C. 1981. High-silica resources in Augusta, Bath, Highland, and Rockbridge Counties, Virginia. *Virginia Division of Mineral Resources Publication* 32: 22.

Sweet, P.C., and G.P. Wilkes. 1986. High-silica resources in Alleghany, Botetourt, Craig, and Roanoke Counties, Virginia. *Virginia Division of Mineral Resources Publication* 67: 22.

Taylor, J.F., Repetski, J.E., Loch, J.D., and S.A. Leslie. 2013. Biostratigraphy and chronostratigraphy of the Cambrian-Ordovician Great American Carbonate Bank. In *The Great American Carbonate Bank: Geology and Economic Resources of the Cambrian-Ordovician Sauk Megasequence of Laurentia*, ed. Derby, J.R., Fritz, R.G., Longacre, S.A., Morgan, W.A. and C.A. Sternbach *American Association of Petroleum Geologists Memoir* 98: 15–36.

Uhler, P.R. 1905. The Niagara Period and its associates near Cumberland, Maryland. *Maryland Academy of Science Transactions* 2: 19–26.

Ulrich, E.O. 1911. Revision of the Paleozoic System. *Geological Society of America Bulletin* 22: 281–680.

Vanuxem, L. 1839. Third annual report of the geological survey of the Third District. *New York Geological Survey, Annual Report* 3: 241–285.

White, W.B., and J.W. Hess. 1982. Geomorphology of Burnsville Cove and the geology of the Butler Cave—Sinking Creek System. *National Speleological Society (NSS) Bulletin* 44(3): 67–77.

Willard, B., Ed. 1939. The Devonian of Pennsylvania. *Pennsylvania Geological Survey, 4th Series, Bulletin* G-19: 481.

Wood, G.H., Jr. 1973. Geologic map of the Pottsville quadrangle, Schuylkill County, Pennsylvania. U.S. Geological Survey Geologic Quadrangle Map GQ-1028: 1 sheet (1:24,000 scale).

Woodward, H.P. 1941. Silurian System of West Virginia. *West Virginia Geological Survey, Report* 14: 326.

Woodward, H.P. 1943. Devonian System of West Virginia. *West Virginia Geological Survey, Report* 15: 655.

Woodward, H.P. 1951. Ordovician System of West Virginia. *West Virginia Geological Survey, Report* 21: 627.

Young, S.A., M.R. Saltzman, and S.M. Bergström. 2005. Upper Ordovician (Mohawkian) carbon isotope (d13C) stratigraphy in eastern and central North America: Regional expression of a perturbation of the global carbon cycle. *Palaeogeography, Palaeoclimatology, Palaeoecology* 222 (1–2): 53–76.

Appendix B
Electronic Map Files

The maps that form the core of this book have been reproduced as .pdf or .jpg files that are accessible electronically. Full scale maps of the larger caves would require very large sheets of paper to display the maps in full detail. Thus we avoid clumsy fold-outs or separate folded sheets and also allow zoom features to inspect the maps in detail. To view these maps see: http://extras.springer.com

Barberry Cave: Nevin Davis map, 2012

Basswood Cave: Nevin Davis map, 2011

Better Forgotten Cave: Nevin Davis map, 2012

Buckwheat Cave: Nevin Davis map, 2011

Butler Cave Sinking Creek System: Although some re-mapping and minor additional mapping have been accomplished in the past several decades, the folio sheets compiled by Les Good in 1985 remain the most complete map available. The sheets have been scanned by Tony Canike to produce the electronic files.

Bvideo Pit: Nevin Davis map, 2011

Caves of the Cove: Regional map showing relationship of caves

Chestnut Ridge Cave System: Tommy Shifflett map, 2014.

Drainage Basins: Figure 17.1 at higher resolution.

Emory Dye Traces: Figure 17.4 at higher resolution.

Helictite Cave: Philip C. Lucas map, 2009.

Rathole 1180 Cave: Les Good and Nevin Davis map, 1979

Water Sinks Caves (Water Sinks Cave, Water Sinks Subway, and Owl Cave): Philip C. Lucas map, 2008.

Wishing Well Cave: Philip C. Lucas map, 2010

W.B. White (ed.), *The Caves of Burnsville Cove, Virginia*,
Cave and Karst Systems of the World, DOI 10.1007/978-3-319-14391-0,

Afterword: The Future

William B. White and Gregg S. Clemmer

Introduction

When Dr. William Halliday revised his classic *Depths of the Earth* ten years after it was first published in 1966, extensive and significant discoveries awaited his updated chronicling of Virginia's Burnsville Cove. Stunned by the documented finds in this corner of Bath and Highland Counties, Halliday opined that "between the head of Butler Cave and Aqua Cave, there is room for twenty times as much cave as in now known. Or one hundred times."

Despite what some might alleged to be "cave hype," Bill never blinked. "The Butler Cave Conservation Society faces one of American caving's most intriguing challenges. This peaceful little Virginia valley and points beyond still may hide the world's largest cave".

Yet as we—the collective "we" of the writers who have contributed to this volume—reach the end of the chronicling, we will simply note that this is not "the end of the story." The narrative is still unfolding. It is simply a matter of drawing a line across the evolving saga and sending the document to the printer. We fully expect some future editor to be staring at a huge pile of maps, photographs, exploration accounts, scientific papers, and theses and wondering how this mass of material is to be compressed between two covers.

May 24, 2008

BCCS had selected May 24, 2008 as marking the 50th anniversary of the discovery of the Butler Cave-Sinking Creek System and as the 40th anniversary of the founding of the Butler Cave Conservation Society. To some extent, this was an arbitrary choice—the exploration and study of Breathing Cave had begun much earlier and was largely complete. But May, 1958 was the point in time when a new large and complex cave had been added to Burnsville Cove by the expedient of removing a few rocks from an air-blowing hole.

Just as May, 1958 wasn't the beginning, May 24, 2008 wasn't the ending. The half-century between the dates saw the exploration and scientific investigation of Butler, the discovery and integration of the caves of the Chestnut Ridge System, the discovery, exploration, and survey of Helictite Cave and the numerous "Pancake Caves." On the anniversary date, the exploration and survey of the newly discovered Subway Section of Water Sinks Cave was underway. In the five years that have passed since the anniversary, there has been more exploration, more digging, and Wishing Well Cave has been added to the large caves of the Cove. As the editor hastens to tidy up the final bits of what has been a long and complex writing and editorial enterprise, he hears, over the hum of his computer, the sound of digging tools, exploding soda straws, and the slosh of buckets as an entrance shaft to Robins Rift is being opened.

Exploration Prospects

The list of caves given in Chap. 1 totals to 71.70 miles (115.44 km). That number was correct in October, 2013. It is doubtless again obsolete. The long-collapsed entrance to Robins Rift has been opened and a re-mapping of the cave is underway. Inspection of the Cove map in either Figs. 1.12 or 24.1 reveals a great

W.B. White (ed.), *The Caves of Burnsville Cove, Virginia*,
Cave and Karst Systems of the World, DOI 10.1007/978-3-319-14391-0,

deal of cave passage. However, even casual inspection of the map also reveals a great deal of blank space and discontinuous drainage lines. As examples:

The streams in the Butler Cave-Sinking Creek System terminate in downstream sumps but these sumps are several miles upstream from the known resurgence in Aqua Cave. This implies several miles of cave passage, perhaps all under water, perhaps with segments of air-filled passage separated by sumps, and perhaps with segments of dry upper levels.

The same could be said for the Cathedral River, last seen in the Blarney Stone Section of the Chestnut Ridge System with more than a mile separating it from Cathedral Spring.

It has been assumed that the downstream reaches of both Butler and Chestnut Ridge caves are sumped deep below regional base levels by the plunging Sinking Creek Syncline and White Oak Syncline. This is a convenient assumption because it places all of the unknowns below water where they cannot be observed. The discovery of the Emerald Pool and Roaring River in the Subway Section raises doubts about the hypothesis. Our knowledge of the cave systems and drainage at their northern terminus in the Bullpasture Gorge is extremely limited. Maybe a drone submarine is the answer.

The Emory River is down there somewhere. The Streamway in Helictite Cave and Sugar Run in Wishing Well Cave give some hints as to what the passage might look like, but a route is needed down to the active drainage level.

Much of the blank space on the map is around and south of Burnsville where there is much limestone exposed but only the small Armstrong and Jackson Caves are known. The rising main structure axes carry the limestone up into the air where it has been eroded away, but some limestone remains and the local drainage is sinking into it.

By-the-Road Cave and Robins Rift both lie east of the main trend lines of caves and may represent fragments of the Blue Spring drainage system trending northeast along the base of Tower Hill Mountain. BCCS now owns Robins Rift. The entrance is now open and exploration proceeds apace. The question of the role of this cave in the overall pattern may soon be answered.

In all of the above, the name of the game is entrances. Although Nature has endowed Burnsville Cove with a richness of caves, she has been very stingy with entrances. The few caves with entrances formed by openings on the hillside were discovered very early. Then came the entrance to Butler, obtained by moving a few rocks. After that, entrance making got harder. There was the endless dig in Backyard Cave, the Big Bucks Pit entrance to Barberry, the dig that gave access to Blarney Stone, the dig into Helictite Cave, and the amazing mining operation that opened Wishing Well Cave. More digs will be required, some turning cavers into hard rock miners. But the caves are down there. It is only a question of how to insert the explorers.

Scientific Prospects

The mineralogy of the Chestnut Ridge System is a problem ripe for the picking. Extensive sampling and analysis would perhaps give a broader picture of the various mineral crusts and small crystals that coat many surfaces in the dry upper level of the cave. We know that aragonite is present in profusion but we don't know why it is present in profusion. Sampling with detailed chemical analysis and petrographic, light, cathodoluminescence, and scanning electron microscopes would help with an interpretation of the quite spectacular speleothem displays.

Our knowledge of the detailed flow paths of today's drainage remains fragmentary. Dye traces link point A with point B but quantitative flow measurements and water balance calculations would place more constraints on the system. In spite of the accomplished tracer work, Cove hydrology would benefit from a more extensive study using modern methods of tracer testing with fluorescence analysis of multiple dyes, automatic water samples, and a more comprehensive use of both water chemistry and flood hydrograph analysis.

Also concerning hydrology, there is the enigmatic dye trace from a swallet on Jack Mountain that shows the water to flow under the Bullpasture River to emerge from Clover Creek Spring on the eastern side. Measurements of travel times, discharge hydrographs, and chemical variations at the spring would perhaps reveal the nature of the flow path that apparently follows the carbonate rocks down the syncline and up the other side. Then one could raise the question of whether similar deep flow paths could be responsible for the hot springs that occur in the Jackson River Valley just to the west.

The geomorphic history of the Cove and the age and sequence of cave development remains largely speculative. The Schwartz-Doctor hypothesis implies a complex history with major drainage reversals far back in time. Cosmogenic isotope dating of the clastic sediments in the different main cave trunks would go a long way in providing anchor points. Nothing limits speculation quite so effectively as a few facts. The geomorphic evolution of the Cove is linked to the larger geomorphic history of the Appalachians and the rich karst resources of the Cove make it an excellent test site for investigation.

The Why of All This

The Butler Cave Conservation Society indeed had (and has) reason to celebrate. By purchasing the Butler Farm, by purchasing the Bobcat Property, by the recent purchase of the Robins Rift property, and by its careful maintenance of cordial relations with other landowners in the Cove, BCCS continues to provide access to known caves and discover digging spots for new ones. The care and upgrading of the Butler homestead provides a base of operations far into the future.

Although a few of the original explorers are still doddering about the place, the hard core discoverers of Bobcat and Burns have passed into middle age, and it is the newest generation of equally hard core explorers that will carry Cove discoveries into the future.

Yet the passion that fuels this continuing exploration and documentation pays little heed to Father Time. If caves seem timeless, perhaps that essence fires the cavers who love them…and who pass it on to the next generation.

In 1923, when George Leigh Mallory was touring the United States to raise money for yet another expedition to Mount Everest, many wondered why he was so fervent to be the first to climb the world's tallest mountain. The Englishman's classic answer appeared in the March 18, 1923 issue of the *New York Times*: "Because it's there."

Cove cavers identify with this…but with a twist. Unlike Mallory, who could always see his goal in the icy, soaring heights before him (and would tragically lose his life on Everest the following year), we never really know if the virgin lead before us saunters to a great gallery, plunges down a "bottomless pit," intersects a stream, or blindly ends at a limestone wall. For in a world where there's never been a photon of light…yet perhaps with just a flash that Bill Halliday might be right, we advance into the unknown because *it might be there.*

MIX
Papier aus verantwortungsvollen Quellen
Paper from responsible sources
FSC® C105338

If you have any concerns about our products,
you can contact us on
ProductSafety@springernature.com

In case Publisher is established outside the EU,
the EU authorized representative is:
Springer Nature Customer Service Center GmbH
Europaplatz 3, 69115 Heidelberg, Germany

Printed by Libri Plureos GmbH
in Hamburg, Germany